Practice and Application of Artificial Intelligence in Built Environment

Practice and Application of Artificial Intelligence in Built Environment

Guest Editors

Junming Chen
Yile Chen
Yanyan Liang

MDPI

Basel • Beijing • Wuhan • Barcelona • Belgrade • Novi Sad • Cluj • Manchester

Guest Editors

Junming Chen
Faculty of Humanities and Arts
Macau University of Science and Technology
Macau
China

Yile Chen
Faculty of Humanities and Arts
Macau University of Science and Technology
Macau
China

Yanyan Liang
Faculty of Innovation Engineering
Macau University of Science and Technology
Macau
China

Editorial Office
MDPI AG
Grosspeteranlage 5
4052 Basel, Switzerland

This is a reprint of the Special Issue, published open access by the journal *Buildings* (ISSN 2075-5309), freely accessible at: https://www.mdpi.com/journal/buildings/special_issues/05NN2349TC.

For citation purposes, cite each article independently as indicated on the article page online and as indicated below:

Lastname, A.A.; Lastname, B.B. Article Title. *Journal Name* **Year**, *Volume Number*, Page Range.

ISBN 978-3-7258-7124-7 (Hbk)
ISBN 978-3-7258-7125-4 (PDF)
https://doi.org/10.3390/books978-3-7258-7125-4

Contents

About the Editors

Junming Chen

Junming Chen has a multidisciplinary background and engages in interdisciplinary research. Backgrounds include architecture, design, computer science, and a Ph.D. in digital media. Since 2017, he has been working at the Macau University of Science and Technology, focusing on the application and improvement of artificial intelligence in digital media, design, and architecture. He serves as a tutor and Assistant Professor for postgraduate students. He excels in using parametric and AI technologies in his design practice, having mentored students to complete multiple projects and won over ten international and national design awards, including the A Design Award and Huacan Awards. He is also a certified senior data analyst by the Ministry of Industry and Information Technology. Additionally, his papers have been published in SCI TOP journals, A&HCI, SSCI, and CCF core journals, with more than 30 articles, and with more than 20 of those as first author/corresponding author. He serves as a reviewer for more than 30 SCI and SSCI journals.

Yile Chen

Yile Chen, Ph.D. in Architecture, is an Assistant Professor at the Faculty of Humanities and Arts, Macau University of Science and Technology. She has published more than 50 papers in SCIE, SSCI, and A&HCI journals. She focuses on architectural technology in high-density urban form and a healthy environment, as well as the protection and renewal of urban and architectural cultural heritage in Macau and Portuguese-speaking regions. She is currently leading the Macau University of Science and Technology Research Fund (FRG) project, focusing on the spatial morphological development of Macau and Timor-Leste. She has also participated in numerous projects, including the Science and Technology Development Fund (FDCT) Science and Technology Innovation Enhancement Program of the Macau Special Administrative Region, the Cultural Affairs Bureau of the Macau Special Administrative Region's Consulting Services, the Macau Special Administrative Region Government Higher Education Fund Research Project on Prevention and Response to Major Infectious Diseases, the China–Portugal Joint Laboratory for Cultural Heritage Conservation "Belt and Road" Open Project, the Zhejiang Provincial Philosophy and Social Sciences Planning Project, the Digital Rural Research Center of Hangzhou City University, and the Guangdong Provincial Department of Education's Key Scientific Research Platform Project for General Universities. She also serves as a peer reviewer for 74 SCIE/SSCI/A&HCI/ESCI journals.

Liang Yanyan

Liang Yanyan, Ph.D., is an Associate Professor and Doctoral Supervisor. He obtained his Ph.D. in Science from Macau University of Science and Technology in 2009. From 2009 to 2013, he served as a Postdoctoral Researcher at the Space Science Institute of MUST, where he was engaged in data analysis and visualization for the Chang'e-1 and Chang'e-2 lunar exploration missions. Since 2014, he has held positions as Assistant Professor and Associate Professor at the Faculty of Information Technology and the Macau Institute of Smart City at MUST. In 2012, he received the Third Prize of the Natural Science Award from the Macau Special Administrative Region (ranked third among awardees). His current research focuses on artificial intelligence, image processing, computer vision, machine learning, data mining, computer architecture, parallel computing, and distributed computing. He has published over 60 papers in high-quality domestic and international journals and major international conferences (including IJCAI, AAAI, CVPR, ICCV, ICML, ACM MM, TPAMI, IJCV, TIP, TIFS, TITS, TCYB, TMM, PR,

AJ, IEEE Access, MVAP, FG, etc.) and authored one monograph. He has led and completed multiple research projects, including eight projects funded by the Macau Science and Technology Development Fund, one key R&D project in Macau (as Sub-project Leader), one key national R&D project under the "Strategic Scientific and Technological Innovation Cooperation" program—a joint project with the Macau Science and Technology Development Fund, one international (regional) cooperation and exchange project of the National Natural Science Foundation of China—a joint project with the Macau Science and Technology Development Fund, and one key R&D project of Guangdong Province. As a key participant (non-Principal Investigator), he has contributed to multiple research projects funded by the Macau FDCT, the National 973 Program, and the National Natural Science Foundation of China.

buildings

MDPI

Editorial

Practice and Application of Artificial Intelligence in the Built Environment

Junming Chen [1], Yile Chen [1,*] and Yanyan Liang [2,*]

1 Faculty of Humanities and Arts, Macau University of Science and Technology, Avenida Wai Long, Tapai, Macau 999078, China; jmchen@must.edu.mo

2 Faculty of Innovation Engineering, Macau University of Science and Technology, Avenida Wai Long, Tapai, Macau 999078, China

* Correspondence: chenyile@must.edu.mo (Y.C.); yyliang@must.edu.mo (Y.L.)

1. Introduction

With the rapid development of science and technology, the integration of artificial intelligence into the entire life cycle of the built environment is accelerating, demonstrating a broad range of potential applications in planning evaluation, design simulation, construction management, and operation and maintenance [1,2]. As a complex system, the built environment has diverse data sources spanning a large range of scales, including near-ground spatial observation data, such as street view images, as well as regional-scale information, such as remote sensing images [3–7]. On this basis, improving edge computing and smart city security governance frameworks, alongside research on energy consumption prediction, interpretable modeling, and digital twins, makes real-time risk identification, performance diagnosis, and decision feedback in the built environment increasingly feasible [8–13]. At the same time, generative artificial intelligence and multimodal large models are driving the exploration of solutions from evaluation to simulation, providing new methodological paths for built environment auditing and spatial governance [14,15].

Driven by the aforementioned trends, this Special Issue of *Buildings* includes 15 papers focusing on key aspects and processes of the built environment, including research on urban perception supported by street view images and interpretable learning, the interpretable assessment of street quality through a street view–remote sensing fusion, a path identification of public open space satisfaction among elderly visitors in World Heritage city centers, an interpretable threshold analysis of environmental safety perception for elderly residents in pocket parks, the identification and configuration optimization of campus landscape elements, the optimization of campus functional zoning considering the differences between "students and visitors", interpretable AI-driven campus building safety assessment and design feedback, heating load prediction for residential buildings, residential unit generation based on multi-relation graph attention, the extraction and updating of facade features in historic districts, an AIGC-LoRA pre-generation and collaborative evaluation framework for updating facades of historic buildings, a modular AI workflow for building facade style transfers, a review of the progress and cutting-edge trends in generative AI research, and a systematic review of the knowledge structure of intelligent building operation and maintenance research.

2. Overview of Contributions

This Special Issue brings together the research of scholars from South Korea, Japan, New Zealand, Malaysia, the UK, and China, primarily covering real-world case studies from South Korea and China, including specific urban studies in Seoul, Guangzhou,

Zhengzhou, Tianjin, and Shantou. These studies encompass both data-driven urban perceptions and street quality assessments centered on street view/remote sensing (e.g., Zhengzhou and Tianjin) and interpretable analyses of age-friendly public spaces (e.g., Tapgol Park in Seoul). This collection provides a reusable methodological framework, as well as case studies for the empirical analysis of AI-assisted evaluation, optimization, and renewal decisions in built environments at different scales.

The methodological framework for artificial intelligence is becoming increasingly sophisticated, covering a wide range of technical approaches from traditional machine learning to deep learning, computer vision, and generative modeling. This Special Issue focuses on application scenarios within the built environment, highlighting the following key models and technical pathways:

- Semantic segmentation and scene parsing (for street view/campus landscape element recognition).
- Object detection (for campus/building safety risk identification).
- Decision trees.
- Light GBM (Optical Gradient Boosting Machine).
- Random forest (RF).
- Geographically weighted random forest (GWRF).
- Elastic net regression.
- Generalized additive model (GAM).
- Shapley interpretable analysis (SHAP) and partial dependency analysis (PDP).
- Generative adversarial networks and conditional generation (GANs/CGANs, such as pix2pix-like frameworks).
- Efficient LoRA parameter fine-tuning (for style consistency and controllable generation for AIGC).
- Multi-relationship graph attention and WGAN generation (graph attention/RWGAN-like methods for apartment type generation).
- Convolutional long short-term memory networks (ConvLSTM) and attention modules (such as channel/temporal attention).
- Configuration analysis (fsQCA for multi-element combination mechanism identification).
- Bibliometrics and Knowledge Graph Analysis (for outlining the research trajectory of operational and generative AI).

By employing these technologies, the authors of this Special Issue offer tangible methods and examples for enhancing urban decision-making using artificial intelligence.

2.1. *Urban Street Perception and Street Quality Assessment*

In Contribution 1 (Zhang et al.), the authors construct an urban perception analysis framework integrating street view images and interpretable machine learning. Using Zhengzhou as a case study, they extract street visual elements based on deep learning technology and systematically analyze the multi-dimensional urban perception formation mechanism using a LightGBM model and the SHAP method. This provides data evidence for refined urban cognition and spatial governance.

In Contribution 2 (Yuan et al.), the authors use the major urban areas of Tianjin as a case study, combining street view images with remote sensing images to construct point-scale environmental indicators. They then conduct interpretable evaluations of multi-dimensional perception results, including comfort, aesthetics, perceived green space, perceived summer heat, and willingness to stay in a location, using elastic network logistic regression, a generalized additive model with logical links, and Shapley addition, providing quantitative references for street renewal and the improvement of environmental quality.

https://doi.org/10.3390/buildings16050926

2.2. Age-Friendly Public Spaces and Safety Perception

Contribution 3 (Wu et al.) focuses on environmental safety issues related to elderly visitors in urban pocket parks in China, quantifying influencing factors by combining panoramic semantic segmentation and interpretable modeling (including SHAP and PDP). The authors highlight the positive effects of pedestrian space and moderate pedestrian activity on safety perception, as well as the significant negative impact of vehicular elements, providing actionable directions for environmental elements in micro-renewal.

In Contribution 4 (Wu et al.), the authors use Tapgol Park in central Seoul as an example, drawing on 477 valid questionnaires collected in and around the park. They use decision tree methods to identify the key paths and hierarchical logic of low-income elderly residents' satisfaction with open spaces, providing interpretive evidence for the governance of age-friendly public spaces in high-density historical urban areas.

The authors of Contribution 5 (Yang et al.) select 149 public spaces in the Yuexiu District, Guangzhou, and use Baidu heatmap data and a geographically weighted random forest (GWRF) model to analyze the impact of the built environment under four temporal scenarios. They find a significant spatiotemporal heterogeneity—building density is more important at night, while residential density is more important during the day, especially on weekends.

2.3. Campus Space Optimization and Campus Building Safety Assessment

In Contribution 6 (Zhuang et al.), the authors address the optimization of landscape composition on university campuses. By combining semantic segmentation, interpretable machine learning, and configuration analysis (fsQCA), they reveal how the combination of landscape elements collectively shapes perceptions of safety, comfort, and belonging, providing a method for configurational optimization of the campus environment.

Contribution 7 (Zhuang et al.) further focuses on the perceptual differences between students and visitors in open campuses. The authors utilize landscape element identification and enhanced decision tree (SHAP) methods to identify key factors and their threshold effects, and use this to divide campus functional zones, providing more targeted evidence for campus spatial governance.

The authors of Contribution 8 (Hou et al.) propose a YOLOv11-Safe interpretable AI framework for identifying and assessing campus building safety risks. This framework translates identification results into feedback suggestions for design optimization, demonstrating an engineering-oriented implementation path of "detection–assessment–optimization".

2.4. Residential Space Generation and Residential Building Load Forecasting

In Contribution 9 (Ye et al.), the authors propose the Graph-RWGAN method, which characterizes the complex relationships between rooms based on a multi-relationship graph attention mechanism. It achieves an automatic generation of residential unit types through iterative generation and global consistency constraints, demonstrating a strong adaptability and flexible generation under weak constraints and diverse layouts. This provides a technical route for controllable space generation and rapid solution derivation.

The authors of Contribution 10 (Zhang et al.) propose a ConvLSTM-DyT-CBAM(T) hybrid framework, replacing the traditional batch normalization layer with a dynamic Tanh (DyT) activation function. The framework utilizes multi-scale time-series data to predict heating loads in residential buildings and enhances the response to sudden load peaks, serving such functions as operation control and energy-saving optimization.

https://doi.org/10.3390/buildings16050926

2.5. Historic Districts and Building Facade Renovation and Style Relocation

Contribution 11 (Yan et al.) addresses the issues of facade degradation and the efficiency of traditional restoration methods in the Shantou Xiaogongyuan historical district. Based on the pix2pix model, the authors develop an intelligent facade generation system and propose a facade renewal method integrating computer vision and generative design, significantly improving design efficiency and providing a replicable path for the AI-assisted systemic renewal of historical urban areas.

Contribution 12 (Duan et al.) discusses the challenge of balancing heritage preservation and modern adaptation in historical conservation areas. The authors propose a AIGC–LoRA framework for the virtual simulation of historical building facades "before renewal" and construct a participatory, iterative "generation–evaluation" system to balance authenticity, coordination, and feasibility. Through AIGC technology, they provide an effective solution for mitigating the risks of continued preservation in historical conservation areas.

In Contribution 13 (Xu & Qu), focusing on architectural facade style transfer, the authors construct a system that integrates Flux Redux and Flux Depth models, using the node-based visual deep learning platform ComfyUI to build a modular AI workflow. By introducing a deep style collaboration process based on node-based platforms such as ComfyUI, the authors provide a reference for reusable and composable workflows in design practice.

2.6. Overview and Knowledge Graphs: Generative AI and Intelligent Operations

The authors of Contribution 14 (Yang et al.) systematically organize and analyze research literature related to use of generative artificial intelligence in architectural design from the Web of Science database, published between 2005 and 2025, using the CiteSpace visualization tool. This analysis outlines the research progress and cutting-edge trends of generative AI in architectural design, providing a general framework for the problem domain, methodological clues, and future directions of "generative design."

In Contribution 15 (Zheng et al.), the authors retrieve 423 publications related to AI-enabled building maintenance from the Web of Science Core Collections, covering the period from 2000 to 2025. Using tools such as bibliometrix v5.0.1 and VOSviewer (v1.6.20), they conduct a systematic bibliometric and scientometric analysis, summarizing the current status and evolutionary hotspots of intelligent building operation and maintenance research. They emphasize that research hotspots are shifting from passive monitoring to proactive prediction and further toward system-level intelligent decision-making and multi-technology integration.

3. Outlook

This Special Issue compiles 15 research articles on artificial intelligence (AI) and its empowerment of the built environment field, and the work presented here will further promote the development of urban and architectural space research and practice and inspire additional related studies. In the future, applications of AI in the built environment field will continue to integrate multi-source data and intelligent algorithms to build more interpretable and transferable assessment and decision-making systems. From optimizing the quality of streets and public spaces to the renewal and generation of campuses and historical buildings, and furthermore to the prediction of residential operation and building operation and maintenance, AI will effectively support the improvement of spatial quality and the optimization of resource efficiency, helping to create a smarter, safer, more inclusive, and more sustainable built environment.

https://doi.org/10.3390/buildings16050926

Author Contributions: Conceptualization, J.C. and Y.C.; methodology, J.C.; software, J.C.; validation, J.C.; formal analysis, J.C.; investigation, J.C.; resources, J.C.; data curation, J.C.; writing—original draft preparation, J.C. and Y.C.; writing—review and editing, J.C. and Y.C.; visualization, Y.C.; supervision, Y.L.; project administration, J.C. and Y.C.; funding acquisition, J.C. and Y.C. All authors have read and agreed to the published version of the manuscript.

Funding: This research was funded by the (1) Faculty Research Grants funded by Macau University of Science and Technology (FRG-MUST grant number: FRG-25-064-FA; FRG-25-067-FA); (2) Science and Technology Development Fund of Macau Project (grant number: 0087/2025/ITP2); (3) Guangdong Provincial Department of Education's key scientific research platforms and projects for general universities in 2023: Guangdong, Hong Kong, and Macau Cultural Heritage Protection and Innovation Design Team (grant number: 2023WCXTD042); (4) Guangdong Provincial Philosophy and Social Sciences Planning 2025 Lingnan Cultural Project (grant number: GD25LN30). The funders had no role in the study conceptualization, data curation, formal analysis, methodology, software, decision to publish, or preparation of the manuscript. There was no additional external funding received for this study.

Acknowledgments: We sincerely thank the authors who contributed to this Special Issue and submitted high-quality manuscripts. We also thank the reviewers for their responsible and rigorous feedback, which has greatly improved each manuscript. Finally, we sincerely thank the editorial team of *Buildings* for their professional support, which ensured the success of this Special Issue.

Conflicts of Interest: The Guest Editors declare that they have no known competing financial interests or personal relationships that could have appeared to influence the work reported in this paper.

List of Contributions:

1. Zhang, M.; Pan, C.; Huang, X.; Zhang, L.; Lee, M. Research on Urban Perception of Zhengzhou City Based on Interpretable Machine Learning. *Buildings* **2026**, *16*, 314. https://doi.org/10.3390/buildings16020314.
2. Yuan, Y.; Tang, F.; Zhou, S.; Zhang, Y.; Li, X.; Wang, S.; Wang, L.; Wang, Q. Interpretable Assessment of Streetscape Quality Using Street-View Imagery and Satellite-Derived Environmental Indicators: Evidence from Tianjin, China. *Buildings* **2026**, *16*, 1. https://doi.org/10.3390/buildings16010001.
3. Wu, S.; Wu, S.; Chen, J.; Pan, C. An Interpretable Machine Learning Approach to Studying Environmental Safety Perception Among Elderly Residents in Pocket Parks. *Buildings* **2025**, *15*, 3411. https://doi.org/10.3390/buildings15183411.
4. Wu, C.; Chen, Y.; Zhang, F.; Zheng, L.; Liang, J.; Yang, S.; Wang, Y. Exploring the Satisfaction of Low-Income Elderly People with Open Space Environment in Tapgol Park of Central Seoul: A Decision Tree Approach to Machine Learning. *Buildings* **2026**, *16*, 172. https://doi.org/10.3390/buildings16010172.
5. Yang, Y.; Lin, X.; Li, X.; Chen, Q.; Sun, X. Day–Night and Weekday–Weekend Heterogeneity in Built Environment Impacts on Public Space Vitality: A GWRF Analysis in Yuexiu District. *Buildings* **2026**, *16*, 523. https://doi.org/10.3390/buildings16030523.
6. Zhuang, X.; Cai, Y.; Tang, Z.; Ding, Z.; Gan, C. Identification and Configuration Optimization of Key Campus Landscape Features Using Augmentation-Based Machine Learning and Configuration Analysis. *Buildings* **2025**, *15*, 3868. https://doi.org/10.3390/buildings15213868.
7. Zhuang, X.; Cai, Y.; Tang, Z.; Ding, Z.; Gan, C. Optimizing University Campus Functional Zones Using Landscape Feature Recognition and Enhanced Decision Tree Algorithms: A Study on Spatial Response Differences Between Students and Visitors. *Buildings* **2025**, *15*, 3622. https://doi.org/10.3390/buildings15193622.
8. Hou, J.; Hu, Y.; Jiang, B.; Chang, Z.; Cao, M.; Wang, B. YOLOv11-Safe: An Explainable AI Framework for Data-Driven Building Safety Evaluation and Design Optimization in University Campuses. *Buildings* **2025**, *15*, 4125. https://doi.org/10.3390/buildings15224125.
9. Ye, Z.; Liu, S.; Tian, Z.; Chen, Y.; Zheng, L.; Chen, J. Graph-RWGAN: A Method for Generating House Layouts Based on Multi-Relation Graph Attention Mechanism. *Buildings* **2025**, *15*, 3623. https://doi.org/10.3390/buildings15193623.

10. Zhang, H.; Gao, X.; Liu, X.; Liu, Z. A ConvLSTM-Based Hybrid Approach Integrating DyT and CBAM(T) for Residential Heating Load Forecast. *Buildings* **2025**, *15*, 3781. https://doi.org/10.3390/buildings15203781.
11. Yan, W.; Wang, T.; Zhang, C. Renewal Design of Architectural Facade Features in the Shantou Xiaogongyuan Historic District Based on Deep Learning. *Buildings* **2025**, *15*, 4404. https://doi.org/10.3390/buildings15244404.
12. Duan, W.; Rao, J.; Zhao, J.; Tao, N.; Chen, J. AI-Based Pre-Renewal Design for Historic Building Facades: An AIGC–LoRA Framework with Collaborative Assessment. *Buildings* **2025**, *15*, 4212. https://doi.org/10.3390/buildings15234212.
13. Xu, C.; Qu, C. A Modular AI Workflow for Architectural Facade Style Transfer: A Deep-Style Synergy Approach Based on ComfyUI and Flux Models. *Buildings* **2026**, *16*, 494. https://doi.org/10.3390/buildings16030494.
14. Yang, Y.; Li, Y.; Bai, X.; Zhang, W.; Chen, S. Research Progress and Frontier Trends in Generative AI in Architectural Design. *Buildings* **2026**, *16*, 388. https://doi.org/10.3390/buildings16020388.
15. Zheng, Y.; Sun, B.; Guan, Y.; Yang, Y. Artificial Intelligence Empowering the Transformation of Building Maintenance: Current State of Research and Knowledge. *Buildings* **2025**, *15*, 4118. https://doi.org/10.3390/buildings15224118.

References

1. Moradi, L.; Biloria, N. Implications of Artificial Intelligence for Assessing the Built Environment. *J. Urban Technol.* **2025**, *32*, 163–191. [CrossRef]
2. Pan, Y.; Zhang, L. Roles of artificial intelligence in construction engineering and management: A critical review and future trends. *Autom. Constr.* **2021**, *122*, 103517. [CrossRef]
3. Biljecki, F.; Ito, K. Street view imagery in urban analytics and GIS: A review. *Landsc. Urban Plan.* **2021**, *215*, 104217. [CrossRef]
4. Li, J.; Hong, D.; Gao, L.; Yao, J.; Zheng, K.; Zhang, B.; Chanussot, J. Deep learning in multimodal remote sensing data fusion: A comprehensive review. *Int. J. Appl. Earth Obs. Geoinf.* **2022**, *112*, 102926. [CrossRef]
5. Yuan, Y.; Shi, X.; Gao, J. Building extraction from remote sensing images with deep learning: A survey on vision techniques. *Comput. Vis. Image Underst.* **2025**, *251*, 104253. [CrossRef]
6. Babun, L.; Denney, K.; Celik, Z.B.; McDaniel, P.; Uluagac, A.S. A survey on IoT platforms: Communication, security, and privacy perspectives. *Comput. Netw.* **2021**, *192*, 108040. [CrossRef]
7. Kong, L.; Tan, J.; Huang, J.; Chen, G.; Wang, S.; Jin, X.; Zeng, P.; Khan, M.K.; Das, S.K. Edge-computing-driven internet of things: A survey. *ACM Comput. Surv.* **2022**, *55*, 1–41. [CrossRef]
8. Sookhak, M.; Tang, H.; He, Y.; Yu, F.R. Security and privacy of smart cities: A survey, research issues and challenges. *IEEE Commun. Surv. Tutor.* **2018**, *21*, 1718–1743. [CrossRef]
9. Fathi, S.; Srinivasan, R.; Fenner, A.; Fathi, S. Machine learning applications in urban building energy performance forecasting: A systematic review. *Renew. Sustain. Energy Rev.* **2020**, *133*, 110287. [CrossRef]
10. Darvishvand, L.; Kamkari, B.; Huang, M.J.; Hewitt, N.J. A systematic review of explainable artificial intelligence in urban building energy modeling: Methods, applications, and future directions. *Sustain. Cities Soc.* **2025**, *128*, 106492. [CrossRef]
11. Fang, W.; Ding, L.; Love, P.E.; Luo, H.; Li, H.; Peña-Mora, F.; Zhong, B.; Zhou, C. Computer vision applications in construction safety assurance. *Autom. Constr.* **2020**, *110*, 103013. [CrossRef]
12. Opoku, D.G.J.; Perera, S.; Osei-Kyei, R.; Rashidi, M. Digital twin application in the construction industry: A literature review. *J. Build. Eng.* **2021**, *40*, 102726. [CrossRef]
13. Abdelrahman, M.; Macatulad, E.; Lei, B.; Quintana, M.; Miller, C.; Biljecki, F. What is a Digital Twin anyway? Deriving the definition for the built environment from over 15,000 scientific publications. *Build. Environ.* **2025**, *274*, 112748. [CrossRef]
14. Khan, A.; Chang, S.; Chang, H. Generative AI approaches for architectural design automation. *Autom. Constr.* **2025**, *180*, 106506. [CrossRef]
15. Jang, K.M.; Kim, J. Multimodal large language models as built environment auditing tools. *Prof. Geogr.* **2025**, *77*, 84–90. [CrossRef]

https://doi.org/10.3390/buildings16050926

buildings

MDPI

Article

Research on Urban Perception of Zhengzhou City Based on Interpretable Machine Learning

Mengjing Zhang [1], Chen Pan [2], Xiaohua Huang [3,*,†], Lujia Zhang [4,*,†] and Mengshun Lee [2]

1 School of Traffic Engineering, Huanghe Jiaotong University, Jiaozuo 454950, China; 2021090602@zjtu.edu.cn
2 Faculty of Innovation and Design, City University of Macau, Macau 999078, China; jourdan29@gdupt.edu.cn (C.P.); lisalee@cityu.edu.mo (M.L.)
3 School of Design, NingboTech University, Ningbo 315100, China
4 School of Architecture, Zhengzhou University, Henan International Joint Laboratory of Eco-community & Innovative Technology, Zhengzhou 450000, China
* Correspondence: hxh@nbt.edu.cn (X.H.); lujzhang2023@zzu.edu.cn (L.Z.)
† These authors contributed equally to this work.

Abstract

Urban perception research has long focused on global metropolises, but has overlooked many cities with complex functions and spatial structures, resulting in insufficient universality of existing theories when facing diverse urban contexts. This study constructed an analytical framework that integrates street scene images and interpretable machine learning. Taking Zhengzhou City as the research object, it extracted street visual elements based on deep learning technology and systematically analyzed the formation mechanism of multi-dimensional urban perception by combining the LightGBM model and SHAP method. The main findings of the research are as follows: (1) The urban perception of Zhengzhou City shows a significant east–west difference with Zhongzhou Avenue as the boundary. Positive perceptions such as safety and vitality are concentrated in the central business district and historical districts, while negative perceptions are more common in the urban fringe areas with chaotic built environments and single functions. (2) The visibility of greenery, the openness of the sky and the continuity of the building interface are identified as key visual elements affecting perception, and their directions and intensifies of action show significant differences due to different perception dimensions. (3) The influence of visual elements on perception has a complex mechanism of action. For instance, the promoting effect of greenery visibility on beauty perception tends to level off after reaching a certain threshold. The research results of this study can provide quantitative basis and strategic reference for the improvement in urban space quality and humanized street design.

Keywords: deep learning; urban perception; street view image; interpretable machine learning; LightGBM; SHAP value; Zhengzhou City; urban design

1. Introduction

Cities are important spatial carriers of human civilization. Their built environment not only supports the production and living activities of residents, but also profoundly influences people's psychological experiences and behavioral patterns [1]. In this complex process of human-land interaction, urban visual landscapes, as the direct presentation of environmental information, play a crucial role in shaping and developing human perceptual responses. Urban perception, which refers to the subjective cognition and comprehensive evaluation of the urban spatial environment by residents based on visual cues [2], has become an important field of interdisciplinary research involving architecture, urban and rural planning, and sociology.

The existing research methods still face significant challenges. Traditional research methods such as questionnaires and in-depth interviews [3,4], although they can obtain relatively in-depth perceptual data, generally have inherent limitations such as limited sample size, high survey costs, and difficulty in achieving continuous spatial coverage [3,4]. In recent years, breakthroughs in deep learning technology have provided a new technical path for large-scale urban perception research [5,6]. The method based on street view images and deep neural networks can achieve efficient and standardized perception prediction, effectively overcoming the shortcomings of traditional methods. However, advanced predictive models often exhibit a "black-box" nature, which means that while they achieve high prediction accuracy, their internal decision-making logic remains opaque [7,8]. This lack of transparency hinders the clear explanation of how predictions are generated and fails to effectively reveal the specific environmental factors that shape urban perception. Consequently, the practical application of such models in urban planning is limited, as planners require not only predictive outcomes but also interpretable insights to inform design interventions. Although traditional statistical methods have good interpretability, they often have deficiencies when dealing with the complex nonlinear correlations between the environment and perception [9,10].

Against this backdrop, the emergence of explainable machine learning methods offers new ideas for addressing the aforementioned challenges [11]. Techniques such as SHAP (Shapley Additive Explanations) have been applied to uncover the nonlinear effects of visual elements on safety perception [12] or to identify key streetscape features influencing the sense of security [13]. Despite these advances, several critical research gaps persist, limiting both theoretical understanding and practical application.

Most studies focus on identifying key visual features that influence perception, but have not systematically analyzed the interactions between different visual elements and their synergistic or antagonistic effects on multidimensional perception. Urban perception is shaped by a combination of environmental features, yet the compound effects and potential trade-offs among elements remain underexplored. Existing research predominantly examines coastal metropolises or cities at specific developmental stages, paying insufficient attention to inland megacities undergoing rapid urbanization and spatial restructuring. The unique spatial patterns, development trajectories, and perceptual dynamics of such cities—characterized by distinct urban forms, historical layers, and development pressures—are not adequately represented in the current literature, limiting the generalizability of findings. While correlations between visual elements and perceptual scores have been established, the causal mechanisms and directional effects linking specific environmental features to particular perceptual dimensions require further validation through more rigorous modeling and interpretability frameworks. The "why" behind the predictions—the explicit pathways through which environmental attributes translate into subjective experience—remains inadequately addressed.

To address these gaps, this study constructs a comprehensive analytical framework integrating street view images, deep learning, and explainable machine learning. We select Zhengzhou City, a major inland transportation hub and a rapidly developing megacity in central China, as an empirical case. Zhengzhou's urban spatial structure shows distinct east–west differences with Zhongzhou Avenue as the boundary: the eastern part is a newly built urban area with larger street scales and higher greening levels, while the western part is the old urban area with high density and complex street interfaces. This spatial duality provides an ideal setting to explore perceptual variations and their environmental drivers.

https://doi.org/10.3390/buildings16020314

This study focuses on the following three research questions: (1) How to measure the multi-dimensional urban perception spatial pattern of Zhengzhou City based on machine learning methods? (2) What are the key environmental elements that influence the formation of urban perception in Zhengzhou City? (3) How do these environmental elements act on different dimensions of urban perception?

By answering these questions, this study aims to advance the methodological application of explainable machine learning in urban perception research and provide a scientific, interpretable evidence base for spatial quality enhancement and human settlement optimization in Zhengzhou and similar inland megacities.

2. Literature Review

2.1. The Concept and Measurement Dimensions of Urban Perception

Urban perception research originated from environmental psychology and humanistic geography, focusing on the interaction between people and the built environment. In "Urban Image", Lynch innovatively proposed that residents form psychological representations of a city through five elements: paths, boundaries, regions, nodes and landmarks [13]. This theory laid the foundation for the study of urban perception. With the deepening of research, the connotation of urban perception has been continuously enriched, expanding from the initial spatial cognition to multiple dimensions such as emotional response and esthetic evaluation. Its research scope has also transcended static spatial images, covering more complex psychological constructions such as emotions and a sense of belonging [14]. In terms of measurement dimensions, existing research has established a diversified evaluation system. The Place Pulse project developed by the MIT Media Lab identified six core perception dimensions—safety, vitality, beauty, affability, repression and boredom—through a large-scale online survey [15]. This framework has been widely adopted due to its systematicness and operability. Sal et al. further verified the intrinsic correlations among these dimensions through factor analysis and found significant positive correlations between dimensions such as safety and prosperity, vitality and beauty [16]. These studies provide a theoretical basis for the quantitative assessment of urban perception. With the popularization of multi-source data and deep learning methods, the measurement dimensions of urban perception are showing a trend of refinement and contextualization. For instance, Liu et al. integrated the physical characteristics of the built environment and subjective perception dimensions in their research to quantify urban vitality [17]; Zhao et al. conducted a more detailed assessment of the visual quality of the built environment by integrating deep learning models [18]. These studies not only verified the classical dimensions but also promoted the deepening and development of the perceptual measurement methodology.

2.2. The Application of Traditional Data and Methods in Urban Perception Research

Traditional urban perception research mainly relies on two types of data sources: location-based abstract data and concrete image data. In terms of abstract data, indicators such as Point of interest (POI) density and land use mix degree are widely used to characterize the functional features of cities. Liu et al. verified the positive correlation between the degree of functional mixing and the perception of block vitality through the calculation of entropy values of POI data [19]. Social media data provides a new perspective for perception research. Ge et al. analyzed the emotional perception characteristics of urban green Spaces using Weibo check-in data [20]. In terms of the application of image data, remote sensing images, due to their wide coverage capability, were widely used in the analysis of urban forms at a macro scale in the early days. Xie et al. combined high-resolution remote sensing images and social perception data to achieve precise identification of urban

https://doi.org/10.3390/buildings16020314

functional areas [18]. However, the top-down perspective of remote sensing images limits their representational ability for visual perception at the street scene level.

Traditional research methods such as questionnaires and in-depth interviews [19,20], although they can obtain relatively in-depth perceptual data, generally have inherent limitations such as limited sample size, high survey costs, and difficulty in achieving continuous spatial coverage [21,22]. Against this backdrop, the emergence of explainable machine learning methods offers new ideas for addressing the aforementioned challenges. Among them, the SHAP (Shapley Additive Explanations) framework is based on the principle of game theory and can quantify the contribution of each feature to the prediction results of the model [23,24], providing an effective tool for understanding the decision-making mechanism of machine learning models.

2.3. Breakthrough Progress in Street View Imaging and Deep Learning

The emergence of street view images has brought about a revolutionary change to urban perception research. Compared with remote sensing images, street view images truly present the built environment from a human perspective and can more directly capture the visual elements that affect perception. Cheng et al. systematically demonstrated the advantages of street view images in visual perception research, including the authenticity of perspective, spatial continuity and temporal comparability [25]. The application of deep learning methods has further promoted the development of this field. Zhang et al. applied deep learning technology to large-scale urban perception research. By training convolutional neural network models, they achieved a quantitative assessment of the urban image of Beijing [26]. Wei et al. employed the transfer learning method to demonstrate the generalization ability of deep learning models among different cities [27]. These studies have established the position of "street view images + deep learning" as a new paradigm in urban perception research.

2.4. The Application and Challenges of Explainable Machine Learning in Urban Research

Although deep learning performs well in prediction accuracy, its "black box" feature limits the transformation of research results into planning practice. To address this challenge, explainable machine learning methods emerged [28]. In the field of urban studies, explainable machine learning methods have demonstrated great potential. Liu et al. analyzed the nonlinear influence of visual elements on safety perception using the SHAP method and found that the improvement effect of tree coverage on safety perception was the most significant within a specific range [29]. In the research on community safety perception in Macao, Pan et al. identified through SHAP values that the tidiness of building interfaces and the permeability of streets are the key factors affecting a sense of security [30].

In conclusion, although existing research has made significant progress in the quantification and mechanism exploration of urban perception, there is still room for further improvement in its theoretical framework and empirical basis. Firstly, most studies focus on identifying the key features that affect perception, but have not systematically analyzed the interactions between different visual elements and their synergistic or antagonistic effects on perception. Secondly, existing research mostly focuses on the types of cities at specific development stages or spatial forms, paying insufficient attention to typical inland metropolises that are in the stage of rapid urbanization and structural reorganization, which limits the universality and representativeness of the theoretical system. Furthermore, the causal relationship between visual elements and specific perceptual dimensions still needs to be verified through more rigorous modeling and interpretability methods.

https://doi.org/10.3390/buildings16020314

Based on the above, this study will advance the relevant work in the following aspects: (1) Construct a complete technical chain of "street view images–deep learning–explainable machine learning", forming a closed-loop analysis process from perception prediction to mechanism explanation; (2) Take Zhengzhou City as a case of a rapidly developing inland megacity, and expand the regional types and sample spectrum of urban perception research; and (3) Introduce SHAP interaction value analysis to deeply reveal the joint action mechanism of multiple visual elements in the process of influencing perception, providing a theoretical basis for refined urban design oriented to perception optimization. By systematically analyzing the visual driving mechanism of urban perception in Zhengzhou City, this study aims to deepen the research on urban perception from "phenomenon description" to "mechanism explanation", and provide methodological support for data-driven urban planning.

3. Data Collection

3.1. Research Area

Zhengzhou, the capital of Henan Province, is located in the southern part of the North China Plain and on the south bank of the lower reaches of the Yellow River. It is an important comprehensive transportation hub and a core city in the central region of China. According to the data from the seventh national census, the permanent resident population of Zhengzhou City has reached 12.6 million, with the urban population accounting for 78.4% [31]. In 2023, the city's gross domestic product (GDP) exceeded 1.3 trillion yuan, forming a modern industrial system dominated by electronic information, automobile manufacturing, modern logistics and other industries. This study selected the main urban area of Zhengzhou City as the research scope, including Jinshui District, Huiji District, Guancheng Hui District, Zhongyuan District, Erqi District and the central administrative area of Zhongmu County (Figure 1). The total area of the research region is approximately 580 square kilometers, and the built-up area exceeds 400 square kilometers. This area concentrates the main administrative, commercial, cultural and residential functions of Zhengzhou City. Its spatial structure presents the development characteristics of "multiple centers and clusters", specifically manifested as a spatial pattern of "two axes, one belt and multiple centers". With Jinshui Road–Zhongyuan Road as the east–west development axis and Huayuan Road–Zijing Mountain Road as the north–south development axis, relying on the Yellow River ecological belt, multiple functional centers such as CBD, Longzi Lake and High-tech Zone have been formed. The street view images in this area are fully covered and can meet the data requirements of large-scale sampling. The area encompasses various spatial types such as traditional commercial districts, modern business districts, industrial heritage areas, and waterfront landscape belts, with rich environmental features. This is conducive to exploring the perception differences under different built environments. The area is currently in the process of rapid urbanization and spatial reconstruction. The research results have direct practical guiding significance for optimizing the quality of urban space.

https://doi.org/10.3390/buildings16020314

Figure 1. Research Area Map. (**a**) Location of Zhengzhou in China; (**b**) Administrative divisions of Zhengzhou main urban area; (**c**) Street view sampling points distribution.

3.2. Street View Image Data

To quantify the visual characteristics of the urban environment, this study adopts the street view image service provided by the Baidu Maps Open Platform. As a leading digital map service provider in China, Baidu Maps' street view data features comprehensive coverage, timely updates, and accurate positioning, and has been widely applied in domestic urban research [14]. Considering the temperate monsoon climate characteristics of Zhengzhou City, there are seasonal variations in vegetation coverage. Image data from 2022 were uniformly collected to reflect the urban landscape during the vegetation growth season and minimize the impact of seasonal factors on perception assessment to the greatest extent [32]. The data collection process includes the following steps:

(1) Road network data acquisition: Obtain vector data of all grades of roads in the study area through Open street map, including expressways, main roads, secondary roads and branch roads;

(2) Sampling point generation: Based on the spatial analysis tool of the Geographic Information System, sampling points were systematically generated along the route elements at intervals of 100 m, and a total of 39,736 initial point positions were obtained. The 100 m interval was selected based on a balance between data density and computational efficiency. This spacing ensures sufficient coverage to capture street-level visual variability while avoiding oversampling redundancy, and aligns with previous urban perception studies using street view imagery [33].

(3) Image acquisition: The Baidu Panorama Static Image API (http://lbs.baidu.com, accessed on 3 February 2022) was called through a Python program 3.9 to implement coordinate conversion to the BD09 geodetic coordinate system. The parameters for simulating human visual perception are set as follows: vertical field of view = 0°, horizontal field of view Angle = 90°, and image resolution = 512 × 512 pixels. Each sampling point collects images from four directions (0°, 90°, 180°, and 270°) to comprehensively record the street environment.

https://doi.org/10.3390/buildings16020314

(4) Data cleaning: Invalid images caused by access restrictions, construction areas or data loss were excluded. Ultimately, 25,862 valid sampling points were obtained, corresponding to 103,448 street view images.

3.3. Perception Label Data

This study adopts the MIT Place Pulse 2.0 dataset as the source of perception labels. This dataset contains 110,988 street view images from cities around the world. Each image was perceived and evaluated by 81,630 participants through pairwise comparisons, covering six dimensions: beauty, boring, depression, vitality, safety and affluence [25]. By adopting the transfer learning strategy and using the ResNet-50 (Residual Network with 50 layers) model pre-trained on this dataset, through the forward propagation calculation of Zhengzhou street view images, the perception scores of six dimensions were obtained. This method has been proven to maintain stable prediction performance in different geographical environments [34]. Thus, the dataset used in this study consists of two components:

(1) Visual element data: 103,448 Baidu street view images from Zhengzhou, from which 19 categories of visual elements were extracted via semantic segmentation.

(2) Perception label data: Six-dimensional perception scores predicted by the ResNet-50 model trained on the MIT Place Pulse 2.0 dataset.

4. Research Methods

This study adopts a comprehensive analysis framework integrating street view images, deep learning and interpretable machine learning. The framework is designed to address the three core research questions outlined in the Introduction: (1) measuring the spatial pattern of multi-dimensional urban perception, (2) identifying key visual elements influencing perception, and (3) elucidating the mechanisms through which these elements affect different perceptual dimensions. The technical route covers three core stages that are successively connected (Figure 2). The first stage is visual element analysis: The SegFormer-BB5 model pre-trained on the Cityscape dataset is used to perform fine semantic segmentation on street view images, and the system extracts and quantifies the composition ratios of 19 key visual elements. The second stage is multi-dimensional perception prediction: Based on the ResNet-50 deep residual network trained on the large-scale crowdsourced dataset MIT Place Pulse 2.0, the perception evaluation is constructed as a binary classification task, and the perception scores of street view images in the six dimensions of beauty, safety, vitality, wealth, depression, and boredom are obtained through transfer learning. And by means of visualization methods, it presents its spatial differentiation pattern. The third stage is the analysis of perceptual elements: Taking 19 types of visual elements as features, the six-dimensional perception score and the comprehensive perception score as target variables, the LightGBM model was constructed, and combined with the SHAP (Shapley Additive Explanations) interpretation framework based on game theory, Quantify the contribution degree and action direction of each visual element in different perception dimensions, thereby identifying the key visual features that affect urban perception.

https://doi.org/10.3390/buildings16020314

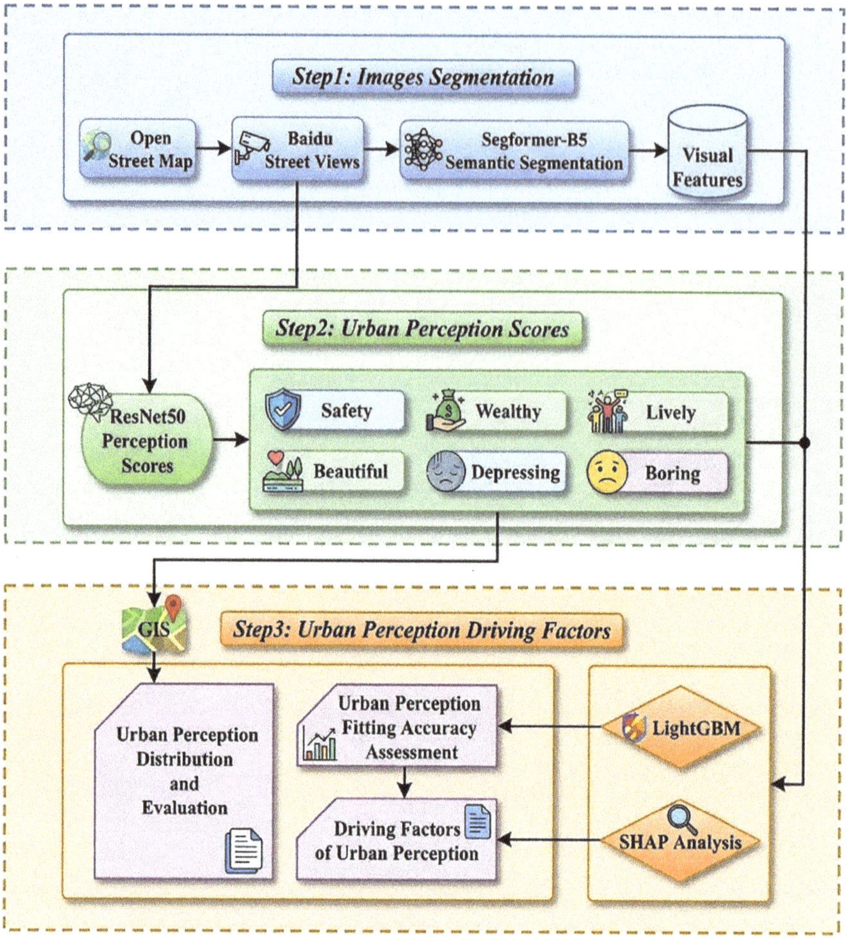

Figure 2. Technology Roadmap.

4.1. Street View Image Semantic Segmentation

This study implemented the Pyramid Scene Parsing Network (PSPNet) architecture (Figure 3), trained on the MIT CSAIL-developed ADE20K dataset, to systematically quantify urban visual components. The ADE20K repository provides semantically rich annotations for 150 object categories spanning built environment features (e.g., buildings, infrastructure) and natural elements (e.g., sky, vegetation) [35]. PSPNet demonstrates particular efficacy in complex urban contexts, achieving 79.73% mean intersection-over-union (mIoU) segmentation accuracy, with its robustness for streetscape analysis empirically validated across multiple urban computing studies [36,37].

$$\mathrm{cS_n^0} = \mathrm{CNN}(I_n) \tag{1}$$

$$\mathrm{S_n^l} = \mathrm{f_{seg}}(\mathrm{MAS}(\mathrm{LN}(\mathrm{s_n^{l-1}}))), \mathrm{l} = 1, 2, \ \cdots, \mathrm{L} \tag{2}$$

$$\mathrm{I_n} = [\frac{0 \text{ in } \mathrm{f_{seg}}(\mathrm{S_n^l})}{\text{total pixels}}, \frac{1 \text{ in } \mathrm{f_{seg}}(\mathrm{S_n^l})}{\text{total pixels}}, \frac{19 \text{ in } \mathrm{f_{seg}}(\mathrm{S_n^l})}{\text{total pixels}}] \tag{3}$$

https://doi.org/10.3390/buildings16020314

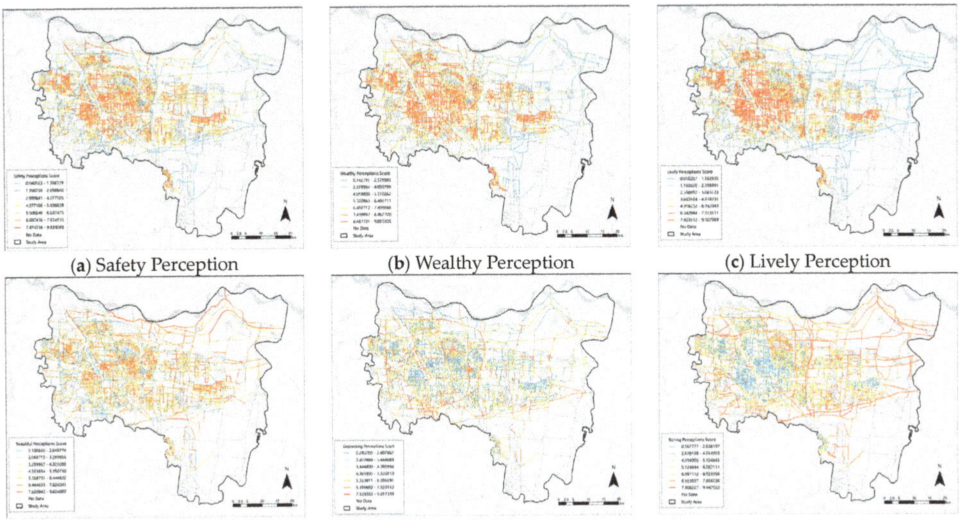

(**a**) Safety Perception (**b**) Wealthy Perception (**c**) Lively Perception

(**d**) Beautiful Perception (**e**) Depressing Perception (**f**) Boring Perception

Figure 3. Spatial distribution of different perception indicators.

Among them, f_{seg} is a multi-layer perceptron with 19 classes, and each ln is a pixel-by-pixel label mapping of in. CNN stands for Convolutional Neural Network. Each label map ln contains the pixel ratio of each object category in the street view image.

4.2. Perceptual Score Based on ResNet-50

This study adopts the ResNet-50 architecture, which is a 50-layer deep residual network proposed by He et al. [38] and uses a bottleneck design. Use its pre-trained form to take advantage of its proven effectiveness in capturing the hierarchical features of urban elements such as building facades and vegetation. To address the inherent uncertainties in perceptual evaluation, the established method is adjusted by reconstructing the task into a binary classification problem [39]. The implementation process consists of four consecutive components: data standardization, transfer learning, spatial quantification, and compound scoring. Our data standardization process involves strictly screening street view images with $\geq$5 annotations, using Q-score classification (mean $\pm$ standard deviation) to eliminate median values and enhance model robustness. Subsequent transfer learning is carried out through a 5:1 training-validation segmentation configuration, and the model performance is improved by iteratively optimizing the network parameters. In the spatial quantization stage, the classification probability is transformed into perception intensity indicators to achieve spatial mapping of subjective perception. The final output generates perception scores in six dimensions: beauty, boredom, depression, vitality, safety, and affluence. These scores are then aggregated into a composite score through weighted averaging. It is worth noting that score inversion is applied to the negative perception dimension to ensure that higher values always correspond to positive perception attributes and maintain the consistency of interpretation for all evaluation indicators.

4.3. Build an Environmental Perception Model Based on LightGBM

This study employs the LightGBM algorithm to model and analyze environmental perception and visual elements. LightGBM (Light Gradient Boosting Machine) is a Boosting method based on Gradient Boosting Decision Tree (GBDT) [40]. This algorithm effectively excludes most of the small Gradient samples by introducing gradient-based One-side

https://doi.org/10.3390/buildings16020314

Sampling (GOSS), and only retains the large gradient samples for information gain calculation. Thus, while ensuring the model accuracy, it reduces the amount of training data. In addition, LightGBM also adopts a leaf-wise strategy with depth constraints, that is, each time, the Leaf with the maximum splitting gain is found among all current leaf nodes for splitting, and this cycle continues. Under the condition of the same number of splits, the Leaf-wise strategy can reduce errors and improve accuracy. The restriction on the maximum tree depth further avoids the overfitting phenomenon.

To comprehensively evaluate the model's performance, three quantitative indicators were implemented: Mean Absolute Error (MAE) quantifies the average prediction deviation, Root mean square error (RMSE) emphasizes larger prediction errors, and Coefficient of Determination (R^2) assesses the explanatory power. The combination of these indicators provides a multi-dimensional assessment of the model's accuracy in simulating human perceptual responses to urban visual stimuli. The calculation formula is as follows:

$$\mathrm{MAE} = \frac{1}{n}\sum\nolimits_{i=1}^{n} |(y_i - \hat{y}_i)| \tag{4}$$

$$\mathrm{RMSE} = \sqrt{\frac{1}{n}\sum\nolimits_{i=1}^{n} (y_i - \hat{y}_i)^2} \tag{5}$$

$$\text{Standard } R^2 = 1 - \frac{\sum_{i=1}^{n} (\hat{y}_i - y_i)^2}{\sum_{i=1}^{n} (\hat{y} - y_i)^2} \tag{6}$$

Among them, $\hat{y}_i$ and y_i respectively represent the predicted value and actual value of the RF model for the sample. $\overline{y}$ represents the mean of the actual values of all samples. n represents the total number of samples.

4.4. An Explanation Based on SHAP

SHAP (Shapley Additive Explanations) is an explainable machine learning framework based on the concept of Shapley values in cooperative game theory [41], providing model-independent explanations for feature contributions in predictive modeling. This game theory approach can operate without understanding the internal model architecture. Instead, it quantifies the positive and negative directional impacts of each variable on individual predictions by analyzing feature interactions and distribution patterns and through conditional expectation calculations. The obtained SHAP values not only reveal the intrinsic data distribution characteristics but also explain the model's decision-making pattern.

In implementation, this interpretive framework assesses the differentiated impact of urban visual elements on different perception dimensions through the Shapley value decomposition system, and the calculation process is formalized in Formula (7) [42]. The magnitude and polarity of these values establish a quantified ranking of feature importance, thereby enabling systematic identification of the key visual determinants that shape human perception and response to the urban environment.

$$g(\acute{Z}) = \varnothing_0 + \sum_{i=1}^{M} \varnothing_i \acute{Z}_i \tag{7}$$

Here, $g(\acute{Z})$ represents the perceived predicted value of the visual element. $\varnothing_0$ represents the average value of environmental perception. M represents the number of visual elements in the perception model. $\varnothing_i$ represents the SHAP value of the i-th visual element.

https://doi.org/10.3390/buildings16020314

$\acute{Z}_i i \in \{0,1\}$ indicates whether the i-th visual element participates in the model prediction. The calculation formula for the SHAP value [43] is as follows:

$$\phi_i = \sum_{S \subseteq F\{i\}} \frac{|S|!(|F| - |S| - 1)!}{|F|!} [f_{S \cup i}(x_{S \cup i}) - f_S(x_S)] \quad (8)$$

Here, ϕ_i represents the SHAP value of the -th visual element. F represents the set of visual elements involved in the prediction. f represents the collection of all visual elements.

In which the perceptual impact of urban visual elements is quantified as a weighted aggregation of score differences across different spatial configurations. This process is mechanically equivalent to calculating the weighted sum of the changes in perception scores under different combinations of visual element proportions, and ensures mathematical rigor through double validation: Formula (8) enforces algorithmic fairness through the principle of symmetrical distribution [44,45]. By processing the visual element dataset with the corresponding perception metrics through this SHAP framework, we have established a direct quantitative representation of the perception impact of environmental features. The obtained values exhibit magnitude dependent significance (absolute values indicate the intensity of influence) and pole-specific interpretation (positive/negative signs indicate directional effects), jointly forming an interpretable mapping between the visual characteristics of a city and its psychological impact. This method, through comparative analysis of their Shapley-derived contribution metrics, can systematically identify key design elements.

5. Results

5.1. Spatial Distribution of Different Perception Indicators

Based on the ArcGIS Pro 3.1.5 platform, the coordinate data of street view image collection points are spatially correlated with their perception scores. The natural breakpoint method is used to divide the scores of six types of perception indicators into seven grades, generating a spatial distribution map (Figure 3). The analysis results show that the four positive perception indicators of safety, prosperity, vitality and beauty present a spatial pattern of "Zhongzhou Avenue as the boundary" in the main urban area of Zhengzhou City. The area east of Zhongzhou Avenue generally shows a relatively high positive perception score, while the old urban area west of Zhongzhou Avenue shows the characteristics of obvious perception differentiation and concentration of local high values. The high-value zones for safety perception (high-value zone > 7.874736), prosperity perception (high-value zone > 8.467721), and vitality perception (high-value zone > 7.923512) have extremely high thresholds and are highly overlapping in space, mainly concentrated in areas such as the CBD of Zhengdong New District and the core business districts of Jinshui District. These places, with their well-developed infrastructure, high-quality building facades and vibrant business atmosphere, have jointly shaped a highly positive urban impression. In contrast, the high-value area of beauty perception (>7.626042) shows a different distribution pattern. Its distribution is not limited to the commercial core, but is more widely distributed along the waterfront green belts such as Jinshui River and Xiong 'er River (>5.358731), as well as newly built landscape areas such as Beilong Lake and Longzi Lake, indicating that natural and artificial landscapes are the main driving forces of beauty perception.

In contrast to positive perception, the two negative perception indicators of suppression and boredom present a spatial pattern of "high in the periphery and specific areas". The high depression perception areas (>7.529553) are mainly distributed in the traditional industrial areas in the west, around railway stations and construction sites on the outskirts of cities. These areas generally have the characteristics of dense buildings and chaotic environment. The high boredom perception area (>7.806227) is concentrated

https://doi.org/10.3390/buildings16020314

in large residential areas with a single architectural style and along the lackluster branch roads. The huge range of its values from the lowest 0.567777 to the highest 9.447562 reveals the extreme unevenness of the visual richness of the city's streets. It is worth noting that the data analysis has revealed a special phenomenon: in the CBD of Zhengdong New District and some waterfront areas, there are regions with a high perception of beauty but also a relatively high perception of boredom. This indicates that although high-quality greening and modern architecture support a high beauty score, the single function of the street interface and the lack of diverse commercial and public activity Spaces have led to insufficient visual richness, thereby triggering a high sense of boredom. This proves that a single esthetic appeal is not enough to eliminate the negative perception brought about by the monotony of functions.

5.2. Spatial Distribution of Different Environmental Indicators

Based on the ArcGIS platform, the coordinate data of street view image collection points are spatially associated with nine types of environmental indicators. The natural breakpoint method is used to divide each indicator into seven levels, generating a spatial distribution map of environmental elements (Figure 4). The analysis results show that different types of built environmental elements present significant spatial differentiation characteristics in Zhengzhou City, and these environmental features provide a physical basis for understanding the spatial differences in urban perception.

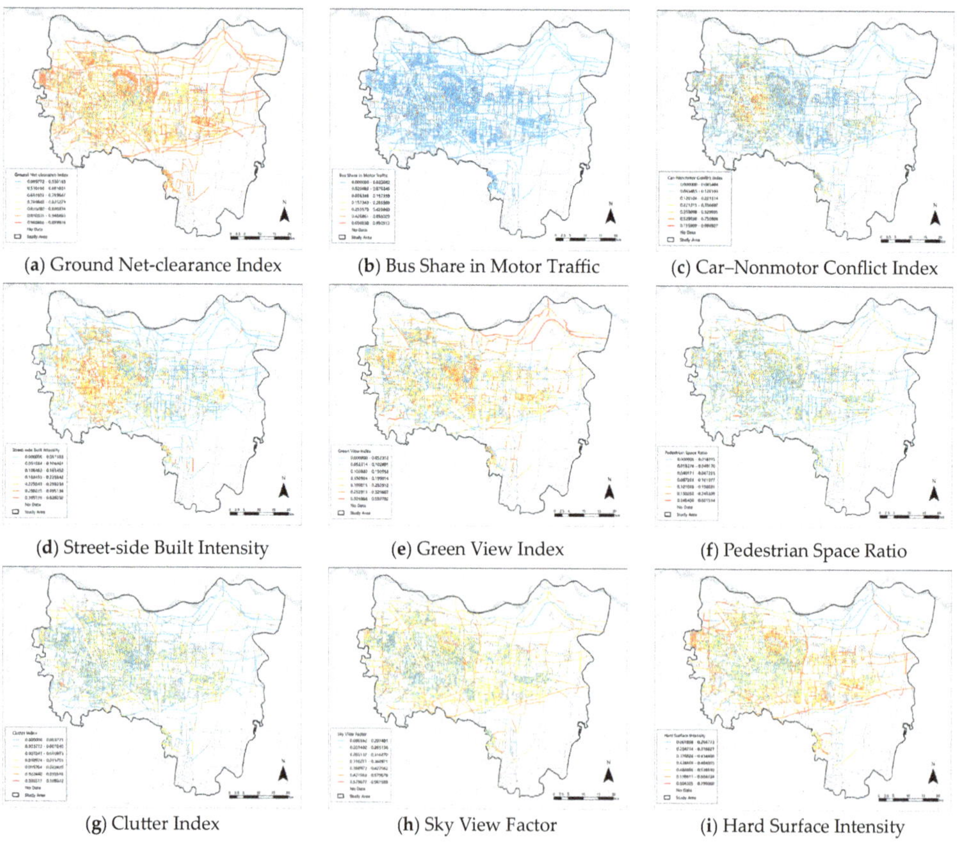

(**a**) Ground Net-clearance Index (**b**) Bus Share in Motor Traffic (**c**) Car–Nonmotor Conflict Index

(**d**) Street-side Built Intensity (**e**) Green View Index (**f**) Pedestrian Space Ratio

(**g**) Clutter Index (**h**) Sky View Factor (**i**) Hard Surface Intensity

Figure 4. Spatial distribution of different environmental indicators.

https://doi.org/10.3390/buildings16020314

In the main urban area of Zhengzhou City, the positive elements characterizing environmental quality show spatial differentiation with Zhongzhou Avenue as the boundary. The high-value area (>0.329868) of the Green View Index is mainly distributed along the waterfront green belts such as Jinshui River and Xiong 'er River, as well as the newly built landscape areas such as Beilong Lake and Longzi Lake in Zhengdong New District, which is highly consistent with the high-value area of beauty perception. The high-value areas of Pedestrian Space Ratio (>0.245400) and Ground Net-clearance index (>0.940466) are mainly concentrated in modern areas such as the CBD of Zhengdong New District and the core business district of Jinshui District. These areas provide material support for the perception of safety and affluence through spacious pedestrian Spaces and clean street interfaces. The high-value areas of the Sky View Factor (>0.579677) are also concentrated in the new urban areas with open architectural layouts, reflecting the sense of spatial openness in these areas. The old urban area west of Zhongzhou Avenue exhibits typical features such as high building density, rich but visually chaotic street interfaces, small block scales, and strong road connectivity, which form a sharp contrast with the newly built urban area.

In contrast, the indicators characterizing environmental pressure present different spatial patterns. The high-value areas of the Car–Nonmotor Conflict Index (>0.755909) are mainly concentrated around railway stations, traditional business districts and intersections of major roads. The traffic flow in these areas is complex and there are significant safety hazards. The high-value areas of Clutter Index (>0.035517) are mainly distributed in the traditional industrial areas in the west, the old urban areas and some urban fringe areas. These areas generally have problems such as disorderly street facilities and messy billboards, corresponding to the perception of poor environmental tidiness. The high-value areas of Hard Surface Intensity (>0.604325) are widely distributed along the core business districts and major traffic arteries of the city, reflecting the insufficiency of ecological regulation capacity in these areas. It is worth noting that, in some areas such as the CBD of Zhengdong New District, a special phenomenon emerged where the Green View Index was relatively high (0.252913–0.329867), but the Pedestrian Space Ratio was relatively low (0.150282–0.245399). Street view image analysis shows that although these areas have high-quality central greenery and modern buildings, the scale of the streets is large, and the convenience and humanization of walking are insufficient, resulting in a mismatch between visual esthetics and walking experience. This might be the environmental cause of the phenomenon where both the perception of beauty and boredom in this area are relatively high simultaneously.

5.3. Correlation Analysis of Different Perception Indicators

To explore the intrinsic connection between urban perception and built environmental elements, this study extracted nine environmental indicators based on street view images and conducted Pearson correlation analysis with six types of perception indicators (Figure 5). The results show that there is a significant correlation between different perception dimensions and specific environmental elements.

The environmental drivers of negative perception are relatively concentrated. Boredom perception was significantly positively correlated with Hard Surface Intensity ($r = 0.48$) and Clutter Index ($r = 0.40$), and strongly negatively correlated with Green View Index ($r = -0.69$). Repressive perception is also highly negatively correlated with the Green View Index ($r = -0.66$), indicating that green scarcity and high-density built environments are prone to induce negative emotions.

https://doi.org/10.3390/buildings16020314

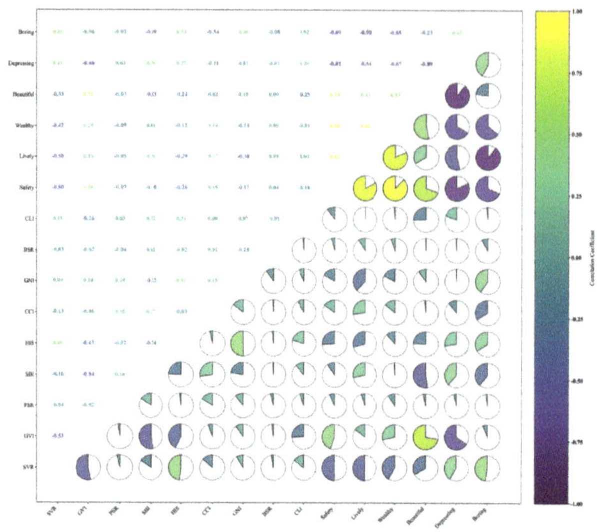

Figure 5. Spatial distribution of different environmental.

The environmental basis for positive perception is more diverse. The Green View Index has the strongest positive correlation with beauty perception (r = 0.72), highlighting the shaping effect of green view landscapes on the esthetic appeal of the city. Affluent perception is highly correlated with the Pedestrian Space Ratio (r = 0.84) and the Ground Net-clearance Index (r = 0.83), indicating the key role of wide and clean streets in affluent perception. Vitality perception also depends on the Pedestrian Space Ratio (r = 0.83), while security perception is positively correlated with the Green View Index (r = 0.45) and the Sky View Factor (r = 0.51), reflecting the importance of an open and transparent visual environment for a sense of security.

Furthermore, CAR-nonmotor Conflict Inde is negatively correlated with safety perception (r = −0.26), while Clutter Index systematically weakens positive perception and enhances negative perception, becoming a key entry point for optimizing urban perception. Relevant analysis provides a clear path for precisely intervening in perceptual experience through environmental elements.

5.4. Regression Analysis of Zhengzhou City

This study conducted a regression analysis of street view images based on the LightGBM algorithm to evaluate the urban perception characteristics of Zhengzhou City in six perception dimensions: beauty, boredom, depression, vitality, safety and prosperity. Taking the perception score predicted by the ResNet-50 model as the dependent variable and nine street view visual elements as features, a regression model of six perception dimensions was constructed. During the training process, a 5-fold cross-validation method was adopted and repeated twice to optimize the model's hyperparameters. The fitting degree and generalization ability of the model were evaluated through the mean square error (MSE) and the coefficient of determination (R^2).

The model evaluation results show (Table 1) that there are differences in the prediction performance of each perception dimension. The esthetic perception model has the strongest explanatory power. The R^2 of the test set reaches 0.6190, indicating that the model can capture the visual features related to esthetics in the image quite well. The model performances of boredom perception and vitality perception are also relatively stable, with R^2 of the test sets being 0.6427 and 0.6434, respectively. In contrast, the models of repressive

https://doi.org/10.3390/buildings16020314

perception (test set $R^2 = 0.5699$) and affluent perception (test set $R^2 = 0.4703$) performed relatively weakly, possibly due to the higher subjectivity of these perception dimensions and the complexity of visual cues. The predictive performance of the safety perception model lies somewhere in between, with the test set R^2 being 0.5001.

Table 1. Regression Analysis Results.

Perception Category	Beautiful	Boring	Depressing	Lively	Safety	Wealthy
MAE	1.0756	0.8724	0.9538	1.3876	1.4986	1.3218
RMSE	1.3213	1.1187	1.2219	1.6852	1.7406	1.5427
R^2	0.6190	0.6427	0.5699	0.6434	0.5001	0.4703

Overall, this series of models has successfully revealed the connection between the street scene visual environment and multi-dimensional perception in Zhengzhou City. The perceptual features of beauty, boredom and vitality dimensions can be effectively captured by the model, while the modeling of suppression and abundance perception suggests that future research needs to incorporate more non-visual features such as socio-economic or functional mix degree to enhance the explanatory power.

5.5. LightGBM Regression Results

Based on the results of the feature importance analysis (Figure 6), the research found that the Green Visibility Index (GVI) is a core element in shaping the positive perception of a city. Its importance in the perception of beauty is as high as 70.53%, and it contributes 22.25% and 14.67%, respectively, in the perception of safety and prosperity, highlighting the fundamental role of greening in enhancing the multi-dimensional perception quality of a city. The Sky visibility factor (SVF) exhibits significant dual characteristics in different perception dimensions: on the one hand, it serves as a key element in the formation of a sense of security (39.79%), while on the other hand, it may exacerbate boredom by causing the absence of street interfaces (21.75%), revealing the complexity of the influence mechanism of built environmental elements. In addition, the proportion of pedestrian space (PSR) dominates in the perception of affability (31.15%), and the ground clearance index (GNI) contributes the most to the perception of vitality (34.67%), all pointing to the importance of humanized design and spatial tidiness for high-quality urban experiences. The research also identified building density (SBI) and traffic conflict (CCI) as the main sources of negative perception, having a significant impact on both boredom and depression perception. These findings indicate that the formation of urban perception depends on the synergistic effect of multiple elements. A single environmental element may play a completely different role in different perception dimensions, which provides a scientific basis for regulating urban perception through precise environmental design and suggests that future urban renewal should focus on the coordinated advancement of the construction of a green system, the control of spatial scale, and the optimization of functional layout.

Multidimensional analysis based on SHAP values reveals the complex influence mechanisms and directions of different visual elements on urban perception (Figure 7). Research has found that the Green Visibility Index (GVI) has a significant positive driving effect on beauty perception. Its high value significantly enhances the evaluation of beauty, reflecting the core position of greening in shaping the esthetic appeal of a city. The Sky visibility factor (SVF) shows a clear positive impact on security perception, indicating that an open visual space helps enhance a sense of security. However, in the perception of boredom, a higher SVF value is associated with negative evaluations, revealing that the same element may play opposite roles in different perception dimensions. Further analysis indicates that the ground clearance Index (GNI), as a key positive element of vitality

https://doi.org/10.3390/buildings16020314

perception, significantly promotes the feeling of street vitality when its improvement is enhanced. However, building density (SBI) and vehicle conflict (CCI) exhibit a relatively strong negative impact in the perception of oppression. It is particularly worth noting that the contribution of the pedestrian space ratio (PSR) to the perception of affluence shows nonlinear characteristics, only generating positive effects within the appropriate range, and the promoting effect weakens after exceeding the threshold. The hard interface ratio (HIS) also shows a positive correlation in the perception of affordance, but a negative correlation with the perception of vitality, suggesting its dual role in different contexts.

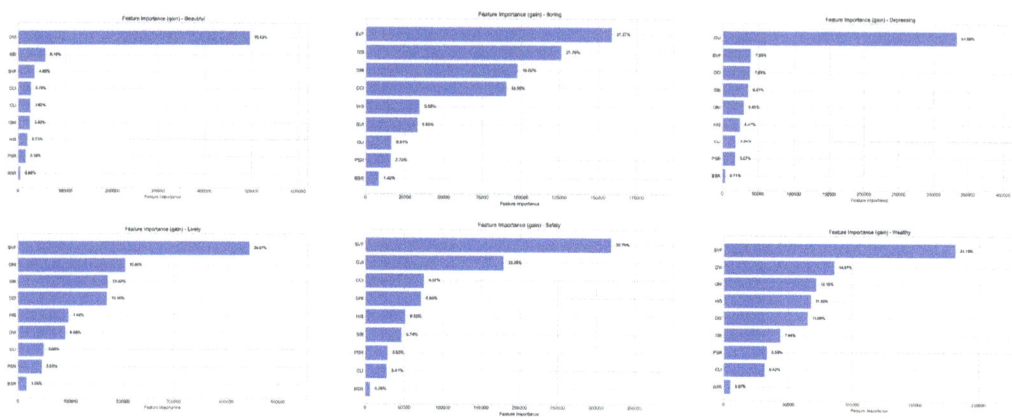

Figure 6. LightGBM Feature Importance Ranking.

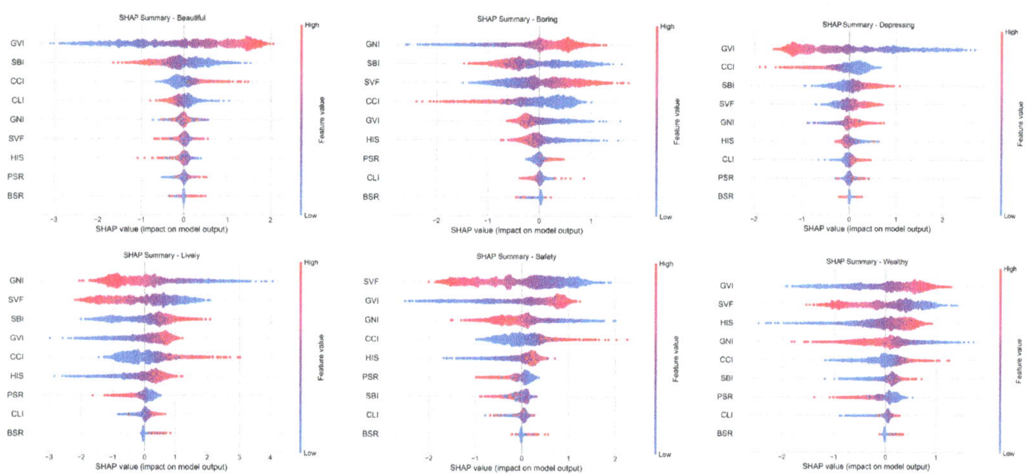

Figure 7. Multidimensional Analysis of SHAP Values.

It is found that from the two dimensions of the direction and intensity of the elements' effects, the understanding of the complex relationship between the urban visual environment and subjective perception has been deepened, providing a quantitative basis for precise urban design. It is suggested that in planning practice, attention should be paid to the coordinated allocation among elements, and differentiated strategies should be adopted for specific perception goals, so as to achieve a systematic improvement in the quality of urban space.

https://doi.org/10.3390/buildings16020314

6. Discussion and Conclusions

6.1. Conclusions

This paper constructs a comprehensive analysis framework integrating street view images, deep learning and interpretable machine learning, and quantitatively measures and explains the mechanism of urban perception in Zhengzhou City. Unlike traditional research, this framework analyzes the contribution and direction of action of visual elements to subjective perception from within the model through the SHAP method, thereby transcending the description of phenomena and precisely revealing the spatial pattern of perception results and the deep environmental driving forces behind their formation.

(1) The urban perception of Zhengzhou City presents a spatial differentiation feature with Zhongzhou Avenue as the boundary. Positive perception (safety, prosperity, vitality, and beauty) has demonstrated a "high center" agglomeration pattern in modern areas such as the CBD of Zhengdong New District and the core business district of Jinshui District. Its high value is closely related to the high-quality environmental elements such as the open Sky View Factor, sufficient Green View Index and spacious pedestrian space in the newly built urban area. The old urban area west of Zhongzhou Avenue, on the other hand, is characterized by a rich variety of historical buildings, a dense flow of people, diverse but visually chaotic street interfaces, small block scales, and high building density. In some areas, due to the chaotic environment and single function, negative perception has accumulated.

(2) The influence of each visual element was quantified through the random forest model and SHAP analysis. The Green View Index was identified as a core element for shaping positive perception (with the importance of beauty perception features reaching 70.53%) and suppressing negative perception. The Sky View Factor and the Pedestrian Space Ratio together constitute the basis of the perception of safety, affability and vitality. The research also reveals the complexity of the role of environmental factors. In the newly built urban area east of Zhongzhou Avenue, due to the large street scale and single building interface, the SVF value is relatively high. Although it enhances the sense of openness, it intensifies the sense of boredom in specific situations, reflecting the nonlinear characteristics of the perceptual influence mechanism.

(3) It is suggested that, in the planning, the characteristics of the built environment on both the east and west sides of Zhongzhou Avenue should be combined to implement differentiated renewal strategies. The visual order and vitality of the old urban area on the west side should be enhanced through interface improvement, functional integration and micro-updates. The newly built urban area on the east side should enhance the details of the building interface, control the scale of the streets and increase the diversity of functions to alleviate the perceptual contradiction of "beautiful but boring". The findings provide a scientific basis and precise entry points for urban design and renewal practices. It is suggested that in the planning, the characteristics of the built environment on both the east and west sides of Zhongzhou Avenue should be combined to implement differentiated renewal strategies.

6.2. Discussion

The perceptual spatial pattern of Zhengzhou City revealed in this study not only confirms the universal law of the "center-edge" gradient of urban perception, but also demonstrates the uniqueness of it as a regional central city. Compared with first-tier cities like Beijing and Shanghai, the high-value areas of beauty perception in Zhengzhou are more prominently distributed along the riverside green belts such as Jinshui River and Xiong 'er River, which is closely related to its construction concept of "Beautiful City by Water". This finding contrasts sharply with Zhang et al. [14]'s research on Beijing, where the perception of beauty is more associated with historical and cultural heritage areas and modern commercial centers. In contrast, Zhengzhou demonstrates the emphasis placed by an emerging inland central city on ecological landscape construction. Meanwhile, as a newly built urban area, Zhengdong New District, with its wide street scale, complete infrastructure and reasonable functional layout, has jointly contributed to the continuous high-value distribution of vitality perception. This feature is relatively rare in the research on Chinese cities by Wei et al. [18], reflecting the practical achievements of urban planning concepts in the new era.

The causes of negative perception are particularly worthy of in-depth exploration. The research finds that the perception of repression in Zhengzhou City presents a "dual-core" distribution feature, which not only occurs in the old urban area with dense buildings, but also exists in some newly built areas. The SHAP analysis results reveal the environmental causes of this phenomenon from the mechanism level: On the one hand, some new districts overly pursue visual effects, leading to an imbalance in spatial scale. Although an excessively high SVF value (>0.65) enhances the sense of openness, it also intensifies the sense of spatial coldness due to the lack of building enclosure. On the other hand, the lack of vertical greening and details on the building facade has prevented the positive effects of the GVI from being fully exerted, intensifying the sense of oppression in the space. The special phenomenon of "beautiful but boring" that has emerged in areas such as the CBD of Zhengdong New District reveals the common problem of "emphasizing landscape over function" in current urban construction. Simple visual beautification is not enough to create a vibrant urban space; it must be combined with diverse commercial facilities and public activity Spaces. This finding corroborates the research conclusion of Liu et al. [25] on the impact of functional mixing degree on vitality perception, further strengthening the key role of functional diversity in the formation of urban perception.

Based on the SHAP analysis results, this study proposes the following targeted optimization suggestions:

(1) In terms of vegetation configuration, it is recommended to adopt a composite greening system of trees, shrubs and ground covers to maintain the road green coverage rate within the optimized range of 25% to 35%. This range has been proven in the SHAP analysis to maximize the positive effects of GVI. At the same time, control the density of trees to maintain visual corridors, ensuring that the SVF remains within the ideal range of 0.4 to 0.6, taking into account both openness and spatial enclosure.

(2) In terms of architectural interface treatment, the visual richness should be enhanced by increasing the advance and retreat relationship, using harmonious material colors, and controlling the length of continuous interfaces (it is recommended not to exceed 50 m). SHAP analysis shows that appropriate changes in the architectural interface can reduce the negative impact of SBI on boredom perception by approximately 23%.

(3) In terms of the organization of street Spaces, emphasis should be placed on functional integration. Especially in waterfront areas where beauty perception is high but boredom is obvious, small-scale commercial functions such as coffee shops, bookstores, and convenience stores, as well as cultural and leisure functions such as exhibitions and performances, should be appropriately added. Research shows that when the functional blending degree is raised above 1.8, the perception of vitality can increase by approximately 35%, effectively alleviating the perceptual contradiction of "beautiful but boring".

These design suggestions based on quantitative analysis provide specific and feasible technical paths for improving the spatial quality of Zhengzhou and other similar cities, and also offer practical examples for data-driven urban design methods.

https://doi.org/10.3390/buildings16020314

6.3. Limitations and Future Research

Despite its contributions, this study has several limitations. Firstly, the reliance on Baidu Street View imagery inherently limited data coverage, excluding some back streets and renewal areas, which may affect the generalizability of the perceptual map. Secondly, while seasonal effects were controlled, the analysis captured a static visual snapshot and did not account for dynamic factors like weather, time of day, or pedestrian flow, which are integral to real-time urban experience. Thirdly, the perception model, though robust, is based on a globally trained dataset; its cultural specificity and adaptability to the inland Chinese context could be further enhanced with a localized training corpus. Additionally, the study primarily assessed the individual impact of visual elements, leaving the interaction effects between different environmental features (e.g., greenery and building interfaces) for future exploration. Finally, urban perception is a multi-sensory process, yet this research focused solely on visual cues, omitting the influences of sound, smell, and tactile experiences. Addressing these limitations in future work, through multi-source data integration, dynamic modeling, and cross-sensory investigation, will lead to a more holistic understanding of human perception in urban environments.

Author Contributions: M.Z., Conceptualization; Writing—original draft; Writing—review and editing. C.P., Writing—original draft; Validation; Investigation. X.H., Writing—review and editing; Visualization. L.Z., Writing—review; Supervision; Validation; Data curation; Resources. M.L., Writing—review and editing; Software. All authors have read and agreed to the published version of the manuscript.

Funding: This research was funded by (1) Zhejiang Provincial Educational Science Planning Project (2025SCG205); (2) 2025 Annual Guangdong Provincial Ordinary Higher Education Institutions Young Innovative Talent Project (2025SCG205); (3) University Scheme of Social Sciences Research Fund of Henan Province (No. 2025-ZZJH-354); (4) Projects of Talents Recruitment of GDUPT (2023rcyj2015); (5) Science and Technology Programme of Maoming of Guangdong Province of China (2025664); and (6) Guangdong Province Maoming City Philosophy and Social Sciences Planning Jointly Built Project (2025GJ10).

Data Availability Statement: The datasets used and/or analyzed during the current study are available from the corresponding author on reasonable request.

Conflicts of Interest: The authors declared no potential conflicts of interest with respect to the research, authorship, and/or publication of this article.

References

1. Lynch, K. *The Image of the City*; MIT Press: Cambridge, MA, USA, 1960.
2. Saleses, P.; Schechtner, K.; Hidalgo, C.A. The collaborative image of the city: Mapping the inequality of urban perception. *PLoS ONE* **2013**, *8*, e68400. [CrossRef]
3. Dubey, A.; Naik, N.; Parikh, D.; Raskar, R.; Hidalgo, C.A. Deep learning the city: Quantifying urban perception at a global scale. In Proceedings of the European Conference on Computer Vision, Amsterdam, The Netherlands, 11–14 October 2016; Springer: Cham, Switzerland, 2016; pp. 196–212.
4. Liu, X.; He, J.; Yao, Y.; Zhang, J.; Liang, H.; Wang, H.; Hong, Y. Classifying urban land use by integrating remote sensing and social media data. *Int. J. Geogr. Inf. Sci.* **2017**, *31*, 1675–1696. [CrossRef]
5. Ge, Y.; Zhang, H.; Liu, X. Assessing sense of place in urban neighborhoods using social media data: A case study of Luoyang, China. *Sustainability* **2020**, *12*, 2134.
6. Xie, L.; Zhang, H.; Wang, C. Identification of urban functional areas based on the multimodal deep learning fusion of high-resolution remote sensing images and social perception data. *Buildings* **2022**, *12*, 556. [CrossRef]
7. Cheng, L.; Chu, S.; Zong, W.; Li, S.; Wu, J.; Li, M. Use of Tencent street view imagery for visual perception of streets. *ISPRS Int. J. Geo-Inf.* **2017**, *6*, 265. [CrossRef]
8. Zhang, F.; Zhou, B.; Liu, L.; Liu, Y.; Fung, H.H.; Lin, H.; Ratti, C. Measuring human perceptions of a large-scale urban region using machine learning. *Landsc. Urban Plan.* **2018**, *180*, 148–160. [CrossRef]

https://doi.org/10.3390/buildings16020314

9. Wei, J.; Yue, W.; Li, M.; Gao, J. Mapping human perception of urban landscape from street-view images: A deep-learning approach. *Int. J. Appl. Earth Obs. Geoinf.* **2022**, *112*, 102886. [CrossRef]
10. Lundberg, S.M.; Lee, S.I. A unified approach to interpreting model predictions. In Proceedings of the Advances in Neural Information Processing Systems, Long Beach, CA, USA, 4–9 December 2017; pp. 4765–4774.
11. Liu, Y.; Chen, M.; Wang, M.; Huang, J.; Thomas, F.; Rahimi, K.; Mamouei, M. A Study on the Interpretable Urban Perception Model Based on Street View Images. *J. Geo-Inf. Sci.* **2022**, *24*, 2045–2057.
12. Office of the Leading Group for the Seventh National Population Census of the State Council. *Main Data of the Seventh National Population Census*; China Statistics Press: Beijing, China, 2021.
13. Yao, Y.; Liang, Z.; Yuan, Z.; Liu, P.; Bie, Y.; Zhang, J.; Wang, R.; Wang, J.; Guan, Q. A human-machine adversarial scoring framework for urban perception assessment using street-view images. *Int. J. Geogr. Inf. Sci.* **2019**, *33*, 2363–2384. [CrossRef]
14. Ma, X.; Ma, C.; Wu, C.; Xi, Y.; Yang, R.; Peng, N.; Zhang, C. Measuring human perceptions of streetscapes to better inform urban renewal: A perspective of scene semantic parsing. *Cities* **2021**, *110*, 103086. [CrossRef]
15. Tang, J.; Long, Y. Measuring visual quality of street space and its temporal variation: Methodology and its application in the Hutong area in Beijing. *Landsc. Urban Plan.* **2019**, *191*, 103436. [CrossRef]
16. Liu, S.; Long, Y.; Zhang, L.; Yang, J.; Dong, W. Quantitative measurement of urban spatial vitality by integrating physical built environment and subjective perception dimensions. *Environ. Plan. B Urban Anal. City Sci.* **2024**, *52*, 131–149. [CrossRef]
17. He, K.; Zhang, X.; Ren, S.; Sun, J. Deep residual learning for image recognition. In Proceedings of the IEEE Conference on Computer Vision and Pattern Recognition, Denver, CO, USA, 3–7 June 2016; pp. 770–778.
18. Han, H.; Yang, K.; Yang, C.; Yang, G.; Xu, L. Influence and Mechanism of a Multi-Scale Built Environment on the Leisure Activities of the Elderly: Evidence from Hefei City in China. *Int. J. Environ. Res. Public Health* **2022**, *19*, 9237. [CrossRef]
19. Pan, C.; Li, H.; Wang, L.; Wu, J.; Guo, J.; Qiu, N.; Liu, X. Data science basis and influencing factors for the evaluation of environmental safety perception in Macau parishes. *Adv. Contin. Discret. Models* **2024**, *2024*, 44. [CrossRef]
20. Chen, C.; Li, H.; Luo, W.; Xie, J.; Yao, J.; Wu, L.; Xia, Y. Predicting the effect of street environment on residents' mood states in large urban areas using machine learning and street view images. *Sci. Total Environ.* **2022**, *816*, 151605. [CrossRef] [PubMed]
21. Hu, S.; Xing, H.; Luo, W.; Wu, L.; Xu, Y.; Huang, W.; Liu, W.; Li, T. Uncovering the association between traffic crashes and street-level built-environment features using street view images. *Int. J. Geogr. Inf. Sci.* **2023**, *37*, 2367–2391. [CrossRef]
22. Ke, G.; Meng, Q.; Finley, T.; Wang, T.; Chen, W.; Ma, W.; Ye, Q.; Liu, T.Y. LightGBM: A Highly Efficient Gradient Boosting Decision Tree. In Proceedings of the Advances in Neural Information Processing Systems, Long Beach, CA, USA, 4–9 December 2017; pp. 3146–3154.
23. Lundberg, S.M.; Erion, G.; Chen, H.; DeGrave, A.; Prutkin, J.M.; Nair, B.; Katz, R.; Himmelfarb, J.; Bansal, N.; Lee, S.I. From local explanations to global understanding with explainable AI for trees. *Nat. Mach. Intell.* **2020**, *2*, 56–67. [CrossRef]
24. Yang, J.; Li, X.; Du, J.; Cheng, C. Exploring the relationship between urban street spatial patterns and street vitality: A case study of Guiyang, China. *Int. J. Environ. Res. Public Health* **2023**, *20*, 1646. [CrossRef]
25. Li, X.; Zhang, C.; Li, W. Does the visibility of greenery increase perceived safety in urban areas? Evidence from the Place Pulse 1.0 dataset. *ISPRS Int. J. Geo-Inf.* **2015**, *4*, 1166–1183. [CrossRef]
26. Biljecki, F.; Ito, K. Street view imagery in urban analytics and GIS: A review. *Landsc. Urban Plan.* **2021**, *215*, 104217. [CrossRef]
27. Xu, G. Study on Spatial Heterogeneity of Urban Agglomeration Network Structure from the Perspective of Multidimensional Flow in Western Ethnic Minority Areas. Ph.D. Thesis, Southwest University, El Paso, TX, USA, 2022.
28. Liu, Y.; Li, Z.; Tian, Y.; Gao, B.; Wang, S.; Qi, Y.; Zou, Z.; Li, X.; Wang, R. A study on identifying the spatial characteristic factors of traditional streets based on visitor perception: Yuanjia Village, Shaanxi Province. *Buildings* **2024**, *14*, 1815. [CrossRef]
29. Quercia, D.; O'Hare, N.K.; Cramer, H. Aesthetic capital: What makes London look beautiful, quiet, and happy? In Proceedings of the 17th ACM Conference on Computer Supported Cooperative Work & Social Computing, Baltimore, MD, USA, 15–19 February 2014; pp. 945–955.
30. Porzi, L.; Rota Bulò, S.; Lepri, B.; Ricci, E. Predicting and understanding urban perception with convolutional neural networks. In Proceedings of the 23rd ACM International Conference on Multimedia, Brisbane, Australia, 26–30 October 2015; pp. 139–148.
31. Chen, B.; Feng, Q.; Niu, B.; Yan, F.; Gao, B.; Yang, J.; Gong, J.; Liu, J. Multi-modal fusion of satellite and street-view images for urban village classification based on a dual-branch deep neural network. *Int. J. Appl. Earth Obs. Geoinf.* **2022**, *109*, 102794.7. [CrossRef]
32. Naik, N.; Philipoom, J.; Raskar, R.; Hidalgo, C. Streetscore-Predicting the Perceived Safety of One Million Streetscapes. In Proceedings of the IEEE Conference on Computer Vision and Pattern Recognition Workshops, Columbus, OH, USA, 23–28 June 2014; pp. 779–785.
33. Liu, Y.; Chen, M.; Wang, M.; Huang, J.; Thomas, F.; Rahimi, K.; Mamouei, M. An interpretable machine learning framework for measuring urban perceptions from panoramic street view images. *Iscience* **2023**, *26*, 106132. [CrossRef]
34. Huang, J.; Qing, L.; Han, L.; Liao, J.; Guo, L.; Peng, Y. A collaborative perception method of human-urban environment based on machine learning and its application to the case area. *Eng. Appl. Artif. Intell.* **2023**, *119*, 105746. [CrossRef]

https://doi.org/10.3390/buildings16020314

35. Suo, W.; Zhao, J. Exploring the Streetscape Perceptions from the Perspective of Salient Landscape Element Combination: An Interpretable Machine Learning Approach for Optimizing Visual Quality of Streetscapes. *Land* **2025**, *14*, 1408. [CrossRef]
36. Lin, M.; Chau, C.K.; Hou, H.C.; Tang, S.K. Development of an interpretable machine-learning model for capturing nonlinear dynamics of multisensory interactions in public open spaces. *Build. Environ.* **2025**, *280*, 113072. [CrossRef]
37. Kuang, B.; Yang, H.; Jung, T. The Impact of Visual Elements in Street View on Street Quality: A Quantitative Study Based on Deep Learning, Elastic Net Regression, and SHapley Additive exPlanations (SHAP). *Sustainability* **2025**, *17*, 3454. [CrossRef]
38. Cao, H.; Weng, M.; Kang, M.; Su, S. Unraveling the relationship between coastal landscapes and sentiments: An integrated approach based on social media data and interpretable machine learning methods. *Trans. GIS* **2024**, *28*, 1065–1089. [CrossRef]
39. Li, Y.; Lu, J.; Meng, Y.; Luo, Y.; Ren, J. Exploring Urban Spatial Quality Through Street View Imagery and Human Perception Analysis. *Buildings* **2025**, *15*, 3116. [CrossRef]
40. Zhang, J.; Yu, Z.; Li, Y.; Wang, X. Uncovering bias in objective map** and subjective perception of urban building functionality: A machine learning approach to urban spatial perception. *Land* **2023**, *12*, 1322. [CrossRef]
41. Zhao, X.; Lu, Y.; Lin, G. An integrated deep learning approach for assessing the visual qualities of built environments utilizing street view images. *Eng. Appl. Artif. Intell.* **2024**, *130*, 107805. [CrossRef]
42. Zhong, W.; Wang, L.; Han, X.; Gao, Z. Spatiotemporal Analysis of Urban Perception Using Multi-Year Street View Images and Deep Learning. *ISPRS Int. J. Geo-Inf.* **2025**, *14*, 390. [CrossRef]
43. Kang, W.; Kang, N.; Wang, P. Predicting Perceived Restorativeness of Urban Streetscapes Using Semantic Segmentation and Machine Learning: A Case Study of Liwan District, Guangzhou. *Buildings* **2025**, *15*, 3671. [CrossRef]
44. Lee, S.; Cho, N.; Park, H. Exploring nonlinear and interactive associations between built environment features and subjective streetscape perceptions. *NPJ Urban Sustain.* **2025**, *5*, 104. [CrossRef]
45. Yu, M.; Chen, X.; Zheng, X.; Cui, W.; Ji, Q.; Xing, H. Evaluation of spatial visual perception of streets based on deep learning and spatial syntax. *Sci. Rep.* **2025**, *15*, 18439. [CrossRef]

https://doi.org/10.3390/buildings16020314

buildings

MDPI

Article

Interpretable Assessment of Streetscape Quality Using Street-View Imagery and Satellite-Derived Environmental Indicators: Evidence from Tianjin, China

Yankui Yuan [1], Fengliang Tang [2], Shengbei Zhou [1], Yuqiao Zhang [3], Xiaojuan Li [3], Sen Wang [4], Lin Wang [3,*] and Qi Wang [3,*]

[1] International School of Engineering, Tianjin Chengjian University, Tianjin 300384, China; yankui_yuan@163.com (Y.Y.); shengbeizhou@163.com (S.Z.)
[2] School of Architecture, Tianjin University, Tianjin 300072, China; tang_fengliang@tju.edu.cn
[3] School of Architecture, Tianjin Chengjian University, Tianjin 300384, China; 18522566933@163.com (Y.Z.); xiaojuanli@tcu.edu.cn (X.L.)
[4] Tianjin Urban Management Research Center, Tianjin 300201, China; rokeboy2008@163.com
* Correspondence: lindasensen@tcu.edu.cn (L.W.); wq93343@163.com (Q.W.)

Abstract

Amid accelerating climate change, intensifying urban heat island effects, and rising public demand for livable, walkable streets, there is an urgent practical need for interpretable and actionable evidence on streetscape quality. Yet, research on streetscape quality has often relied on single data sources and linear models, limiting insight into multidimensional perception; evidence from temperate monsoon cities remains scarce. Using Tianjin's main urban area as a case study, we integrate street-view imagery with remote sensing imagery to characterize satellite-derived environmental indicators at the point scale and examine the following five perceptual outcomes: comfort, aesthetics, perceived greenness, summer heat perception, and willingness to linger. We develop a three-step interpretable assessment, as follows: Elastic Net logistic regression to establish directional and magnitude baselines; Generalized Additive Models with a logistic link to recover nonlinear patterns and threshold bands with Benjamini–Hochberg false discovery rate control and binned probability calibration; and Shapley additive explanations to provide parallel validation and global and local explanations. The results show that the Green View Index is consistently and positively associated with all five outcomes, whereas Spatial Balance is negative across the observed range. Sky View Factor and the Building Visibility Index display heterogeneous forms, including monotonic, U-shaped, and inverted-U patterns across outcomes; Normalized Difference Vegetation Index and Land Surface Temperature are likewise predominantly nonlinear with peak sensitivity in the midrange. In total, 54 of 55 smoothing terms remain significant after Benjamini–Hochberg false discovery rate correction. The summer heat perception outcome is highly imbalanced: 94.2% of samples are labeled positive. Overall calibration is good. On a standardized scale, we delineate optimal and risk intervals for key indicators and demonstrate the complementary explanatory value of street-view imagery and remote sensing imagery for people-centered perceptions. In Tianjin, a temperate monsoon megacity, the framework provides reproducible, actionable, design-relevant evidence to inform streetscape optimization and offers a template that can be adapted to other cities, subject to local calibration.

Keywords: streetscape quality; street-view imagery; urban perception; remote sensing imagery; temperate monsoon city

1. Introduction

Urban streets are core elements of the public realm. They serve as the fundamental units of circulation and the most immediately experienced public spaces in everyday life [1,2]. Systematic reviews and empirical studies consistently show that higher street visual quality is associated with greater psychological comfort and more physical activity. Within mainstream frameworks in environmental behavior and environmental psychology, visual elements shape place evaluations and travel choices through mediating factors such as perceived safety, environmental legibility, and perceptual load [3,4]. In line with this evidence, high-quality streetscapes enhance comfort and support active mobility [5,6]. These patterns accord with the 'perceived is felt' proposition, whereby visual boundaries and perceived safety strongly condition walking behavior [7]. Beyond individual experience, streetscape quality also shapes the perceived attractiveness of routes and neighborhoods, residents' quality of life, and the commercial vitality of urban areas [8,9]. As climate change intensifies and urban heat island effects worsen, street-scale greenery configurations and open spaces play an increasingly important role in regulating thermal comfort [8,10,11]. Accordingly, assessing urban street quality is both timely and necessary [12]. However, despite growing interest in streetscape perception, much of the literature still centers on morphological or ecological indicators. Systematic quantification of multidimensional subjective perceptions and climate-specific adaptive thresholds remains limited, so there is still a shortage of operational quantitative evidence to support streetscape design and management. Operationalizing these theoretical pathways into measurable evidence requires appropriate data and metrics. Contemporary studies, therefore, rely primarily on two complementary sources, street-view imagery (SVI), which captures the pedestrian perspective, and remote sensing imagery (RSI), which provides aerial observations.

Within this framework, most studies quantify streetscape quality with two complementary data sources, SVI and RSI, yet each on its own struggles to capture both fine-scale perception and wide area structure [13]. SVI is sensitive to visual experience at the human scale and, through semantic segmentation, yields segment-level phenotypic metrics such as the Green View Index (GVI), Sky View Factor (SVF), and Building Visibility Index (BVI). These visual variables are widely used to quantify and interpret urban perceptual qualities [14,15], to detect increases in walking propensity, to reveal interactions between facade colors and street functions, to relate street spatial quality to attributes of the built environment, and to investigate housing market responses to street greenery [16–19]. Recent work has constructed GVI datasets across cities and seasons and combined them with RSI to produce detailed green coverage maps, with street-scale correlations as high as 0.918, underscoring the value of a pedestrian viewpoint for fine scale perception [20]. Taken together, these findings show that SVI reflects street quality as seen by people and is especially informative for explaining walking behavior, visual experience, and perceptual preferences. At the macro level, RSI characterizes the ecological and thermal context of a city, and key indicators including the Normalized Difference Vegetation Index (NDVI), Land Surface Temperature (LST), and Normalized Difference Built-up Index (NDBI) are standard inputs in urban environmental analysis [13]. NDVI and LST provide stable evidence for urban heat island (UHI) assessment and are widely used to analyze UHI intensity and thermal risk [21,22]. Studies consistently report that NDVI robustly indicates blue and green cover and is negatively associated with thermal exposure, LST depicts urban thermal risk, time series track the evolution of green space patterns, and multi-index combinations reveal coupling between built-up intensity and the thermal environment [23,24]. These strengths make RSI a practical and scalable tool for urban ecological planning, includ-

https://doi.org/10.3390/buildings16010001

ing resource-constrained and high-throughput applications that require broad coverage and reusability [25–27]. Accordingly, aligning micro visual SVI information and macro ecological RSI layers on the same spatial anchors is a necessary step to translate theory into measurable evidence and defines the data motivation for this study.

Despite the strengths of SVI and RSI, single-source approaches have notable limitations for assessing streetscape quality. SVI captures detailed features from the pedestrian viewpoint, but its spatial representativeness is constrained by sampled road segments and image availability, with limited coverage of small streets and secondary roads [28]. In addition, operational definitions and computation of GVI and SVF vary across studies, which hampers cross-study comparability [29]. By contrast, RSI relies on overhead observations with spatial resolution on the order of 10 to 30 m. NDVI, LST, and NDBI stably indicate greenness, heat exposure, and building intensity [21], yet are insensitive to shading, interface continuity, and visibility at the street level [30], making it difficult to directly map residents' sensory experience and differences in thermal comfort in street scenes [31–33]. Comparative work also shows that visible streetscape features explain feelings of safety, aesthetics, and vitality by 7 to 15 percentage points more than RSI features alone [34]. Moreover, neglecting scale linkages can produce supply–demand mismatches. In Hangzhou, 81 percent of the area showed a mismatch in thermal risk, and even where some streets improved visibility, overall thermal risk intensified due to weak integration between macro ecological patterns and population vulnerability [35]. In Singapore, high-NDVI zones are often inside parks, and limited accessibility means they do not benefit high-traffic street segments that face more urgent thermal exposure and landscape needs [36]. In sum, SVI is stronger for micro-scale perceptual detail and RSI is stronger for macro-scale ecology and the thermal environment, but either source alone reveals only one side. These limitations directly motivate joint modeling of SVI and RSI anchored to the same spatial points and inform the multi-scale evaluation framework proposed in this study.

Building on the above theory and evidence, integrating SVI and RSI within a single evaluation framework shows clear potential in three dimensions, namely identification, interpretation, and application and transferability. For identification, combining RSI with SVI improves the discrimination of complex urban features and street elements, and studies in Jakarta report higher identification accuracy with stable outputs [37]. For interpretation, linking visual elements such as streetscape canopy and openness with remote sensing metrics such as NDVI and LST enables the quantification of ecological services and heat exposure mechanisms and yields actionable chains of evidence at the street level [38,39]. For application and transferability, interdisciplinary reviews indicate that integrating SVI and RSI within the same assessment framework helps address residents' comprehensive needs across visual perception, thermal comfort, and health and has become a common direction in urban planning and environmental perception research [40–42].

Despite the above evidence for convergent pathways, established studies still show three common limitations. First, single-source, single-scale, single-dependent-variable modeling remains prevalent [43]. Pairwise regressions that relate one dependent variable to isolated predictors such as LST, NDVI, and GVI using ordinary least squares (OLS), geographically weighted regression (GWR), or multi-scale geographically weighted regression (MGWR) struggle to span both micro-scale perception and macro structure or to capture differences in multidimensional subjective perceptions [44]. Second, weakly specified assumptions and limited diagnostics constrain mechanism discovery [41]. Linear models rarely recover nonlinear relationships or threshold effects [45], whereas machine learning models can fit complex patterns but often do not disclose functional form or provide calibrated diagnostics, with many studies reporting only coefficients or accuracy metrics without reliability calibration by bins or control for multiple comparisons [46]. Third,

https://doi.org/10.3390/buildings16010001

spatial anchors and scales are inconsistent, since the road location and orientation in SVI do not align with the spatial (10–30 m) and temporal resolutions of RSI, which undermines cross-scale and cross-study comparability and introduces structural underestimation and representativeness bias [41,43,47].

Most sample cities in existing studies are in tropical or humid climate zones, leaving megacities with temperate monsoon climates underrepresented and limiting model transferability across climates. Accordingly, we select Tianjin, a megacity in northern China, as the study area. It has a temperate monsoon climate, has experienced more frequent summer extreme heat in recent years, and exhibits a pronounced UHI [48]. The urban fabric mixes historic and newly built neighborhoods with diverse street forms and complex functions, which makes a single scale or a single indicator inadequate to represent residents' perceptions of streetscapes. In the context of an aging population, streets must provide ecological and thermal comfort functions while also meeting residents' needs for visual comfort and environmental perception [49]. Therefore, in non-tropical cities, there is a clear need for an integrated modeling approach based on multidimensional perceptual outcomes that brings together indicators of visual experience, temperature and heat exposure, and spatial interfaces and morphology. Such an integrated framework linking visual, thermal, and spatial morphology can explain residents' multidimensional street-level perceptions and support more balanced streetscape design and management.

Motivated by these gaps and by the characteristics of the study area, this study integrates SVI and RSI at the point scale to build a unified, standardized multidimensional database and to assess the following five perceptual outcomes: comfort, aesthetics, perceived greenness, summer heat perception, and willingness to linger. Our aim is to identify design relevant optimal and risk intervals for key indicators on a standardized scale, to clarify the main streetscape elements and mechanisms associated with these five perceptions, and to provide a quantitative basis for street-level planning and design. To address limitations related to single dependent variables, linear assumptions, and climate zone transferability, we adopt a three-step interpretability chain that combines Elastic Net logistic regression (ENet–Logit), Generalized Additive Models with a logistic link (GAM–Logit), and SHAP-based explanations. Accordingly, we address the following three research questions:

(1) What are the spatial patterns and overall levels of Tianjin's main streets from the street view and remote sensing perspectives?
(2) What are the directions and strengths of association between multidimensional elements and the five perceptual outcomes, and do nonlinearities and threshold intervals exist?
(3) How can the three-step interpretability chain produce reviewable optimal and risk intervals that locate key management elements and mechanisms?

Compared with previous work on visual environment perception and combined SVI–RSI approaches, this study, therefore, contributes in three ways. First, at the data level, we build a point-level multi-source database that aligns street-view imagery, satellite-derived environmental indicators, and five perceptual outcomes within a unified sampling frame in a temperate monsoon megacity. Second, at the methodological level, we apply a three-step interpretability chain, ENet–Logit, GAM–Logit, and SHAP, to recover nonlinear relationships and threshold intervals across multiple outcomes under explicit calibration and control for multiple comparisons. Third, at the substantive level, we derive climate-specific optimal and risk ranges for key indicators such as GVI, SVF, BVI, and LST and interpret these as design-relevant guidance for balancing comfort, aesthetics, perceived greenness, heat exposure, and willingness to linger.

https://doi.org/10.3390/buildings16010001

2. Study Area and Data

2.1. Study Area

Tianjin is situated in the northeastern North China Plain, bounded by the Bohai Sea to the east and the Yanshan Mountains to the north, within 38°34′–40°15′ N and 116°43′–118°04′ E, as shown in Figure 1. It has a temperate monsoon climate with hot humid summers, cold dry winters, and recorded extreme temperatures up to 39.6 °C. The city is highly urbanized, with about 13 million residents, a dense road network, and an urban area that is largely built up, dominated by multistory residential buildings and high-rise commercial and office buildings, alongside historic districts and new developments that yield diverse street forms and functions. Green space is generally scarce and unevenly distributed, with continuous street tree canopy in central districts and small isolated patches in newer areas. These climatic and morphological characteristics, together with a pronounced urban heat island, make Tianjin a representative setting for point-scale assessment and provide a stable spatial basis for constructing indicators and analyzing mechanisms under unified standards.

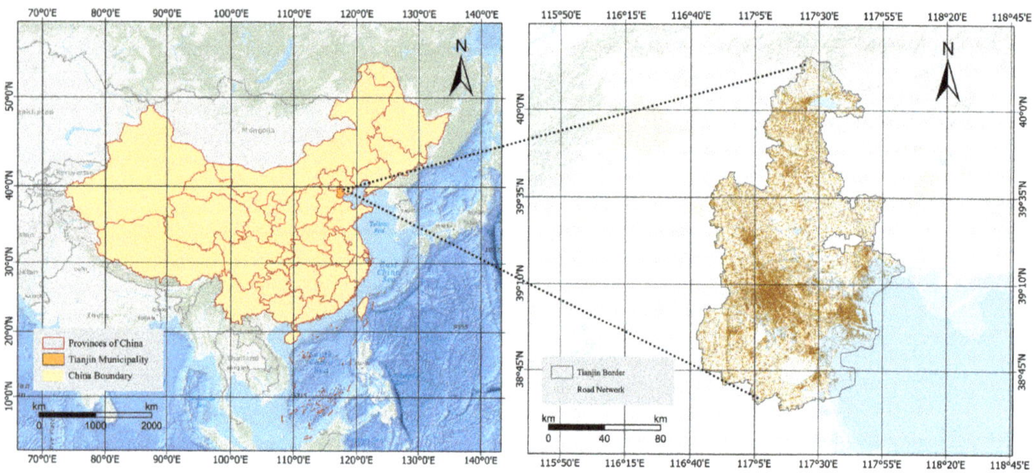

Figure 1. Location of the study area and road network context in Tianjin, China.

2.2. Analytical Framework

This study establishes an interpretable point-scale workflow that proceeds through data preparation, an ENet–Logit linear baseline, GAM–Logit nonlinear modeling with threshold interval identification, and SHAP interpretation. To limit post-selection bias, subsequent stages use the same set of independent variables on a unified z-score standardization scale. Data preparation aligns SVI and RSI sampling points, removes missing and anomalous values, and normalizes units to build a comparable point-scale database. ENet–Logit provides directional and magnitude references for each variable on the logit scale and reports log loss and AUC as lower-bound performance metrics without pruning variables. GAM–Logit fits smooth functions and uncertainties using the same inputs, identifies nonlinear patterns, and extracts threshold intervals when internal extrema are present. SHAP additively decomposes GAM–Logit predictions via KernelExplainer, reporting global importance and sample-level dependence for parallel comparison and consistency checks with preceding results. The workflow is illustrated in Figure 2.

https://doi.org/10.3390/buildings16010001

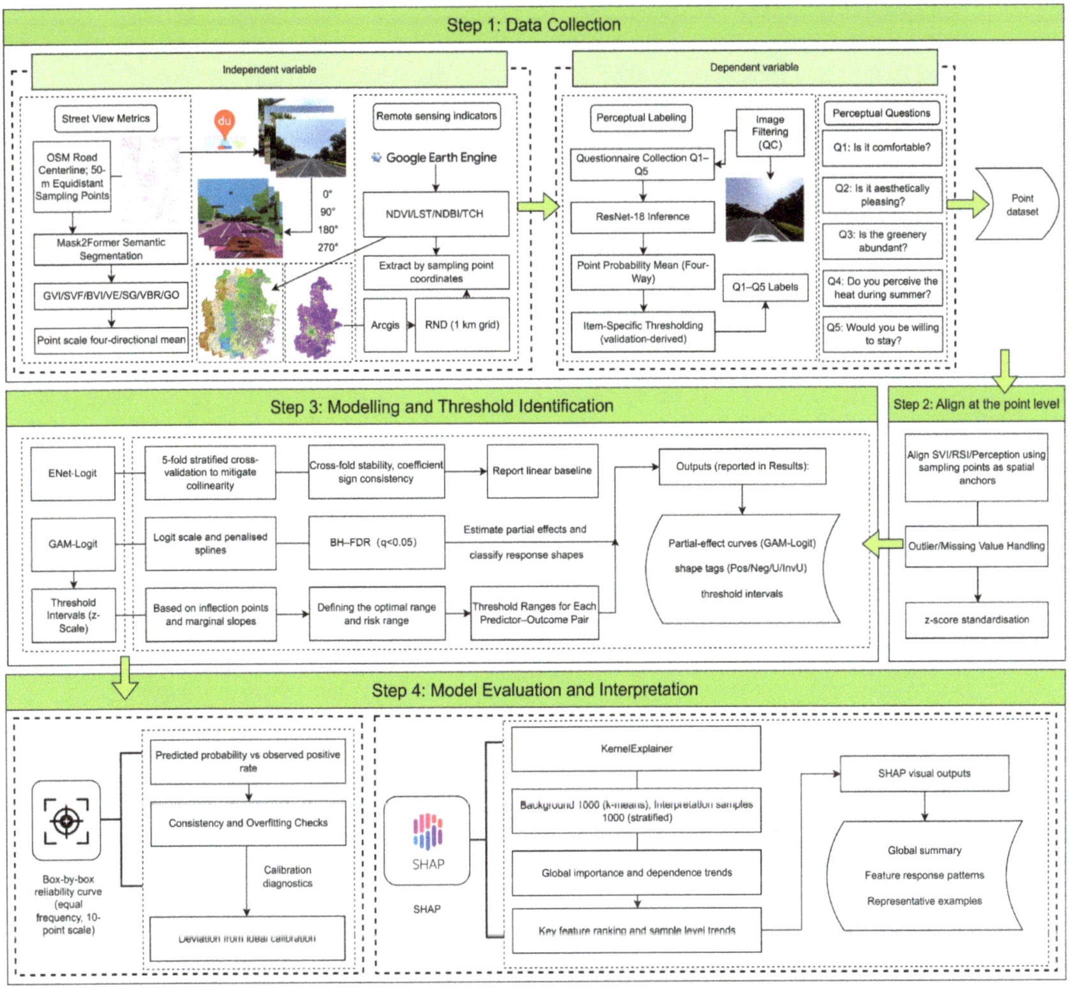

Figure 2. Flowchart.

2.3. Data Sources

At the data level, the analytical framework organizes the variables into the following three groups of environmental features: (i) visual streetscape indicators derived from SVI, capturing greenery, openness–enclosure and façade qualities; (ii) satellite-derived environmental indicators such as NDVI, LST, and NDBI; and (iii) basic spatial and contextual controls, including road configuration and building coverage. Together, these variables are used to explain the five perceptual outcomes of comfort, aesthetics, perceived greenness, summer heat perception, and willingness to linger so that both the pedestrian perspective and the overhead ecological context are represented on the same point-level sampling frame (Figure 2). The specific data sources for each group are as follows.

To ensure comparability and consistent inputs, we restrict the analysis to road space within Tianjin's built-up area that is covered by SVI. Using OpenStreetMap (OSM) road centerlines, we placed sampling points at 50 m intervals and retained only locations with valid streetscape images in four directions. At these points, we computed seven SVI and five RSI indicators and attached five binary perception labels (Q1–Q5) derived from a questionnaire based on 1000 sample images. All variables are fused at the point scale, with SVI describing the pedestrian visual interface, RSI characterizing vegetation structure and the thermal environment, and road network indicators complementing spatial morphology and accessibility. To ensure temporal consistency, RSI layers were processed for June to August 2023 with cloud masking and projection harmonization and then extracted at the SVI points. All indicators and labels are aligned to the same SVI point set to reduce structural underestimation and representativeness bias, and extrapolation is limited to road segments with SVI coverage rather than areas without streetscape imagery. Maps are displayed in Web Mercator (EPSG:3857) for visualization only and are not used in quantitative calculations. The full workflow is shown in Figure 2, and data sources and statistical settings are summarized in Table 1.

Table 1. Data sources, spatial resolution, and feature dimensions.

Data Type	Data Source	Resolution	Unit	Feature Dimension
SVI	Baidu Street View API (1)	1024 × 512 pixels	Point	Visual interface
NDVI (RSI)	Landsat 8 C2 L2 via GEE (2)	30 m	Point	Vegetation cover
LST (RSI)	Landsat 8 C2 L2 via GEE (2)	30 m	Point	Thermal environment
NDBI (RSI)	Landsat 8 C2 L2 via GEE (2)	30 m	Point	Built up intensity
TCH (RSI)	10 m Canopy Height 2020 (3)	10 m	Point	Vegetation structure
RND	OSM road vectors (4)	1 km grid	Point	Road network density
Q1–Q5	Questionnaire	Inference	Point	Perception

Table note: (1) Baidu Maps Open Platform Street View API (Baidu, Beijing, China): https://lbsyun.baidu.com/ (accessed on 13 June 2025); (2) Google Earth Engine (Google LLC, Mountain View, CA, USA): https://earthengine.google.com/ (accessed on 14 June 2025). Landsat 8 Collection 2 Level 2 data were obtained and processed on this platform; (3) GEE Community Catalog, Canopy Height 2020: https://gee-community-catalog.org/projects/canopy (accessed on 14 June 2025); and (4) OpenStreetMap (OpenStreetMap Foundation, Cambridge, UK): https://www.openstreetmap.org/ (accessed on 1 June 2025). All variables are aligned to the same SVI point set, the analysis unit is the point, and inference is limited to road space with SVI coverage. The TCH time window is 2020, while all other RSI layers use June to August 2023. Landsat spatial resolution is 30 m and TCH resolution is 10 m.

Street-view images were obtained through the Baidu Street View API, and Python 3.10 (PSF, Beaverton, OR, USA) scripts batch retrieved the latest available imagery. The sampling frame followed OpenStreetMap (OSM) road centerlines, with points placed at 50 m intervals; at each point, four directional images were captured at 0°, 90°, 180°, and 270° aligned with the road orientation. Original images at 1024 × 800 pixels were cropped to 1024 × 512 pixels to support semantic segmentation and feature extraction. In total, approximately 979,676 images were collected. This uniform high-density sampling yielded about 6 to 10 representative streetscape images within a 500 m radius of a typical neighborhood, providing adequate coverage of mixed function areas and dense street networks [50].

To quantify ecological and thermal environments at the street scale, we acquired and preprocessed multisource RSI on Google Earth Engine (GEE). NDVI, LST, and NDBI were derived from Landsat 8 Collection 2 Level 2, with imagery covering June to August 2023 selected for the study area. The Fmask algorithm was applied to mask clouds and shadows, and seasonal means were computed to characterize vegetation and thermal conditions during peak summer. Tree canopy height (TCH) was taken from the Canopy Height 2020 raster at 10 m resolution to represent urban canopy structure. In parts of the study area, this product is labeled NoData by the provider. We did not interpolate these gaps; during raster to point extraction, NoData locations were recorded with 0 as a placeholder in the data tables and flagged as missing in the metadata, where 0 does not indicate 0 m. Road network density (RND) was calculated on a 1 km by 1 km grid from OSM road vectors by summing total road length and expressing it as kilometers of road per square kilometer as

https://doi.org/10.3390/buildings16010001

a background measure of accessibility and spatial morphological complexity. All rasters were reprojected and resampled in GEE and clipped to the built-up area of Tianjin. Finally, point values were extracted at the SVI sampling coordinates so that the RSI indicators NDVI, LST, NDBI, TCH, and RND were aligned to the same SVI point set.

Crowd perception data were derived from online questionnaire annotations using SVI. To enhance the representativeness of visual composition, we first computed the pixel proportions of four primary elements, vegetation, road, sky, and building, from semantically segmented street-view images. Based on these proportions, images were categorized into low, medium, and high, forming 81 strata as the sampling frame. At most, 25 images were sampled per stratum; when fewer were available, all images in the stratum were retained and supplemented from adjacent strata. To minimize perspective variation, only longitudinal front-facing road views oriented at 90° and 180° were kept, yielding a candidate set of about 2025 images, from which 1000 images were confirmed after quality control. To improve response quality and balance visual exposure, images were divided into 50 task units of 20 images each, and a secondary stratification at the task level ensured an overall even distribution of the four elements. Each participant was randomly assigned two task units, with a randomized image order within each unit. To reduce attentional bias, minimum response durations and attention checks were implemented, and only one valid response was collected for each image. The questionnaire produced a five-item binary response matrix for 1000 images, and coded labels were aggregated to the corresponding SVI point locations to generate five binary perceptual dependent variables at the point scale.

It should be noted that because this study relies on online questionnaire annotation, the total sample is small relative to the full image database of approximately 970,000 images. Each image was evaluated by a single participant, which may introduce individual subjective bias and reduce label stability. Although the sample achieved stratified balance in visual characteristics, its spatial distribution may be biased toward central urban districts, with relatively limited coverage in peripheral neighborhoods, which can introduce structural bias. To mitigate these risks, subsequent modeling incorporated cross validation and bootstrap resampling to strengthen the robustness of estimated variable effects and the consistency of model interpretations.

2.4. Independent Variables

2.4.1. Street View Data

To characterize street environments from a pedestrian perspective, each metric is computed at the image level and then averaged across the four directions at the point scale. Semantic outputs are summarized into seven scene elements, including vegetation, sky, building, road, pavement, water, and others. Based on this classification, we calculate seven visual indicators (GVI, SVF, BVI, VE, SG, VBR, and GO) to quantify visibility and interface structure from multiple dimensions. For core metrics, GVI and SVF represent the visible proportions of vegetation and sky in the imagery and serve as key parameters for assessing greening level, shading potential, and spatial openness [51,52]. Prior work shows that these metrics are significantly associated with pedestrian thermal comfort and environmental satisfaction [51]. BVI quantifies facade enclosure, reflecting perceived confinement and ventilation potential, and has been widely applied in studies of human settlements and urban design [53]. In addition, VE measures the visible information content across elements, and existing studies report a positive association between streetscape and perceived complexity [54].

Within the set of foundational indicators, we further derive three metrics to capture element interactions and interface coordination. Recent work using panoramic street-view

https://doi.org/10.3390/buildings16010001

imagery and deep learning-based semantic segmentation shows that, even at comparable levels of overall greenness, different configurations of trees, shrubs, and grass can produce markedly different ecological and experiential outcomes [55], highlighting the need for structure-sensitive indicators in addition to aggregate measures such as GVI. SG integrates GVI and SVF to represent combined green cover and geometric openness. SVF serves as a proxy for street canyon geometry. Prior studies show that street canyon geometry and green elements act jointly to regulate thermal comfort, and the effects of different geometry and green volume combinations on radiative exposure are not simply additive [56]. In a Tianjin case study, the interaction between vegetation cover and street canyon geometry significantly influenced thermal comfort [57]. Building on this evidence, we define SG to characterize this composite effect. VBR measures interface coordination as the ratio of GVI to BVI, consistent with findings that vegetation can mitigate the negative effects of high facade enclosure and is a key determinant of perceived street quality [58,59]. GO is computed from the pixel ratio of road to pavement to indicate ground openness and accessibility, in line with the Road Area Index based on road and pavement pixels for assessing street spatial quality [60] and supported by the SWAUR framework, which treats pavement as a core dimension of walkability [61]. All metrics are computed at the image scale using pixel ratios or entropy formulas, and point-scale values are obtained by averaging across the four directions at each point. Formal definitions and formulas are provided in Table 2.

Table 2. Definitions and formulas for SVI visual indicators.

Indicator	Abbrev.	Description	Formula
Green View Index	GVI	Share of vegetation pixels, reflecting greening level and shading potential	$GVI = \frac{N_{green}}{N_{total}}$
Sky View Factor	SVF	Share of sky pixels, reflecting spatial openness	$SVF = \frac{N_{sky}}{N_{total}}$
Building Visibility Index	BVI	Share of building pixels including wall class, reflecting facade enclosure	$BVI = \frac{N_{building+wall}}{N_{total}}$
Visual Entropy	VE	Shannon entropy based on seven semantic pixel proportions, reflecting landscape complexity	$VE = -\sum_{k=1}^{7} p_k \log_2(p_k)$
Spatial Balance	SG	Product of GVI and 1 − SVF, representing the composite effect of greenness and openness	$SG = GVI \times (1 - SVF)$
Vegetation to Building Ratio	VBR	Ratio of GVI to BVI, characterizing interface coordination	$VBR = \frac{GVI}{BVI+\epsilon}$
Ground Openness	GO	Share of road and sidewalk pixels, reflecting ground openness and accessibility	$GO = \frac{N_{road}+N_{sidewalk}}{N_{total}}$

Table note: (1) All indicators are computed at the image scale and averaged across the four directions at the point scale. (2) N_{green}, N_{sky}, $N_{building}$, N_{road}, $N_{sidewalk}$ denote the pixel counts of the corresponding semantic classes, and N_{total} is the total pixel count per image. (3) p_k is the proportion of the k semantic class in the image, with seven core classes used in this study. (4) ϵ is a small constant to avoid a zero denominator.

2.4.2. Remote Sensing Data

To quantify macro ecological and spatial morphological characteristics, we use foundational RSI indicators. NDVI measures macro-level green coverage and vigor and can be compared with GVI to assess cross-scale consistency or discrepancies in perceived greenness; evidence from multiple cities shows that NDVI does not fully align with human visual perception of urban green space, underscoring the need to combine macro and micro indicators [47]. LST represents summer surface thermal exposure and is included as an

https://doi.org/10.3390/buildings16010001

environmental stress dimension, with recent work linking LST and green space exposure to subjective outcomes such as perceived heat and sleep quality [62]. NDBI characterizes the intensity of built and impervious surface cover and serves as a macro-scale remote sensing indicator of the built environment that complements BVI, with the former reflecting surface coverage and the latter capturing facade visibility. Recent reviews identify NDBI as a core index for monitoring built up areas.

For extended indicators, TCH uses a global 10 m canopy height product to represent the three-dimensional structure of vegetation, supplementing the two-dimensional limitations of NDVI and GVI. Comparative evaluations have validated multiple global canopy height datasets [63,64]. RND is computed as the total length of OSM roads within a regular 1 km grid and expressed as kilometers of road per square kilometer, describing road network complexity and the accessibility context. Recent research on urban health and walkability emphasizes the influence of road structure on human experience and access, making RND a suitable macro-level covariate for this study. Accordingly, we employ five RSI indicators, NDVI, LST, NDBI, TCH, and RND. Sampling points coincide with the SVI capture locations. Except for RND, indicators are extracted at the point scale from their native raster (30 m for NDVI, LST, and NDBI; 10 m for TCH), whereas RND is assigned to points by 1 km nearest neighbor mapping. Formal definitions and calculation formulas are provided in Table 3.

Table 3. Definitions and formulas for RSI indicators.

Indicator	Abbrev.	Description	Formula
Normalized Difference Vegetation Index	NDVI	Vegetation cover and vigor	$NDVI = \frac{NIR-Red}{NIR+Red}$
Land Surface Temperature	LST	Summer surface thermal exposure	$LST = \overline{T_K} - 273.15$
Normalized Difference Built-up Index	NDBI	Intensity of built and impervious cover	$NDBI = \frac{SWIR1-NIR}{SWIR1+NIR}$
Tree Canopy Height	TCH	Vertical vegetation structure canopy height	External raster product
Road Network Density	RND	Road density and morphological complexity	$RND = \frac{L_{road}}{A_{grid}}$

Table note: (1) $NIR, Red, SWIR1$ denote reflectance in the near-infrared, red, and shortwave infrared 1 bands. (2) $\overline{T_K}$ is the seasonal mean of per scene land surface temperature in degrees Celsius. (3) TCH is a 10 m global canopy height product, clipped and resampled only. (4) L_{road} is the total road length within a 1 km grid, and A_{grid} is the grid area.

2.5. Dependent Variables

Perceptual data were derived from the following five binary questionnaire items applied to each image: whether the scene was comfortable (Q1), whether it was aesthetically pleasing (Q2), whether it was rich in greenery (Q3), whether it was perceived as hot in summer (Q4), and whether one would be willing to linger (Q5). Responses of "yes" and "no" were encoded as 1 and 0, respectively. To address class imbalance, training used weighted binary cross entropy with a positive class weight, with the objective function defined as follows:

$$\mathcal{L}_k = -\frac{1}{N}\sum_{i=1}^{N}[\alpha_k y_{i,k}\log\sigma(\ell_{i,k}) + \alpha_k(1-y_{i,k})\log(1-\sigma(\ell_{i,k}))] \quad (1)$$

Here, k indexes the task, N is the number of training samples, $y_{i,k} \in \{0,1\}$ is the image-level label, $\ell_{i,k}$ is the model logit, $\sigma(x) = 1/(1+e^{-x})$ is the sigmoid function, and α_k is the positive class weight defined as the ratio of negative to positive samples. The classifier backbone is ResNet-18 initialized with ImageNet pretrained weights. Inputs are

https://doi.org/10.3390/buildings16010001

(384 × 384), with standard augmentation including random cropping, horizontal flipping, and color perturbation. AUC is the primary model selection metric, and F1 is referenced for threshold determination. Validation performance is reported in Table 4.

Table 4. Final selected models and validation performance for the five questionnaire items.

Item	Selected Epochs	AUC	Optimal Threshold	F1 at Threshold (Validation)
Q1 Comfort	80	0.721	0.450	0.585
Q2 Aesthetics	80	0.721	0.400	0.583
Q3 Vegetation	80	0.768	0.300	0.627
Q4 Summer heat	50	0.780	0.350	0.880
Q5 Stay	100	0.770	0.500	0.515

Table note: AUC = area under the ROC curve. "Optimal threshold" and "F1 at threshold" are computed on the validation set using a grid search over thresholds 0.05–0.95 (step 0.05).

At inference, the model first computes image-level probabilities, then averages the multi-orientation predictions at the point scale as follows:

$$p_{i,k} = \sigma(l_{i,k}), p_{s,k} = \frac{1}{M_s} \sum_{i \in \Omega_s} p_{i,k} \tag{2}$$

In this expression, $p_{i,k}$ is the image-level probability, $p_{s,k}$ is the mean probability at the point scale, Ω_s is the set of multi-orientation images for point s, and M_s is the number of orientations for that point, at most four.

At the point level, the dependent variable is defined as follows:

$$Y_{s,k} = 1(p_{s,k} \geq \tau_k) \tag{3}$$

Here, τ_k is the task-specific threshold selected on the validation set. During inference, the script loads this threshold with the corresponding task model and applies it to the probabilities.

After the above processing, SVI, RSI, and the five perceptual outcomes share the same coordinate system and form a point-level database. This unified point scale dataset provides the common input for the subsequent modeling and interpretation.

3. Methods

Building on this harmonized point-level dataset, the methods proceed in three main stages. We first derive visual streetscape indicators from SVI via semantic segmentation, then estimate ENet–Logit and GAM–Logit models to quantify associations between the predictors and the five perceptual outcomes, and finally use SHAP to interpret the GAM–Logit model and decompose feature contributions at both the global and local levels.

3.1. Semantic Segmentation Model

This study uses Mask2Former for street-view semantic segmentation with publicly available pretrained weights trained on Mapillary Vistas [65,66] (official implementation repository: https://github.com/facebookresearch/Mask2Former; accessed on 20 June 2025). Mapillary Vistas offers broad coverage of street-view imagery across multiple countries, seasons, and meteorological conditions; it contains 25,000 images spanning 66 object categories and provides instance-level annotations for 37 classes, enabling stable recognition of urban scene elements. The required metrics GVI, SVF, BVI, and related indicators depend on reliable identification and accurate boundary delineation of vegetation, sky, buildings, roads, pavements, and vehicles. In urban scenes with complex occlusions and substantial scale variation, Mask2Former shows a stable performance and supplies dependable inputs for subsequent metric computation. Semantic masks are mapped to the defined feature sets, and pixel proportions for each feature form the basis for metric calculation. Metric definitions and aggregation are detailed in Table 2.

Mask2Former uses mask-based prediction as its fundamental unit. During decoding, masked attention restricts focus to candidate mask boundaries, enhancing discriminative features within those regions. The Transformer architecture further supports modeling of cross-scale correlations. Compared with the per-pixel classification paradigm, this mechanism yields segmentation outputs with more consistent contours and coherent region features. It reduces category confusion under severe occlusion while preserving recognition fidelity for slender or porous structures such as gaps in tree canopies, skies beneath bridges, and continuous street interfaces. Public benchmarks show that Mask2Former achieves a leading performance on semantic segmentation, instance segmentation, and panoramic segmentation tasks. Its stable cross-task and cross-dataset capabilities make it suitable as a general-purpose engine for extracting urban scene elements.

At inference, a fixed resolution, cropping window, and sliding window overlap ratio were used, with a fixed random seed and batch inference. Automated quality screening for brightness, sharpness, and exposure was applied alongside anomaly detection in category proportions based on sample distributions. Severely occluded instances and invalid outputs were discarded according to predefined thresholds. All settings were held constant across the entire dataset to ensure comparability between result batches.

3.2. Linear Baseline: ENet–Logit

This study integrates SVI and RSI features at the point scale, treating the two information types as complementary rather than interchangeable across representational dimensions and spatial scales. Prior work identifies street-view imagery as a key urban data source and highlights strong scale dependence and spatial heterogeneity between street-view green coverage and remote sensing green coverage [13]. At the same time, urban greenness, surface thermal conditions, and built-up intensity systematically covary, producing widespread correlations among the merged predictors. Under such correlational designs, LASSO model selection can fail to meet representability conditions and, thus, yield unstable variable sets and inconsistent effect directions [67]. To address multicollinearity while capturing group effects, we adopt Elastic Net logistic regression (ENet–Logit), which combines L1 and L2 regularization to provide a robust linear baseline for variable direction and relative strength [68].

Twelve candidate predictors were used, seven derived from SVI and five from RSI. All predictors were standardized to z-scores before modeling. Each dependent variable (Q1–Q5) was modeled separately using stratified five-fold cross validation for out-of-sample evaluation; when any class had insufficient cases, the number of folds was reduced automatically, but never below two. Elastic Net regularization mixing coefficients and penalty strengths followed the regularization path and were held constant across tasks [69]. Log loss and AUC are reported as benchmark metrics for the linear baseline; because units and effect scales differ, these metrics are not compared directly with GAM partial effects or SHAP contributions. Stability was assessed by cross-fold consistency: a coefficient was considered a stable linear signal if it was nonzero and displayed a consistent sign in at least three folds. For presentation, coefficients with absolute standardized values below 0.10 were recorded as zero. For each predictor we report the sum of absolute standardized coefficients across tasks and the number of effective dependent variables. Tree canopy height (TCH) exhibits structural missingness in parts of the study area; it is retained in

https://doi.org/10.3390/buildings16010001

ENet–Logit for completeness of presentation but excluded from subsequent GAM–Logit and SHAP analyses and is not used for causal interpretation.

This step provides linear baseline estimates of direction and magnitude for each predictor, enabling direct comparison with the nonlinear shapes and threshold intervals derived from GAM–Logit, as well as SHAP contribution rankings and sample-level explanations. To avoid post-selection bias, the same set of independent variables is carried forward across models. The ENet–Logit baseline, therefore, serves as the starting point and constraint in the evidence chain, supporting downstream checks on the consistency of threshold interval interpretations and mechanism directionality [68].

3.3. Nonlinear Modeling: GAM–Logit

This study requires variable specific nonlinear shapes and uncertainty bearing threshold intervals for five dependent variables at the point scale, using the same set of independent variables on a unified logit scale to preserve comparability across outcomes. After aligning SVI and RSI, relying on post hoc effect visualization from tree ensembles or deep learning models to recover shapes and thresholds risks interpolation artifacts and bias in the presence of correlated predictors [70]. GAM–Logit directly estimates a smooth function for each predictor and its uncertainty on the log odds scale, producing fitted curves and confidence bands that enable more robust definition of threshold ranges while maintaining comparability across dependent variables on the same scale [71].

Within a log likelihood framework, penalized spline smoothing terms were specified for each predictor using thin plate regression splines as basic functions, with smoothing parameters selected by restricted maximum likelihood (REML). To preserve input consistency and interpretability, no interaction terms were included and all features were standardized to z scores prior to model fitting. The GAM logistic specification is, therefore, as follows:

$$logit(p) = \beta_0 + \sum_{j=1}^{k} f_j(x_j) \tag{4}$$

Here, p denotes the probability of the perceptual outcome and $f_j(\cdot)$ is the smooth function for predictor x_j on the logit scale.

On the standardized scale, relationship patterns are classified by the monotonicity of the partial effect curve and by the presence or absence of a single internal extremum. For figure annotation, we use the labels Monotonic Increase (Pos), Monotonic Decrease (Neg), U-Shaped (U), and inverted-U-Shaped (InvU). Significance is assessed with Benjamini–Hochberg false discovery rate control (BH-FDR) at $q < 0.05$. In plots, significant curves are shown with solid lines and non-significant curves with dashed lines; in tables, $q < 0.05$ is marked with $*$ and $q \geq 0.05$ with *n.s.* Threshold intervals are reported only when internal extrema exist; their left and right bounds are determined by predefined rules that account for the effect direction and its uncertainty. Calibration is evaluated with bin reliability curves using equal frequency deciles, comparing mean predicted probability per bin with the observed positive rate to assess calibration and detect overfitting. For dependent variables with extreme prior imbalance, interpretation emphasizes calibration consistency rather than discrimination metrics. Independent variables with near-zero variance or with structural missingness in the study area are retained for data alignment and reproducibility but excluded from bias effect plots, threshold interval reporting, and SHAP presentations. All GAM effect curves, shape labels, significance flags, threshold bands, and calibration diagnostics are cross-referenced with the ENet–Logit linear baseline for direction and relative strength and with SHAP global- and individual-level interpretations for consistency checks.

https://doi.org/10.3390/buildings16010001

3.4. Model Interpretation: SHAP

To complement ENet–Logit and GAM–Logit, we apply SHAP to additively decompose model outputs and report feature contributions at the global and individual levels. Grounded in Shapley values from cooperative game theory, SHAP attributes a model prediction to the marginal contribution of each feature, producing locally consistent and globally additive importance scores. Compared with conventional regression coefficients or generic feature importance measures, SHAP offers a unified additive interpretation that improves comparability across features and dependent variables and supports sample-level explanation of model behavior [72].

In this study, the explanatory target is the prediction function of the GAM–Logit model, for which we use the kernel SHAP explainer provided by the SHAP library. Because GAM–Logit is neither a tree model nor a deep model, TreeExplainer and DeepExplainer are not suitable. KernelExplainer is agnostic to model class and applies to any callable prediction function, which makes it an appropriate choice [72]. To balance representativeness and computational efficiency, k-means clustering sampled 1000 points from the full dataset as the background set, and stratified sampling selected 1000 points as the interpretation sample. SHAP explanations are limited to describing statistical associations within the current model and sample structure at two levels, global ranking and grouped summarization. Feature-level importance is reported as the mean absolute SHAP value, indicating each variable's average contribution to the prediction.

4. Results

The empirical findings are organized in line with this three-stage design. The results first present the distributions, spatial patterns, and correlations of the five perceptual outcomes and the environmental indicators, then report the ENet–Logit linear baseline and the nonlinear partial effects and threshold intervals from the GAM–Logit models, and finally summarize SHAP-based feature importance and dependence patterns as complementary evidence.

4.1. Distribution of Dependent Variables

At the point scale, the five perceived dependent variables (Q1–Q5) exhibited markedly skewed distributions. Q4 was nearly saturated, with a median of 1.00, an interquartile range (IQR) of 1.00–1.00, and a positive class proportion of 94.2%, indicating that this variable was predominantly positive across the study area. Q5 exhibited extremely low values, with a median of 0.00, an IQR of 0.00–0.00, and a positive class proportion of only 2.9%. The remaining dependent variables predominantly exhibited low values; among these, Q1 had a median of 0.25, an IQR of 0.00–0.50, and a positive class proportion of 17.1%. Overall, the positive class proportion ranged from 2.9% to 94.2%, spanning approximately 91 percentage points. See Table 5 and Figure 3a for details.

Table 5. Median, interquartile range (IQR), and positive class proportion (%) for the five perceived dependent variables.

Dependent Variable	Median Probability	IQR	Positive Class Proportion (%)
Q1	0.25	0.00–0.50	17.10
Q2	0.00	0.00–0.25	7.40
Q3	0.00	0.00–0.25	5.80
Q4	1.00	1.00–1.00	94.20
Q5	0.00	0.00–0.00	2.90

Table note: Positive class proportion is calculated using a threshold of 0.5. Median and IQR are descriptive statistics of the predicted probabilities at the point scale.

https://doi.org/10.3390/buildings16010001

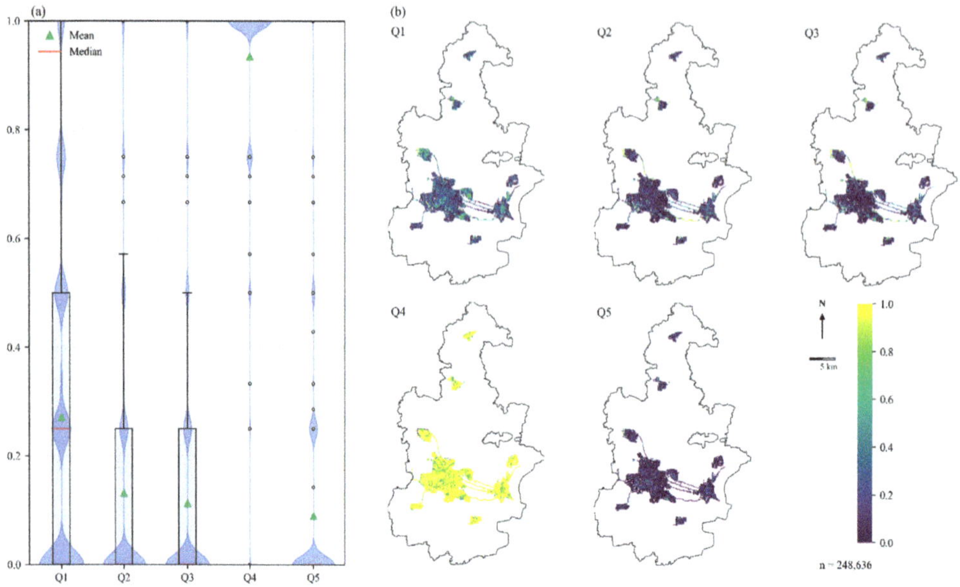

Figure 3. Probability distributions and spatial patterns of the five perceptual dependent variables. (**a**) Violin plots of predicted probability distributions, with green triangles indicating the mean and red horizontal lines showing the median; (**b**) spatial distribution of predicted probabilities aggregated into 500 m grids, with blank cells representing areas without street-view coverage or insufficient samples ($n < 5$).

Spatially, the patterns of high values differ markedly across the dependent variables, as illustrated in Figure 3b. High values for Q4 form continuous belts along the main urban area and arterial roads, with scattered high values in some peripheral areas. High values for Q1 are fewer and exhibit a fragmented distribution. Q2 and Q3 are predominantly point-like and discrete, whereas Q5 is the rarest. Subsequent comparisons are restricted to areas with street-view coverage.

4.2. Distribution and Correlation of Independent Variables

At the point scale, the distributions and missingness of all independent variables were generally acceptable. Missingness rates for NDVI and NDBI were 0.032% each, whereas LST exhibited a rate of 8.169%, primarily due to cloud masks and invalid pixels. TCH shows structural missingness within the study area; its distribution and correlations were not statistically analyzed. Medians and interquartile ranges (IQRs) for all variables are provided in Table 6. Most variables exhibited concentrated distributions with low skewness, although certain indicators displayed positive skewness with right tails, most notably VBR followed by RND, while LST was nearly symmetric. Overall, correlations were weak to moderate, as shown in Figure 4a. GVI was highly positively correlated with SG, NDVI showed a significant negative correlation with NDBI, and BVI exhibited a strong negative correlation with VBR. Other correlations were generally weak to moderate. Boxplots and scatterplots indicate stable medians and IQRs across variables, with few outliers exerting limited influence on overall patterns in Figure 4b. Figure 4c compares the density distributions of the positively skewed, right-tailed variables in their original scale versus after log1p transformation. The transformation was applied solely for visualizing skewness and tail morphology, whereas all models were estimated using the original scale. Overall, variable distributions show a coexistence of centralization and skewness,

https://doi.org/10.3390/buildings16010001

with correlations concentrated in a few variable pairs, while remaining variables preserve distinctiveness, collectively providing complementary signals.

Table 6. Descriptive statistics of independent variables.

Variable	Median	IQR	Missing%
GVI	0.125	0.053–0.222	0.000
SVF	0.290	0.182–0.393	0.000
BVI	0.121	0.039–0.262	0.000
VE	2.016	1.864–2.146	0.000
SG	0.084	0.034–0.169	0.000
VBR	1.779	0.386–9.776	0.000
GO	0.271	0.211–0.314	0.000
NDVI	0.325	0.225–0.437	0.032
LST	40.744	38.934–42.424	8.169
NDBI	−0.087	−0.148–−0.032	0.032
RND	21.535	15.384–28.493	0.000
TCH	—	—	—

Table note: Medians and interquartile ranges (IQRs) for all variables are calculated and presented in their original units. The unit of RND is km·km^{-2}, and the unit of LST is °C. Missingness is reported as a percentage. TCH exhibits structural missingness and is, therefore, excluded from descriptive statistics and correlation analyses.

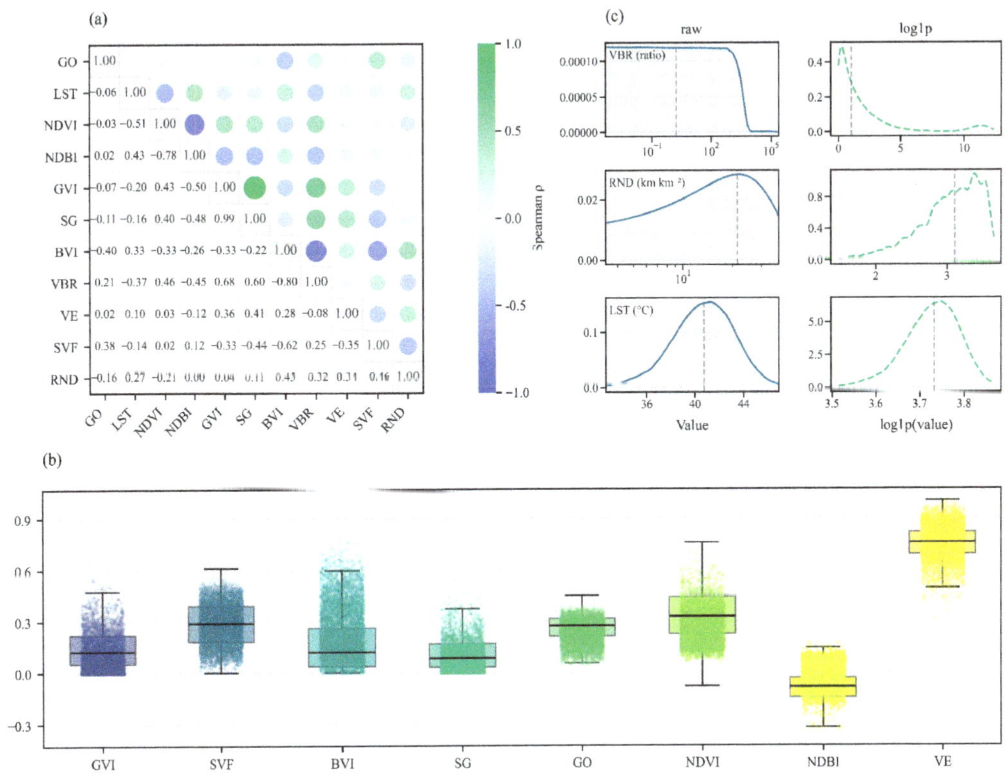

Figure 4. Distribution and correlation of independent variables. (**a**) Spearman correlation matrix; the upper triangle shows bubbles with area proportional to |ρ| and color indicating sign, while the lower triangle displays numeric values; (**b**) boxplots with jittered points for proportion and index variables; boxes indicate interquartile range (IQR), the red line shows the median, and jittered points are uniformly light blue; and (**c**) three representative distributions: VBR (positively skewed with a long right tail), RND (mildly positively skewed), and LST (approximately symmetric); log1p transformation is applied for visualization only, while models are fitted on the original scale.

https://doi.org/10.3390/buildings16010001

4.3. Linear Baseline (ENet–Logit)

On the logit scale, standardized coefficients from ENet–Logit provide a baseline for variable direction and relative strength. In aggregate, the seven SVI indicators sum to about 41.41, or roughly 92% of the total signal strength, while the five RSI indicators sum to 3.58, about 8% in Table 7. These values reflect only baseline linear signals and are on a different scale from GAM–Logit partial effects or SHAP contributions, so they are not directly comparable. At the variable level, GVI consistently provides the dominant positive signal across the five outcomes, whereas SG is the dominant negative signal, followed by SVF, BVI, and VE; VBR and GO show limited intensity. Within RSI, RND totals 1.29, concentrated mainly in Q5, while LST, NDVI, and NDBI are weaker. TCH is structurally missing in parts of the study area; it is listed for completeness but not interpreted. By outcome, Q1, Q2, and Q3 show positive GVI and negative SG; Q4 is dominated by a positive SVF, with GVI turning negative under multivariate control and BVI, VE, and GO being mildly positive. Q5 keeps GVI positive and SG negative, with RND stably positive. Cross outcome comparisons indicate GVI and SG have consistent directions for most variables, while the positive RND signal is concentrated in Q5. Q4's signal structure differs markedly from the others in Figure 5a. Overall intensity rankings appear in Figure 5b, with full values in Table 7.

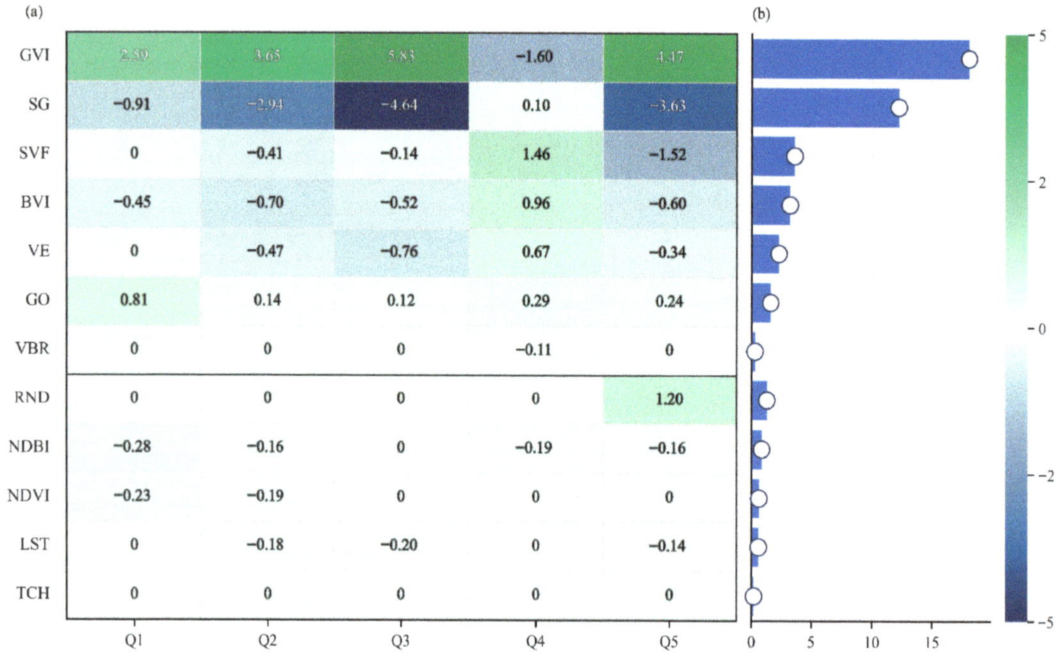

Figure 5. ENet–Logit linear baseline. (**a**) Heatmap of standardized coefficients; entries with absolute standardized values < 0.10 are set to 0. The color scale is standardized across all dependent variables, and horizontal dividers separate SVI from RSI; (**b**) bar chart of total strength by variable. Total strength is the sum of absolute standardized coefficients across the five dependent variables and describes linear signal under the screening criterion. TCH is shown for completeness only and is not interpreted; it is excluded from GAM–Logit and SHAP analyses.

https://doi.org/10.3390/buildings16010001

Table 7. Standardized coefficients from ENet–Logit.

Variable	Q1	Q2	Q3	Q4	Q5	Total$\|\beta\|$	Count ($\geq$0.10)
GVI	2.59	3.65	5.83	−1.60	4.47	18.14	5
SG	−0.91	−2.94	−4.64	0.10	−3.63	12.22	5
SVF	−0.08	−0.41	−0.14	1.46	−1.52	3.61	4
BVI	−0.45	−0.70	−0.52	0.96	−0.60	3.23	5
VE	−0.06	−0.47	−0.76	0.67	−0.34	2.30	4
VBR	−0.08	−0.04	0.06	−0.11	0.02	0.31	1
GO	0.81	0.14	0.12	0.29	0.24	1.60	5
SVI subtotal						41.41	
LST	−0.03	−0.18	−0.20	0.05	−0.14	0.60	3
NDBI	−0.28	−0.16	−0.06	−0.19	−0.16	0.85	4
NDVI	−0.23	−0.19	−0.02	−0.10	−0.10	0.64	4
RND	0.01	−0.03	−0.04	−0.01	1.20	1.29	1
TCH	0.01	−0.07	0.00	−0.06	−0.06	0.20	0
RSI subtotal						3.58	

Table note: All predictors were standardized to z-scores before modeling; coefficients represent the logit scale effect per 1 standard deviation. Total $|\beta|$ is the sum of absolute standardized coefficients across the five outcomes for each variable. Count ($\geq$0.10) is the number of outcomes with absolute standardized coefficients of at least 0.10 and is used to describe stability, not statistical significance. For presentation, absolute standardized coefficients less than 0.10 are recorded as 0. TCH has structural missingness in the study area and is listed for completeness only; it is not interpreted and is excluded from GAM–Logit and SHAP.

4.4. Nonlinear Partial Effects and Threshold Intervals (GAM–Logit)

We use GAM–Logit to estimate smooth terms and partial effects for predictors of Q1–Q5. Significance is controlled with BH-FDR at ($q < 0.05$). Except for NDVI in Q4, which is not significant, most smooth terms are significant and predominantly nonlinear, as shown in Table 8. Bias effect curves are steeper with narrower confidence bands in densely sampled regions and flatter with greater uncertainty in sparsely sampled regions, as Figure 6a illustrates. Calibration is satisfactory, with bin reliability curves close to the identity line and predicted probabilities matching observed positive rates, as shown in Figure 6b. Given Q4's 94.2% positive class proportion, interpretation should focus on calibration consistency and avoid attributing prior driven distributional features to discriminative advantage. Following predefined rules, threshold intervals are reported only for relationships that exhibit internal extrema on the standardized scale; these are labeled as optimal or risk intervals and their bounds are set by the bias effect curves and associated uncertainties. No thresholds are reported for monotonic relations. Reported threshold positions align with the inflection points and slope changes in Figure 6a, as shown in Figure 6c.

Table 8. Variable Form.

Variable	Q1	Q2	Q3	Q4	Q5
GVI	Pos *	Pos *	Pos *	Pos *	Pos *
SG	Neg *	Neg *	Neg *	Neg *	Neg *
SVF	InvU *	InvU *	InvU *	Pos *	Neg *
BVI	Neg *	InvU *	InvU *	Pos *	InvU *
NDVI	Neg *	Neg *	InvU *	n.s.	InvU *
LST	InvU *	InvU *	U *	InvU *	U *
VE	U *	Neg *	Neg *	Pos *	InvU *
VBR	U *	U *	U *	InvU *	U *
GO	Pos *	U *	U *	U *	U *
NDBI	U *	U *	InvU *	U *	U *
RND	Pos *	Pos *	Pos *	Pos *	Pos *

Table note: Form labels are defined as follows: Pos denotes monotonic increase; Neg denotes monotonic decrease; U denotes U-shaped; InvU denotes inverted-U-shaped; n.s. denotes not significant. An asterisk * marks $q < 0.05$ after Benjamini–Hochberg FDR control within the corresponding dependent variable. TCH is excluded due to structural missingness.

https://doi.org/10.3390/buildings16010001

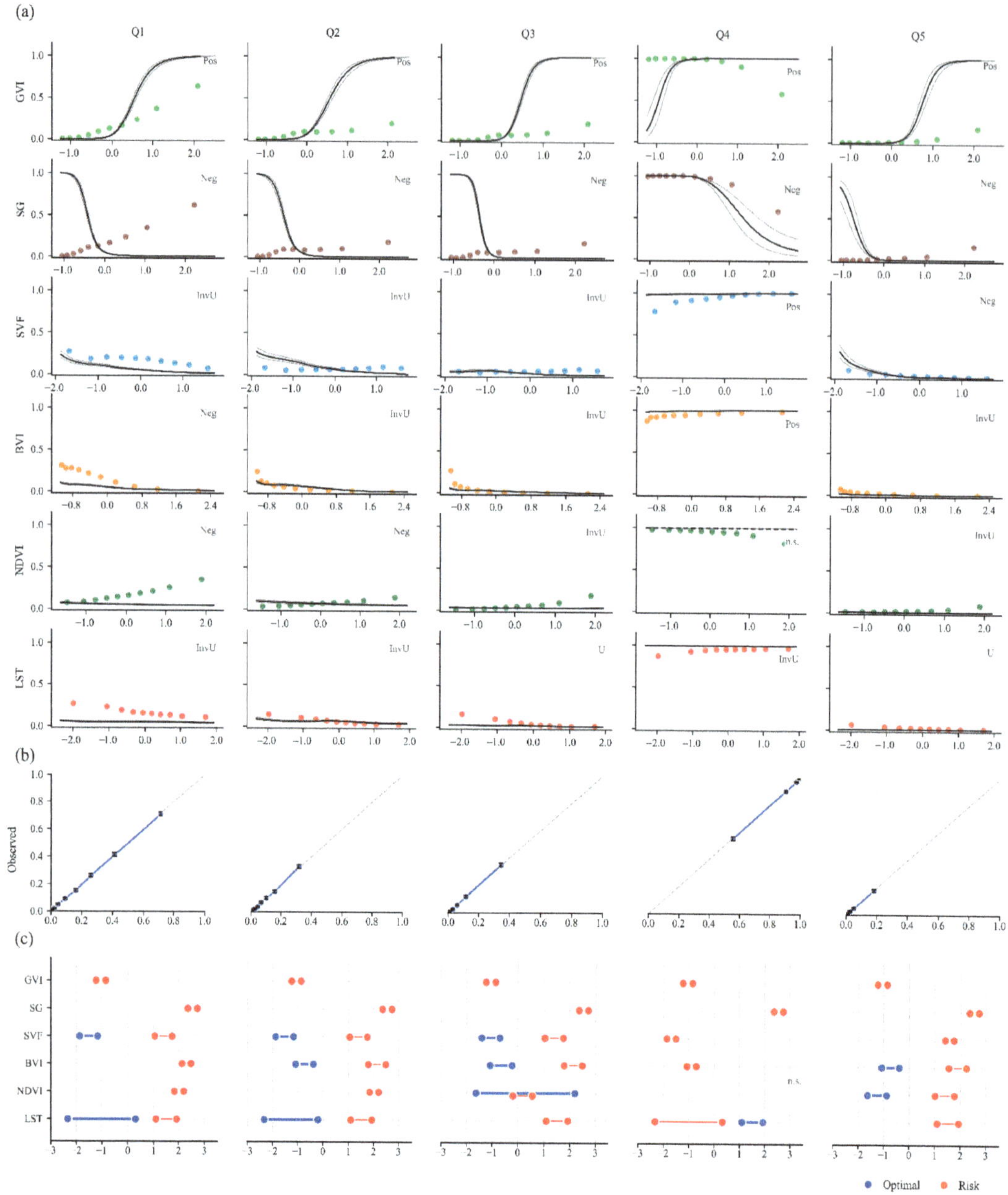

Figure 6. GAM–Logit bias effects, calibration, and threshold bands. (**a**) Bias effects: x axis is z-score and y axis is the positive probability $Pr(Y = 1)$; points show observed positive class proportions in equal deciles; the black line is the fitted curve and the gray band is the 95% confidence interval; significance is controlled by BH–FDR at $q < 0.05$, and non significant results $q \geq 0.05$ are shown with dashed lines; (**b**) calibration: polylines connect bin means and points display 95% confidence intervals; the gray dashed line denotes ideal calibration; (**c**) threshold bands: optimal and risk intervals are shown only when an internal extremum exists; "*n.s.*" denotes not significant; TCH is not displayed.

https://doi.org/10.3390/buildings16010001

Across outcomes, the primary variables show clear and stable patterns. GVI is positive for Q1–Q5, with higher GVI associated with higher positive probabilities, indicating a stable signal. SG is negative throughout, with stronger negative effects in the medium-to-high range that attenuate at the upper tail. SVF shows an inverted-U shape in Q1–Q3 and turns positive in Q4 and negative in Q5, with peaks concentrated in the medium-to-high range, suggesting moderate optimality. BVI is negative in Q1, inverted-U in Q2 and Q3, positive in Q4, and again inverted-U in Q5. NDVI is negative in Q1 and Q2, inverted-U in Q3 and Q5, and not significant in Q4. LST mainly exhibits U or inverted-U forms, with marginal effects flattening at both extremes. In general, most variables reach peak sensitivity around the median range, with relatively gradual margins at the extremes. VBR predominantly shows U shapes with high stability; RND is monotonic increasing; NDBI is mainly U, with some outcomes showing inverted-U; and VE and GO switch patterns across outcomes, as shown in Table 8. Per the methodological convention, TCH is excluded from bias effect and threshold displays due to structural missingness.

4.5. SHAP Importance and Marginal Effects for Each Dependent Variable

As Figure 7 and Table 9 show, rankings are stable across Q1–Q5. GVI and SG form the top tier for all outcomes, SVF and BVI occupy an intermediate tier, and VE and GO are smaller in magnitude with consistent variation. RSI indicators are generally low in average absolute SHAP value, though LST, NDBI, and RND display discernible increases. Dependency trends align with the GAM–Logit bias effects: GVI rises through the medium-to-high range and then stabilizes at the upper end; SG remains negative over the full range, with larger magnitude in the medium-to-high range and stabilization at the top; SVF peaks near the median; BVI is positive in the medium-to-high range and slightly negative at low values; NDVI shows small overall magnitude and is not significant for Q4; and LST is negative at high values, as shown in Figure 7b. The Cleveland plot in Figure 7c indicates that GVI and SG have the highest average absolute SHAP values with minimal variation across Q1–Q5. SVF and BVI are intermediate, while most remote sensing indicators cluster at lower levels without notable outliers. TCH is excluded from interpretation due to structural missingness.

Table 9. SHAP global importance for each dependent variable.

Feature	Q1	Q2	Q3	Q4	Q5	Mean Across
GVI	0.295	0.260	0.263	0.197	0.210	0.245
SG	0.188	0.226	0.230	0.243	0.196	0.216
SVF	0.055	0.049	0.035	0.019	0.051	0.042
BVI	0.037	0.028	0.015	0.022	0.015	0.023
VE	0.008	0.035	0.024	0.013	0.007	0.018
VBR	0.002	0.001	0.002	0.002	0.001	0.002
GO	0.047	0.011	0.004	0.006	0.007	0.015
NDVI	0.006	0.006	0.003	0.002	0.003	0.004
LST	0.003	0.010	0.006	0.003	0.005	0.006
NDBI	0.007	0.005	0.006	0.004	0.003	0.005
RND	0.017	0.009	0.004	0.003	0.001	0.007

Table note: Entries are mean absolute SHAP values. The last column reports the mean across Q1–Q5. No significance testing is conducted.

https://doi.org/10.3390/buildings16010001

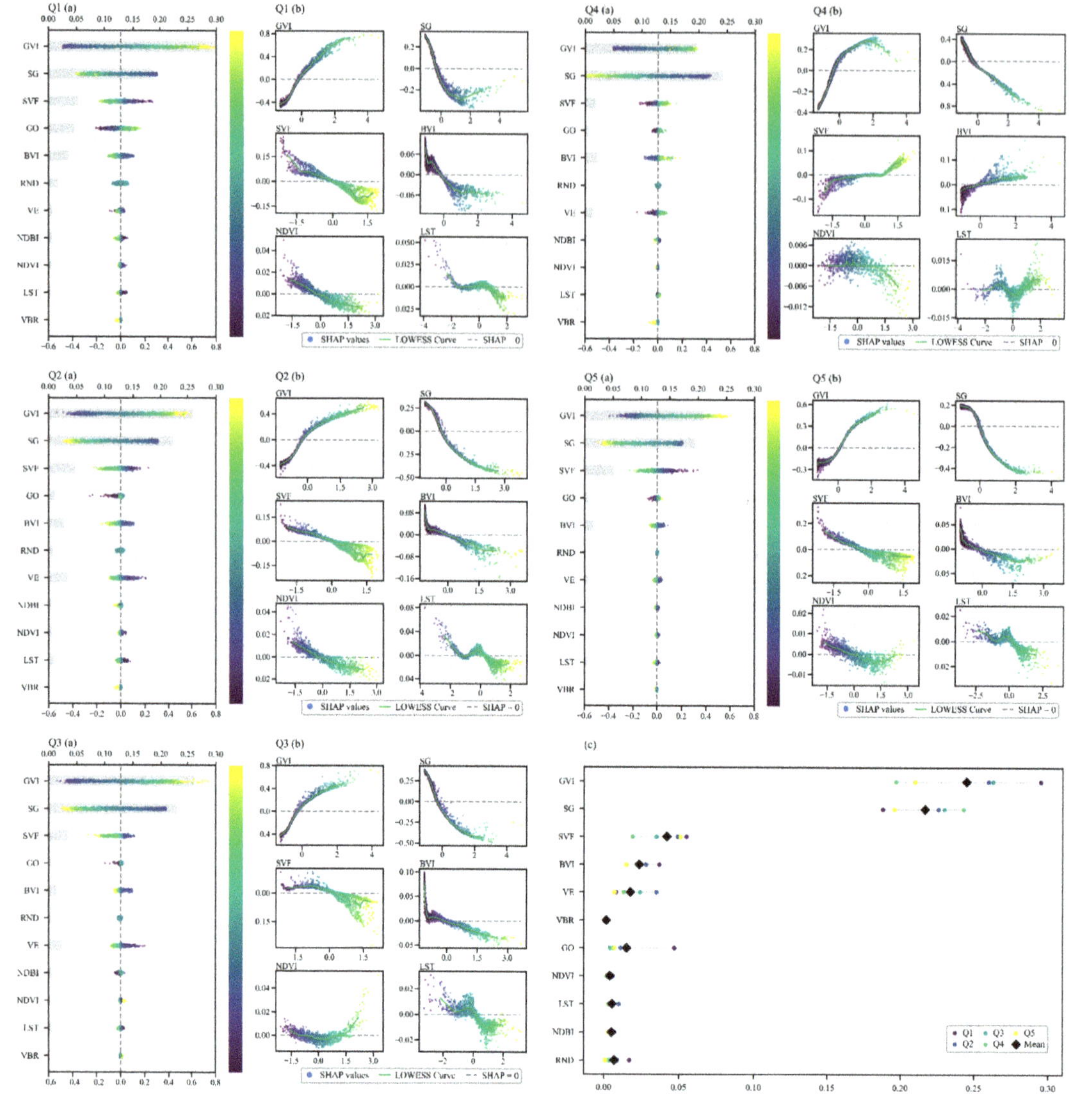

Figure 7. SHAP explanations for the five dependent variables (**Q1–Q5**). (**Q1**(**a**)–**Q5**(**a**)) Swarm plots summarizing global importance and the distribution of sample level contributions. (**Q1**(**b**)–**Q5**(**b**)) Dependency plots showing marginal trends for GVI, SG, SVF, BVI, NDVI, and LST. (**c**) Cleveland plot comparing mean absolute SHAP values across Q1–Q5 on a unified scale, with black diamonds marking the Q1–Q5 mean. TCH is excluded due to structural missingness in the study area.

5. Discussion

5.1. Key Findings

This subsection synthesizes how the three-stage interpretability chain connects the multi-source environmental indicators and the five perceptual dimensions to the three research questions (RQ1–RQ3) posed in the introduction. Addressing the limitations of single data sources and linear-only assumptions in current streetscape research, this study systematically analyzed how 12 environmental factors shape five perceptual dimensions through a design that integrates data from multiple sources, evaluates multiple perceptual outcomes, and implements an interpretable chain of evidence with three stages. The core finding is that, within Tianjin at the point scale, streetscape perception is governed primarily by nonlinear relationships and threshold effects, showing pronounced moderation rather than simple monotonic "higher is better" patterns. Specifically, SG shows significant negative associations with Q1, Q2, and Q5, with stronger marginal effects at medium-to-high levels; SVF exhibits a clear inverted-U-shaped relation with Q2 and Q3 that delineates optimal ranges but is monotonically positive for Q4; and GVI maintains a stable positive effect with marked diminishing returns at high levels. Overall, most variables display peak sensitivity near moderate levels, with marginal effects that flatten toward both extremes. These results align with recent evidence on nonlinear and threshold patterns identified using street view and multiple data sources [73,74].

First, the ENet–Logit baseline confirms the dominant role of SVI and shows that SVI and RSI are complementary rather than substitutable for representing green cover and perception [75]. At the same time, remote sensing indicators provide stable supplements in the thermal exposure and built-up intensity dimensions, underscoring the value of combining multiple data sources. Second, GAM–Logit reveals pervasive nonlinear relationships and multidimensional perceptual conflicts, highlighting the need for multivariate assessment. As shown in Table 8, except for NDVI in Q4, most smooth terms are statistically significant at ($q < 0.05$) and are predominantly inverted-U-shaped or U-shaped. The stable inverted-U-shaped relationship of SVF for Q1–Q3 delineates optimal ranges for enhancing visual comfort, whereas SVF becomes monotonically increasing for Q4. This accords with evidence that a higher sky view factor increases MRT and PET under summer conditions; while greater openness may improve legibility and brightness, it can exacerbate thermal discomfort in summer [73,76]. Consequently, optimizing a single outcome can lead to one-sided or even conflicting recommendations. The directional contradictions of the same environmental factor across different perceptual dimensions demonstrate the prevalence of perceptual tradeoffs and show that focusing on a single dependent variable risks partial or misleading conclusions.

Crucially, SHAP analysis provides independent validation of the conclusions from a model agnostic perspective. At the global importance level, SHAP rankings closely align with ENet–Logit and GAM–Logit, jointly identifying GVI and SG as the most influential core variables. At the sample level, SHAP reproduces the key nonlinear patterns revealed by GAM–Logit from a distinct methodological basis, including diminishing returns for GVI and an inverted U shaped effect for SVF. This cross-validation strengthens the robustness of the principal findings and provides an interpretable measure of feature contributions grounded in cooperative game theory.

Ultimately, the synergistic use of the three-step explanatory chain advances the analysis from identifying phenomena to uncovering mechanisms. Considering all 12 predictors together reveals a clear hierarchy of influence: GVI and SG, with exceptionally high total intensity and stability, emerge as the primary drivers of perception; SVF, BVI, and LST play critical contextual roles through nonlinear patterns; and the remaining variables show weaker or unstable effects. This is not a limitation, but rather reflects the underlying dominance structure of complex systems, and it provides quantitative grounds for prioritizing resources in planning practice. In sum, the principal findings establish a coherent analytic paradigm in which rigorous understanding and effective improvement of streetscape quality require integrated attention to the following three principles: the complementarity of multiple data sources, the multidimensional balancing of perceptions, and the interplay of nonlinear influence zones. These three principles provide an integrated answer to RQ1–RQ3 in our case study of Tianjin.

https://doi.org/10.3390/buildings16010001

5.2. Dialog with Existing Research: Evidence from Temperate Monsoon City

Within the temperate monsoon context, a central finding is that SG exerts a persistent negative effect on comfort, aesthetics, and willingness to linger. This contrasts with studies suggesting that openness and green quantity can act synergistically to enhance experience. Prior research reports a curvilinear relationship between openness and street vitality, with an optimal zone in the middle range and adverse effects at both extremes [77]. Our results do not show such synergy across multiple perceptual dimensions. This divergence indicates strong site and climate specificity in morphological effects, implying that the influence of form on perception is not universal and must be recalibrated by climate and season. Under cold winters and hot summers, high uniformity and convergent forms can jointly restrict summer ventilation and reduce winter sunlight, diminishing overall experience. By contrast, moderate morphological heterogeneity is more conducive to balancing thermal conditions and daylight across seasons [73], offering clear guidance for the refined remediation of urban street segments.

Our findings indicate that SVF has an optimal range within the visual dimension, whereas in summer heat perception, greater openness may increase discomfort. External studies likewise show a positive association between higher SVF and higher PET [73]. GVI exerts an overall positive effect but displays diminishing returns at high levels. However, studies that decompose GVI into tree, shrub, and grass components using panoramic street-view imagery and deep learning semantic segmentation show that structural differences within similar levels of greenness can lead to markedly different ecological and experiential outcomes [55], suggesting that structure-sensitive indicators are needed alongside aggregate greenness. Work on street-view analysis and interpretable learning has identified nonlinear relationships and thresholds between green quantity and psychological stress [74] and between streetscape greenery and walking propensity or active travel [78,79]. Recent health-oriented studies in semi-arid cities likewise report nonlinear effects of urban green space on older adults' physical activity, with threshold patterns varying by obesity status [80]. From the perspective of nonlinearity and thresholds, the effects of GVI, SVF, and related morphological indicators on public perception and behavior can amplify or even reverse once certain thresholds are reached, so simply increasing quantity does not guarantee improved experience. Broader health research similarly reports threshold relationships between green quantity and health outcomes [81].

SVI provides a more direct and stable measure of multidimensional perception, whereas RSI contributes independent information on thermal environments and built-up intensity. The two are complementary rather than mutually exclusive, consistent with recent work showing that GVI aligns more closely with visible greenery and perception, while NDVI reflects horizontal coverage [82]. At the macro level, we observe a negative correlation between LST and NDVI and a positive correlation between LST and NDBI, supplying independent signals for heat-related outcomes, consistent with evidence across multiple cities [43]. From an indicator standpoint, BVI and GO display contextual rather than universal dominance in visual outcomes. External studies likewise report nonlinear relationships and thresholds between visual complexity and perception [83]. Regarding multi-source consistency, complementary relationships and systematic differences between NDVI and GVI are supported by independent evidence, indicating divergent explanatory pathways and greater combined utility for these indicator types [84].

In summary, our comparative analysis provides multidimensional perceptual evidence from Tianjin as a temperate monsoon city, demonstrating contextual heterogeneity and interval thresholds among greenery, openness, and interface characteristics. These patterns are likely to be informative for other temperate monsoon settings, but they should be tested and recalibrated in additional cities and climatic contexts. Taken together, these findings

https://doi.org/10.3390/buildings16010001

support an approach based on intervals rather than direction alone. Moreover, integrating street-level imagery with remote sensing is more effective for explaining perceptual variation and improving comparability across cities, consistent with the use of multi-source and explainable learning to reveal nonlinear and synergistic effects [85].

5.3. Planning and Design Implications

Building on the nonlinear relationships and threshold intervals identified here, effective streetscape optimization hinges on regulation based on intervals for key environmental elements. We propose three priority levels. First, ensure GVI lies within the mid-to-high portion of its optimal range and use it as a foundational indicator for street renewal. This can be achieved by connecting visible greenery and canopy continuity, for example, by planting continuous rows of trees and linking pocket green spaces, while maintaining adequate branch clearance to avoid excessive density that pushes GVI into a zone of high risk with diminishing returns. Second, strictly avoid the zone of high risk for SG by correcting unsuitable combinations of green coverage and openness; use pruning, increased branch clearance, and staggered vegetation placement to provide shade while creating rhythmic light and dark patterns, thereby mitigating its negative effects. Third, apply mid-to-high range control for SVF, maintaining openness within the optimal range for Q2 and Q3; introduce intermittent widenings to prevent excessive enclosure while remaining mindful of its positive association with Q4.

During implementation, context-sensitive design and coordination should integrate street functions with the macro thermal environment. Commercial pedestrian streets should enforce tighter controls on the SG risk zone and improve GO to prioritize visual comfort and dwell time; on major thoroughfares, the upper limit of SVF may be relaxed moderately while thermal regulation remains a priority. For typical heat risk segments that show high LST and NDBI with low NDVI, a synergistic system of continuous shading, enhanced evapotranspiration, and improved ventilation should be established, consistent with multi-city evidence that targeted greening can mitigate outdoor thermal disparities [86]. Heat-tolerant, dense canopy trees should be planted and combined with permeable paving and materials with medium albedo to reduce heat absorption while avoiding glare. Street valley airflow should be organized along prevailing summer winds and maintain ventilation gaps through staggered planting. All interventions should account for potential side effects. When increasing GVI, its shading effect on SVF should be concurrently evaluated to avoid greening that increases heat; when adjusting SVF, its dual effects on visual openness and thermal exposure should be balanced. By mapping each element's position within diagrams based on thresholds, differentiated one-street, one-strategy approaches should be formulated. Ultimately, a dynamic maintenance regime should be established that intensifies irrigation and shading in summer and applies moderate pruning in winter to ensure sunlight penetration. This approach shifts street design from extensive remodeling toward a precision-oriented and adaptive paradigm grounded in quantitative evidence and aimed at multidimensional perceptual objectives.

5.4. Limitations

Although this study advances SVI and RSI fusion and the identification of nonlinear mechanisms, several limitations merit further work. On the data side, perception labels derived from online questionnaires have relatively small sample sizes and potential subjective bias. The temporal resolution of SVI does not fully capture seasonal vegetation dynamics, and structural gaps in TCH limit a complete analysis of vertical vegetation structure. Future research could combine mobile sensing with crowdsourcing to expand geographic coverage and obtain repeated measures while preserving the visual representativeness of samples, thereby improving the stability and reliability of perception labels. In addition, integrating multi-source three-dimensional data such as LiDAR would allow more precise characterization of vegetation's vertical features. Methodologically, while the GAM–Logit model effectively reveals nonlinear relationships for individual variables, it does not explicitly model interactions, which may overlook complex synergistic or antagonistic effects. This study also did not model spatial autocorrelation in street network structure, which could bias estimates of local clustering. Finally, for inference and transferability, future work should validate and calibrate the proposed threshold ranges across different climate zones, urban morphologies, and street types to build a more generalizable design knowledge system based on intervals. Studies on rail transit, property prices and urban vitality have already shown that built-environment effects are strongly place- and time-varying [87–89], suggesting that the threshold ranges identified here are also likely to shift across contexts and over time. Coupling these analyses with evaluations conducted before and after implementation and physiological monitoring would further test the practical applicability of the thresholds in engineering contexts and move beyond statistical association toward causal mechanisms.

6. Conclusions

This study developed a point-anchored assessment framework that integrates street-view imagery with satellite-derived environmental indicators for use in Tianjin, a temperate monsoon megacity, and offers a structure that can be adapted to other cities subject to local calibration. The research centered on the following five perceptual dimensions: comfort, aesthetic appeal, perceived greenness, summer heat perception, and willingness to linger, and it used an interpretable evidence chain of ENet–Logit, GAM–Logit, and SHAP to examine the nonlinear effects of 12 environmental factors. Core findings showed that streetscape perception is governed primarily by nonlinear relationships and threshold effects. GVI is the most robust facilitator of multidimensional perceptions, whereas SG shows a consistent negative effect, revealing local design tradeoffs. SVF and BVI commonly exhibit optimal ranges and multidimensional conflicts. Accordingly, the study shifts the focus from directional effects to effect shapes and sensitive intervals. Based on these mechanisms, we distill the following three actionable implications for planning and design: prioritize continuous canopy connections so that GVI lies within the optimal gain interval; regulate SVF within the mid-to-high portion of its optimal range to balance visual comfort with thermal needs; and avoid excessive enclosure or sparseness as reflected by BVI to seek perceptual balance. The study also confirms the complementary value of SVI and RSI, providing a multidimensional data foundation for refined design.

The findings apply primarily to road spaces in Tianjin with street-view coverage. To enhance transferability and practical applicability, future work will conduct observations across seasons to capture perceptual dynamics, develop spatially dependent models to deepen mechanism discovery, quantify the benefits of fusing SVI and RSI through controlled experiments, and validate the framework across diverse climate zones and urban morphologies to build a more generalizable streetscape assessment methodology.

Author Contributions: Conceptualization, Y.Y., F.T. and L.W.; methodology, Y.Y.; software, Y.Y.; validation, Y.Y., F.T. and L.W.; formal analysis, Y.Y.; investigation, Y.Z. and Y.Y.; resources, L.W., Q.W. and X.L.; data curation, Y.Y. and S.Z.; writing—original draft, Y.Y., F.T., S.Z. and Y.Z.; writing—review and editing, Y.Y., F.T., L.W. and Q.W.; visualization, Y.Y., S.Z. and Y.Z.; supervision, L.W. and X.L.; project administration, Y.Y., F.T., L.W., Q.W. and S.W.; funding acquisition, L.W., Q.W. and S.W. All authors have read and agreed to the published version of the manuscript.

Funding: This research received no external funding.

https://doi.org/10.3390/buildings16010001

Data Availability Statement: The original contributions presented in the study are included in the article, further inquiries can be directed to the corresponding author.

Conflicts of Interest: The authors declare no conflicts of interest.

References

1. Fuller, M.; Moore, R. *An Analysis of Jane Jacobs's the Death and Life of Great American Cities*; Macat Library: London, UK, 2017.
2. Wang, F.; Tan, C.; Li, M.; Gu, D.; Wang, H. Research on Traffic Design of Urban Vital Streets. *Sustainability* **2022**, *14*, 6468. [CrossRef]
3. Nguyen, P.-Y.; Astell-Burt, T.; Rahimi-Ardabili, H.; Feng, X. Green Space Quality and Health: A Systematic Review. *Int. J. Environ. Res. Public Health* **2021**, *18*, 11028. [CrossRef] [PubMed]
4. Ewing, R.; Handy, S. Measuring the Unmeasurable: Urban Design Qualities Related to Walkability. *J. Urban Des.* **2009**, *14*, 65–84. [CrossRef]
5. Hematian, H.; Ranjbar, E. Evaluating Urban Public Spaces from Mental Health Point of View: Comparing Pedestrian and Car-Dominated Streets. *J. Transp. Health* **2022**, *27*, 101532. [CrossRef]
6. Dai, L.; Zheng, C.; Dong, Z.; Yao, Y.; Wang, R.; Zhang, X.; Ren, S.; Zhang, J.; Song, X.; Guan, Q. Analyzing the Correlation Between Visual Space and Residents' Psychology in Wuhan, China Using Street-View Images and Deep-Learning Technique. *City Environ. Interact.* **2021**, *11*, 100069. [CrossRef]
7. Tavolari, B. Jane Jacobs: Contradições e Tensões | Jane Jacobs: Contradictions and Tensions. *Rev. Bras. Estud. Urbanos Reg.* **2019**, *21*, 13. [CrossRef]
8. Huang, Z.; Duan, L.; Xu, Y.; Yang, S.; Lin, Z.; Yue, H.; Yang, J. Exploring the Influence of Urban Green Space and Urban Morphology on Urban Heat Islands Using Street View and Satellite Imagery. *Sci. Rep.* **2025**, *15*, 23759. [CrossRef]
9. Zhou, S.; Ji, Q.; Zhang, L.; Wu, J.; Li, P.; Zhang, Y. Exploration of Differences in Housing Price Determinants Based on Street View Imagery and the Geographical-XGBoost Model: Improving Quality of Life for Residents and Through-Travelers. *ISPRS Int. J. Geo-Inf.* **2025**, *14*, 391. [CrossRef]
10. Chen, C.; Lin, X.; Wu, Y.; Xie, J.; Liu, J. How Do Residents Perceive Ecosystem Service Benefits Received from Urban Streets? A Case Study of Guangzhou, China. *J. Clean. Prod.* **2024**, *449*, 141554. [CrossRef]
11. Yang, J.; Li, X.; Du, J.; Cheng, C. Exploring the Relationship Between Urban Street Spatial Patterns and Street Vitality: A Case Study of Guiyang, China. *Int. J. Environ. Res. Public Health* **2023**, *20*, 1646. [CrossRef]
12. Lee, A.; Jordan, H.; Horsley, J. Value of Urban Green Spaces in Promoting Healthy Living and Wellbeing: Prospects for Planning. *Risk Manag. Healthc. Policy* **2015**, *8*, 131–137. [CrossRef]
13. Biljecki, F.; Ito, K. Street View Imagery in Urban Analytics and GIS: A Review. *Landsc. Urban Plan.* **2021**, *215*, 104217. [CrossRef]
14. He, N.; Li, G. Urban Neighbourhood Environment Assessment Based on Street View Image Processing: A Review of Research Trends. *Environ. Chall.* **2021**, *4*, 100090. [CrossRef]
15. Ogawa, Y.; Oki, T.; Zhao, C.; Sekimoto, Y.; Shimizu, C. Evaluating the Subjective Perceptions of Streetscapes Using Street-View Images. *Landsc. Urban Plan.* **2024**, *247*, 105073. [CrossRef]
16. Huang, G.; Yu, Y.; Lyu, M.; Sun, D.; Zeng, Q.; Bart, D. Using Google Street View Panoramas to Investigate the Influence of Urban Coastal Street Environment on Visual Walkability. *Environ. Res. Commun.* **2023**, *5*, 65017. [CrossRef]
17. Zhai, Y.; Gong, R.; Huo, J.; Fan, B. Building Façade Color Distribution, Color Harmony and Diversity in Relation to Street Functions: Using Street View Images and Deep Learning. *ISPRS Int. J. Geo-Inf.* **2023**, *12*, 224. [CrossRef]
18. Li, S.; Ma, S.; Tong, D.; Jia, Z.; Li, P.; Long, Y. Associations Between the Quality of Street Space and the Attributes of the Built Environment Using Large Volumes of Street View Pictures. *Environ. Plan. B Urban Anal. City Sci.* **2022**, *49*, 1197–1211. [CrossRef]
19. Ji, Q.; Zhou, S.; Zhang, L.; Yuan, Y.; Wu, L.; Tang, F.; Wu, J.; Meng, Y.; Zhang, Y. Urban Greening Strategies and Ecosystem Services: The Differential Impact of Street-Level Greening Structures on Housing Prices. *Forests* **2025**, *16*, 1713. [CrossRef]
20. Ma, Y.; Chen, P.; Gong, M.; Cai, Y.; Jian, I.Y. The First Seasonal Green View Index Mapping Dataset Across Chinese Cities Powered by Deep Learning. *Sci. Data* **2025**, *12*, 1356. [CrossRef]
21. Zhao, L.; Fan, X.; Hong, T. Urban Heat Island Effect: Remote Sensing Monitoring and Assessment—Methods, Applications, and Future Directions. *Atmosphere* **2025**, *16*, 791. [CrossRef]
22. Pritipadmaja; Garg, R.D.; Sharma, A.K. Assessing the Cooling Effect of Blue-Green Spaces: Implications for Urban Heat Island Mitigation. *Water* **2023**, *15*, 2983. [CrossRef]
23. Kara, Y.; Yavuz, V.; Lupo, A.R. Multi-Index Assessment of Surface Urban Heat Island (SUHI) Dynamics in Samsun Using Google Earth Engine. *Atmosphere* **2025**, *16*, 712. [CrossRef]
24. Wang, H.; Lin, C.; Ou, S.; Feng, Q.; Guo, K.; Wei, X.; Xie, J. Multilevel Change of Urban Green Space and Spatiotemporal Heterogeneity Analysis of Driving Factors. *Sustainability* **2024**, *16*, 4762. [CrossRef]

https://doi.org/10.3390/buildings16010001

25. Zhang, J.; Yuan, S. Urban and Rural Planning Combined with Environmental Remote Sensing Technology for Planning and Design of Ecological City. *Preprints* **2024**, 2024072090. [CrossRef]
26. Lima, L.B.; Franca Rocha, W.J.S.; Souza, D.T.M.; Lobão, J.S.B.; De Santana, M.M.M.; Cambui, E.C.B.; Vasconcelos, R.N. Urban Quality: A Remote-Sensing-Perspective Review. *Urban Sci.* **2025**, *9*, 31. [CrossRef]
27. Wu, Y. Research Progress of Urban Ecological Evaluation Methods Based on Remote Sensing Indexes. *Highlights Sci. Eng. Technol.* **2024**, *108*, 27–33. [CrossRef]
28. Fan, Z.; Feng, C.-C.; Biljecki, F. Coverage and Bias of Street View Imagery in Mapping the Urban Environment. *Comput. Environ. Urban Syst.* **2025**, *117*, 102253. [CrossRef]
29. Yin, D.; Hirata, T. Comprehensive Comparative Analysis and Innovative Exploration of Green View Index Calculation Methods. *Land* **2025**, *14*, 289. [CrossRef]
30. Wu, S.; Wang, D.; Wei, Y.; Wang, X.; Lou, R.; Duan, J. Bridging the Key Components of Urban Ecosystem Services Assessment: Links and Feedbacks Between "Typology-Multifunctionality-Supply and Demand". *Landsc. Ecol.* **2025**, *40*, 67. [CrossRef]
31. Wang, L.; Han, X.; He, J.; Jung, T. Measuring Residents' Perceptions of City Streets to Inform Better Street Planning Through Deep Learning and Space Syntax. *ISPRS J. Photogramm. Remote Sens.* **2022**, *190*, 215–230. [CrossRef]
32. Tang, L.; He, J.; Peng, W.; Huang, H.; Chen, C.; Yu, C. Assessing the Visibility of Urban Greenery Using MLS LiDAR Data. *Landsc. Urban Plan.* **2023**, *232*, 104662. [CrossRef]
33. Xu, F.; Yan, J.; Heremans, S.; Somers, B. Pan-European Urban Green Space Dynamics: A View from Space Between 1990 and 2015. *Landsc. Urban Plan.* **2022**, *226*, 104477. [CrossRef]
34. Larkin, A.; Gu, X.; Chen, L.; Hystad, P. Predicting Perceptions of the Built Environment Using GIS, Satellite and Street View Image Approaches. *Landsc. Urban Plan.* **2021**, *216*, 104257. [CrossRef]
35. Chen, H.; Zheng, Y.; Xing, J.; Zhou, C. Urban Planning Priorities Analysis Based on Urban Heat Risk Supply—Demand Mismatch: A Case Study of Hangzhou, China. *Sustain. Cities Soc.* **2025**, *130*, 106502. [CrossRef]
36. Ye, Y.; Richards, D.; Lu, Y.; Song, X.; Zhuang, Y.; Zeng, W.; Zhong, T. Measuring Daily Accessed Street Greenery: A Human-Scale Approach for Informing Better Urban Planning Practices. *Landsc. Urban Plan.* **2019**, *191*, 103434. [CrossRef]
37. Najmi, A.; Gevaert, C.M.; Kohli, D.; Kuffer, M.; Pratomo, J. Integrating Remote Sensing and Street View Imagery for Mapping Slums. *ISPRS Int. J. Geo-Inf.* **2022**, *11*, 631. [CrossRef]
38. Barbierato, E.; Bernetti, I.; Capecchi, I.; Saragosa, C. Integrating Remote Sensing and Street View Images to Quantify Urban Forest Ecosystem Services. *Remote Sens.* **2020**, *12*, 329. [CrossRef]
39. Zhang, T.; Lin, Z.; Wang, L.; Zhang, W.; Zhang, Y.; Hu, Y. The Heterogeneous Effects of Microscale-Built Environments on Land Surface Temperature Based on Machine Learning and Street View Images. *Atmosphere* **2024**, *15*, 549. [CrossRef]
40. Ren, Z.; Zhao, H.; Fu, Y.; Xiao, L.; Dong, Y. Effects of Urban Street Trees on Human Thermal Comfort and Physiological Indices: A Case Study in Changchun City, China. *J. For. Res.* **2022**, *33*, 911–922. [CrossRef]
41. Ni, Y.; Lin, L.; Xia, H.; Wang, X. Urban Street Greening and Resident Comfort: An Integrated Approach Based on High-Precision Shadow Distribution and Facade Visual Assessment. *Sustainability* **2025**, *17*, 1026. [CrossRef]
42. Wang, G.-Y. Integrating Street Views, Satellite Imageries and Remote Sensing Data into Economics and the Social Sciences. *Soc. Sci. Comput. Rev.* **2024**, *42*, 326–351. [CrossRef]
43. Cetin, M.; Ozenen Kavlak, M.; Senyel Kurkcuoglu, M.A.; Bilge Ozturk, G.; Cabuk, S.N.; Cabuk, A. Determination of Land Surface Temperature and Urban Heat Island Effects with Remote Sensing Capabilities: The Case of Kayseri, Türkiye. *Nat. Hazards* **2024**, *120*, 5509–5536. [CrossRef]
44. Shu, B.; Chen, Y.; Zhang, K.; Dehghanifarsani, L.; Amani-Beni, M. Urban Engineering Insights: Spatiotemporal Analysis of Land Surface Temperature and Land Use in Urban Landscape. *Alex. Eng. J.* **2024**, *92*, 273–282. [CrossRef]
45. Song, Y.; Zhou, M.; Tan, J.; Cheng, J.; Wang, Y.; Feng, X.; Yu, H. Association Between Street Greenery and Physical Activity among Chinese Older Adults in Beijing, China. *Sci. Rep.* **2025**, *15*, 19509. [CrossRef] [PubMed]
46. Chen, C.; Wang, J.; Li, D.; Sun, X.; Zhang, J.; Yang, C.; Zhang, B. Unraveling Nonlinear Effects of Environment Features on Green View Index Using Multiple Data Sources and Explainable Machine Learning. *Sci. Rep.* **2024**, *14*, 30189. [CrossRef]
47. Teeuwen, R.; Milias, V.; Bozzon, A.; Psyllidis, A. How Well Do NDVI and OpenStreetMap Data Capture People's Visual Perceptions of Urban Greenspace? *Landsc. Urban Plan.* **2024**, *245*, 105009. [CrossRef]
48. Wei, T.; Li, W.; Tang, J. Decoding Tianjin: How Does Urban Form Shape the Diurnal Cycle of Surface Temperature? *Sustainability* **2024**, *16*, 10241. [CrossRef]
49. Ahern, J. Urban Landscape Sustainability and Resilience: The Promise and Challenges of Integrating Ecology with Urban Planning and Design. *Landsc. Ecol.* **2013**, *28*, 1203–1212. [CrossRef]
50. Zou, Y.; Chen, W.; Li, S.; Wang, T.; Yu, L.; Xu, M.; Singh, R.P.; Liu, C.-Q. Spatio-Temporal Changes in Vegetation in the Last Two Decades (2001–2020) in the Beijing–Tianjin–Hebei Region. *Remote Sens.* **2022**, *14*, 3958. [CrossRef]
51. Liu, Y.; Yu, Z.; Song, Y.; Yu, X.; Zhang, J.; Song, D. Psychological Influence of Sky View Factor and Green View Index on Daytime Thermal Comfort of Pedestrians in Shanghai. *Urban Clim.* **2024**, *56*, 102014. [CrossRef]

52. Wang, S.; Feng, S.; Liang, S.; Yan, L.; Yin, H.; Zhang, J. A Study of the Effects of Sky View Factor on Street Pedestrians' Sensations & Comforts Against Thermal Environments and Air Qualities. *J. Eng. Res.* 2025, *in press*. [CrossRef]
53. Zhang, S.; Lu, J.; Guo, R.; Yang, Y. Exploring the Relationship Between Visual Perception of the Urban Riverfront Core Landscape Area and the Vitality of Riverfront Road: A Case Study of Guangzhou. *Land* **2024**, *13*, 2142. [CrossRef]
54. Florio, P.; Leduc, T.; Sutter, Y.; Brémond, R. Visual Complexity of Urban Streetscapes: Human vs Computer Vision. *Mach. Vis. Appl.* **2024**, *35*, 7. [CrossRef]
55. Zhang, L.; Wang, L.; Wu, J.; Li, P.; Dong, J.; Wang, T. Decoding Urban Green Spaces: Deep Learning and Google Street View Measure Greening Structures. *Urban For. Urban Green.* **2023**, *87*, 128028. [CrossRef]
56. Peng, M.; Huang, H. The Synergistic Effect of Urban Canyon Geometries and Greenery on Outdoor Thermal Comfort in Humid Subtropical Climates. *Front. Environ. Sci.* **2022**, *10*, 851810. [CrossRef]
57. Wang, L.; Chen, T.; Yu, Y.; Wang, L.; Zang, H.; Cang, Y.; Zhang, Y.; Ma, X. Impacts of Vegetation Ratio, Street Orientation, and Aspect Ratio on Thermal Comfort and Building Carbon Emissions in Cold Zones: A Case Study of Tianjin. *Land* **2024**, *13*, 1275. [CrossRef]
58. Kuang, B.; Yang, H.; Jung, T. The Impact of Visual Elements in Street View on Street Quality: A Quantitative Study Based on Deep Learning, Elastic Net Regression, and SHapley Additive exPlanations (SHAP). *Sustainability* **2025**, *17*, 3454. [CrossRef]
59. Switalski, M.; Torres, M.G.; Grêt-Regamey, A. The 3P's of Place-Making: Measuring Place-Making Through the Latent Components of Person, Procedures and Place. *Landsc. Urban Plan.* **2023**, *238*, 104817. [CrossRef]
60. Gao, W.; Qian, Y.; Chen, H.; Zhong, Z.; Zhou, M.; Aminpour, F. Assessment of Sidewalk Walkability: Integrating Objective and Subjective Measures of Identical Context-Based Sidewalk Features. *Sustain. Cities Soc.* **2022**, *87*, 104142. [CrossRef]
61. Rui, Q.; Cheng, H. Quantifying the Spatial Quality of Urban Streets with Open Street View Images: A Case Study of the Main Urban Area of Fuzhou. *Ecol. Indic.* **2023**, *156*, 111204. [CrossRef]
62. Beele, E.; Aerts, R.; Reyniers, M.; Somers, B. Urban Green Space, Human Heat Perception and Sleep Quality: A Repeated Cross-Sectional Study. *Environ. Res.* **2024**, *263*, 120129. [CrossRef] [PubMed]
63. Lang, N.; Jetz, W.; Schindler, K.; Wegner, J.D. A High-Resolution Canopy Height Model of the Earth. *Nat. Ecol. Evol.* **2023**, *7*, 1778–1789. [CrossRef]
64. Jing, X.; Li, Z.; Chen, H.; Zhang, C. "Is What We See Always Real?" A Comparative Study of Two-Dimensional and Three-Dimensional Urban Green Spaces: The Case of Shenzhen's Central District. *Forests* **2024**, *15*, 983. [CrossRef]
65. Cheng, B.; Misra, I.; Schwing, A.G.; Kirillov, A.; Girdhar, R. Masked-Attention Mask Transformer for Universal Image Segmentation. In Proceedings of the IEEE/CVF Conference on Computer Vision and Pattern Recognition (CVPR 2022), New Orleans, LA, USA, 19–24 June 2022; pp. 12809–12819.
66. Neuhold, G.; Ollmann, T.; Bulo, S.R.; Kontschieder, P. The Mapillary Vistas Dataset for Semantic Understanding of Street Scenes. In Proceedings of the 2017 IEEE International Conference on Computer Vision (ICCV), Venice, Italy, 22–29 October 2017; IEEE: Venice, Italy, 2017; pp. 5000–5009.
67. Zhao, P.; Yu, B. On Model Selection Consistency of Lasso. *J. Mach. Learn. Res.* **2006**, *7*, 2541–2563.
68. Zou, H.; Hastie, T. Regularization and Variable Selection via the Elastic Net. *J. R. Stat. Soc. Ser. B Stat. Methodol.* **2005**, *67*, 301–320. [CrossRef]
69. Friedman, J.; Hastie, T.; Tibshirani, R. Regularization Paths for Generalized Linear Models via Coordinate Descent. *J. Stat. Softw.* **2010**, *33*, 1–22. [CrossRef]
70. Apley, D.W.; Zhu, J. Visualizing the Effects of Predictor Variables in Black-Box Supervised Learning Models. *J. R. Stat. Soc. Ser. B* **2020**, *82*, 1059–1086. [CrossRef]
71. Wood, S.N. *Generalized Additive Models: An Introduction with R*, 2nd ed.; Chapman and Hall/CRC: New York, NY, USA, 2017; ISBN 978-1-315-37027-9.
72. Lundberg, S.; Lee, S.-I. A Unified Approach to Interpreting Model Predictions. *arXiv* **2017**, arXiv:1705.07874. Available online: https://arxiv.org/abs/1705.07874v2 (accessed on 8 September 2025).
73. Solanki, P.; Amirtham, L.R.; Deb, C. Optimizing Sky View Factor and Vegetation to Mitigate Urban Heat in Hot-Humid Climates. *Discov. Cities* **2025**, *2*, 79. [CrossRef]
74. Zhang, T.; Wang, L.; Zhang, Y.; Hu, Y.; Zhang, W. Assessing the Nonlinear Impact of Green Space Exposure on Psychological Stress Perception Using Machine Learning and Street View Images. *Front. Public Health* **2024**, *12*, 1402536. [CrossRef]
75. Torkko, J.; Poom, A.; Willberg, E.; Toivonen, T. How to Best Map Greenery from a Human Perspective? Comparing Computational Measurements with Human Perception. *Front. Sustain. Cities* **2023**, *5*, 1160995. [CrossRef]
76. Sun, C.; Lian, W.; Liu, L.; Dong, Q.; Han, Y. The Impact of Street Geometry on Outdoor Thermal Comfort Within Three Different Urban Forms in Severe Cold Region of China. *Build. Environ.* **2022**, *222*, 109342. [CrossRef]
77. Shao, J.; Long, Y.; Liu, X.; Zheng, Y.; Song, Y.; Wang, J.; Liu, B.; Yang, J.; Chen, Y.; Zhang, F. Machine Learning-Based Study on Factors Influencing Street Vitality in Urban Fringe Commercial Districts: A Case of Wuhan. *Front. Archit. Res.* **2025**, *in press*. [CrossRef]

https://doi.org/10.3390/buildings16010001

78. Yang, L.; Ao, Y.; Ke, J.; Lu, Y.; Liang, Y. To Walk or Not to Walk? Examining Non-Linear Effects of Streetscape Greenery on Walking Propensity of Older Adults. *J. Transp. Geogr.* **2021**, *94*, 103099. [CrossRef]
79. Yang, L.; Yang, H.; Yu, B.; Lu, Y.; Cui, J.; Lin, D. Exploring Non-Linear and Synergistic Effects of Green Spaces on Active Travel Using Crowdsourced Data and Interpretable Machine Learning. *Travel Behav. Soc.* **2024**, *34*, 100673. [CrossRef]
80. Qiu, H.; Tang, H.; Yang, Y. The Non-linear Effects of Urban Green Space on Promoting Physical Activity of Old Adults at Different Obesity Status in Semi-Arid Area: A Case Study of Lanzhou, China. *Geohealth* **2025**, *9*, e2024GH001273. [CrossRef] [PubMed]
81. Navarrete-Hernandez, P.; Kiarostami, N.; Yang, D.; Ozcakir, A. Green Enough? A Dose-Response Curve of the Impact of Street Greenery Levels and Types on Perceived Happiness. *Landsc. Urban Plan.* **2024**, *251*, 105130. [CrossRef]
82. Ma, Y.; Chen, P.; Qin, Y.; Yang, Z.; Li, S. From Sky to Ground: Monitoring Visible Street Greenery via Multisource Remote Sensing Imagery with Deep Learning. *Urban For. Urban Green.* **2025**, *109*, 128866. [CrossRef]
83. Zhu, J.; Wang, S.; Ma, H.; Shan, T.; Xu, D.; Sun, F. Nonlinear Effect of Urban Visual Environment on Residents' Psychological Perception—An Analysis Based on XGBoost and SHAP Interpretation Model. *City Environ. Interact.* **2025**, *27*, 100202. [CrossRef]
84. Huang, Y.; Sanatani, R.P.; Liu, C.; Kang, Y.; Zhang, F.; Liu, Y.; Duarte, F.; Ratti, C. No "True" Greenery: Deciphering the Bias of Satellite and Street View Imagery in Urban Greenery Measurement. *Build. Environ.* **2025**, *269*, 112395. [CrossRef]
85. Xie, Y.; Zhang, J.; Li, Y.; Zhu, Z.; Deng, J.; Li, Z. Integrating Multi-Source Urban Data with Interpretable Machine Learning for Uncovering the Multidimensional Drivers of Urban Vitality. *Land* **2024**, *13*, 2028. [CrossRef]
86. Tang, H.; Qiu, H.; Yang, Y. Thermal Environment Disparities across 11 Cities in the Greater Bay Area: Nonlinear Relationships and Urban Greening-Based Solutions. *Sustain. Cities Soc.* **2025**, *133*, 106867. [CrossRef]
87. Chen, Y.; Yu, B.; Shu, B.; Yang, L.; Wang, R. Exploring the Spatiotemporal Patterns and Correlates of Urban Vitality: Temporal and Spatial Heterogeneity. *Sustain. Cities Soc.* **2023**, *91*, 104440. [CrossRef]
88. Yang, L.; Yu, B.; Liang, Y.; Lu, Y.; Li, W. Time-Varying and Non-Linear Associations Between Metro Ridership and the Built Environment. *Tunn. Undergr. Space Technol.* **2023**, *132*, 104931. [CrossRef]
89. Yang, L.; Chen, Y.; Xu, N.; Zhao, R.; Chau, K.W.; Hong, S. Place-Varying Impacts of Urban Rail Transit on Property Prices in Shenzhen, China: Insights for Value Capture. *Sustain. Cities Soc.* **2020**, *58*, 102140. [CrossRef]

https://doi.org/10.3390/buildings16010001

buildings

MDPI

Article

An Interpretable Machine Learning Approach to Studying Environmental Safety Perception Among Elderly Residents in Pocket Parks

Shengzhen Wu [1], Sichao Wu [2,*], Jingru Chen [1] and Chen Pan [3]

1 College of Arts and Design, Jimei University, Xiamen 361000, China; 202361000218@jmu.edu.cn (S.W.); 202261000138@jmu.edu.cn (J.C.)
2 Xiamen Academy of Arts and Design, Fuzhou University, Xiamen 361000, China
3 Architecture and Civil Engineering Institute, Guangdong University of Petrochemical Technology, Maoming 525000, China; jourdan29@gdupt.edu.cn
* Correspondence: wusichao@fzu.edu.cn

Abstract: This research explores the environmental safety challenges faced by pocket parks in the context of urban aging within Chinese cities. It systematically analyzes visual elements that influence the elderly's perception of environmental safety by applying interpretable machine learning techniques. By integrating panoramic image semantic segmentation and explainable AI models (e.g., SHAP and PDP), the study transforms subjective environmental perception into measurable indicators and constructs an environmental safety perception model using the LightGBM algorithm. Results indicate that sufficient pedestrian areas and moderate crowd activities significantly enhance safety perception among the elderly. Conversely, the presence of cars emerges as the most substantial adverse factor. Natural elements, such as vegetation and grass, exhibit nonlinear effects on safety perception, with an optimal threshold range identified. The research further elucidates the intricate synergies and constraints among visual elements, underscoring that the highest perceived safety arises from the synergistic combination of positive factors. This study deepens the understanding of environmental perception among the elderly and offers a data-driven framework and practical guidelines for urban planners and designers. It holds significant theoretical and practical implications for advancing the refined and human-centered renewal of urban public spaces.

Keywords: urban aging; pocket parks; environmental safety perception; visual elements; machine learning

1. Introduction

As the most populous country in the world, China faces an increasingly aging population. While the World Health Organization defines senior citizens as citizens aged 65 and above, the Law of the People's Republic of China on the Protection of the Rights and Interests of the Elderly defines senior citizens as those aged 60 and above. Despite recent continuous economic growth and rapid urban renewal, the built environment of many urban public spaces still fails to meet the basic needs of the elderly. With declining physical and mental health, the elderly require a more supportive living environment to enhance their quality of life [1]. Urban public spaces, particularly parks, serve as key locations for the elderly to engage in outdoor activities and interact socially within urban environments. However, these spaces often struggle to meet the needs of elderly users, not only compromising their ability to engage effectively but also exposing them to safety risks.

Many urban parks exhibit facility deterioration and uneven terrain, heightening the risk of falls among older adults due to inadequate ground maintenance and facility upkeep [2]. The significant noise problem in the park adversely impacts the recreational experience of elderly individuals [3]. Research indicates that elders' perceptions of urban park safety are shaped by factors such as cleanliness and the condition of facilities [4]. Relative to large parks, pocket parks exhibit the characteristics of a small scale, flexible layout, and geographical proximity to neighborhoods. Therefore, they are more likely to function as critical venues for the daily activities of elderly residents. Pocket parks, typically smaller than one hectare, can provide functional and practical open space through thoughtful design and strategic layout [5]. As a key element of urban micro-renewal, the environmental design of pocket parks plays a essential role in enhancing the psychological well-being of the elderly population. Elderly individuals aged between 66 and 75 are frequent visitors to pocket parks, where the landscaped environment contributes significantly to psychological restoration [6]. The green coverage, spatial arrangement, and facility designs of pocket parks collectively shape both the elderly's willingness to use these spaces and their experiential quality during activities [7]. There is a gap in the current research on the quantitative analysis of pocket park environments' effects on elderly safety perception based on visual elements [8]. Prior research suggests that factors such as greening and spatial enclosure influence elderly spatial perception. However, systematic quantitative analyses on how visual elements specifically impact older adults' safety perception remain underdeveloped. Enhancing the environmental safety of pocket parks and fulfilling the functional needs of elderly individuals for secure and comfortable public spaces have emerged as critical priorities in contemporary urban renewal initiatives.

To investigate the behaviors and interactions of elderly individuals in urban public spaces, it is significant to analyze the critical role of environmental safety perception in shaping their experiences. The perception of environmental safety is a key psychological element for the elderly in urban public spaces. This perception is influenced by several theoretical frameworks, notably the Stress Reduction Theory (SRT) and the Attention Restoration Theory (ART). SRT posits that exposure to natural environments can foster positive psychological experiences, which in turn alleviates stress. Similarly, ART suggests that natural settings offer the restorative benefits necessary for individuals to recover their attentional capacity from the demands of daily stressors [9]. Safety perception exhibits significant group differences and environmental dependence. Zhou et al., through crowdsourcing research, revealed significant differences in the perception of criminal- and disability-related security between men and women, highlighting the necessity to precisely define perception and incorporate demographic variables into modeling [10]. By focusing on a female-centered perspective, Chen et al. revealed that environmental indicators such as sky openness, the visibility of greenery, and road visibility significantly influence women's sense of security [11,12]. These theories establish a decisive theoretical framework for elucidating how the elderly evaluate the safety of built environment. This study defines environmental safety perception as the elderly's subjective evaluation of their sense of security in pocket parks. This perception is shaped by two sets of factors: the physical characteristics of pocket parks (e.g., greening, spatial layout, and facility configuration) and the individual characteristics of the elderly (e.g., experience, knowledge, and emotional state) [13].

With the rapid development of computer vision and deep learning technologies, SVI and deep learning models have increasingly been employed by researchers to quantify and analyze human perception of the built environment, particularly in environmental safety perception. Larkin et al. leveraged Amazon Mechanical Turk to crowdsource annotations and trained deep models via transfer learning to predict variables such as natural environment quality, aesthetics, relaxation, and security. The model demonstrated an explanatory power of up to 77.6%. This method not only accelerates data acquisition efficiency but also enhances the objectivity and reproducibility of perceptual measurement [14]. Machine learning has been extensively applied in urban spatial data analysis, notably for evaluating how urban green spaces affect residents' mental health. However, the black-box characteristic of machine learning restricts its interpretability, thereby undermining its credibility and limiting its application to scenarios requiring explanation. Recent developments in Explainable Artificial Intelligence (XAI) offer a solution by enabling the interpretation of machine learning outcomes [15]. For instance, SHAP (SHapley Additive exPlanation) and PDP (Partial Dependency Plot) are two popular model interpretation techniques. The SHAP method, introduced by Lundberg et al. and grounded in the Shapley value from cooperative game theory, provides feature-level explanations by quantifying the contribution of each feature to the model's output for a given prediction [16]. This method enables a deeper understanding of the decision-making mechanisms in complex models, thereby enhancing their interpretability and establishing greater confidence in their predictions [17]. The SHAP method combines global and local interpretations within a unified framework. Its findings align with human perception, making it robust and intuitive for interpretation [18]. Partial dependence plots (PDP) visualize the marginal effect of individual features on the output of a machine learning model and are particularly valuable for exploring non-monotonic relationships in high-dimensional, multi-faceted data structures [19]. SHAP and PDP can be leveraged to determine the most effective intervention by quantifying feature importance and visualizing domain-specific outcomes.

Existing studies have laid a significant theoretical foundation for understanding the environmental perception of the elderly; however, several notable limitations remain. Empirical research is predominantly based on traditional questionnaires or limited field observations, thereby hindering large-scale, efficient quantitative analysis of urban public spaces. Although machine learning models have been introduced in some studies, their inherent "black box" characteristics obscure the intricate relationships underlying their prediction mechanisms. This limitation complicates the translation of research findings into practical and actionable design principles. Based on the above considerations, this study uses 29 pocket parks located on Xiamen Island, Xiamen City, Fujian Province, China as examples. Semantic segmentation technology was employed to extract visual elements from panoramic images. Subsequently, a road safety perception dataset was constructed by combining these visual elements with safety perception scores. An environmental safety perception model was established using the LightGBM algorithm. The influence mechanism of visual factors on environmental safety perception was analyzed using SHAP and PDP interpretation algorithms. The key elements and their influence intensity were analyzed.

2. Research Methodology

2.1. Research Area and Data Collection

2.1.1. Research Area

This study examines the pocket parks on Xiamen Island in Xiamen City, Fujian Province, China. During the study, 29 pocket parks (with an area of less than 1 hectare) were found to have been constructed on Xiamen Island, comprising 19 in Siming District and 10 in Huli District. As the core area of Xiamen City, Xiamen Island occupies a highly desirable geographical position between 118°04′ and 118°23′ east longitude and 24°23′ and 24°55′ north latitude and has a total area of approximately 158 square kilometers. Xiamen Island exhibits a relatively high population density and a pronounced aging trend. Based on 2022 statistics, 18.5% of the island's population was aged 60 or above, exceeding the national average. This provides an ideal case study for analyzing the spatial needs of the elderly in urban public spaces [20].

The 29 pocket parks examined in this study are distributed across various administrative districts on the island, such as Siming and Huli Districts. They encompass diverse urban environments, including high-density residential zones, mixed commercial areas, and older residential communities. Most of these pocket parks were constructed post-2010 as part of urban micro-renewal initiatives. They typically span an area of 200 to 500 square meters and are characterized by a high vegetation coverage rate—averaging over 65%—as well as a variety of facilities, including seating arrangements, walking paths, and fitness equipment.

Xiamen Island was chosen as a study site since it epitomizes the common issues faced by China's rapidly urbanizing coastal regions. Firstly, rapid economic growth has spurred urban renewal. Secondly, approximately 40% of the local communities are aging, and the discrepancy between the existing built environment and the elderly's needs is pronounced. Consequently, pocket parks have emerged as a vital experimental platform to enhance outdoor activity safety for the elderly. Xiamen Island's climate is mild, with an average annual temperature of 21 °C, and its annual precipitation is moderate at roughly 1200 mm. These conditions facilitate year-round outdoor activity research and ensure stable data collection (Figure 1).

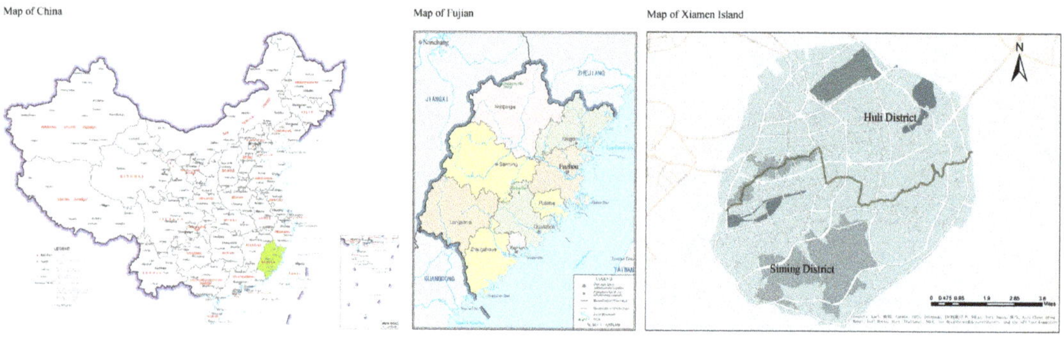

Figure 1. Research Scope.

2.1.2. Data Collection

Panoramic image data: In elderly individuals, the typical visual field spans approximately 500 m [21]. While pocket parks typically occupy areas less than one hectare with simple internal road layouts, their scenery is characterized by variety and staggered visual impact. To ensure comprehensive coverage of walking paths and scenic elements within the parks, strategic placement of collection points is essential. Collection guidelines are as follows:

1. No points are required for sections shorter than 5 m.
2. Points should be placed at the midpoint of sections ranging from 5 to 10 m.
3. For sections exceeding 10 m, points should be spaced at 20 m intervals.
4. Additionally, points should be positioned at every intersection to ensure coverage.

A total of 497 collection points and corresponding panoramic images were selected for this study. These images were captured using the Insta 360X4 panoramic camera (Shenzhen Insta360 Innovation Technology Co., Ltd., Shenzhen, China), which automatically records metadata for each photo, including geospatial coordinates, shooting time, and environmental lighting conditions. This functionality ensures data consistency and traceability. After acquisition, all 497 images underwent standardized preprocessing. Denoising and brightness normalization were applied using the OpenCV Python 3.14 library to mitigate weather-related interference (e.g., rain or harsh lighting) during visual element extraction. Images were uniformly resized to 1024 × 512 pixels to optimize input for subsequent deep learning models.

Environmental safety perception data of the elderly: This study examined the environmental safety perception of elderly individuals in pocket parks on Xiamen Island, with data collected over a 15-month period. The research team randomly selected 2000 panoramic images across multiple pocket parks and invited 88 volunteers (an equal gender distribution of 44 males and 44 females) to evaluate their environmental safety perception using a 1-to-10 semantic differential scale. The volunteers were primarily elderly residents aged 60 and above, which established the robustness and representativeness of the sample. To ensure the reliability of the scores, the research team selected volunteers with frequent use of pocket parks. Based on the 2021 Xiamen City Seventh National Population Census Report, the city's population aged 60 and above is 490,600, constituting 9.56%. The study resulted in 176,000 valid scoring data points, derived from 88 volunteers independently evaluating 2000 images each. Each environmental stimulus was independently assessed 88 times, providing adequate data repetition to ensure statistical accuracy and stability in the mean scores of scene perception. Additionally, all volunteers completed the scoring of all images independently, thereby ensuring uniformity in evaluation criteria and minimizing between-group variability. Ultimately, 88 volunteers completed the scoring of all sample pictures, producing the environmental security score dataset for the study.

2.2. Extraction and Quantification of Visual Elements

2.2.1. Extraction of Visual Variables

The explanatory variables are visual elements primarily extracted via image semantic segmentation. Semantic segmentation constitutes a core task in computer vision, aiming to assign every individual pixel in an image to a predefined semantic category [22]. Unlike traditional image classification, which only identifies objects in the image, semantic segmentation categorizes each pixel, thereby providing a pixel-level understanding [23]. By applying semantic segmentation techniques to panoramic imagery, pixel-level urban features including roads, buildings, and green spaces can be systematically classified. This classification enables the systematic assessment of urban safety and comfort based on the identified features [24]. Street view image segmentation was conducted using the MIT ADE20K model with a ConvNeXt backbone [25], as well as a supplementary model based on segformer and trained with the Mapillary Vistas dataset [26]. ADE20K is a large-scale dataset for semantic segmentation, containing over 20,000 images divided into training, validation, and test sets. These images are densely annotated with 100 'thing' categories (e.g., vehicles, pedestrians) and 50 'stuff' categories (e.g., roads, sky), covering 89% of the pixels. Mapillary Vistas is a street view image dataset comprising 25,000 images, split into training, validation, and test sets. These images are densely annotated with 28 'stuff' categories and 37 'thing' categories, covering 98% of the pixels [27]. This study examines several key element categories, including the sky, trees, buildings, motor vehicles, and pedestrian areas [28]. The segmentation results for the datasets demonstrate these element categories and thereby satisfy the study's requirements (Figure 2).

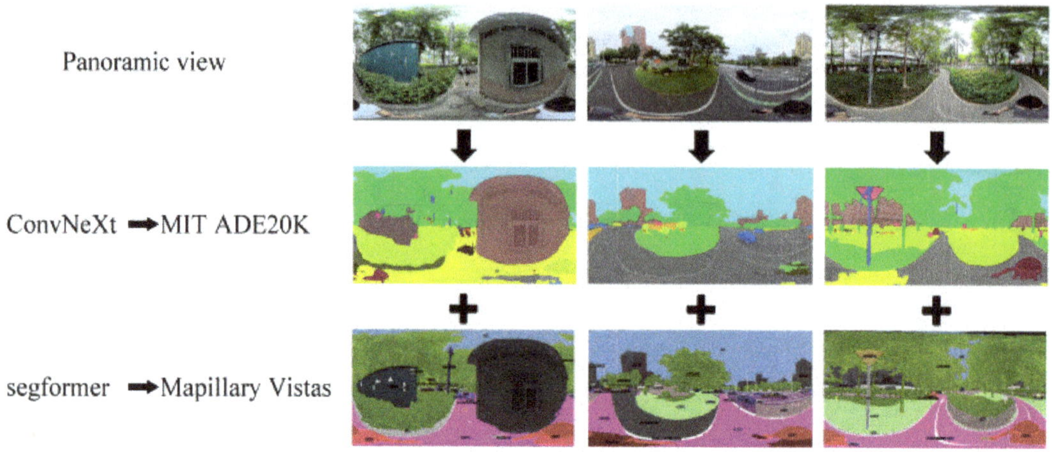

Figure 2. Extraction of visual elements.

2.2.2. Quantification of Visual Variables

This study centers on the environmental safety perception score and employs a hybrid method that combines manual assessment and machine learning to conduct quantitative analysis on large-scale data. Deep learning technologies are employed to train models, thereby improving computational efficiency and resource utilization. Through its self-attention mechanism, as a deep learning model, the Transformer architecture can effectively capture long-distance dependencies in images, leading to significant improvements in segmentation accuracy. This architecture demonstrates superior performance in handling complex scenes and high-resolution images, further illustrating its effectiveness and robustness [29]. We constructed a dataset with 10,000 image comparison pairs, of which 8000 were allocated for training and 2000 for testing. The ConvNeXt model, built on the Transformer architecture, was selected for training. Training hyperparameters included a learning rate of 0.001, the Adam optimizer, a batch size of 20, and a training duration of 120 epochs. Cross-Entropy Loss was utilized as the loss function [30]. Once the training phase concluded, the model was applied to the test set, achieving an accuracy of 89.2%. Subsequently, all panoramic images of pocket parks were processed to compute their safety perception scores.

2.3. Construction of Environmental Safety Perception Model

This study utilized the LightGBM algorithm to investigate the associations between environmental safety perception and visual cues [31]. We used a learning rate of 0.001, the Adam optimizer, a batch size of 20, and trained for 120 epochs. The validation split was 20% of the dataset. LightGBM is a highly efficient gradient boosting decision tree algorithm that achieves faster training by optimizing data processing efficiency. It prioritizes samples with large gradient variations while deemphasizing those with small variations, thus preserving model accuracy without unnecessary computational overhead. The algorithm employs a leaf growth strategy, where each iteration selects and splits the leaf node with the highest gain until the maximum number of splits is achieved. By capping the tree depth, LightGBM prevents overfitting and enhances model stability. Compared to traditional algorithms, LightGBM demonstrates superior performance in terms of training speed, memory efficiency, and accuracy, positioning it as one of the most efficient frameworks for gradient boosting tasks. Its high efficiency originates from optimized histogram algorithms and leaf growth strategies, which work together to substantially increase both training efficiency and memory utilization. Through its gradient boosting mechanism, LightGBM incrementally improves predictive performance by iteratively refining residuals. The algorithm excels in handling large-scale data, supports parallel computing and distributed learning, and is particularly well-suited for high-dimensional data scenarios [32]. Hyperparameter tuning of LightGBM, including learning rate, tree depth, and regularization parameters, significantly impacts the model performance [33].

2.4. Quantify the Impact of Visual Elements on Environmental Safety Perception Based on SHAP and PDP

SHAP (SHapley Additive exPlanations) is a method rooted in Shapley values from cooperative game theory, enabling the analysis and interpretation of model predictions [34]. SHAP quantifies the contribution of each feature to the model's prediction and determines whether these contributions are positive or negative. PDP is employed to visually illustrate the marginal effect of one or two variables on the model's predictions. This technique involves computing the marginal effect of a specified variable by systematically varying its values while holding other variables constant. SHAP combines both global and local interpretability by providing insights into the overall model behavior and individual prediction contributions. This study integrates SHAP and PDP to assess how visual elements influence environmental safety perception. Because they provide both global and local interpretability, which is crucial for understanding the complex relationships between visual elements and safety perception. Other methods like LIME were considered but were less suitable for our high-dimensional data and complex models.

3. Results

3.1. Spatial Distribution and Assessment of Perception of Environmental Safety

The urban development of Xiamen Island features a 'dual-core' structure: Siming District as the traditional central urban area and Luhui District as the emerging development zone. This development model has significantly shaped the urban fabric and public space quality of both areas, particularly evident in the elderly residents' ratings of park environment security.

The safety awareness scores of elderly residents were assessed in 29 pocket parks across Xiamen Island (Figure 3). Notably, elderly residents in Siming District exhibited consistently lower scores compared to those in Hulin District. Siming District, as Xiamen Island's historic core, comprises numerous aging residential communities. Parks 5, 8, 9, and 10, which scored between 4.60 and 5.83, are typically found within such communities. These low scores reflect persistent deficiencies in the hardware quality and management systems of these aging communities. In contrast, Hulin District, a rapidly developing area with new high-density residential and commercial mixed-use developments, contains high-scoring parks (e.g., 19, 20, 22–25, 28, and 29; scores of 6.60–7.17), illustrating successful urban planning outcomes. These high-scoring parks benefit from modern urban development principles, including well-designed hardware infrastructure and professional management, which together create high-quality and secure public activity spaces for elderly residents. These findings highlight the importance of urban planning in addressing the specific needs of elderly residents through well-designed facilities and effective management systems.

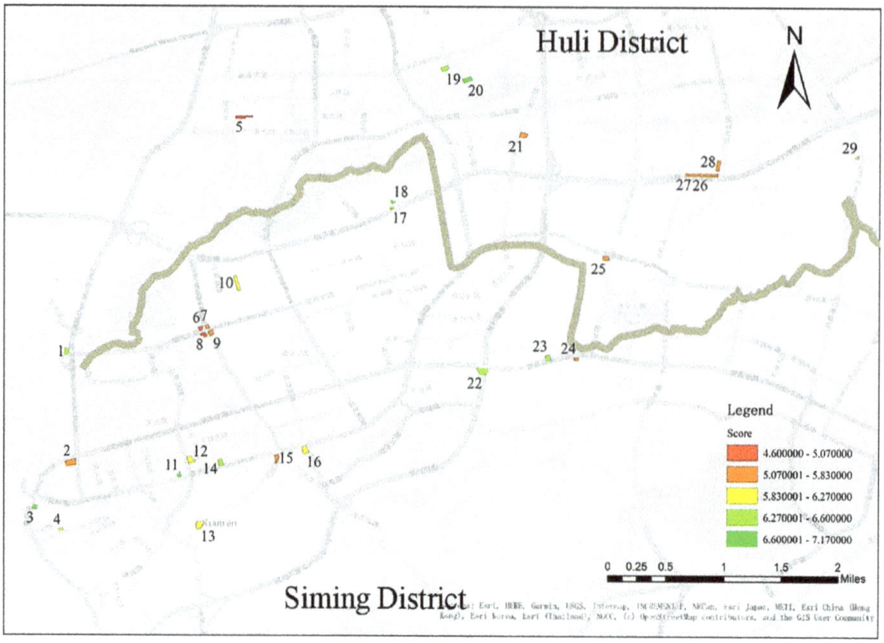

Figure 3. Environmental Safety Perception Distribution Characteristics of Elderly People's Pocket Parks on Xiamen Island.

3.2. SHAP Feature Contribution Results

Figure 4a demonstrates the effect of distinct visual elements on elderly individuals' environmental safety perception. The elements are ranked by their importance, with the horizontal axis representing the SHAP value (i.e., influence weight) of each element. Areas with larger width and higher density indicate stronger influence. The dots denote pocket park samples. Darker shades of red correspond to higher safety perception, while darker shades of blue indicate lower safety perception. The SHAP values for pedestrian areas exhibit a broad distribution, with a primary concentration in the positive region, suggesting that it exerts a strong positive effect on environmental safety perception. Augmenting walkable spaces and enhancing walking infrastructure can improve safety perception among elderly individuals [35]. Research on urban design indicates that narrow sidewalks are likely to negatively influence pedestrian perception, such as by restricting visual accessibility and movement comfort. Conversely, wider sidewalks are associated with an enhanced perception of space due to improved comfort and safety [36]. The car exhibits a significant influence in the graph, primarily through negative SHAP values. This suggests that the presence of vehicles significantly diminishes the environmental safety perception of the elderly. Vehicles moving outside pocket parks may induce anxiety among the elderly, who may have concerns about traffic accidents and safety [37]. Vehicle noise and exhaust emissions may negatively impact the health of the elderly, particularly their respiratory and cardiovascular systems, thereby reducing their comfort and sense of security in their environment [38,39]. Furthermore, vehicles parked in pedestrian zones can limit the mobility of the elderly, increasing their risk of collisions and subsequently diminishing their perceived environmental safety [40]. The SHAP values for walls exhibit a dispersed distribution, with some values in the positive range and others in the negative range. This dual distribution reflects the dual role of walls in shaping environmental safety perception. Specifically, low fences or flower beds that provide visual screening and psychological reassurance can enhance the elderly's sense of security.

In contrast, tall walls may block visibility, reduce spatial transparency, and thereby induce feelings of oppression and insecurity among the elderly [41]. Person presence correlates with improved environmental safety perception, as reflected by predominantly positive SHAP values. A low visitor count may indicate insufficient natural surveillance, allowing potential offenders to operate undetected and thus heightening feelings of insecurity, particularly among elderly individuals [42]. Excessive human gatherings can cause visual information overload, impairing elderly individuals' ability to interpret environmental cues and thus compromising their judgment of environmental safety [43]. Furthermore, excessive crowd density can induce psychological distress in elderly individuals, which may, in turn, impair their perceptual abilities [44]. Excessive crowd density may induce anxiety and unease in the older population, potentially compromising their perceptual abilities. Significantly, the positive effects of signage features suggest that they can enhance the older population's perception of environmental safety. Well-planned notice boards, which provide clear directional cues and location information, serve as effective tools for guiding older adults in navigating and utilizing spaces, ultimately boosting their sense of security [45]. Most of the SHAP values for parterre and vegetation are positive, indicating their relatively favorable contribution to environmental safety perception. While their overall contribution may be modest in comparison to other factors, their effect remains positive, contributing to the enhancement of environmental safety perception. A well-designed flower bed layout and appropriate vegetation density not only enhance the visual appeal of the environment but also contribute to psychological comfort. Furthermore, their isolating and buffering functions mitigate external disturbances, thereby improving the sense of security among elderly residents [46,47]. The SHAP values of elements such as grass, motorcycles, roads, fences, sky, buildings, streetlights, trees, and bicycles are relatively small, suggesting a minimal contribution to environmental safety perception. Greening, however, positively reinforces the environmental safety of the elderly [48,49]. Trees on pedestrian paths enhance the perceived attractiveness and safety of the walking environment, underscoring the significance of infrastructure quality. However, the presence of trees may not always directly enhance perceived safety, depending on specific circumstances [50]. Specifically, the SHAP values of grass and streetlight are predominantly positive, suggesting they exert a mild positive influence on safety perception. Conversely, the SHAP values of motorcycles and roads often display negative contributions, potentially undermining safety perception. Furthermore, overgrown vegetation may conceal hazards such as uneven terrain or obstacles, thereby raising the likelihood of accidents.

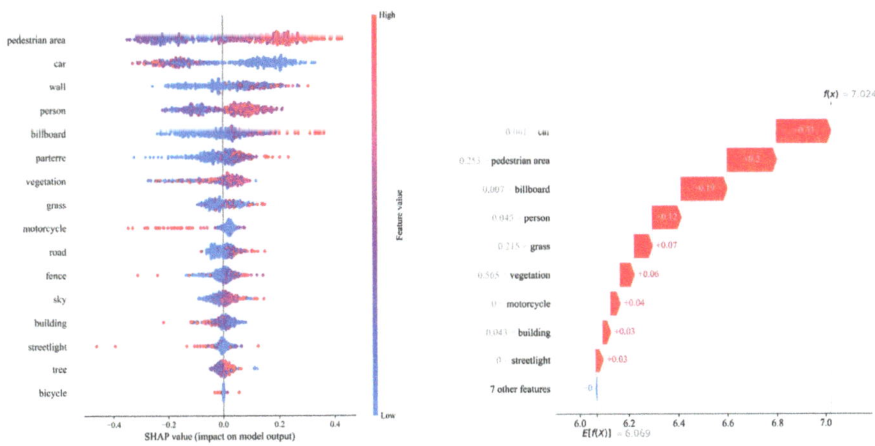

(a) SHAP Summary Plot (b) SHAPwaterfall plot

Figure 4. The influence of visual elements on the environmental safety perception of the elderly.

Figure 4b illustrates the specific contributions of each visual element to the model output, reflecting their impact on environmental safety perception. The model output value f(x) = 7.024, which reflects a relatively high environmental safety perception score for the pocket park. The contribution values of each element, listed in the figure, indicate their impact on the model output: positive values represent a positive influence, whereas negative values signify a negative influence. While the contribution value of the car to the model output is +0.23 (denoted by the red arrow in the figure), this positive value represents the absolute magnitude of its negative impact, indicating that vehicles significantly reduce perceived environmental safety. This aligns with the global interpretation that vehicles negatively affect safety perception [51]. The pedestrian area contributes +0.2 to the model output, suggesting that its width and accessibility significantly enhance the environmental safety perception of the elderly. This finding is consistent with the global interpretation that pedestrian zones positively affect safety perception, underscoring their importance in pocket park design. The billboard contributes +0.19 to the model output, likely due to its clear and well-designed informational content, which effectively enhances environmental safety perception among the elderly. The person contributes +0.12 to the model output, increasing social interaction, vitality, and perceived safety among the elderly. Grass (+0.07) and vegetation (+0.06) positively influence environmental safety perception due to their appropriate quantity and well-planned layout. This aligns with the global interpretation that grassland and vegetation positively influence safety perception [52]. While elements such as motorcycles, buildings, and streetlights have smaller contribution values (e.g., streetlights contribute +0.03), they still positively influence environmental safety perception, particularly by enhancing nighttime or low-light security [53].

This study focuses on the analysis of the internal environment of pocket parks, particularly examining the impact of visual elements on the environmental safety perception of the elderly. Based on relevant literature and on-site investigations, six key visual elements were identified for in-depth research: pedestrian area, car, person, billboard, vegetation, and grass. The vegetation includes various plant types above the ground, such as trees, shrubs, grasslands, hedges, and shrub forests. They have a certain obstructive effect on the vision of the elderly. In contrast, grassland is mainly composed of low-growing herbaceous plants and covers the ground surface, and its effect on vision obstruction is not significant. Therefore, in this study, it is divided into two different indicators.

3.3. *Univariate Key Visual Elements Contribute to Shaping Elderly Individuals' Perception of Environmental Safety*

Figure 5a–f illustrates the quantitative characterization of the main effects of six key visual elements using partial dependency plots (PDPs). All curves lie within a 95% confidence interval, with the horizontal axis denoting the pixel proportion of the elements and the vertical axis denoting the safety perception score of the elderly. The blue solid line, representing the Partial Dependence Curve (PDC), illustrates the partial dependence of the model's predictions on a specific feature. It demonstrates how the model's predictions are affected by variations in the feature. The blue shaded area indicates the 95% confidence interval, reflecting the statistical uncertainty of the curve. A narrow shaded area indicates high consistency in the model's response, while a wide shaded area suggests significant variability, likely due to feature interactions or data sparsity. The red dots indicate the distribution of actual sample data, with their positions aligned with the feature values in the dataset. Observing their density helps identify regions where sufficient data exists to support reliable model predictions. For instance, a dense cluster of red dots in a specific interval implies higher reliability of model predictions within that area. The pedestrian area exerts a significant nonlinear effect on the prediction results. Once the area exceeds a certain threshold (approximately 0.18–0.22), the larger the area, the higher the prediction result, indicating an optimal area range or a positive impact after the critical point. Car count is inversely related to the prediction results; a small number of vehicles leads to a marked decrease in results, which remain low. Overall, vehicles negatively impact the target variable. Person exhibits a threshold effect on the prediction results. When density reaches approximately 0.040, the results significantly improve and stabilize at a high level, suggesting better performance of the target variable at this density. Billboard count may positively impact the prediction results up to a point, but exceeding approximately 1.0% may lead to a decline, indicating an optimal number of signboards. Vegetation density has a nonlinear influence, with an optimal range at approximately 0.30. Too low or too high densities may negatively impact the results. The grass area may initially decrease the prediction results but turn positive once the area reaches approximately 0.07, indicating a beneficial effect after a certain scale.

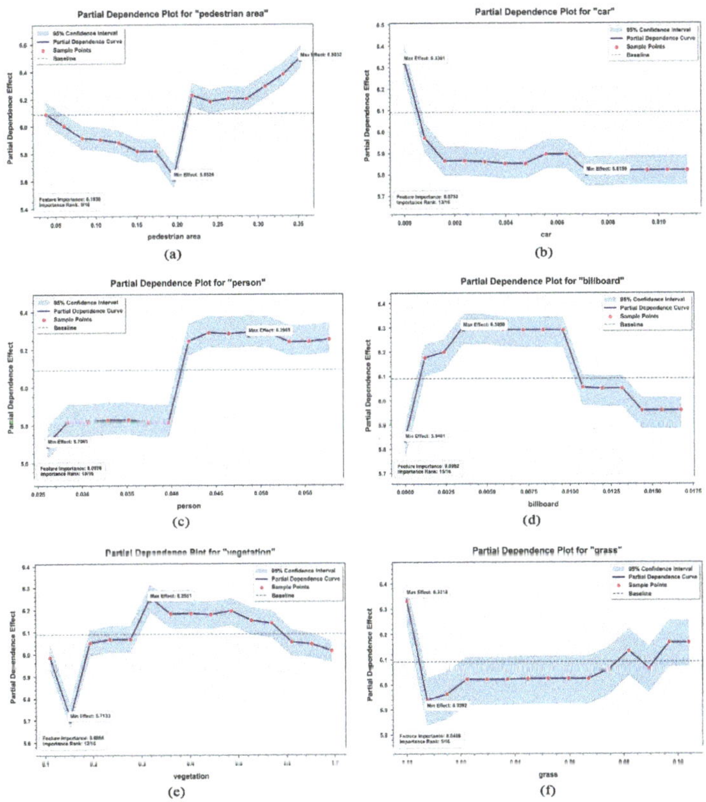

Figure 5. Main effect analysis of univariate PDP on the environmental safety perception of the elderly. (**a**) Partial Dependence Plot for pedestrian area; (**b**) Partial Dependence Piot for car; (**c**) Partial Dependence Plot for person; (**d**) Partial Dependence Piot for billboard; (**e**) Partial Dependence Plot for vegetation; (**f**) Partial Dependence Plot for grass.

3.4. Bivariate Key Visual Elements Contribute to Shaping Elderly Individuals' Perception of Environmental Safety

Figure 6a–o depicts pairwise analyses of six key visual elements. The color intensity indicates the elderly's safety perception scores, with the color gradient reflecting the vertical axis values (or color bar). The darker the color, the higher the perceived safety in the corresponding area, while the lighter the color, the lower the perceived safety. Therefore, the optimal combination range and risk areas are visually distinguishable based on color variations. According to the magnitude of partial dependence, the 15 visual element combinations are categorized into three influence magnitudes: high, medium, and low, as presented in Table 1. The classification was based on the magnitude of the partial dependence range, which was calculated as the difference between the maximum and minimum partial dependence values for each element combination. The specific ranges for high-, medium-, and low-impact amplitudes are as follows:

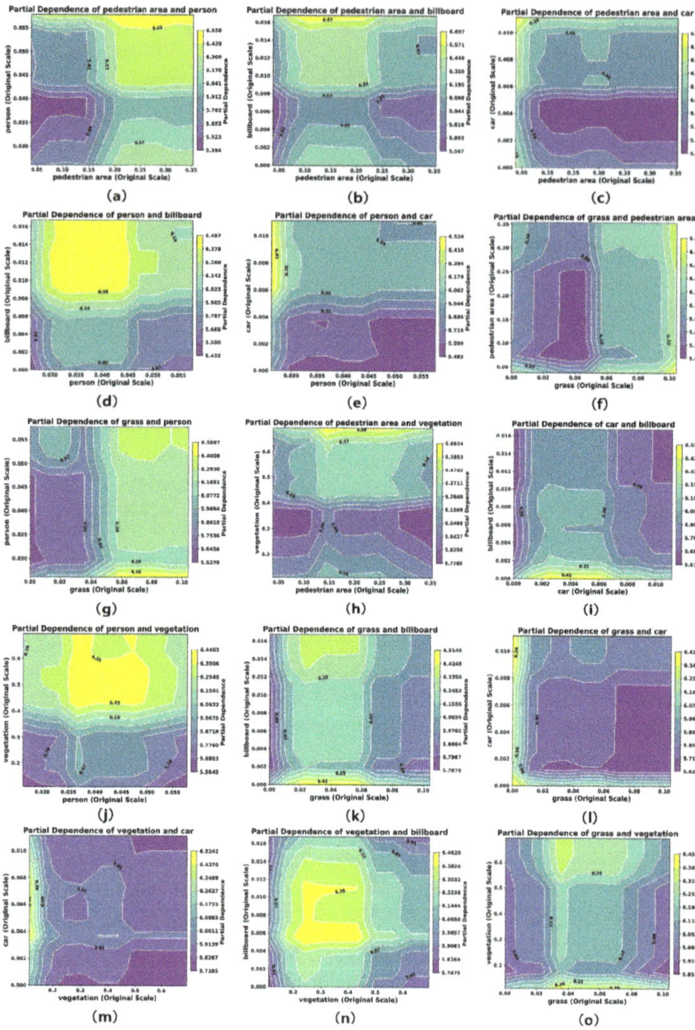

Figure 6. Analysis of the regulatory mechanism of interactive PDP on the environmental safety perception of the elderly. (**a**) Partial Dependence of pedestrian area and person; (**b**) Partial Dependence of pedestrian area and billboard; (c) Partial Dependence of pedestrian area and car; (**d**) Partial Dependence of person and billboard; (**e**) Partial Dependence of person and car; (**f**) Partial Dependence of grass and pedestrian area; (**g**) Partial Dependence of grass and person; (**h**) Partial Dependence of pedestrian area and vegetation; (**i**) Partial Dependence of car and billboard; (**j**) Partial Dependence of person and vegetation; (**k**) Partial Dependence of grass and billboard; (**l**) Partial Dependence of grass and car; (**m**) Partial Dependence of vegetation and car; (**n**) Partial Dependence of vegetation and billboard; (**o**) Partial Dependence of grass and vegetation.

Table 1. Key Visual Elements Grouped by Adjusted Interaction Influence.

Feature Pair	Calculation Range (Maximum PD–Minimum PD)	Classification of Impact Amplitude
grass_vegetation	0.6021	Low
vegetation_billboard	0.7145	Low
vegetation_car	0.7847	Low
grass_car	0.801	Medium
grass_billboard	0.8075	Medium
person_vegetation	0.8618	Medium
car_billboard	0.9519	Medium
pedestrian area_vegetation	0.9639	Medium
grass_person	0.9708	Medium
grass_pedestrian area	1.004	High
person_car	1.043	High
person_billboard	1.065	High
pedestrian area_car	1.127	High
pedestrian area_billboard	1.13	High
pedestrian area_person	1.164	High

High-Impact Amplitude: ≥0.98;

Medium-Impact Amplitude: 0.79 to 0.97;

Low-Impact Amplitude: <0.79.

The high-impact amplitude is shown in Figure 6a–f. Feature combinations in this group demonstrate the most significant impact on safety perception in model predictions, characterized by the largest partial dependence range (≥0.98). These combinations possess the strongest capacity to influence the safety perception of elderly individuals [54]. Significant differences in elderly safety perception are observed under varying feature combinations, aligning closely with the model's predictive outcomes. The features of the high-bias dependent range group, which primarily centers on pedestrian areas and person, form a strong synergy with other positive features such as grass, car, and billboard. This synergy indicates that when these elements coexist, they mutually reinforce positive outcomes on elderly safety perception. Among these combinations, optimal feature combinations are typically characterized by high intensities of relevant features. This strongly supports the "Street Eye" theory and highlights the importance of promoting the creation of active [55], accessible, and green urban environments. When these elements coexist and are well managed, their synergistic potential can enhance positive influences to maximize perceived safety. Conversely, the risk areas associated with these combinations tend to occur when all related characteristics remain at low levels. This indicates that a lack of human activities, restricted pedestrian space, and inadequate green infrastructure or urban vitality can result in a significant decrease in perceived safety [56]. This phenomenon is particularly pronounced among the elderly, whose physical and mental vulnerabilities heighten their sensitivity to environmental risks.

The influence amplitude, as visualized in Figure 6g–l, demonstrates that the characteristics within this group exert a moderate yet statistically significant impact on safety perception. This relationship exhibits a partial dependence range of 0.79 to 0.98. Combinations within this moderate dependence range reflect diverse interaction patterns, including synergistic and compensatory effects. For example, optimal safety perception often emerges when positive features such as grass, person, car, and billboard are present at relatively high levels. These interactions effectively enhance perceived safety; however, their impact typically remains less pronounced than that of the high-range group. For instance, the interaction between grass and car illustrates that when grass levels are low, the presence of cars contributes significantly to safety perception. This occurs as cars act as compensatory elements by signaling urban activity and accessibility, thereby enhancing the sense of security in undergreened settings [57]. However, the compensatory effect of vegetation does not universally enhance safety perception. In certain scenarios, dense vegetation—particularly in areas with limited human activity or pedestrian functionality—can negatively influence safety perception. This highlights that vegetation density and management quality significantly impact safety perception; overly dense or poorly maintained vegetation can become a safety hazard. This phenomenon is especially prevalent among the elderly, given their heightened sensitivity to environmental safety due to physical and cognitive limitations. Consequently, public space design and management should holistically integrate these factors to enhance the safety perception of elderly individuals.

Figure 6m–o illustrates the relative influence of low-amplitude factors on safety perception. These combinations, which often involve vegetation, demonstrate counterintuitive negative trends under specific conditions. Vegetation's impact on safety perception is complex: its optimal contribution occurs at low or medium levels when paired with high-level urban features (e.g., billboards or cars), potentially enhancing urban vitality and visibility. Conversely, high vegetation levels coupled with low levels of other urban features may create risk zones. This highlights that excessive vegetation, especially without complementary positive safety signals, can be perceived as a safety hazard, significantly reducing perceived safety [58]. This indicates that in urban design, the types, densities, and layouts of vegetation must be carefully controlled to mitigate their adverse effects on perceived safety. Evidence shows that vegetation density significantly influences people's perceived tendency to engage in activities. A more pronounced sense of community hierarchy in vegetation layouts can reduce overly intense behavioral tendencies and encourage a broader range of human activities. Furthermore, integrating vegetation communities with rest facilities enhances positive emotions among users and fosters greater social engagement. In urban design, vegetation should strike a balance between fulfilling ecological functions and addressing its influence on human behavior and emotional responses [59].

4. Discussion

This study employed an interpretable machine learning model to quantify and investigate the influence of pocket parks' visual elements on the environmental safety perception of the elderly. The results confirmed previous theories while revealing complex nonlinear relationships and interaction mechanisms between the elements. These findings contribute to a scientific basis for urban micro-space planning and design, particularly in settings involving aging populations.

4.1. The Nonlinear Impact Dynamics of Key Visual Elements

Unlike traditional research that often simplifies elements like greening and human flow into single positive or negative factors, this study identifies the complex and threshold-dependent nature of their contributions using SHAP and PDP analysis. The core roles of "safe space" and "social vitality" are as follows: Research indicates that pedestrian area and person are the two most significant factors for improving the sense of safety among elderly pedestrians. This aligns with Jacobs' "Eyes on the Street" theory: moderate crowd activities provide natural surveillance, thereby enhancing the sense of security in the space. The spacious and clear pedestrian area not only ensures ease of mobility and potential escape routes for the elderly but also promotes a psychological perception of "controllability" and "accessibility," thereby significantly reducing anxiety caused by uncertainty. PDP analysis further quantified its threshold effect: when the proportion of pedestrian area exceeds approximately 0.20 and the density of human vision exceeds 0.04, the sense of security rises to a relatively high and stable level. This suggests that successful pocket park design must prioritize ensuring sufficient walking and activity spaces, while creating a social environment that attracts moderate human activity.

The overwhelming negative impact of traffic threats: Cars are identified as the strongest negative contributor to reduced safety perception. Even if the visual proportion of cars is relatively low, their presence leads to a sharp decline in safety perception. This indicates that the elderly, as a vulnerable group in traffic environments, are highly sensitive to potential traffic accident risks. The noise, exhaust fumes, and visual intrusion generated by vehicles jointly disrupt the peaceful and safe atmosphere of pocket parks, contradicting the restorative environment advocated by the Stress Relief Theory (SRT). This finding highlights that effective physical isolation or visual buffering measures are essential in pocket park site selection and design to minimize traffic disruption as much as possible.

The "double-edged sword" effect of natural elements: Research confirms the positive role of vegetation and grass in safety perception, aligning with the Attention Restoration Theory (ART), which highlights the natural environment's contribution to psychological recovery. However, this study reveals the nonlinear nature of these elements' influence. The PDP curve shows that vegetation has an optimal density range (approximately 0.30). If density is too low, ecological benefits remain limited; if it is too high, it may create a sense of insecurity by obstructing views and providing hiding spots. This challenges the traditional notion that "the more greenery, the better." Designers must balance aesthetic value, ecological benefits, and safety requirements by, for example, adopting plant configurations with optimal spacing, controlling shrub height, and ensuring visibility within forest understories.

4.2. The Interaction, Coordination and Restraint Mechanism of Visual Elements

The interplay of visual elements was analyzed via bivariate PDP analysis in this study. By moving beyond single-element limitations, the analysis revealed the synergistic amplification effect and the intricate trade-offs of mutual restraint among elements, representing a key research insight.

Synergistic Amplification of High-Impact Combinations: Combinations like pedestrian area with pedestrian activity and pedestrian area with car presence show strong synergistic effects among positive elements. When spacious walkable spaces are paired with moderate social activity, safety perception peaks. This highlights how physical space design and social vitality interact to form the foundation of elderly safety perception. Notably, a limited vehicular presence in such zones—reflecting accessibility or regional vitality—exerts minimal negative impact due to the buffering effect of positive factors.

Complex Trade-Offs of Medium- and Low-Impact Combinations: The interactions between vegetation and other elements display the greatest complexity. For example, in sparsely used zones, high-density vegetation correlates with reduced safety perception. By contrast, in densely populated areas, vegetation density has negligible negative effects. This underscores the importance of natural surveillance, which mitigates insecurity caused by visual obstructions. These findings provide implications for park management: sparsely used zones should prioritize vegetation permeability and maintenance, while crowded zones may benefit from increased greenery to enhance ecological comfort.

4.3. The Theoretical and Practical Significance of the Research

This study advances both methodological frameworks and practical insights.

Methodological Contribution: The research establishes a closed-loop framework of "data-driven modeling—interpretive analysis—design optimization." By combining panoramic imagery, deep learning semantic segmentation, and explainable AI models (SHAP, PDP), this study transforms subjective environmental perceptions into measurable and interpretable indicators. This approach enhances research rigor and provides a robust analytical framework for studying the humanistic design of future urban environments.

The findings can be directly translated into design guidelines for pocket parks:

(1) Optimize pedestrian pathways: Create wide, continuous, and barrier-free walking areas using slip-resistant materials like EPDM rubber floors. Design interconnected paths to form a circular or networked layout, minimizing narrow or sharp turns for safety.
(2) Restrict vehicular access: Establish no-parking zones with clear signage and surveillance. Install safety barriers, such as greenbelts or bollards, to shield parks adjacent to roads, reducing vehicle threats and promoting low-carbon travel.
(3) Enhance social space: Place rest seats and social facilities in safe, sheltered areas to accommodate the physical needs and social preferences of users. These arrangements foster interaction and community engagement.
(4) Design effective signage: Position notice boards at prominent locations (e.g., entrances and along walkways) to convey park maps, facility details, and event schedules. Include health-related information boards to meet educational needs while maintaining visual clarity.
(5) Select appropriate vegetation: Plant tall trees for shade and opt for low-growing, non-toxic plants. Prioritize layered vegetation layouts with transparency for safety. Regular maintenance ensures cleanliness and aesthetic appeal.
(6) Plan grassland areas: Design grassy zones based on park scale and functionality. Use durable, soft grass species (e.g., ryegrass) and incorporate seating or pathways along edges to enhance accessibility.

4.4. The Limitations of the Research and Future Prospects

Despite meaningful findings, this study has limitations. The research focuses on Xiamen City, a unique coastal region with distinct cultural, climatic, and urban characteristics [60]. These factors may influence the generalizability of the conclusions. The applicability of research findings may vary across countries with differing levels of economic development, public safety, political systems, and population well-being. Future research will therefore aim to address these limitations and extend the applicability of the findings to diverse contexts. Secondly, the study employs static panoramic images as the primary environmental evaluation tool. While intuitive, this approach fails to account for dynamic environmental changes, such as atmosphere variations across time of day, weather conditions, and seasonal transitions. Additionally, it excludes multi-sensory environmental elements like sound and smell, which could significantly impact the perception of safety among the elderly. The aging population is heterogeneous, encompassing differences in age, physical and mental health, socioeconomic status, life experiences, and education. Such diversity likely complicates the mechanisms influencing their safety perception [61]. However, this study provides only limited insight into these variations.

The study focuses on pocket parks within urban micro-renewal initiatives due to their significant role in enhancing the mental well-being of elderly residents. While pocket parks lack the expansive open spaces, low population density, and walkability of larger parks, their compact size and high accessibility create unique challenges and opportunities. As integral components of urban green spaces, pocket parks directly influence the quality of life of urban residents, particularly the elderly, through their layout and design. Pocket parks serve dual functions: they provide urban residents with access to nature while fostering community interaction, mitigating the urban heat island effect, and enhancing urban aesthetics. Although smaller in scale, pocket parks are indispensable elements of urban renewal and community building due to their high-density distribution and accessibility.

Given the aforementioned limitations, future research could explore the following directions in greater depth for operational feasibility:

(1) Cross-regional and cross-cultural comparative studies could be conducted by selecting cities or regions with distinct characteristics. This would allow for a more rigorous examination of the universality and boundary conditions of the current findings.
(2) Dynamic video data and time-series image analysis could be employed to systematically capture and analyze environmental dynamics, including changes in pedestrian and vehicular flows, diurnal lighting variations, and seasonal landscape shifts. This would enhance the understanding of real-time safety perceptions among the elderly.
(3) A multi-sensory environmental perception model could be developed by integrating multi-source data collection techniques. These techniques include in-depth field questionnaire surveys, continuous behavioral observations, physiological index measurements (e.g., heart rate variability and skin conductance response), and wearable sensor data. This integrated approach would help reveal the complex relationships between environmental characteristics and perceived safety.
(4) Differentiated studies targeting subgroups of the elderly (e.g., advanced-age vs. young-old populations, those with varying health conditions, and individuals from diverse socio-economic backgrounds) could be conducted. This would enable a deeper analysis of their distinct safety concerns and environmental preferences, thereby offering robust scientific evidence and design insights for developing inclusive, tailored, and diversified age-friendly environments.

5. Conclusions

This study examines environmental safety issues in pocket parks within urban aging contexts, employing interpretable machine learning methods for a systematic quantitative analysis of visual elements influencing older adults' environmental safety perception.

Our findings uncover three core principles governing elderly safety perception, derived from field investigations and data analysis:

(1) Key positive factors include sufficient walkable areas, which ensure safe and barrier-free mobility for the elderly, and moderate group activities, which foster community vitality and enhance perceived safety.
(2) The relationship between natural elements and safety perception is nonlinear. While moderate greening improves environmental quality and reduces stress, excessive vegetation can obstruct vision, diminish path clarity, and introduce safety risks.
(3) Complex synergistic relationships exist between visual elements. For instance, the combination of spacious pedestrian areas, vibrant social activity, and optimal greening yields the highest safety perception among the elderly.

This research not only offers insights into older adults' environmental perception but also provides an interpretable analytical framework and practical guidance for urban planners and designers. By emphasizing the importance of overall environmental coordination, the study highlights the need to transition from isolated optimization of design elements to the creation of comprehensive, people-oriented spaces that encourage social interaction, manage natural elements effectively, and minimize traffic threats. Under the dual trends of globalization and aging populations, these findings carry significant theoretical and practical value for advancing the refinement and humanization of urban public spaces.

Author Contributions: S.W. (Shengzhen Wu), Conceptualization; Supervision; Validation; Writing—original draft; Writing—review and editing. S.W. (Sichao Wu), Writing—original draft; Validation; Investigation. J.C., Writing—original draft; Data curation; Resources; Validation. C.P., Writing—original draft; Investigation; Software. All authors have read and agreed to the published version of the manuscript.

Funding: This research was funded by (1) Fujian Provincial education System Philosophy and Social Science Research Project (JAS24049). (2) Research Project of Higher Education Science Research Laboratory of Fujian Higher Education Association in 2024 (24FJSYZD038).

Institutional Review Board Statement: Ethical review and approval were waived for this study, due to this study only invited the elderly to rate the panoramic images, but did not involve questionnaires or human experiments.

Informed Consent Statement: This study only invited 88 elderly individuals to assist in the scoring process. It did not involve human experiments, so no informed consent form was required.

Data Availability Statement: The datasets used and/or analyzed during the current study are available from the corresponding author on reasonable request.

Conflicts of Interest: The authors declared no potential conflicts of interest with respect to the research, author-ship, and/or publication of this article.

References

1. Guo, N.; Xia, F.; Yu, S. Enhancing elderly well-being: Exploring interactions between neighborhood-built environment and outdoor activities in old urban area. *Buildings* **2024**, *14*, 2845. [CrossRef]
2. Onose, D.A.; Onose, D.A.; Iojă, I.C.; Niță, M.R.; Vânău, G.O.; Popa, A.M. Too old for recreation? How friendly are urban parks for elderly people? *Sustainability* **2020**, *12*, 790. [CrossRef]
3. Shan, W.; Xiu, C.; Ji, R. Creating a healthy environment for elderly people in urban public activity space. *Int. J. Environ. Res. Public Health* **2020**, *17*, 7301. [CrossRef] [PubMed]
4. Kimic, K.; Polko, P. The use of urban parks by older adults in the context of perceived security. *Int. J. Environ. Res. Public Health* **2022**, *19*, 4184. [CrossRef]
5. Kerishnan, P.B.; Maruthaveeran, S. Factors contributing to the usage of pocket parks—A review of the evidence. *Urban For. Urban Green.* **2021**, *58*, 126985. [CrossRef]
6. Wang, X.; Li, G.; Pan, J.; Shen, J.; Han, C. The difference in the elderly's visual impact assessment of pocket park landscape. *Sci. Rep.* **2023**, *13*, 16895. [CrossRef]
7. Lu, S.; Oh, W.; Ooka, R.; Wang, L. Effects of environmental features in small public urban green spaces on older adults' mental restoration: Evidence from Tokyo. *Int. J. Environ. Res. Public Health* **2022**, *19*, 5477. [CrossRef]
8. Yang, Y.; Lu, Y.; Yang, H.; Yang, L.; Gou, Z. Impact of the quality and quantity of eye-level greenery on park usage. *Urban For. Urban Green.* **2021**, *60*, 127061. [CrossRef]
9. Meng, L.; Wen, K.H.; Zeng, Z.; Brewin, R.; Fan, X.; Wu, Q. The impact of street space perception factors on elderly health in high-density cities in Macau—Analysis based on street view images and deep learning technology. *Sustainability* **2020**, *12*, 1799. [CrossRef]
10. Zhou, H.; Wang, J.; Wilson, K.; Widener, M.; Wu, D.Y.; Xu, E. Using street view imagery and localized crowdsourcing survey to model perceived safety of the visual built environment by gender. *Int. J. Appl. Earth Obs. Geoinf.* **2025**, *139*, 104421. [CrossRef]
11. Chen, S.; Lin, S.; Yao, Y.; Zhou, X. Urban public space safety perception and the influence of the built environment from a female perspective: Combining street view data and deep learning. *Land* **2024**, *13*, 2108. [CrossRef]
12. Fan, C.; Li, S.; Liu, Y.; Jin, C.; Zhou, L.; Gu, Y.; Gai, Z.; Liu, R.; Qiu, B. Using social media text data to analyze the characteristics and influencing factors of daily urban green space usage—A case study of Xiamen, China. *Forests* **2023**, *14*, 1569. [CrossRef]
13. Zeng, E.; Dong, Y.; Yan, L.; Lin, A. Perceived safety in the neighborhood: Exploring the role of built environment, social factors, physical activity and multiple pathways of influence. *Buildings* **2023**, *13*, 2. [CrossRef]
14. Larkin, A.; Krishna, A.; Chen, L.; Amram, O.; Avery, A.R.; Duncan, G.E.; Hystad, P. Measuring and modelling perceptions of the built environment for epidemiological research using crowd-sourcing and image-based deep learning models. *J. Expo. Sci. Environ. Epidemiol.* **2022**, *32*, 892–899. [CrossRef] [PubMed]
15. Behera, R.K.; Bala, P.K.; Rana, N.P. Creation of sustainable growth with explainable artificial intelligence: An empirical insight from consumer packaged goods retailers. *J. Clean. Prod.* **2023**, *399*, 136605. [CrossRef]
16. Lundberg, S.M.; Lee, S.-I. A unified approach to interpreting model predictions. In Proceedings of the 31st International Conference on Neural Information Processing Systems, Long Beach, CA, USA, 4–9 December 2017.
17. Lan, Y.; Li, Z.; Lin, W. "Why Should I Trust You?": Exploring Interpretability in Machine Learning Approaches for Indirect SHM. *e-J. Nondestruct. Test. Ultrason.* **2024**, *29*. [CrossRef]
18. Hassija, V.; Chamola, V.; Mahapatra, A.; Singal, A.; Goel, D.; Huang, K.; Scardapane, S.; Spinelli, I.; Mahmud, M.; Hussain, A. Interpreting black-box models: A review on explainable artificial intelligence. *Cogn. Comput.* **2024**, *16*, 45–74. [CrossRef]

19. Rui, J. Exploring the association between the settlement environment and residents' positive sentiments in urban villages and formal settlements in Shenzhen. *Sustain. Cities Soc.* **2023**, *98*, 104851. [CrossRef]
20. Lin, X.; Yusoff, W.F.M.; Mat, M.K.A. Assessment of Xiamen Community Parks to Promote Physical Activity for Older Adults. *J. Arch. Urban Des.* **2024**, *1*, 1–13. [CrossRef]
21. Song, Y.; Wang, X.; Liao, M.; Baldwin, A.S.; Liu, L. Binocular function in the aging visual system: Fusion, suppression, and stereoacuity. *Front. Neurosci.* **2024**, *18*, 1360619. [CrossRef]
22. Mo, Y.; Wu, Y.; Yang, X.; Liu, F.; Liao, Y. Review the state-of-the-art technologies of semantic segmentation based on deep learning. *Neurocomputing* **2022**, *493*, 626–646. [CrossRef]
23. Guo, Y.; Nie, G.; Gao, W.; Liao, M. 2d semantic segmentation: Recent developments and future directions. *Future Internet* **2023**, *15*, 205. [CrossRef]
24. Wang, X.; Zhang, Y.; Xu, W.; Wang, H.; Cai, J.; Qin, Q.; Wang, Q.; Zeng, J. Construction of Multi-Scale Fusion Attention Unified Perceptual Parsing Networks for Semantic Segmentation of Mangrove Remote Sensing Images. *Appl. Sci.* **2025**, *15*, 976. [CrossRef]
25. Ueda, Y.; Wada, M.; Adachi, M.; Miyamoto, R. Dataset Creation for Segmentation to Enhance Visual Navigation in a Targeted Indoor Environment. In Proceedings of the 2023 17th International Conference on Signal-Image Technology & Internet-Based Systems (SITIS), Bangkok, Thailand, 8–10 November 2023. [CrossRef]
26. Sánchez, I.A.V.; Labib, S. Accessing eye-level greenness visibility from open-source street view images: A methodological development and implementation in multi-city and multi-country contexts. *Sustain. Cities Soc.* **2024**, *103*, 105262. [CrossRef]
27. Elharrouss, O.; Al-Maadeed, S.; Subramanian, N.; Ottakath, N.; Almaadeed, N.; Himeur, Y. Panoptic segmentation: A review. *arXiv* **2021**, arXiv:2111.10250. [CrossRef]
28. Shafiq, M.; Gu, Z. Deep residual learning for image recognition: A survey. *Appl. Sci.* **2022**, *12*, 8972. [CrossRef]
29. Du, F.; Wu, S. ECMNet: Lightweight Semantic Segmentation with Efficient CNN-Mamba Network. *arXiv* **2025**, arXiv:2506.08629. [CrossRef]
30. Woo, S.; Debnath, S.; Hu, R.; Chen, X.; Liu, Z.; Kweon, I.S.; Xie, S. Convnext v2: Co-designing and scaling convnets with masked autoencoders. In Proceedings of the IEEE/CVF Conference on Computer Vision and Pattern Recognition, Vancouver, BC, Canada, 18–22 June 2023. [CrossRef]
31. Hajihosseinlou, M.; Maghsoudi, A.; Ghezelbash, R. A novel scheme for mapping of MVT-type Pb–Zn prospectivity: LightGBM, a highly efficient gradient boosting decision tree machine learning algorithm. *Nat. Resour. Res.* **2023**, *32*, 2417–2438. [CrossRef]
32. Özgür, S.; Orman, M. Application of deep learning technique in next generation sequence experiments. *J. Big Data* **2023**, *10*, 160. [CrossRef]
33. Ramani, R.; Mary, A.R.; Raja, S.E.; Shunmugam, D.A. Optimized data management and secured federated learning in the Internet of Medical Things (IoMT) with blockchain technology. *Biomed. Signal Process. Control* **2024**, *93*, 106213. [CrossRef]
34. Nohara, Y.; Matsumoto, K.; Soejima, H.; Nakashima, N. Explanation of machine learning models using shapley additive explanation and application for real data in hospital. *Comput. Methods Programs Biomed.* **2022**, *214*, 106584. [CrossRef]
35. Croff, R.; Aron, S.; Wachana, A.; Fuller, P.; Mattek, N.; Towns, J.; Kaye, J. Walking and social reminiscence in gentrifying neighborhoods: Feasibility and impact on cognitive, physical, and mental health among older black adults in the SHARP study. *Gerontologist* **2024**, *64*, gnae019. [CrossRef]
36. Kim, Y.; Choi, B.; Choi, M.; Ahn, S.; Hwang, S. Enhancing pedestrian perceived safety through walking environment modification considering traffic and walking infrastructure. *Front. Public Health* **2024**, *11*, 1326468. [CrossRef]
37. Williams, T.G.; Logan, T.M.; Zuo, C.T.; Liberman, K.D.; Guikema, S.D. Parks and safety: A comparative study of green space access and inequity in five US cities. *Landsc. Urban Plan.* **2020**, *201*, 103841. [CrossRef]
38. Ferguson, L.A.; Taff, B.D.; Blanford, J.I.; Mennitt, D.J.; Mowen, A.J.; Levenhagen, M.; White, C.; Monz, C.A.; Francis, C.D.; Barber, J.R.; et al. Understanding park visitors' soundscape perception using subjective and objective measurement. *PeerJ* **2024**, *12*, e16592. [CrossRef] [PubMed]
39. Ren, X.; Wei, P.; Wang, Q.; Sun, W.; Yuan, M.; Shao, S.; Zhu, D.; Xue, Y. The effects of audio-visual perceptual characteristics on environmental health of pedestrian streets with traffic noise: A case study in Dalian, China. *Front. Psychol.* **2023**, *14*, 1122639. [CrossRef]
40. Park, Y.; Garcia, M. Pedestrian safety perception and urban street settings. *Int. J. Sustain. Transp.* **2020**, *14*, 860–871. [CrossRef]
41. Ren, W.; Zhan, K.; Chen, Z.; Hong, X.C. Research on Landscape Perception of Urban Parks Based on User-Generated Data. *Buildings* **2024**, *14*, 2776. [CrossRef]
42. Cozens, P.M.; Saville, G.; Hillier, D. Crime prevention through environmental design (CPTED): A review and modern bibliography. *Prop. Manag.* **2005**, *23*, 328–356. [CrossRef]
43. Gooren, J. The logic of CPTED for public space or the social potential of physical security. *Crime Law Soc. Change* **2023**, *79*, 417–436. [CrossRef]
44. Kim, Y.-A.; Kim, J. Examining the effects of physical environment and structural characteristics on the spatial patterns of crime in Daegu, South Korea. *Crime Delinq.* **2024**, *70*, 2166–2194. [CrossRef]

45. Aceves-González, C.; Ekambaram, K.; Rey-Galindo, J.; Rizo-Corona, L. The role of perceived pedestrian safety on designing safer built environments. *Traffic Inj. Prev.* **2020**, *21* (Suppl. S1), S84–S89. [CrossRef]
46. Ariffin, N.D.M.; Ibrahim, R.; Maulan, S.; Bakar, S.A. Manicured Versus Naturalistic Landscape Style: Public Preference of Urban Park's Landscape in Kuala Lumpur, Malaysia. *Int. J. Acad. Res. Bus. Soc.* **2024**, *14*, 1184–1201. [CrossRef]
47. Lis, A.; Pardela, Ł.; Iwankowski, P. Impact of vegetation on perceived safety and preference in city parks. *Sustainability* **2019**, *11*, 6324. [CrossRef]
48. Kim, D. The transportation safety of elderly pedestrians: Modeling contributing factors to elderly pedestrian collisions. *Accid. Anal. Prev.* **2019**, *131*, 268–274. [CrossRef]
49. Veitch, J.; Flowers, E.; Ball, K.; Deforche, B.; Timperio, A. Designing parks for older adults: A qualitative study using walk-along interviews. *Urban For. Urban Green.* **2020**, *54*, 126768. [CrossRef]
50. Basu, N.; Oviedo-Trespalacios, O.; King, M.; Kamruzzaman, M.; Haque, M.M. The influence of the built environment on pedestrians' perceptions of attractiveness, safety and security. *Transp. Res. Part F Traffic Psychol. Behav.* **2022**, *87*, 203–218. [CrossRef]
51. Wu, H.; Chen, Y.; Zhang, Z.; Jiao, J. The impact of street characteristics on older pedestrians' perceived safety in Shanghai, China. *J. Transp. Land Use* **2020**, *13*, 469–490. [CrossRef]
52. Yin, Y.; Shao, Y.; Wang, Y.; Wu, L. Developing a pocket park prescription program for human restoration: An approach that encourages both people and the environment. *Int. J. Environ. Res. Public Health* **2023**, *20*, 6642. [CrossRef]
53. Lv, M.; Wang, N.; Yao, S.; Wu, J.; Fang, L. Towards healthy aging: Influence of the built environment on elderly pedestrian safety at the micro-level. *Int. J. Environ. Res. Public Health* **2021**, *18*, 9534. [CrossRef] [PubMed]
54. Beneduce, C.; Lepri, B.; Luca, M. Urban Safety Perception Through the Lens of Large Multimodal Models: A Persona-based Approach. *arXiv* **2025**, arXiv:2503.00610. [CrossRef]
55. Jacobs, J. *The Death and Life of Great American Cities*; Random House: New York, NY, USA, 1961. [CrossRef]
56. Tchinda, P.E.; Kim, S.-N. The paradox of "eyes on the street": Pedestrian density and fear of crime in Yaoundé, Cameroon. *Sustainability* **2020**, *12*, 5300. [CrossRef]
57. Aruona, V.A.; Harris, N.; Munera, M.; Cifuentes, C.A. Reimagining Assistive Walkers: An Exploration of Challenges and Preferences in Older Adults. *arXiv* **2025**, arXiv:2504.18169. [CrossRef]
58. Zhu, W.; Chen, Z.; Wang, X.; Hu, F. Research on the influence of the visual perception characteristics of fitness trail landscape space based on psychological perception: A case study of Hunnan District, Shenyang, China. *Front. Psychol.* **2025**, *16*, 1595451. [CrossRef]
59. Sezavar, N.; Pazhouhanfar, M.; Van Dongen, R.P.; Grahn, P. The importance of designing the spatial distribution and density of vegetation in urban parks for increased experience of safety. *J. Clean. Prod.* **2023**, *403*, 136768. [CrossRef]
60. Wei, D.; Wang, Y.; Jiang, Y.; Guan, X.; Lu, Y. Deciphering the effect of user-generated content on park visitation: A comparative study of nine Chinese cities in the Pearl River Delta. *Land Use Policy* **2024**, *144*, 107259. [CrossRef]
61. Wang, R.; Jiang, Y.; Liu, D.; Peng, H.; Cao, M.; Yao, Y. Is perceived safety a prerequisite for the relationship between green space availability, and the use and perceived comfort of green space? *Wellbeing Space Soc.* **2025**, *8*, 100247. [CrossRef]

buildings

MDPI

Article

Exploring the Satisfaction of Low-Income Elderly People with Open Space Environment in Tapgol Park of Central Seoul: A Decision Tree Approach to Machine Learning

Chunhong Wu [1], Yile Chen [2,*], Fenrong Zhang [1], Liang Zheng [2,*], Jingwei Liang [2], Shuai Yang [3] and Yinqi Wang [4]

1 Department of Design, Chung-Ang University, 84 Heukseok-ro, Dongjak-gu, Seoul 06974, Republic of Korea; wuch0142@cau.ac.kr (C.W.); zhangfenrong@cau.ac.kr (F.Z.)
2 Faculty of Humanities and Arts, Macau University of Science and Technology, Avenida Wai Long, Taipa, Macau 999078, China; 2250000402@student.must.edu.mo
3 School of Art and Archaeology, Hangzhou City University, No. 51 Huzhou Street, Gongshu District, Hangzhou 310015, China; samyang@zju.edu.cn
4 Graduate School of Design, Kyushu University, 4-9-1 Shiobaru, Minami-ku, Fukuoka-shi, Fukuoka 815-8540, Japan; wang.yinqi.058@s.kyushu-u.ac.jp
* Correspondence: chenyile@must.edu.mo (Y.C.); zhengliang@must.edu.mo (L.Z.)

Abstract

In urban design, public open spaces (POS) are essential for enhancing health and well-being across the lifetime. High-quality public open spaces facilitate the maintenance of optimal physical and mental health in older individuals by encouraging activities like physical exercise and social engagement. Preserving the physical and mental well-being of elderly individuals is a fundamental concern for aging policy. Nevertheless, urbanization presents considerable problems with the provision of public open spaces for activities aimed at the elderly. South Korea has more significant issues than other nations globally. This study, based on data from 477 valid questionnaires collected in and around Tapgol Park in Jung-gu, Seoul, employed a decision tree approach to identify key factors and paths that influence overall satisfaction. The goal was to identify decision paths that improve satisfaction while ensuring interpretability, thereby providing a scientific basis for urban space design and renovation. The results show that: (1) The decision tree of this study presents a hierarchical logic of quietness first, then accessibility and cleanliness, and finally price and vitality, which is consistent with the high frequency of use of Tapgol Park by the elderly and the diverse facilities in the surrounding area. (2) The key to improving the management and satisfaction of Tapgol Park in Seoul is the quietness of the site. (3) When the park is not quiet, users are most sensitive to bottom-line factors, such as commercial supply, evacuation safety, transportation accessibility, price perception, barrier-free, and anti-slips. (4) When the park is quiet, basic comfort factors such as smooth walking, all-day opening, sunlight, and no odor constitute the minimum condition set for entering the comfort zone. (5) Water experience, waterfront accessibility, proximity to cultural resources, and moderate business and community-oriented leisure facilities are key plus points. Methodologically, this study is among the first to apply a decision tree approach to low-income elderly using a small public open space in a historic city center, clarifying the nonlinear and hierarchical relationships among environmental factors within these low-income elderly groups. This provides empirical support and reference for the aging-friendly urban space in world heritage cities and other historical and cultural cities.

Keywords: world heritage; historic built environment; low income; elderly people; open space for outdoor activities; satisfaction questionnaire analysis; decision tree; machine learning

 https://doi.org/10.3390/buildings16010172

1. Introduction

1.1. Research Background

Public open spaces are a crucial component of healthy aging [1]. In urban planning, public open spaces are a crucial tool for improving health and well-being throughout the lifespan [2,3]. High-quality public open spaces can help older adults maintain excellent physical and mental health by promoting activities such as physical exercise and social interaction. Maintaining the physical and mental health of older adults is a top priority for aging policies. However, urbanization poses substantial obstacles to the planning of public open spaces for elderly activities [4]. South Korea faces even greater challenges than other countries worldwide. In Korea, over 40% of adults aged 65 years and older live in relative poverty, representing the highest old-age poverty rate among OECD countries [5]. According to the *2025 Senior Citizens Statistics* published by Statistics Korea, the number of people aged 65 and over will exceed 20% of the total population, making South Korea a super-aged country [6]. According to data from the Seoul Metropolitan Government, the number of people aged 65 and over in Seoul will be particularly high, accounting for 19% of the population in 2025, totaling 1,860,632 [7]. Seoul's population is on the brink of entering a super-aged era. A super-aged society is accompanied by the problem of elderly poverty. In 2022, South Korea's relative poverty rate for the elderly was 39.7%, making it the highest poverty rate globally [8]. By 2025, the number of elderly people aged 65 and over living in poverty in South Korea will rise from 3.5 million to 4.6 million, with nearly one in every two people living in poverty being an elderly person [9]. The difficulty for elderly people living in poverty [10] to access high-quality public open spaces exacerbates their mental health issues [11]. Loneliness is particularly prominent among older South Koreans [12]. Moreover, the mental health of older adults depends to a large extent on the support provided by public open spaces, and this dependence is particularly pronounced among older men [13].

According to the *Seoul Welfare Foundation's Age-Friendly Seoul City 2.0 report (2025)*, Jongno-gu is one of Seoul's main activity areas for adults aged 60 and older [14]. As a central park for impoverished elderly men, Tapgol Park (탑골공원), located near Insa-dong in Jongno-sam-ga, Nakwon Musical Instrument Arcade, and Jongmyo Shrine in central Seoul, is a representative gathering place for seniors [15,16]. It is also recognized by the Seoul Metropolitan Government as a priority governance area for age-friendly public open spaces [17]. Tapgol Park is built around the historical site of Wongaksa Temple. It is also close to Seoul's representative coffee culture district and the World Heritage Site Jongmyo Shrine and has rich historical and cultural resources. Rich in historical and cultural resources, Tapgol Park holds significant importance for seniors in Korea. Heritage-led redevelopment in 2000 (the "Tapgol Park Sacred Domain") [18] was criticized for excluding older users [19]. Amid persistent alcohol-related disorder (public urination, fights) [19], Jongno-gu designated Tapgol Park and nearby alleys alcohol-free in April 2026 [20], reigniting debate on social equity in low-income elderly people's use of public open spaces [21]. Tapgol Park and its surrounding facilities offer a variety of options for seniors including many low-cost barbershops, affordable restaurants, and free meal services [22]. However, they face the risk of encroachment on their space. Due to factors such as heritage preservation and commercial development, the amount of open space previously reserved for seniors and the number of facilities and spaces accessible to them are shrinking. Despite this, Tapgol Park continues to attract many elderly men, particularly those from low-income families. Nearly a thousand people visit Tapgol Park

https://doi.org/10.3390/buildings16010172

daily, predominantly low-income elderly people, and the park's cultural environment fosters an atmosphere that is accessible to men [23]. At the Wongaksa Temple (원각사) free meal service site in Tapgol Park, low-income elderly people queue from 6:00 a.m.; the daily peak reaches approximately 290 diners [24]. As one of Seoul's most representative gathering places for low-income elderly people [25], Tapgol Park is selected as this study site.

According to the research team's observations, low-income elderly people primarily spend their time outdoors in the streets, buildings, and parks bounded by Jongno, Supyo-ro, Samil-daero, Donhwamun-ro. Overall, the open spaces centered around Tapgol Park in the city center (including the park and surrounding roads) and its surrounding facilities constitute a prime public open space for these elderly men to spend time outdoors. However, with rapid urbanization, the planning and construction of open spaces in the city center face significant challenges, particularly a lack of research and practical support for optimizing spaces specifically for low-income elderly people. This research gap urgently needs to be addressed, as optimizing open spaces for this group not only promotes emotional well-being [26] and enhances subjective well-being [27], but also holds significant implications for promoting social equity and achieving healthy aging. Furthermore, studies have shown that public open spaces that are safe, well maintained, and offer a high level of landscape comfort [28], as well as high-quality [26] public open spaces characterized by good sky visibility [29], high walkability and free access [30], have positive effects on the emotional well-being and subjective well-being of low-income elderly people [27]. These spaces are also of critical importance for promoting social equity [31] and achieving healthy aging. A recent literature review further emphasized that the accessibility, availability, and spatial quality of public open spaces such as green areas exert a clear protective effect on the mental health of low-income populations [32]. At the same time, higher subjective satisfaction with public open spaces can elicit more positive emotions among older adults and enhance their visual comfort [28]. Deficiencies in the facilities, functional diversity, and accessibility of public open spaces in old urban districts significantly reduce older adults' subjective satisfaction [33], whereas positive perceptions of the accessibility, safety, and comfort of these spaces help promote the mental and social health of low-income elderly people [34]. In high-density city centers, better walking accessibility, overall environmental quality [10], and safety- and lighting-related features [35] in public open spaces are associated with higher satisfaction among older adults and more active use. Therefore, this study provides theoretical references and policy implications for the future planning and design of a wider range of age-friendly open spaces and healthy aging from the perspective of low-income elderly people by providing evidence from South Korea.

1.2. Literature Review

As mobility declines with aging, outdoor open spaces become crucial for seniors to connect with nature and participate in daily activities [28]. Given the importance of outdoor activities and access to nature for seniors' emotional well-being, the design of these open spaces should meet their needs. Previous research has shown that the emotional impact of outdoor open spaces on low-income seniors varies. For example, low-income seniors experience positive emotions in open spaces near green spaces and historic buildings, but the impact of public transportation is not as uniformly positive as traditionally assumed [36]. They also prefer outdoor open spaces that are safe, less prone to falls, have useful wayfinding systems, are aesthetically pleasing, and are easily accessible [37]. A United States study found that low-income Chinese American seniors prefer safe, accessible public open spaces that offer cultural activities [38]. Although recent studies have shown that high-quality public open spaces for outdoor activities are more likely to elicit positive emotions among older adults [26], that higher levels of landscape

 https://doi.org/10.3390/buildings16010172

and overall comfort in such spaces are associated with better emotional well-being [28], and that good sky visibility can alleviate psychological anxiety and improve the psychological health of older adults [39], research on the open space preferences of low-income elderly people is urgently needed.

Gender differences also exist in preferences for public open spaces for outdoor activities. Women generally prefer indoor environments, while men primarily use city streets [40]. Furthermore, women place greater importance on the safety of public open spaces for outdoor activities than men do [41,42]. The openness of views and the cleanliness of public spaces influence women's safety perceptions [43]. Ref. [44] also found that many elderly Korean women advocate for fences, gates, and continuous surveillance in parks, as well as for improving safety with diverse materials and facilities. Meanwhile, elderly Korean men are more concerned about fall risks, expressing concern about uneven sidewalks and sloped trails and suggesting the addition of handrails. Older women, who also differ in the distance they use public open spaces, tend to prefer informal public open spaces near their homes (such as street corners) for brief social interactions compared to men [45]. This suggests the need for tailored interventions to address these demographic differences [42]. However, research on public open space satisfaction among low-income elderly people for outdoor activities remains insufficient. Given the high prevalence of mental health issues among elderly in South Korea, research on open space satisfaction among these elderly is crucial for the comprehensive advancement of the WHO's "Decade of Healthy Aging" initiative [46].

With the widespread application of machine learning technologies, researchers are increasingly adopting these advanced methods to analyze the factors influencing senior citizen satisfaction. Although recent studies examining the factors influencing environmental satisfaction among older adults have employed machine-learning approaches such as random forests [47] and non-linear models including ordered Logit [48], the use of decision tree classification models in the analysis of senior citizen satisfaction remains relatively new. Research on the impact of the environment on senior citizens' emotions has primarily utilized methods such as deep learning, space syntax [49], and street view images [50]. Studies on the impact of open space on outdoor activities have primarily used gradient-based decision trees (GBDTs) [51], or random forests [52]. In a study of open space satisfaction among adult residents in a historic heritage site, ref. [53] used a decision tree classification model to reveal key factors influencing open space satisfaction, including Largo square layout, commercial colors and materials, the location of historic buildings, and the legibility of commercial signage. However, no study has yet used this method to conduct an in-depth study on open space satisfaction among low-income elderly people. The decision tree model can be presented in a tree structure with clear logic, is easy to understand and explain, and can clearly identify the key factors and optimal paths that affect the satisfaction of the elderly [53], which is more conducive to the formulation of clear and sustainable open space planning and design strategies.

1.3. Problem Statement and Objectives

The primary objective of this study was to conduct an in-depth analysis of the environmental satisfaction of low-income elderly people visiting Tapgol Park in central Seoul. Tapgol Park in central Seoul holds unique reference value due to its long-standing appeal to elderly residents and its diverse surrounding environment. This study employed a questionnaire survey and decision tree methodology to analyze and understand the complex factors influencing low-income elderly people's satisfaction with the content dimensions of the environmental characteristics of Tapgol Park in central Seoul. At the methodological level, translating survey data into context-specific decision rules through the decision tree

https://doi.org/10.3390/buildings16010172

model mitigates core limitations of conventional linear, variable-by-variable approaches and enables the systematic identification of interaction and configuration effects among multiple environmental factors [53]. This will offer important advice to policymakers and urban planners, helping them develop more effective strategies and policies to improve the life satisfaction of low-income elderly people in their home-based care and ensure the sustainable development of open spaces in central Seoul.

2. Study Area and Methodology

2.1. Study Area: Tapgol Park and Its Surrounding Areas

Jongno 3-ga Tapgol Park and its surrounding streets in Seoul are the main geographical scope of this study (Figure 1). The primary area of activity for seniors at Tapgol Park is concentrated within the area bounded by Jongno-gu, Supyo-ro, Samil-dae-ro, and Donhwamun-ro 11-ga. The area around Tapgol Park and Supyo-ro is particularly popular for socializing with seniors. Tapgol Park, located in Seoul's Jongno-gu district, is near Jongno 3-ga Station on Seoul Subway Lines 1, 3, and 5. It covers an area of 19,599 square meters and is adjacent to Insadong, Jongmyo Park, and Gyeongbokgung Palace.

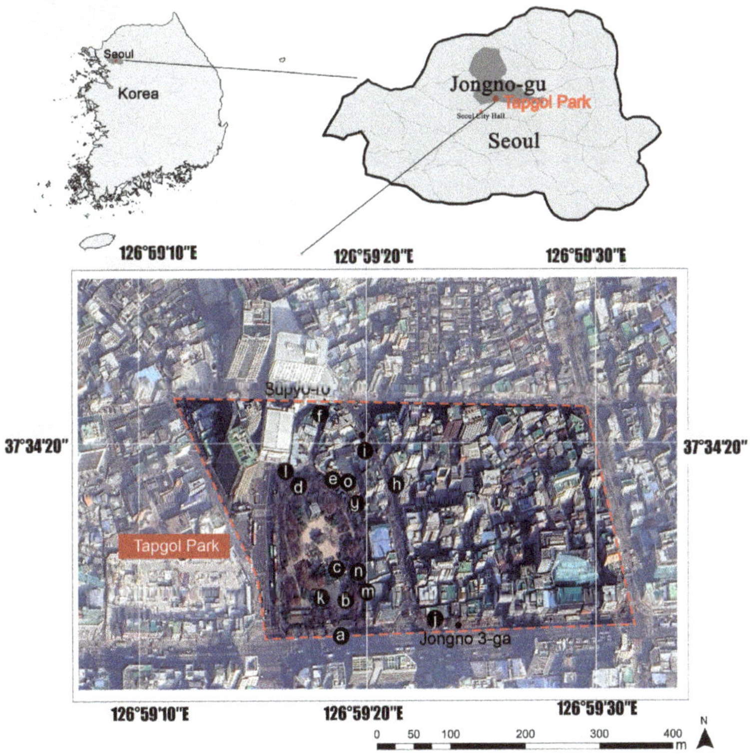

Figure 1. Study Area: Tapgol Park and Its Surrounding Areas.

According to the *Seoul Welfare Foundation's Age-Friendly Seoul City 2.0 report (2025)*, Jongno-gu is a key activity area for adults aged 60+, and Tapgol Park is designated by the Seoul Metropolitan Government as a priority age-friendly public open space. The park's protected area is 15,051 m^2 ($\approx$0.015 km^2) [54], only about 0.0025% of Seoul [55] and 0.06% of Jongno-gu [56]. Although Jongno-gu has about 140,000 residents, its concentration of government, commercial, and cultural functions attracts a daytime population of roughly

https://doi.org/10.3390/buildings16010172

2 million. As a typical downtown open space, Tapgol Park exhibits distinct place-making characteristics (Figure 2). The park provides essential leisure and social activities and fosters a significant agglomeration effect due to the diverse surrounding amenities. In particular, the abundance of traditional food stalls, street food vendors, affordable amenities (such as barber shops and shoe repair shops), free canteens, chess studios, cultural centers, and various commercial facilities surrounding Tapgol Park makes it particularly attractive to older people. Within Tapgol Park's public open spaces, areas in front of snack shops and small businesses, intersections, benches, stone borders along flower beds, plazas, pavilions, and spaces under eaves serve as key venues for seniors to interact and linger, forming a typical gathering place for older people. Convenient transportation accessibility, surrounding commercial and cultural amenities, diverse recreational facilities, and familiar social networks all contribute to Tapgol Park's place-making qualities. Furthermore, the surrounding streets and neighborhood facilities further enhance this appeal, making it a unique hub for seniors' daily activities within the urban fabric.

(**a**) Tapgol Park (main entrance)

(**b**) Elderly people are queuing up to receive free lunch boxes

(**c**) Elderly people chatting in the park

(**d**) Elderly people chatting in the park

(**e**) Elderly people eating for free, buying coffee and chatting on the path at the side entrance of the park

(**f**) Elderly people eating for free, buying coffee and chatting on the path at the side entrance of the park

(**g**) Elderly people eating for free, buying coffee and chatting on the path at the side entrance of the park

(**h**) Elderly people walking on Supyo-ro

(**i**) Street snacks popular among the elderly people

Figure 2. *Cont.*

https://doi.org/10.3390/buildings16010172

(j) On the sidewalk of Jongno 3rd Street, elderly people are selling things and passing by.

(k) Welfare facilities for the elderly around Tapgol Park: facilities inside the park that provide free ice water

(l) There is a welfare information center and a free ice water distribution facility at the north gate of the park, as well as a musical instrument mall (the blue building at the back).

(m) Welfare Center for People with Disabilities

(n) Elderly Welfare Center

(o) Waste recycling station

Figure 2. The research team investigated the situation around Tapgol Park. (**a**) Tapgol Park (main entrance); (**b**) Elderly people are queuing up to receive free lunch boxes; (**c**,**d**) Elderly people chatting in the park; (**e**–**g**) Elderly people eating for free, buying coffee and chatting on the path at the side entrance of the park; (**h**) Elderly people walking on Supyo-ro; (**i**) Street snacks popular among the elderly people; (**j**) On the sidewalk of Jongno 3rd Street, elderly people are selling things and passing by; (**k**) Welfare facilities for the elderly around Tapgol Park: facilities inside the park that provide free ice water; (**l**) There is a welfare information center and a free ice water distribution facility at the north gate of the park, as well as a musical instrument mall (the blue building at the back). (**m**) Welfare Center for People with Disabilities; (**n**) Elderly Welfare Center; (**o**) Waste recycling station. The Korean text in the photos mainly refers to shop signs, road signs, and directions.

2.2. Five Dimensions of Seoul's Central Environment

This study developed measurement items for five dimensions of public open spaces (accessibility, convenience, diversity, safety, and comfort) by adapting two Korean-language questionnaires on age-friendly community public spaces and elderly-friendly walking environments [37,57]. First, we extracted from the original instruments all items that were conceptually related to outdoor public open spaces and everyday walking environments (e.g., path conditions, greenery, lighting, street furniture, nearby facilities, and social spaces), while items referring to indoor housing conditions, employment, or medical systems were excluded. Second, conceptually overlapping items were consolidated and reorganized into the five dimensions above, and several context-specific items were added to reflect the spatial and functional characteristics of Tapgol Park and its surrounding streets (such as access to transit, nearby commercial and leisure facilities, historic monuments, and street vendors). This process yielded a final pool of 41 items (Q1-1~Q1-6 for accessibility, Q2-1~Q2-7 for convenience, Q3-1~Q3-7 for interesting/variety, Q4-1~Q4-7 for safety, and Q5-1~Q5-9 for comfort), plus one item measuring overall satisfaction (OS), as summarized in Appendix A. The questionnaire was administered in Korean using a five-point Likert

 https://doi.org/10.3390/buildings16010172

scale (1 = strongly disagree, 5 = strongly agree); the English wording reported in this article is a translation of the original items for the purpose of international dissemination.

The questionnaire used a five-point Likert scale to measure seniors' satisfaction with the characteristics of Tagu Park's public open spaces. A 30-day field observation process also identified the primary areas of seniors' outdoor activities. In the field observations, we referred to Gehl's classification of behavioral types in public spaces (necessary activities, social activities, and optional activities) and his methods of recording [58]. We observed, during the same time periods on different days, the locations, numbers, types, and durations of activities of older around Tapgol Park. Seniors' overall satisfaction served as the dependent variable (a continuous variable), while 41 attributes across the five dimensions of public open spaces served as independent variables.

To clarify the on-site observation procedure, this study describes three consecutive full-day observations conducted in Tapgol Park. Repeated observations were carried out at each fixed observation point during the same time periods on different days. A manual hand-held tally counter was used to record attendance. Each entry corresponded to a clearly defined time segment, primarily based on 10-min scan intervals (e.g., 10:10–10:20; 11:15–11:25) to capture ongoing activities. "Location" was operationalized as a fixed observation point (observation zone). "Quantity" was operationalized as the number of older people engaged in observable activities within the visible boundary of the observation zone during each time segment (recorded as 0 when none were present), with sex composition recorded simultaneously. "Type" was categorized following Gehl's three behavioral types, including necessary activities (e.g., queuing to receive meals or lunch boxes), social activities (e.g., conversation, playing chess, shared eating), and optional activities (e.g., walking, watching, taking photos, seated resting). "Duration" was operationalized using the start and end times of each time segment (duration = end time − start time). In total, approximately 70 observation time segments were obtained from Tapgol Park, covering key periods in the morning, afternoon, and early evening. These observation data were used to identify primary activity areas and high-frequency activity repertoires of elderly people and to provide contextual evidence for interpreting the public open space attributes assessed in the subsequent questionnaire, thereby enhancing the traceability and validity of the study.

In 2018, UN-Habitat defined outdoor public open spaces as spaces that provide social interaction, such as parks, streets, green spaces, squares, and sidewalks, as well as recreational areas like playgrounds and markets, and marginal spaces between buildings or along roads [59]. Streets, parks, and squares are the primary venues for outdoor activities [58]. The outdoor activities of the elderly mainly include leisure activities, daily activities, and social activities, specifically walking activities, sports activities, participating in gatherings and interest groups, and going out shopping [60–62].

Walkability, convenient travel routes, and accessibility to commercial facilities, transit stations, parks, and recreational areas significantly influence older adults' outdoor physical activity [26,63]. Furthermore, familiar and accessible locations facilitate intergenerational physical activity among older adults [63]. Furthermore, proximity to transit stations and high-quality public open spaces makes it easier for older adults to participate in community activities, fostering social interaction, particularly among older men [42].

On the other hand, the leisure, daily activities, and social interactions of older adults outdoors are primarily influenced by amenities within public open spaces, such as roads, hospitals, bus stops, grocery stores, and subways [64]. Adequate exercise space, nearby parks, fitness centers, and sports equipment in the community effectively support older adults' outdoor activities [63]. Furthermore, research from South Korea has shown that

https://doi.org/10.3390/buildings16010172

amenities and social spaces along walking paths can positively promote walking among older adults [65].

Available recreational facilities around public open spaces can create more attractions for older adults and promote social interaction. For example, clubs and social activity centers provide structured opportunities for older adults to meet and interact with peers, thereby strengthening their social networks [66]. Furthermore, the safety of pedestrian environments and pedestrian spaces, as well as the safety of cycling paths (sidewalks, bike lanes, and pedestrian zones), directly impact the participation of older adults in outdoor physical activities [26,65]. At night, lighting levels and uniformity are crucial factors influencing outdoor activities among older adults [35]. Furthermore, comfort factors significantly influence the experience of older adults in outdoor activities. Convenient and safe transportation facilities, along with good weather conditions, can enhance the participation of older adults in intergenerational physical activities [63]. Furthermore, micro-environmental factors in outdoor open spaces, such as path flatness, shade, seating arrangements, and surface materials, significantly influence older adults' walking and other physical activities [67,68]. Furthermore, the perception of green areas positively impacts the comfort of older adults experiencing outdoor activities at night [35].

Because low-income communities often offer fewer public open spaces suitable for outdoor activities and have less accessible walkability than higher-income communities [69,70], outdoor public open spaces are often crucial for low-income urban residents [57]. Furthermore, significant differences in preferences between older men and women exist when it comes to using outdoor public open spaces: women prioritize safety more than men, while men prioritize the amenities of outdoor public open spaces more. This phenomenon suggests the need for tailored interventions to address these demographic differences [42].

2.3. Data Collection and Questionnaire Distribution

The research team surveyed elderly people aged 65 and older who visited Tapgol Park on Jongno 3-ga, Seoul, from 15–20 June 2025, and from 5–26 August 2025. The researchers selected five dimensions of age-friendly urban public open spaces: accessibility, convenience, safety, and comfort. Data were collected on-site using a paper questionnaire. We created a questionnaire using digital tools provided by wjx.cn to facilitate digital recording and organization. The questionnaire questions are provided in Appendix A. The Institutional Review Board (IRB) of Hangzhou City University reviewed and approved research involving human subjects. Before completing the questionnaire, the participants' legal guardians or next of kin provided written informed consent for this study (Approval Number 2025-0308). The content structure and questions of the questionnaire are in Appendix A.

The researchers distributed a total of 500 questionnaires, eliminating incomplete or incorrectly completed ones. However, a total of 477 valid questionnaires were collected, representing 95.4% of the total. In terms of age distribution, the survey results showed that 115 women, or 24.1%, represented most participants, while 362 men, or 75.9%, comprised the majority (Table 1). The 65–70 age group had the largest number of participants, with 180 participants, or 37.73%. Conversely, the 80-plus age group had the smallest number of participants, with only 139, or 29.1%. The 71–80 age group had 158 participants, or 33.1%. Regarding occupational distribution, the largest number of elderly individuals selected other occupations, totaling 200, or 41.9%. The smallest number of participants were company employees, with 62, or 13%, representing the majority. Managers comprised 67, or 14%, and business owners comprised 78, or 16.4%. There were 70 people working in the service industry, accounting for 14.7% of the total. Regarding educational background, the majority had a high school education or less, totaling 172 people, or 36.1%. High school graduates accounted for 135 people, or 28.3%. Vocational college graduates accounted

https://doi.org/10.3390/buildings16010172

for 84 people, or 17.6%. General university graduates accounted for 86 people, or 18%. Seniors with an average monthly income of less than 1 million South Korean Won (KRW) were the largest group, totaling 191, or 40% of the total. The next largest group was those with an average monthly income between 1 and 2 million won (KRW), totaling 84, or 17.6%. The group with an average monthly income between 2.01 and 3 million won (KRW) accounted for 64, or 13.4% of the total. And those with an average monthly income between 3.01 and 4 million won (KRW), totaling 69, or 14.5%. There are 69 elderly people with an average monthly income of over 4 million won, representing 14.5% of the total population. According to the Ministry of Health and Welfare's 2025 standards, the median monthly income for a single person is approximately 2,392,013 South Korean Won. Those earning less than 50% of the median (₩ 1,196,006) are considered low-income and eligible for education subsidies. Those earning less than 40% of the median (₩ 956,805) are eligible for medical subsidies, and those earning less than 30% of the median (₩ 717,604) are eligible for living allowances.

Table 1. Basic information of the respondents.

Survey Items	Interval Results	Frequency	Percentage (%)
Gender	Female	115	24.1
	Male	362	75.9
Age	65–70 years old	180	37.7
	71–80 years old	158	33.1
	>80 years old	139	29.1
Marital status	Single	137	28.7
	Married	239	50.1
	Other	101	21.2
Housing ownership	Owned	204	42.8
	Rented	171	35.8
	Other	102	21.4
Education	High school graduate or below	172	36.1
	High school graduate	135	28.3
	Vocational college	84	17.6
	University graduate	86	18.0
Occupation before retirement	Company employees	62	13.0
	Managers	67	14.0
	Service industry workers	70	14.7
	Business owners	78	16.4
	Other	200	41.9
Average monthly income entry	Under 1 million KRW	62	13.0
	1 million–2 million KRW	67	14.0
	3 million–4 million KRW	70	14.7
	4 million–5 million KRW	78	16.4
	5 million or more	200	41.9

Data source: the author compiled the data from the questionnaire.

https://doi.org/10.3390/buildings16010172

2.4. Research Process and Decision Tree Method

This study uses a decision tree method to identify key factors and paths that affect overall satisfaction based on questionnaire data from Tapgol Park and its surrounding areas in Jung-gu, Seoul. The goal is to provide a decision path that can improve satisfaction while ensuring interpretability. Tapgol Park has long attracted elderly people to gather, and the surrounding facilities are diverse and representative, making it suitable for conducting path identification research from environmental factors to satisfaction. The questionnaire scale of this study includes five dimensions: accessibility, convenience, interest/variety, safety, and comfort, with a total of 36 items as the characteristic value variables of the model. Overall satisfaction is used as the target value variable of the model. The research process of this study is constructed with reference to the common research steps in existing literature [53], from data preprocessing, data set division, model construction and parameter optimization, model training and evaluation, to visualization and rule extraction, and is adjusted according to the purpose of this study to ensure the scientificity and appropriateness of the method (Figure 3).

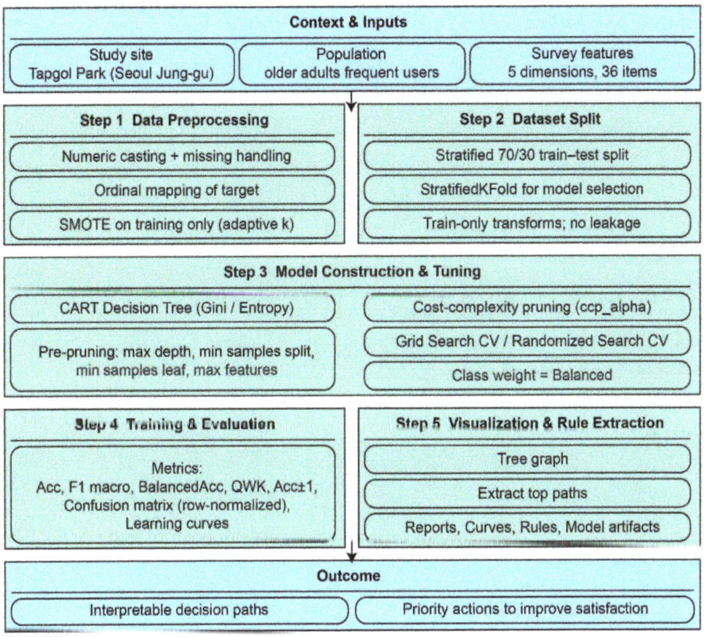

Figure 3. The research process in this study.

(1) Data Preprocessing. To ensure data quality and model comparability, raw questionnaire data were standardized. Independent variables were numericalized [71], with missing or infinite values imputed using out-of-domain constants to maintain training stability. The target variable (overall satisfaction) was mapped to ordered integers. Feature selection was conducted using both mutual information and tree-based embedded importance to filter high-contribution features. To address class imbalance, the Synthetic Minority Oversampling Technique (SMOTE) was applied to the training set, with the number of neighbors adaptively set based on the minority class sample size.

(2) Dataset Partitioning. A stratified random sampling method (70:30 ratio) was used to split data into training and test sets, strictly preserving the distribution of satisfaction levels. Stratified k-fold cross-validation with shuffling was employed within the training set

https://doi.org/10.3390/buildings16010172

for hyperparameter tuning [72]. To prevent data leakage, all preprocessing steps, including feature selection and SMOTE, were strictly isolated within the training folds.

(3) Model Construction and Optimization. The study employed the CART decision tree framework [73]. Overfitting was mitigated through pre-pruning (constraints on max depth, min split/leaf samples) and post-pruning using Cost Complexity Pruning (CCP). Hyperparameters were optimized via GridSearchCV or RandomizedSearchCV [74], maximizing the Macro-average F1 score with class weights balanced. The optimal tree structure was determined by selecting the CCP alpha that yielded the best generalization score on the cross-validation curve [75].

(4) Training and Evaluation. Performance was assessed using multiple indicators: Overall Accuracy (Acc), F1 score, and Balanced Accuracy. Given the ordinal nature of satisfaction, Quadratic Weighted Kappa (QWK) and Accuracy within one error (Ac@$\pm$1) were included. Additionally, an ordinal judgment rule based on the cumulative median of posterior probabilities was implemented to reduce classification errors. Model diagnostics included row-normalized confusion matrices to visualize error structures and learning curves to assess bias and variance.

(5) Visualization and Rule Extraction. The final converged tree model was visualized to display discriminant paths. Decision rules were extracted and simplified into interval expressions, ranked by a joint score of credibility and coverage. All artifacts, including variable mappings, rule sets, and model files, were archived to ensure reproducibility.

The above process thoroughly evaluates the questionnaire scale and sample characteristics, systematically addresses category imbalance, model complexity, and overfitting risks, and utilizes various evidence such as ordered judgment, learning curves, and rule screening to clearly demonstrate the significant role of segmented characteristic value items in assessing target value satisfaction.

3. Results

3.1. Questionnaire Reliability and Validity Analysis Results

3.1.1. Questionnaire Reliability Analysis

This survey aimed to identify the environmental satisfaction of low-income older men living in Tapgol Park in central Seoul. The survey consisted of 37 items divided into five dimensions: accessibility, convenience, interest or variety, safety, and comfort. After collecting the questionnaire data, the researchers conducted a reliability analysis.

According to the reliability analysis in Table 2, the internal consistency coefficients (Cronbach's alpha coefficients) for the five dimensions were 0.953, 0.903, 0.915, 0.949, and 0.967, respectively. The Cronbach's alpha coefficient for the entire scale was 0.955. Cronbach's alpha coefficients for the five dimensions and the entire scale were all greater than 0.8, indicating satisfactory internal consistency and reliability for the five dimensions and the entire scale [76,77].

Table 2. Cronbach's α coefficient.

Dimension Information	Number of Items	Cronbach's Alpha	Subitems
Accessibility	6	0.953	Q1-1 (Smooth and easy to navigate); Q1-2 (Barrier-free); Q1-3 (Water-friendly environment); Q1-4 (Convenient transportation); Q1-5 (Open all day); Q1-6 (Approach satisfaction rating)

 https://doi.org/10.3390/buildings16010172

Table 2. *Cont.*

Dimension Information	Number of Items	Cronbach's Alpha	Subitems
Convenience	7	0.903	Q2-1 (Adequate rest facilities); Q2-2 (Complete sunshade facilities); Q2-3 (Complete public facilities); Q2-4 (Convenient and easy to find restrooms); Q2-5 (Adequate welfare facilities); Q2-6 (Complete sports facilities); Q2-7 (Satisfaction rating of amenities)
Interest/variety	7	0.915	Q3-1 (Abundant shopping district); Q3-2 (Abundant entertainment venues); Q3-3 (Proximity to cultural heritage sites); Q3-4 (Convenient food options); Q3-5 (Leisure facilities for interesting spaces); Q3-6 (Number of small shops); Q3-7 (Satisfaction with interesting)
Safety	7	0.949	Q4-1 (No illegal parking); Q4-2 (Appropriate speed); Q4-3 (Nighttime lighting); Q4-4 (Non-slip road surface); Q4-5 (Clear warning signs); Q4-6 (Complete evacuation facilities); Q4-7 (Satisfaction with safety)
Comfort	9	0.967	Q5-1 (Satisfied with greenery); Q5-2 (Satisfied with sunlight); Q5-3 (No noise); Q5-4 (Open view); Q5-5 (Good hygiene); Q5-6 (Private and comfortable); Q5-7 (No odor); Q5-8 (Reasonable prices); Q5-9 (Satisfaction with comfort)
Overall satisfaction	36	0.955	The total of the above

Data source: the author compiled the data from the questionnaire.

3.1.2. Construct Validity of the Questionnaire

The researchers conducted an exploratory factor analysis on the questionnaire using SPSS software (version 29.0.2.0). Principal component analysis (PCA) was used for factor extraction, with varimax rotation performed. As shown in Table 3, the results of the KMO and Bartlett's test of sphericity (STS) showed a KMO value of 0.944, greater than 0.6, and a significance of less than 0.001 for the Bartlett test of sphericity, indicating that the questionnaire was suitable for factor analysis and factor extraction. Table 4 shows the total variance explained by the exploratory factor analysis. The results showed that the factor analysis extracted five factors, with a cumulative explained variance of 74.197%, exceeding 60%, indicating that the extracted factors explained 74.197% of the original 36 items. The first principal component explained 39.767% of the variance, less than 40% and less than half of the total explained variance of 74.197%, indicating that the questionnaire did not contain common method bias.

Table 3. KMO and Bartlett's test of sphericity.

KMO Sampling Suitability Measure		0.944
Bartlett's test of sphericity	Approximate chi-square	16,131.16
	Degrees of freedom	630
	Significance	<0.001

Data source: the author compiled the data from the questionnaire.

https://doi.org/10.3390/buildings16010172

Table 4. Total variance explained (extraction method: principal component analysis).

Element	Initial Eigenvalues			Extracted Sum of Squares of Loadings			Rotated Sum of Squares of Loadings		
	Total	Percentage of Variance	Cumulative (%)	Total	Percentage of Variance	Cumulative (%)	Total	Percentage of Variance	Cumulative (%)
1	14.316	39.767	39.767	14.316	39.767	39.767	7.289	20.248	20.248
2	4.841	13.448	53.215	4.841	13.448	53.215	5.462	15.171	35.419
3	3.282	9.118	62.333	3.282	9.118	62.333	4.727	13.129	48.548
4	2.412	6.701	69.034	2.412	6.701	69.034	4.686	13.017	61.565
5	1.858	5.162	74.197	1.858	5.162	74.197	4.547	12.631	74.197
6	0.638	1.772	75.969						
7	0.568	1.577	77.546						
8	0.536	1.489	79.035						
9	0.514	1.428	80.463						
10	0.497	1.381	81.844						
11	0.473	1.315	83.159						
12	0.454	1.26	84.419						
13	0.428	1.188	85.606						
14	0.4	1.112	86.719						
15	0.375	1.041	87.76						
16	0.354	0.983	88.743						
17	0.346	0.962	89.705						
18	0.313	0.871	90.575						
19	0.29	0.805	91.381						
20	0.281	0.781	92.162						
21	0.27	0.749	92.911						
22	0.268	0.744	93.654						
23	0.258	0.716	94.37						
24	0.237	0.657	95.028						
25	0.23	0.638	95.666						
26	0.193	0.537	96.203						
27	0.184	0.511	96.713						
28	0.18	0.5	97.213						
29	0.167	0.464	97.677						
30	0.144	0.399	98.077						
31	0.142	0.393	98.47						
32	0.127	0.354	98.824						
33	0.122	0.339	99.164						
34	0.109	0.303	99.467						
35	0.102	0.284	99.751						
36	0.09	0.249	100						

Data source: the author compiled the data from the questionnaire.

Table 4 shows the total variance explained by the exploratory factor analysis. The results indicate that the factor analysis extracted five factors with a cumulative explained variance of 74.197%, exceeding 60%, indicating that the extracted factors account for 74.197% of the original 36 items. The first principal component explained 39.767% of the variance, less than 40%, indicating that the questionnaire was free of common method bias.

Table 5 shows the rotated component matrix obtained through exploration factor analysis. The factors for the nine measurement items in the fifth dimension (comfort) of

https://doi.org/10.3390/buildings16010172

Tapgol Park's public open space are all concentrated on the first principal component. The factor loadings for the seven dimensions of the safety measurement item are all concentrated on the second principal component. The factor loadings for the seven dimensions of the Interest or Variety measurement item are all concentrated on the third principal component. The factor loadings for the six dimensions of the accessibility measurement item are all concentrated on the fourth principal component. The factor loadings for the seven dimensions of the convenience measurement item are all concentrated on the fifth principal component. The structure of the rotated component matrix is consistent with the questionnaire design, indicating that each dimension has excellent structural validity.

Table 5. Rotated component matrix for exploratory factor analysis (extraction method: principal component analysis; rotation method: Nishizawa normalization maximum variance method).

Measurement Item	Element				
	1	2	3	4	5
Q5-1 (Satisfied with greenery)	**0.85**	0.2	0.076	0.143	0.186
Q5-2 (Satisfied with sunlight)	**0.798**	0.188	0.06	0.238	0.139
Q5-3 (No noise)	**0.826**	0.241	0.12	0.194	0.124
Q5-4 (Open view)	**0.847**	0.154	0.097	0.188	0.13
Q5-5 (Good hygiene)	**0.82**	0.185	0.066	0.232	0.15
Q5-6 (Private and comfortable)	**0.827**	0.194	−0.019	0.226	0.196
Q5-7 (No odor)	**0.855**	0.146	0.046	0.256	0.127
Q5-8 (Reasonable prices)	**0.829**	0.23	0.075	0.159	0.134
Q5-9 (Satisfied with comfort)	**0.809**	0.145	0.113	0.194	0.163
Q4-1 (No illegal parking)	0.267	**0.854**	0.057	0.109	0.045
Q4-2 (Appropriate speed)	0.095	**0.835**	0.022	0.188	0.034
Q4-3 (Nighttime lighting)	0.154	**0.828**	0.103	0.056	0.086
Q4-4 (Non-slip road surface)	0.148	**0.811**	0.105	0.156	0.117
Q4-5 (Clear warning signs)	0.204	**0.834**	0.128	0.207	0.123
Q4-6 (Complete evacuation facilities)	0.288	**0.828**	0.121	0.123	0.05
Q4-7 (Satisfaction with safety)	0.363	**0.775**	0.09	0.27	0.044
Q3-1 (Abundant shopping district)	0.041	0.219	**0.771**	0.261	0.176
Q3-2 (Abundant entertainment venues)	0.044	0.054	**0.777**	0.103	0.203
Q3-3 (Proximity to cultural heritage sites)	0.087	0.215	**0.755**	0.138	0.186
Q3-4 (Convenient food options)	0.048	−0.045	**0.782**	0.074	0.194
Q3-5 (Leisure facilities for interesting spaces)	−0.017	0.032	**0.785**	0.102	0.258
Q3-6 (Number of small shops)	0.122	0.066	**0.764**	0.168	0.187
Q3-7 (Satisfaction with interesting)	0.164	0.107	**0.736**	0.036	0.273
Q1-1 (Smooth and easy to navigate)	0.322	0.2	0.148	**0.809**	0.168
Q1-2 (Barrier-free)	0.314	0.197	0.142	**0.804**	0.107
Q1-3 (Water-friendly environment)	0.28	0.207	0.185	**0.773**	0.144
Q1-4 (Convenient transportation)	0.24	0.148	0.158	**0.817**	0.08

 https://doi.org/10.3390/buildings16010172

Table 5. *Cont.*

Measurement Item	Element				
	1	2	3	4	5
Q1-5 (Open all day)	0.321	0.189	0.189	**0.776**	0.115
Q1-6 (Approach satisfaction rating)	0.243	0.202	0.153	**0.832**	0.233
Q2-1 (Adequate rest facilities)	0.093	0.1	0.223	0.175	**0.746**
Q2-2 (Complete sunshade facilities)	0.13	0.048	0.175	0.099	**0.781**
Q2-3 (Complete public facilities)	0.162	0.019	0.209	0.086	**0.749**
Q2-4 (Convenient and easy to find restrooms)	0.188	0.107	0.29	0.125	**0.712**
Q2-5 (Adequate welfare facilities)	0.142	0.145	0.189	0.039	**0.735**
Q2-6 (Complete sports facilities)	0.16	−0.005	0.188	0.065	**0.717**
Q2-7 (Satisfaction rating of amenities)	0.17	0.076	0.205	0.155	**0.744**

Data source: the author compiled the data from the questionnaire.

3.1.3. Convergent and Discriminant Validity of the Questionnaire

In addition, the researchers conducted a confirmatory factor analysis on the questionnaire using AMOS software (version 29) to determine its convergent and discriminant validity. Convergent validity refers to the internal convergence of the five items within each dimension; discriminant validity refers to the degree of discrimination between different dimensions. Figure 4 and Table 6 details the confirmatory factor analysis model.

Table 6. Confirmatory factor analysis fit index.

CFI	TLI	IFI	RFI	NFI	RMSEA
0.922	0.916	0.922	0.881	0.89	0.067

Data source: the author compiled the data from the questionnaire.

3.1.4. Convergent Validity of the Questionnaire

Convergent validity refers to the internal convergence of the five items within each dimension. As shown in Table 7, the standardized loadings of the measurement items corresponding to the five dimensions ranged from 0.699 to 0.919. With the exception of the "Complete Sports Facilities" measurement item, which had a standardized loading of 0.699 (very close to 0.7), all other measurement items were greater than 0.7. The composite reliability (CR) values for the five dimensions of accessibility, safety, convenience, comfort, and diversity were 0.953, 0.903, 0.915, 0.949, and 0.967, respectively. All values were greater than 0.7. The average variance extracted (AVE) values for the five dimensions were 0.772, 0.572, 0.607, 0.726, and 0.766, respectively, all exceeding 0.5. Overall, the standardized loadings, composite reliability, and average variance extracted all met the standards, indicating that the questionnaire possesses excellent convergent validity.

https://doi.org/10.3390/buildings16010172

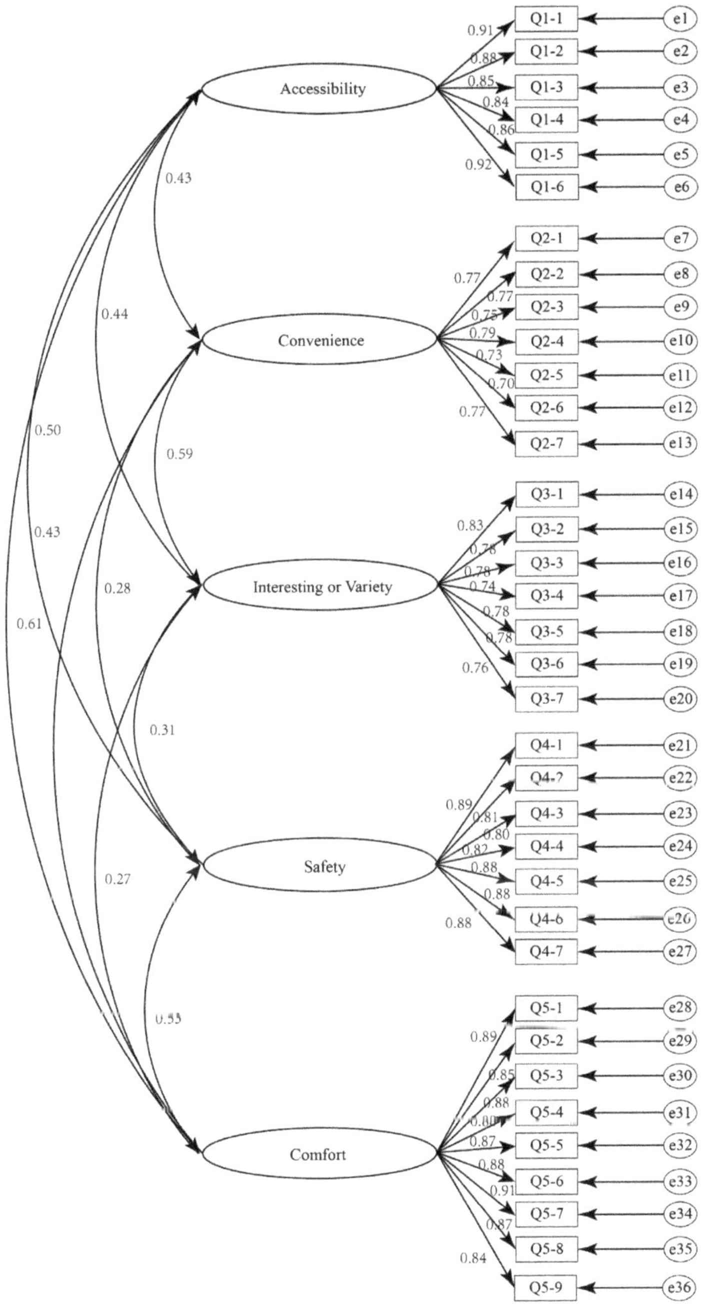

Figure 4. Confirmatory factor analysis model.

https://doi.org/10.3390/buildings16010172

Table 7. Results of convergent validity analysis.

Dimension Information	Item	Unstandardized Load	Standardized Load	Standard Error	*p*	CR	AVE
Accessibility	Q1-1 (Smooth and easy to navigate)	1	0.914			0.953	0.772
	Q1-2 (Barrier-free)	0.98	0.884	0.032	***		
	Q1-3 (Water-friendly environment)	0.816	0.855	0.029	***		
	Q1-4 (Convenient transportation)	0.893	0.839	0.033	***		
	Q1-5 (Open all day)	0.883	0.859	0.031	***		
	Q1-6 (Approach satisfaction rating)	0.973	0.919	0.029	***		
Convenience	Q2-1 (Adequate rest facilities)	1	0.773			0.903	0.572
	Q2-2 (Complete sunshade facilities)	0.959	0.774	0.054	***		
	Q2-3 (Complete public facilities)	0.953	0.754	0.056	***		
	Q2-4 (Convenient and easy to find restrooms)	1.029	0.787	0.057	***		
	Q2-5 (Adequate welfare facilities)	0.926	0.727	0.056	***		
	Q2-6 (Complete sports facilities)	0.879	**0.699**	0.056	***		
	Q2-7 (Satisfaction rating of amenities)	0.973	0.775	0.055	***		
Interesting/Variety	Q3-1 (Abundant shopping district)	1	0.83			0.915	0.607
	Q3-2 (Abundant entertainment venues)	0.856	0.776	0.044	***		
	Q3-3 (Proximity to cultural heritage sites)	0.835	0.784	0.042	***		
	Q3-4 (Convenient food options)	0.797	0.74	0.044	***		
	Q3-5 (Leisure facilities for interesting spaces)	0.873	0.78	0.044	***		
	Q3-6 (Number of small shops)	0.868	0.782	0.044	***		
	Q3-7 (Satisfaction with interesting)	0.786	0.757	0.042	***		
Safety	Q4-1 (No illegal parking)	1.024	0.879	0.037	***		
	Q4-2 (Appropriate speed)	1.024	0.879	0.037	***		
	Q4-3 (Nighttime lighting)	0.982	0.883	0.035	***		

https://doi.org/10.3390/buildings16010172

Table 7. *Cont.*

Dimension Information	Item	Unstandardized Load	Standardized Load	Standard Error	*p*	CR	AVE
	Q4-4 (Non-slip road surface)	0.861	0.818	0.036	***		
	Q4-5 (Clear warning signs)	0.937	0.803	0.04	***		
	Q4-6 (Complete evacuation facilities)	0.913	0.809	0.039	***		
	Q4-7 (Satisfaction with safety)	1	0.886			0.949	0.726
Comfort	Q5-1 (Satisfied with greenery)	1.105	0.884	0.043	***		
	Q5-2 (Satisfied with sunlight)	1.071	0.882	0.042	***		
	Q5-3 (No noise)	1.082	0.872	0.043	***		
	Q5-4 (Open view)	1.069	0.882	0.042	***		
	Q5-5 (Good hygiene)	1.189	0.906	0.044	***		
	Q5-6 (Private and comfortable)	1.056	0.87	0.042	***		
	Q5-7 (No odor)	1	0.842			0.967	0.766
	Q5-8 (Reasonable prices)	1	0.846	0.042	***		
	Q5-9 (Satisfaction with comfort)	1.177	0.892	0.045	***		

Data source: the author compiled the data from the questionnaire. (In SPSS, an asterisk after a T-value usually indicates the level of significance. Three asterisks (***) correspond to the 1% significance level, and the *p* value is less than 0.01).

3.1.5. Discriminant Validity

As shown in Table 8, the values on the diagonal are the square roots of the AVE values of each dimension, indicating the strength of correlation between the dimensions, and the other values are the correlation coefficients between the dimensions. The results in Table 9 show that the diagonal values (currently bold) are larger than the other values, indicating that the questionnaire has strong discriminant validity.

Table 8. Results of discriminant validity analysis (the bold words on the diagonal line are the squares of the AVE values, the other values are the correlation coefficients between the dimensions).

Dimension Information	Comfort	Safety	Interesting/Variety	Convenience	Accessibility
Comfort	**0.79**				
Safety	0.481 **	**0.78**			
Interesting/Variety	0.282 **	0.332 **	**0.737**		
Convenience	0.48 **	0.32 **	0.641 **	**0.776**	
Accessibility	0.564 **	0.447 **	0.424 **	0.444 **	**0.852**
AVE	0.625	0.608	0.543	0.602	0.726

Data source: the author compiled the data from the questionnaire. Two asterisks (**) correspond to significance at the 5% level, and the *p* value is between 0.01 and 0.05.

https://doi.org/10.3390/buildings16010172

3.2. Descriptive Analysis of the Characteristics of Open Spaces for Outdoor Activities in Downtown Seoul Based on Five Dimensions

The questionnaire used a five-point Likert scale to measure respondents' agreement with various aspects of the open space for outdoor activities in Tapgol Park in downtown Seoul. Results indicated that the average scores for the six measurement items in the accessibility dimension were 2.92, 3.12, 2.96, 3.1, 3.18, and 3.21, respectively (Table 9). The average scores for "barrier-free", "transportation convenience", and "open all day" exceeded 3 points, while the average scores for "smooth and easy to navigate" and "water-friendly environment" exceeded 2.9. This indicates that respondents generally agreed that most of the aspects mentioned in these six items reflect the importance of open space for outdoor activities in the city center. Respondents also had high approval ratings for safety and comfort, with scores exceeding 3 points. None of the five dimensions exceeded 4 points, indicating that the overall environment is important to low-income seniors. The average scores for convenience and diversity were all below the 3-point threshold, with average scores for convenience of 2.92, 2.9, 2.93, 2.91, 2.87, 2.84, and 2.98, respectively. The average scores for diversity were 2.71, 2.65, 2.82, 2.68, 2.65, 2.59, and 2.78, respectively. This data suggests that low-income seniors do not agree that the above dimensions influence their satisfaction with open spaces for outdoor activities in downtown Seoul. Therefore, when studying open spaces for outdoor activities in downtown Seoul, we considered abandoning the dimensions of convenience and diversity and focusing on the dimensions of accessibility, safety, and comfort.

Table 9. Descriptive analysis of open space characteristics.

Item or Dimension (Variable)		Mean Value	Standard Deviation
Accessibility	Q1-1 (Smooth and easy to navigate)	2.92	1.41
	Q1-2 (Barrier-free)	**3.12**	1.43
	Q1-3 (Water-friendly environment)	2.96	1.23
	Q1-4 (Convenient transportation)	**3.1**	1.37
	Q1-5 (Open all day)	**3.18**	1.32
	Q1-6 (Approach satisfaction rating)	**3.21**	1.36
Convenience	Q2-1 (Adequate rest facilities)	2.92	1.56
	Q2-2 (Complete sunshade facilities)	2.9	1.5
	Q2-3 (Complete public facilities)	2.93	1.53
	Q2-4 (Convenient and easy to find restrooms)	2.91	1.58
	Q2-5 (Adequate welfare facilities)	2.87	1.54
	Q2-6 (Complete sports facilities)	2.84	1.52
	Q2-7 (Satisfaction rating of amenities)	2.98	1.52
Interesting/Variety	Q3-1 (Abundant shopping district)	2.71	1.28
	Q3-2 (Abundant entertainment venues)	2.65	1.17
	Q3-3 (Proximity to cultural heritage sites)	2.82	1.13
	Q3-4 (Convenient food options)	2.68	1.14
	Q3-5 (Leisure facilities for interesting spaces)	2.65	1.19
	Q3-6 (Number of small shops)	2.59	1.18

 https://doi.org/10.3390/buildings16010172

Table 9. *Cont.*

Item or Dimension (Variable)		Mean Value	Standard Deviation
	Q3-7 (Satisfaction with interesting)	2.78	1.1
Safety	Q4-1 (No illegal parking)	**3.07**	1.44
	Q4-2 (Appropriate speed)	**3.08**	1.44
	Q4-3 (Nighttime lighting)	**3.06**	1.49
	Q4-4 (Non-slip road surface)	2.96	1.34
	Q4-5 (Clear warning signs)	**3.04**	1.42
	Q4-6 (Complete evacuation facilities)	2.96	1.49
	Q4-7 (Satisfaction with safety)	**3.09**	1.47
Comfort	Q5-1 (Satisfied with greenery)	2.91	1.64
	Q5-2 (Satisfied with sunlight)	2.99	1.47
	Q5-3 (No noise)	**3.01**	1.56
	Q5-4 (Open view)	2.69	1.51
	Q5-5 (Good hygiene)	**3.03**	1.54
	Q5-6 (Private and comfortable)	2.92	1.51
	Q5-7 (No odor)	**3.05**	1.63
	Q5-8 (Reasonable prices)	2.9	1.51
	Q5-9 (Satisfaction with comfort)	**3.05**	1.48

Data source: the author compiled the data from the questionnaire.

3.3. *Analyzing Differences in Open Space Characteristics*

Given that different respondent attributes may influence their perceptions of open space, it is necessary to further analyze the differences in open space characteristic scores for downtown outdoor activities based on respondent attributes. The researchers considered respondent attributes such as age group, occupation, marital status, education level, and housing status.

3.3.1. Analysis of Differences Between Age Groups

Table 10, based on one-way ANOVA, shows no significant differences ($p > 0.05$) in scores across the five dimensions of open space for outdoor activities among low-income elderly people individuals based on age.

Table 10. Analysis of differences in age groups.

Variables	65–70 Years Old (N = 180)	71–80 Years Old (N = 158)	80 Years and Above (N = 139)	*p*-Value
Accessibility Total Score	18.81 ± 7.34	17.47 ± 6.94	19.26 ± 7.60	0.083
Convenience Total Score	20.88 ± 8.33	20.39 ± 8.75	19.61 ± 8.57	0.419
Interesting/Variety Total Score	19.41 ± 6.84	18.77 ± 6.30	18.35 ± 6.87	0.36
Safety Total Score	21.71 ± 8.60	19.97 ± 8.51	22.14 ± 9.33	0.074
Comfort Total Score	26.87 ± 12.55	25.25 ± 12.17	27.55 ± 12.18	0.248

Data source: the author compiled the data from the questionnaire.

https://doi.org/10.3390/buildings16010172

3.3.2. Gender Difference Analysis

Table 11 shows that based on the one-way analysis of variance, there is no significant difference ($p > 0.05$) in the scores of low-income elderly men and women on the five dimensions of open space for outdoor activities in the city center, including accessibility, convenience, interest or variety, safety, and comfort.

Table 11. Gender difference analysis.

Variables	Female (N = 115)	Male (N = 362)	p-Value
Accessibility Total Score	18.30 ± 7.78	18.56 ± 7.16	0.738
Convenience Total Score	19.29 ± 8.49	20.69 ± 8.54	0.126
Interesting/Variety Total Score	18.95 ± 6.87	18.87 ± 6.62	0.911
Safety Total Score	20.10 ± 9.03	21.63 ± 8.74	0.107
Comfort Total Score	26.04 ± 12.26	26.68 ± 12.36	0.629

Data source: the author compiled the data from the questionnaire.

3.3.3. Analysis of Intergroup Differences in Marital Status

Table 12 shows that there is no significant difference ($p > 0.05$) in the marital status (married, single, or other) of low-income elderly people on the five dimensions of open space for outdoor activities in the city center, including accessibility, convenience, interest or variety, safety, and comfort.

Table 12. Analysis of Intergroup Differences in Marital Status.

Variables	Single (N = 137)	Married (N = 239)	Other (N = 101)	p-Value
Accessibility Total Score	18.47 ± 7.03	18.70 ± 7.67	18.04 ± 6.85	0.749
Convenience Total Score	20.01 ± 8.20	20.99 ± 8.72	19.30 ± 8.51	0.215
Interesting/Variety Total Score	18.41 ± 6.62	19.39 ± 6.89	18.34 ± 6.20	0.251
Safety Total Score	21.47 ± 8.44	21.03 ± 9.00	21.53 ± 8.97	0.843
Comfort Total Score	25.39 ± 12.62	27.03 ± 12.39	26.89 ± 11.76	0.442

Data source: the author compiled the data from the questionnaire.

3.3.4. Analysis of Differences Between Housing Ownership Groups

Table 13 shows that there is no significant difference ($p > 0.05$) in the housing ownership status (owned, rented, or other) of low-income elderly people in the open space for outdoor activities in the city center, including accessibility, convenience, safety and comfort. However, there are significant differences in the dimensions of interest or variety ($p = 0.025 < 0.05$).

Table 13. Analysis of Intergroup Differences in Housing Ownership Status.

Variables	Owned (N = 204)	Rented (N = 171)	Other (N = 102)	p-Value
Accessibility Total Score	18.53 ± 7.04	18.59 ± 7.57	18.26 ± 7.46	0.935
Convenience Total Score	21.08 ± 8.40	20.38 ± 8.53	18.84 ± 8.72	0.097
Interesting/Variety Total Score	19.67 ± 6.64	18.78 ± 6.52	17.49 ± 6.82	0.025
Safety Total Score	21.46 ± 8.86	21.88 ± 8.77	19.81 ± 8.76	0.157
Comfort Total Score	26.50 ± 11.98	27.21 ± 12.58	25.44 ± 12.63	0.518

Data source: the author compiled the data from the questionnaire.

3.3.5. Analysis of Differences in Educational Attainment

Table 14 shows that there is no significant difference ($p > 0.05$) in the educational attainment (high school graduate or below, high school graduate, vocational college, or university graduate) of low-income elderly people in the open space for outdoor activities

https://doi.org/10.3390/buildings16010172

in the city center, including accessibility, safety and comfort. However, there are significant differences in the two dimensions of convenience ($p = 0.029 < 0.05$) and interest or variety ($p = 0.017 < 0.05$).

Table 14. Analysis of Differences in Educational Attainment.

Variables	High School Graduate or Below (N = 172)	High School Graduate (N = 135)	Vocational College (N = 84)	University Graduate (N = 86)	*p*-Value
Accessibility Total Score	18.92 ± 7.23	19.28 ± 7.19	17.92 ± 7.04	16.97 ± 7.75	0.091
Convenience Total Score	20.40 ± 8.39	21.97 ± 8.41	18.94 ± 8.48	19.09 ± 8.79	0.029
Interesting/Variety Total Score	19.20 ± 6.89	19.90 ± 6.44	18.46 ± 6.24	17.08 ± 6.74	0.017
Safety Total Score	21.31 ± 8.45	22.21 ± 9.21	19.82 ± 8.77	21.07 ± 8.93	0.278
Comfort Total Score	26.82 ± 12.05	27.84 ± 12.27	24.90 ± 12.69	25.47 ± 12.56	0.292

Data source: the author compiled the data from the questionnaire.

3.3.6. Analysis of Occupational Differences

Table 15 shows that through one-way analysis of variance, there are differences in the evaluation of the five dimensions of open space for outdoor activities among low-income elderly people, company employees, managers, service industry workers, business owners, and other occupational groups, and the differences between the scores of each dimension are not significant ($p > 0.05$).

Table 15. Analysis of differences in occupation before retirement.

Variables	Company Employees (N = 62)	Managers (N = 67)	Service Industry Workers (N = 70)	Business Owners (N = 78)	Other (N = 200)	*p*-Value
Accessibility Total Score	16.60 ± 7.07	18.21 ± 7.81	19.11 ± 7.31	18.01 ± 7.33	19.15 ± 7.15	0.15
Convenience Total Score	19.58 ± 8.31	19.76 ± 8.84	21.00 ± 8.93	18.94 ± 8.38	21.11 ± 8.41	0.293
Interesting/Variety Total Score	18.85 ± 6.50	18.42 ± 7.32	17.71 ± 7.22	18.97 ± 6.32	19.43 ± 6.44	0.431
Safety Total Score	19.55 ± 8.62	20.79 ± 9.05	21.40 ± 9.29	20.95 ± 8.29	22.02 ± 8.84	0.391
Comfort Total Score	25.68 ± 12.30	25.79 ± 12.57	27.77 ± 11.93	24.26 ± 12.02	27.49 ± 12.47	0.276

Data source: the author compiled the data from the questionnaire

3.3.7. Analysis of Monthly Income Difference

Table 16 shows that there is no significant difference ($p > 0.05$) in the monthly income (under 1 million KRW, 1 million–2 million KRW, 3 million–4 million KRW, 4 million–5 million KRW, 5 million or more) of low-income elderly people in the open space for outdoor activities in the city center, including accessibility, convenience, interest or variety and comfort. However, there are significant differences in the dimensions of safety ($p = 0.006 < 0.05$).

The analysis of characteristics such as gender, age group, marital status, suggests that these factors do not influence the availability of open space for outdoor activities in the city center, and there is no direct or inevitable connection between them.

Table 16. Analysis of Monthly Income Difference. (Income is in Korean Won).

Variables	Under 1 Million KRW (N = 127)	1 Million–2 Million KRW (N = 26)	3 Million–4 Million KRW (N = 3)	4 Million–5 Million KRW (N = 1)	5 Million or More (N = 2)	*p*-Value
Accessibility Total Score	20.02 ± 7.15	18.85 ± 8.16	27.00 ± 3.61	25.00 ± NA	10.50 ± 6.36	0.129
Convenience Total Score	21.13 ± 7.48	23.23 ± 8.50	32.00 ± 1.73	26.00 ± NA	28.00 ± 7.07	0.068

https://doi.org/10.3390/buildings16010172

Table 16. *Cont.*

Variables	Under 1 Million KRW (N = 127)	1 Million–2 Million KRW (N = 26)	3 Million–4 Million KRW (N = 3)	4 Million–5 Million KRW (N = 1)	5 Million or More (N = 2)	*p*-Value
Interesting/Variety Total Score	20.50 ± 6.47	20.31 ± 8.21	24.67 ± 11.85	25.00 ± NA	15.00 ± 8.49	0.591
Safety Total Score	23.02 ± 8.18	18.65 ± 8.32	31.33 ± 4.04	25.00 ± NA	9.50 ± 0.71	0.006
Comfort Total Score	28.18 ± 10.86	27.00 ± 13.54	41.33 ± 2.52	42.00 ± NA	28.00 ± 16.97	0.221

Data source: the author compiled the data from the questionnaire. In the table, "NA" indicates that the sample size for the corresponding income group (4–5 million Korean won) is only 1 (N = 1). The standard deviation is a statistic used to measure the dispersion of a set of data. The dispersion cannot be calculated from a single data point; the standard deviation requires at least two samples to be calculated. Therefore, the standard deviation result cannot be provided here and is marked as "NA (Not Applicable)."

3.4. Model Training Result

This study divided questionnaire features into six comparable feature sets based on information hierarchy and usage scenarios to further interpret model training results and explore the role of various inputs in predicting overall satisfaction (OS, ordinal). As shown in Table 17 (For a complete list of specific variables included in each set, please refer to Table A2 in the Appendix C), Group A consists solely of the respondent's user profile variables, which tests the independent explanatory power of demographic information on overall satisfaction (OS) and establishes its value as a lower bound baseline, indicating how well the model can perform without contextual information. Group B uses aggregated scores across five dimensions as input to assess the main effects of macro-perception dimensions on OS, facilitating hierarchical comparison with detailed items. Group C builds on Group B by adding user profiles to estimate the marginal contribution of user profiles to discriminative performance after the inclusion of macro-level dimensions, while also examining whether there is an increase in variance and overfitting risk caused by high-cardinality weak signals. Group D uses all 36 detailed items to directly measure the explanatory power of micro-experience cues on OS and provide a practical entry point for subsequent rule extraction. Group E superimposes user profiles on top of Group D to examine whether the profiles still provide valuable insights or are largely redundant when the subdivisions already cover key information. Group F integrates subdivisions, the five dimensions, and the profiles as a comparison of the upper limit of complexity, helping to determine whether simple, information-dense inputs are superior to stacking all variables.

Table 17. Feature sets and analytical intent for decision tree modeling.

Variable Set	Description	Group A	Group B	Group C	Group D	Group E	Group F
Respondent	Gender, Age, Income, etc. (7 items)	✓		✓		✓	✓
Major Category	Approach, Amenities, Safety, etc. (5 items)		✓	✓			✓
Subcategory	Specific indicators for environment & facilities (31 items)				✓	✓	✓

Source: Author's statistics. In the table, "✓" means that the "Group (Group A–F)" corresponding to that column was included as the input feature of the "Variable Set (Variable Set)" corresponding to that row when building the decision tree model; no "✓" means that the variable set was not included in the model analysis of the corresponding group.

The main source of information driving the prediction of overall satisfaction comes from the comparison results of the feature sets of the six groups (Table 18), specifically the segmented items: (1) From the best test values, Group D reached 0.855 in Acc, F1 macro, and BalancedAcc, which is the best decision tree model. Its QWK is 0.936, and Acc@±1 is 0.962, indicating that not only is the overall accuracy the highest, but the errors are mostly small

 https://doi.org/10.3390/buildings16010172

misalignments of adjacent levels. (2) In comparison, Group B's Acc is 0.722, which is 0.133 lower than Group D, and its QWK is 0.895, which is also 0.041 lower, indicating that merging the segmented experiences into five major categories will significantly lose separable information. However, Group B's Acc@±1 has reached 0.957, indicating that its errors are mainly concentrated in adjacent levels, and the macro dimension still has a stable directional signal. (3) Group A has the lowest scores in all indicators, with Acc and BalancedAcc being only 0.406, F1 macro being 0.404, QWK being only 0.411, and Acc@±1 being 0.705. This group can be considered the lower bound baseline of demographic information, meaning that it is difficult to characterize the level of environmental satisfaction based solely on population structure.

Table 18. Summary of test scores for different groups.

Group	Acc	F1 Macro	BalancedAcc	QWK	Acc@±1
A	0.406	0.404	0.406	0.411	0.705
B	0.722	0.721	0.722	0.895	0.957
C	0.620	0.621	0.620	0.826	0.915
D	0.855	0.855	0.855	0.936	0.962
E	0.850	0.850	0.851	0.939	0.974
F	0.774	0.771	0.774	0.926	0.974

Source: Author's statistics.

It is worth mentioning that adding user profile variables did not bring significant gains but weakened the generalization performance of the model. (1) After superimposing user profiles on the macro category to form group C, Acc dropped from 0.722 to 0.620, F1 macro dropped simultaneously to 0.621, QWK dropped from 0.895 to 0.826, and Acc@±1 dropped from 0.957 to 0.915. This shows that adding demographic variables when the macro dimension is already used as input does not bring effective gains but instead introduces noise and splitting bias, weakening the generalization performance. (2) When the baseline is group D of the subdivided items, superimposing the profiles to obtain group E, the strictly judged Acc and F1 macro dropped slightly to 0.850, and the balanced Acc was 0.851, but QWK increased slightly from 0.936 to 0.939, and Acc@±1 increased from 0.962 to 0.974. Comparison of this set of features shows that, given that the segmented items already cover the primary information, user profile variables have a neutral or negative impact on strict hits, and overall, their marginal impact is minimal. Quantitative results suggest that user profile data alone is insufficient to explain satisfaction and that demographic structure alone is weakly predictive.

The fully integrated Group F further demonstrates that simply stacking all variables does not guarantee better generalization. Compared to the best-performing Group D, Group F's Acc dropped to 0.774, its F1 macro dropped to 0.771, its BalancedAcc remained at 0.774, and its QWK dropped to 0.926. However, its Acc@±1 rose to 0.974, on par with Group E, demonstrating more conservative predictions and fewer long-range misalignments within a tolerance of one level of error. This phenomenon typically arises from the increased complexity and inclusion of weak signals after feature stacking, which reduces the model's ability to accurately predict hits while maintaining a high level on the ordered distance metric, showing a tendency to converge toward intermediate levels. Another noteworthy detail is that the BalancedAcc and Acc values across all groups are essentially consistent, indicating that recall across all levels of the test set is relatively balanced after stratification, without any unilateral bias caused by class imbalance. This situation also makes the macro-level conclusions more robust.

https://doi.org/10.3390/buildings16010172

As shown in Figure 5, the confusion matrix of the six groups was tested: the rows are the actual grades, and the columns are the predicted grades. The higher the main diagonal, the better the hit. It can be observed that: (1) Group D has the fullest diagonal and very little long-range misalignment, which is the best in the field; (2) Group B has strong directionality, with high hits at the endpoint grades but still confusion in the middle grades; (3) Group F shrinks towards the median, with strict hits decreasing while the ordered metrics remain high; (4) Group A has a generally low diagonal and spreads to grades 3–4, indicating insufficient information and center drift.

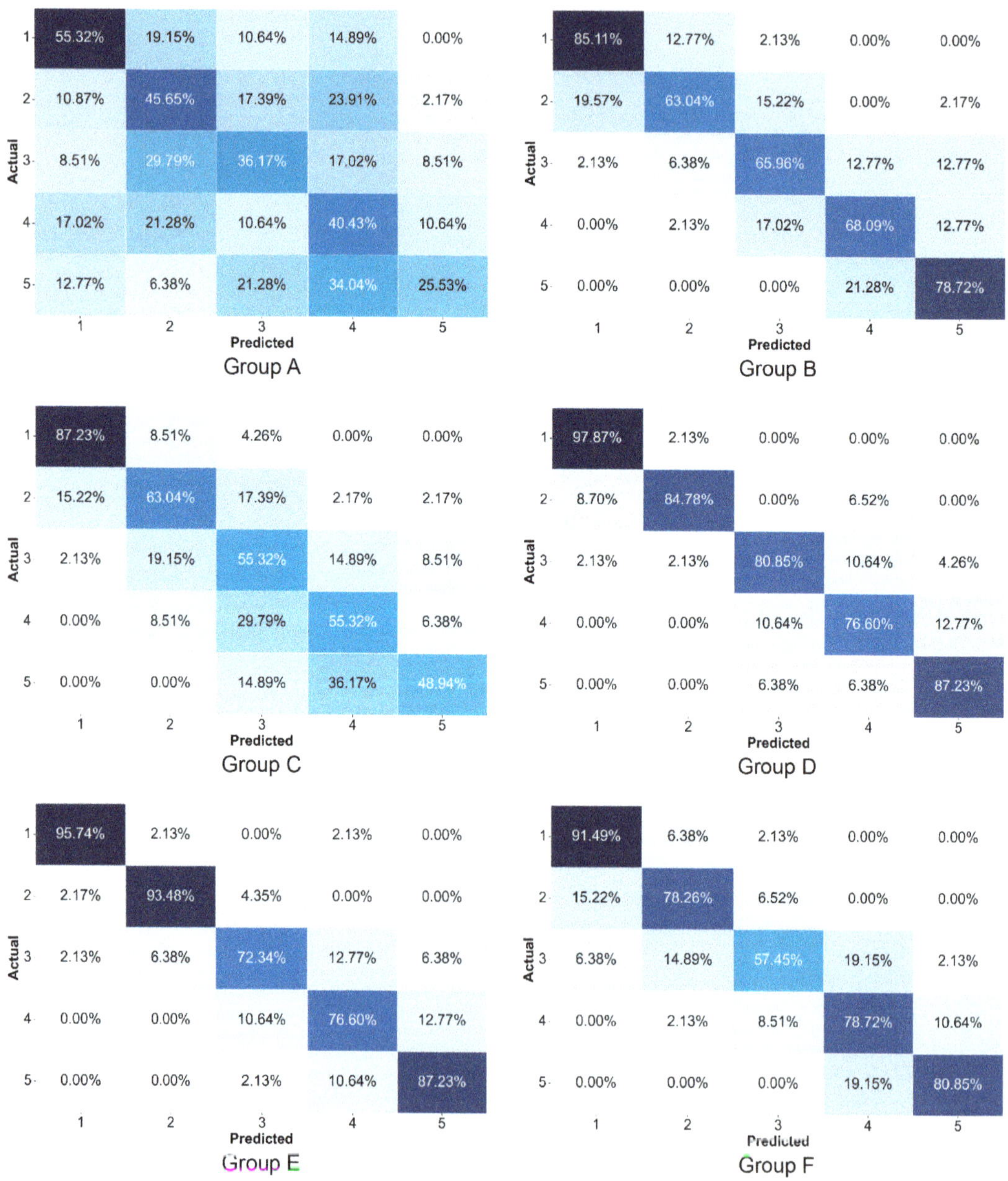

Figure 5. Confusion matrix test of different groups. (Image source: drawn by the author).

https://doi.org/10.3390/buildings16010172

The confusion matrix test shows that: (1) Group A is difficult to distinguish satisfaction levels based on demographic information alone; (2) Group B can effectively distinguish high and low satisfaction levels, with errors mainly due to adjacent files; (3) Group C degrades overall after adding portraits to Group B, reflecting the risk of increased noise and variance; (4) Group D has the most sufficient information on subdivided items, with strict indicators and ordered indicators leading simultaneously, making it the most suitable model; (5) Group E and Group D are similar in form, with a slight decrease in strict hits but a slight increase in QWK and Acc@±1, which is a mild smoothing of boundary samples; (6) Group F stacks all variables, resulting in increased complexity and more centering judgments, and a lower strict hit than Group D. In summary, the results support the use of Group D as the main model, with Group B and Group F as the level and complexity comparison, and Group E and Group C as negative comparisons.

As shown in Figure 6, the learning curve uses the F1 macro as the vertical axis and the training sample size as the horizontal axis to show how the fitting and generalization of the six groups of models change with the sample size: (1) Group A has a low overall ceiling, and both the training and cross-validation curves rise slowly at a low level, showing high bias. Increasing the sample pairs will only improve performance. (2) The cross-validation curve of Group B rises steadily to about 0.71 as the sample size increases, and the gap between training and validation narrows, indicating that the macro dimension provides a clear signal but still has moderate variance. (3) On this basis, Group C, which adds user portraits, has a low training curve and a lower validation curve than Group B. The gap widens, showing an increase in variance and slight degradation caused by noise. (4) The training curve of Group D quickly reaches about 0.9, and the cross-validation curve rises monotonically and stabilizes at a high level of about 0.84, with the smallest gap, indicating that the subdivided items have the best separability and the strongest generalization. (5) In Group E and Group D the morphology is similar, and the validation curve is slightly lower but remains stable, reflecting the mild smoothing of the image for border samples rather than substantial gains. (6) The validation curve of group F is between groups B and D, stabilizing at about 0.80, indicating that feature stacking increases complexity but does not bring corresponding generalization benefits. In general, as the number of samples exceeds about 300, all groups except Group A enter the region of diminishing returns. The optimal solution is to focus on Group D and combine it with moderate regularization and pruning (cost-complexity pruning). For groups C and F, it is necessary to focus on preventing and controlling the variance and overfitting risks caused by high-dimensional weak signals.

The learning curve with the accuracy within one (Acc@±1) as the vertical axis shows that each group quickly reaches a high level and converges after the number of samples increases, indicating that most of the errors in the model are adjacent-level misalignments (Figure 7). (1) Group A performed the worst, with the validation curve hovering in the range of about 0.60–0.70 for a long time and a significant gap with the training curve, reflecting that demographic information is difficult to support the stable improvement of orderly judgment. (2) Group B's validation curve has crossed 0.90 and stabilized at about 0.93–0.95 when there are about 150–200 samples. (3) Group C is slightly lower than Group B, indicating that the superposition of portraits does not bring obvious benefits of tolerating one level of caliber. (4) Groups D and E are the most robust, with training and validation almost matching, converging to 0.96–0.97, and E slightly rising relative to Group D, confirming that the portrait mainly plays a role in smoothing boundary samples when the subdivision items are sufficient. (5) Group F can also reach 0.96–0.97, but the training curve is closer to the ceiling, and the validation curve rises more slowly, suggesting that feature stacking increases complexity but does not bring about a generalization advantage over Group D. Overall, after the sample size exceeds about 300, each group

https://doi.org/10.3390/buildings16010172

enters the region of diminishing returns. The high convergence of Acc@±1 is consistent with the aforementioned confusion matrix, emphasizing that the focus should be on subdividing items and controlling complexity through moderate regularization and pruning (cost-complexity pruning), rather than simply expanding the feature set.

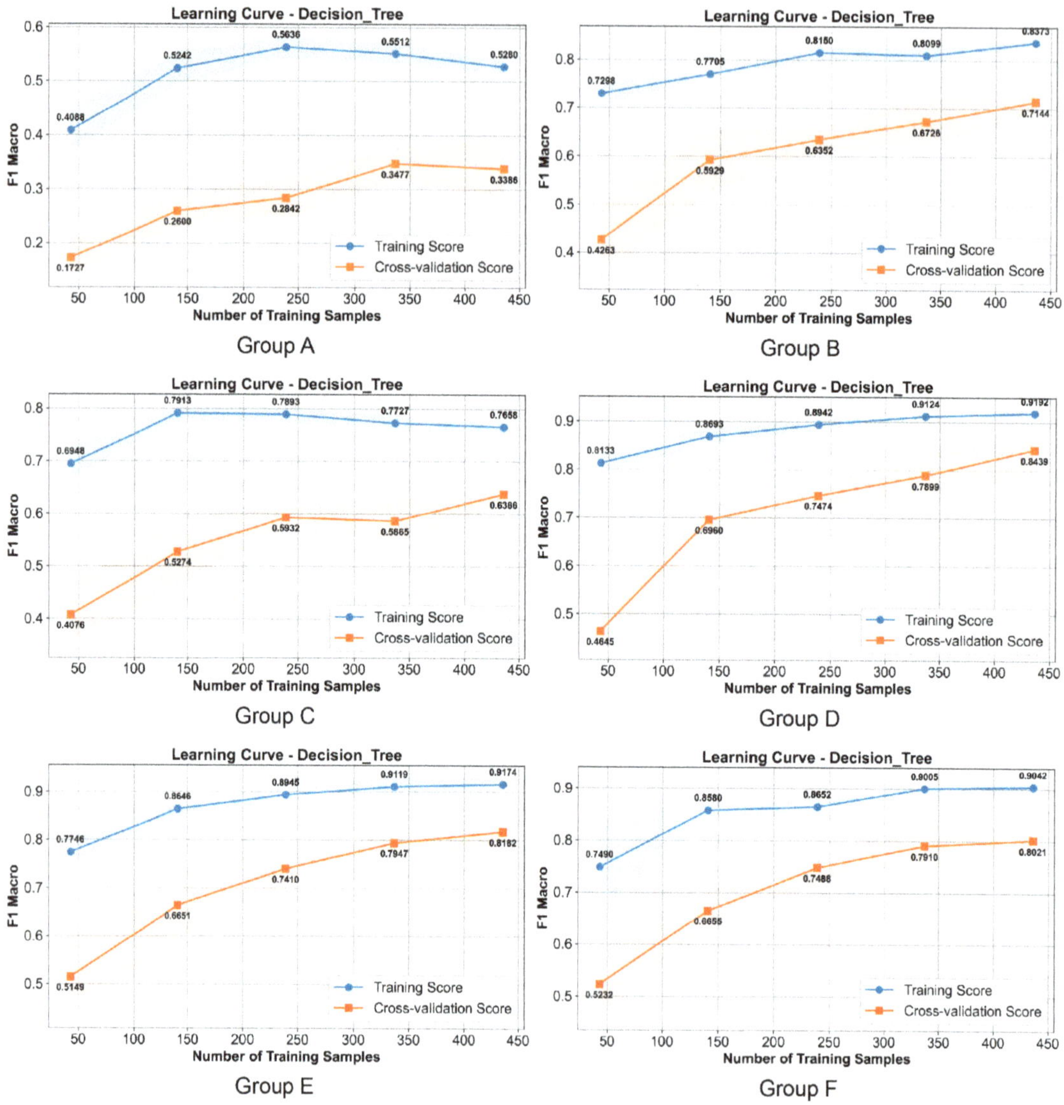

Figure 6. Learning curves of different groups. (Image source: drawn by the author).

The learning curve with quadratic weighted kappa (QWK) as the vertical axis shows (Figure 8): (1) Group D converges steadily to the highest range at the cross-validation end, close to 0.95, with the smallest gap between training and validation, indicating that under the metric that penalizes more distant misalignments, fine-grained features are most effective in reducing cross-grade errors. (2) Group E, which superimposes user portraits on this basis, is slightly lower than Group D but is also smooth and stable, with a validation end of approximately 0.92–0.93, indicating that the boundary samples are gently corrected

https://doi.org/10.3390/buildings16010172

but the overall benefit is limited. (3) The validation curve of Group B quickly increases from approximately 0.86 and converges to around 0.88–0.90. The result shows that the macro dimension can significantly reduce long-range misalignment, but the boundaries are still not sharp enough. (4) After superimposing the user portraits, Group C and Group F finally stay at around 0.85, and the gap between training and verification is larger, suggesting that the addition of weak signals and complexity has led to an increase in variance rather than an improvement in consistency. (5) Only the verification end of User Portrait Group A is always at a low level, far lower than that of other groups, making it difficult to provide effective orderly consistency.

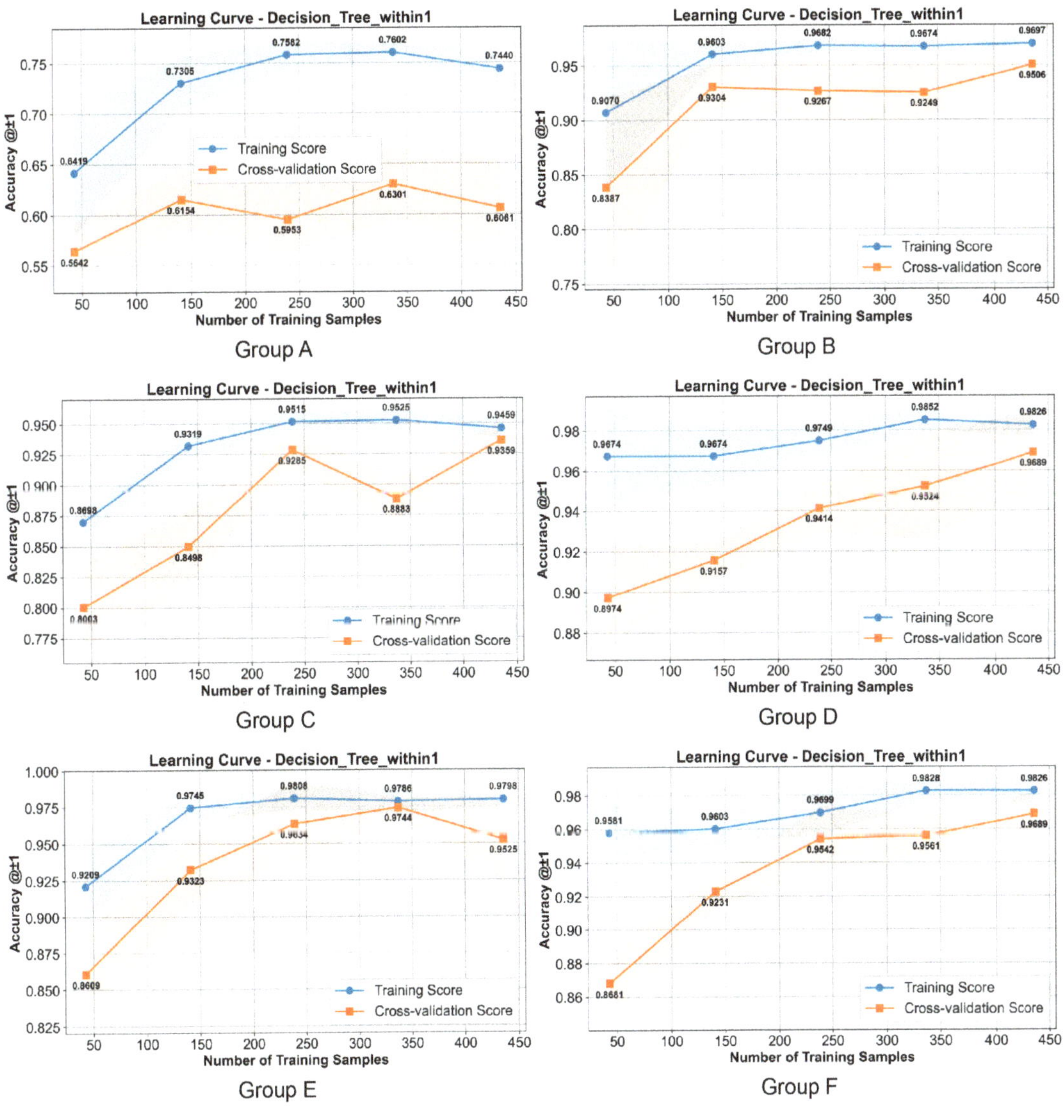

Figure 7. Acc@±1 of different groups. (Image source: drawn by the author).

https://doi.org/10.3390/buildings16010172

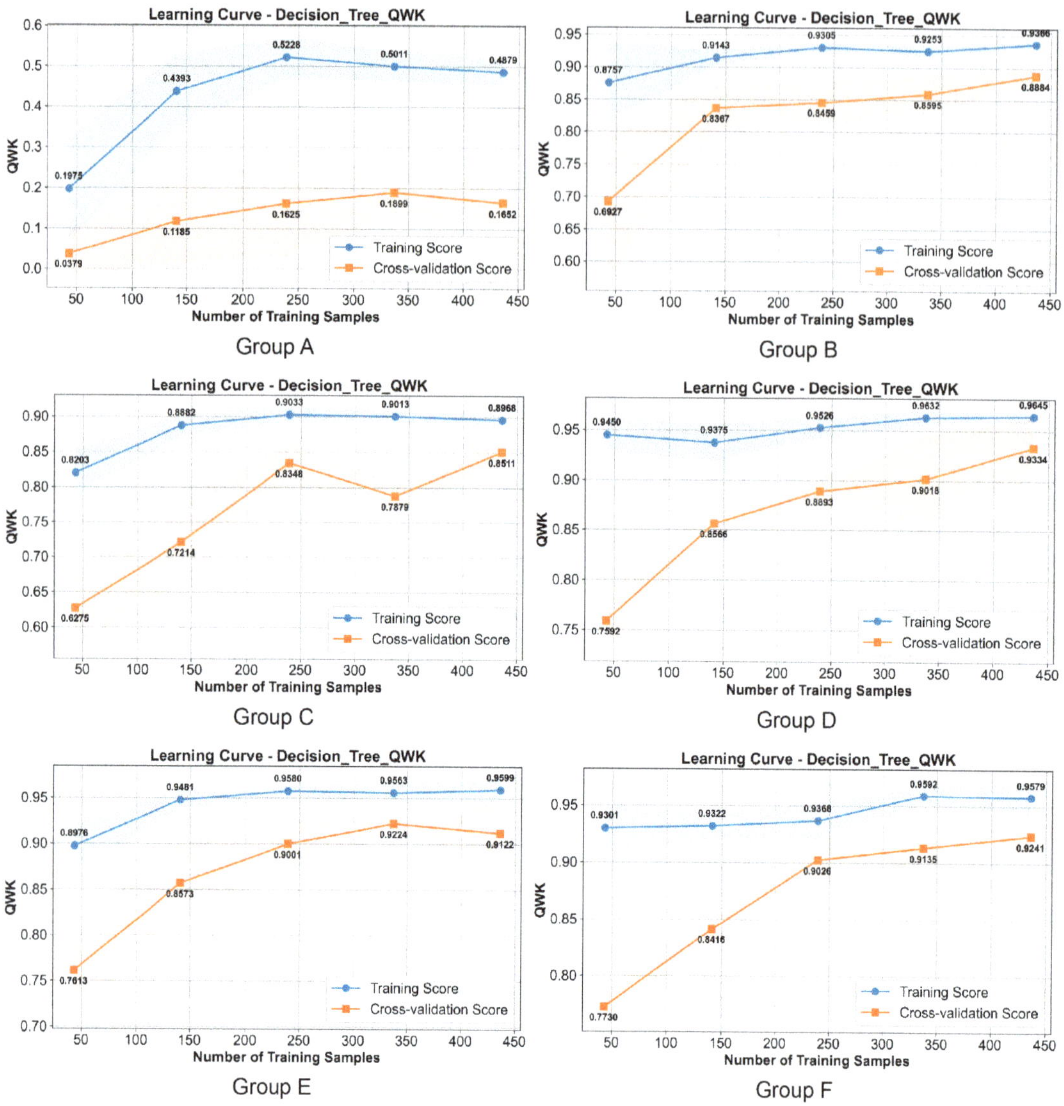

Figure 8. QWK of different groups. (Image source: drawn by the author).

The six curves with the cost-complexity pruning path (ccp_alpha) as the horizontal axis and the cross-validation score as the vertical axis show (Figure 9): (1) The optimal ccp_alpha all fall within a small range close to zero, showing that the scores first rise slightly or remain flat after mild pruning, then decline monotonically as the regularization strength increases and quickly enter the underfitting range. The confidence bands are significantly widened in the large alpha region, indicating that the model variance and uncertainty have increased significantly. (2) Groups D and E have the highest and most stable platform near small alpha, and the flat top area near the optimal point is narrow, indicating that fine-grained features have already provided adequate structure and what is needed is a mild and precise regularization. (3) Group B has a moderate performance level and a mild downward slope, which is consistent with the characteristic of the blunt boundary of macro features. (4) Groups C and F enter the attenuation zone earlier, indicating that weak signals

 https://doi.org/10.3390/buildings16010172

and increased complexity will amplify pruning sensitivity. (5) Group A has a lower overall score but also benefits from mild pruning. In general, heavy pruning is not recommended for all groups. The optimal strategy is to make fine adjustments within a small range of alpha to maintain a balance between structural simplicity and discriminative power.

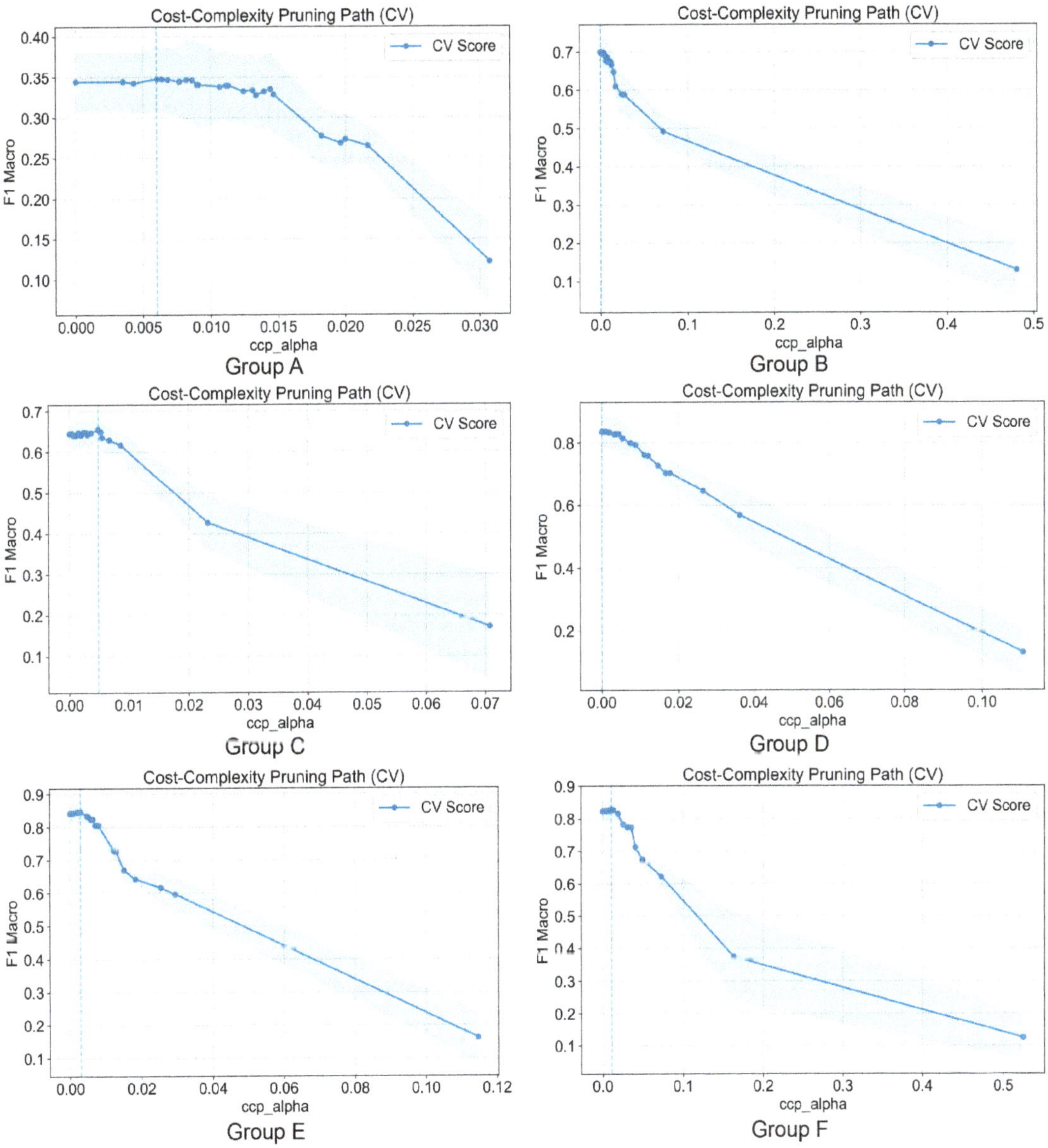

Figure 9. Cost-complexity pruning path for different groups. (Image source: drawn by the author).

In summary, based on metrics such as Acc, F1 macro, BalancedAcc, QWK, and Acc@±1, as well as evidence from confusion matrices, learning curves, and ccp_alpha curves, Group D achieved the best results in both strict hit and ordered consistency (Acc $\approx$ 0.855, QWK $\approx$ 0.936, and Acc@±1 $\approx$ 0.962). The training-validation gap was minimal, and the optimal pruning occurred within a small range, indicating a clear discrimination boundary

https://doi.org/10.3390/buildings16010172

and stable generalization. To emphasize the principal findings, this study performed additional analysis on Group D, showcasing the pruned tree diagrams and decision paths. The decision tree diagrams for the remaining feature combinations have been moved to Appendix B for reference, and the paths and diagrams are not expanded.

4. Discussion

4.1. Decision Tree Analysis

To directly address the spatial management goals of elderly frequenters at Tapgol Park, this section will focus on the actual splitting order of the decision tree for Group D, as well as the sample proportions and category distributions at each node. Figure 10 shows the core information of the first three layers of the decision tree results. The detailed results are shown in Figure A1 in Appendix B. The first splitting variable of the tree is quietness (no noise), with a threshold of 2.4. This attribute splits the sample into two almost even branches: the upper branch, representing 49.8% of the sample with low quietness, has a node distribution dominated by category 1 (overall satisfaction score of 1). The lower branch, representing 50.2% of the sample with relative quietness, has a node distribution dominated by category 4–5 (overall satisfaction score of 4–5). This step directly widens the gap in the baseline satisfaction conditions: quietness determines whether the sample will be subsequently managed to mitigate low satisfaction or to optimize towards high satisfaction.

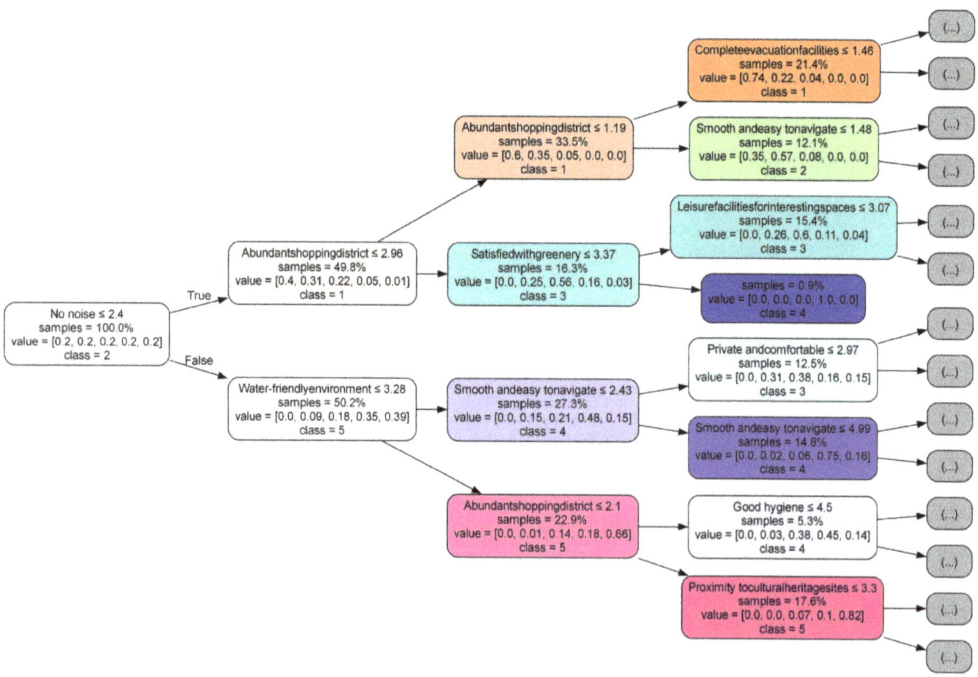

Figure 10. The core information of the first three layers of the decision tree results. (Image source: drawn by the author).

The second layer of the lower branch of the decision tree is dominated by water-friendly environments (threshold of 3.28). It is then further subdivided into conditions such as smooth and easy to navigate, abundant shopping district, satisfied with sunlight, open all day, no odor, and satisfactory hygiene, forming two stable channels mainly at levels 4 and 5.

 https://doi.org/10.3390/buildings16010172

The second layer of the upper branch of the decision tree is dominated by shopping resource richness (threshold of 2.96) and satisfaction with greenery and quickly falls to baseline items such as complete evacuation facilities, convenient transportation, reasonable prices, and barrier-free access, forming a downward channel dominated by levels 1–2 and a small level 3 transition zone.

The low level of satisfaction is characterized by multiple bottom-line indicators failing to meet standards at the same time, and the extreme deficiency of any one of them will significantly lower the score. (1) In the upper branch of the decision tree, the nodes corresponding to extremely low shopping resource richness (threshold of 1.19) are mainly level 1 (33.5% of the samples, about 60% of the samples). (2) Under the same path, when the evacuation facilities are lower than 1.46, the proportion of level 1 rises to about 74% (21.4% of the samples). (3) When the convenient transportation is lower than 1.98, it is almost all level 1 (about 93% of the samples, 15.0% of the samples). (4) When the reasonable price is lower than 1.90, it is a pure level 1 node (12.8% of the samples). (5) When the barrier-free is lower than 1.60, it is also a pure level 1 node. (6) If some of the bottom lines are improved but still not comfortable, the model will give a level 2. For example, when the walking smoothness is lower than 1.48 or the anti-slip property is poor (non-slip road surface, threshold of 3.33), the proportion of level 2 rises to 57% and 79%. This chain shows that when noise is difficult to control and consumption and travel are inconvenient, an extremely low score for any single item is enough to bring the overall satisfaction level to the lowest level. Governance should prioritize rigid bottom lines such as price and accessibility, barrier-free and evacuation safety.

The transition zone to the middle tier is primarily defined by environmental perceptions, such as greenery, lighting, privacy, and comfort, reflecting a baseline level of usability but falling short of comfort. Typical milestones include nighttime lighting around 3.89, which is predominantly rated 3 (approximately 75% of samples, representing 10.3% of the sample); odor-free ratings below 2.5, which are almost exclusively rated 3 (5.3% of the sample); greenery satisfaction around 3.37, which is 56% (16.3% of the sample); and privacy and comfort around 2.97, which is 38% (12.5% of the sample). This statistic suggests that in noisy or crowded environments, even if prices and transportation options improve, if direct experiences such as olfaction and lighting remain subpar, the overall rating will struggle to reach the high-end level.

The two main pathways to high satisfaction are clear and complementary. One is the Level 4 pathway, emphasizing accessibility and basic comfort. Within the context of quietness, as long as walking comfort reaches a moderate-to-high level (thresholds of approximately 2.43 and 4.99, with Level 4 percentages of approximately 48% and 75%, respectively), sunlight is good (threshold of 4.38, corresponding to 85% of Level 4), and accessibility is convenient and 24/7 (threshold of 2.21, corresponding to 96% of Level 4), coupled with an absence of odor and high hygiene scores, it will steadily fall into the Level 4 range. The other is the Level 5 pathway, emphasizing vitality and affordability. Given its quietness, Level 5 nodes with a high abundance of shopping resources are predominantly Level 5 (66% of the sample, representing 22.9% of the sample). This percentage rises further to 82% (17.6% of the sample) near cultural heritage sites, while pure Level 5 nodes emerge when affordability is high (threshold of 4.06, representing 8.6% of the sample). A deeper divide could lead to the formation of pure Level 5 nodes again, with further refinements in shopping resources. The two channels together indicate that to elevate a space that already has a quiet background from comfortable to optimal, on the one hand, it is necessary to ensure that it is accessible and clean and odorless; on the other hand, the willingness to stay and consume should be enhanced through price-friendly and commercial cultural provision.

https://doi.org/10.3390/buildings16010172

Based on the above structure, three representative and operational paths can be given to directly guide the governance sequence:

(1) Path 1, stop-loss type: When quietness does not meet the standard and transportation is inconvenient, prices are high, or accessibility and evacuation are insufficient, the model will almost certainly fall to level 1. The priority should be affordability and accessibility, then accessibility and anti-slip evacuation. The goal is to raise the level 1 sample to level 2–3 as soon as possible.

(2) Path 2, steady-state improvement type: Under the premise of relative quiet, through systematic improvement of smooth walking, all-day openness, and sanitation and odorlessness, it can stably enter level 4 channel. This path is suitable for rapid and effective facility maintenance and management.

(3) Path 3, transition type: After quietness and basic comfort have been met, the combination of reasonable prices and proximity to shopping resources and culture can be added to push the leaf node to level 5, which is suitable for price guidance and business optimization.

Overall, this study's decision tree clearly demonstrates a hierarchical progression: quietness first, accessibility and cleanliness second, and finally price and vitality. This aligns with the high frequency of elderly use of Tapgol Park and the diverse surrounding facilities. Based on this, we recommend prioritizing the baseline within a limited budget, then ensuring stability at Level 4 through pedestrian-friendliness and cleanliness and finally raising the level to Level 5 through a combination of price and business formats, thereby expanding the coverage of high satisfaction.

4.2. Decision Path Analysis

Unlike whole-tree analysis, which focuses on earlier-split features and information-level organization, decision path analysis directly addresses the controllable conditions that result in a certain satisfaction level, thereby transforming statistical findings into a governance checklist. Paths with high p-values (p represents confidence level; all paths below are selected as 1.00) ensure high hit execution certainty, while high support values (support represents coverage; the top five decision paths with the highest coverage are selected below) ensure that the benefits of an action are not too limited. Together, these two factors constitute the trade-off between precision and coverage in policy design.

As shown in Table 19, three levels of usage scenarios are clearly defined: (1) The stop-loss zone with low satisfaction corresponds to high noise and mismatched baseline elements. (2) In a relatively quiet environment, the stable comfort zone formed by pedestrian-friendly and all-day openness corresponds to level 4. (3) In a quiet environment with a high satisfaction zone formed by the linkage of waterfront, commercial, and cultural resources, corresponding to level 5. The thresholds are all based on the 1–5 scale of the questionnaire, with lower values indicating more dissatisfaction and higher values indicating more satisfaction.

Path 1, which leads to Level 1 overall satisfaction, represents a typical baseline gap: high noise (no noise $\leq$ 2.397) combined with insufficient commercial facilities, inadequate evacuation facilities, inconvenient transportation, and perceived unreasonable prices. When all five factors are simultaneously low, the sample inevitably falls into the lowest category, with the highest proportion (Support = 12.821%). This indicates that for the elderly, safe evacuation, accessibility, and affordability are rigid requirements. Any extremely low value will drag down the overall rating. Prioritizing price guidance and optimizing public transportation connections, while also addressing evacuation and pedestrian safety, is crucial. Path 4, corresponding to Level 2, is a secondary form of Path 1. While there are some shops, walking is difficult, the road surface is not slip-resistant, and

 https://doi.org/10.3390/buildings16010172

there is odor, making it difficult to reach medium satisfaction. The key focus here is road improvements, cleaning and odor removal. Detailed improvements to accessibility and continuous walking can help elevate Level 2 to Level 3.

Table 19. Top 5 decision paths.

Path	Decision Path	p	Support
1	IF No noise ≤ 2.397 AND Abundant shopping district ≤ 1.185 AND Complete evacuation facilities ≤ 1.460 AND Convenient transportation ≤ 1.979 AND Reasonable prices ≤ 1.898 THEN class = 1	1.00	12.821%
2	IF No noise > 2.397 AND Water-friendly environment > 3.281 AND Abundant shopping district > 2.104 AND Proximity to cultural heritage sites > 3.305 AND Leisure facilities for interesting spaces > 3.014 THEN class = 5	1.00	8.608%
3	IF No noise ∈ (2.397, 4.993] AND Water-friendly environment ≤ 3.281 AND Smooth and easy to navigate ∈ (2.428, 4.993] AND Satisfied with sunlight ≤ 4.385 AND Open all day > 2.212 THEN class = 4	1.00	7.875%
4	IF No noise ≤ 2.397 AND Abundant shopping district ∈ (1.185, 2.962] AND Smooth and easy to navigate ∈ (1.480, 2.923] AND Non-slip road surface ≤ 3.325 AND No odor ≤ 2.714 THEN class = 2	1.00	4.762%
5	IF No noise ≤ 2.397 AND Abundant shopping district > 2.962 AND Satisfied with greenery ≤ 3.372 AND Leisure facilities for interesting spaces ≤ 3.067 AND Nighttime lighting > 3.891 AND No odor ≤ 2.500 THEN class = 3	1.00	4.762%

Source: Author's statistics.

Path 5 defines the median zone for overall satisfaction level 3: While noise levels remain high, there's a good supply of shops and adequate nighttime lighting, but insufficient greenery and leisure facilities, and the presence of odors ultimately leads to medium satisfaction. This graph reveals the tension between bustle and comfort: vitality can offset some of these negative factors, but a lack of greenery and accessible spaces, coupled with negative olfactory stimulation, suppresses overall satisfaction. Countermeasures include adding greenery, providing reliable seating areas, and reducing noise and odors, ultimately shifting this demographic toward level 4.

Path 3 provides a stable path to Level 4 overall satisfaction: In a relatively quiet environment, even if water accessibility is average, as long as walking is smooth to medium-to-high levels, sunlight exposure is not too low, and the space is open all day, it can steadily fall into Level 4. The key here is the walking experience and open access. This means that for elderly frequent users, a continuous, smooth, and all-day pedestrian system is key to raising the space to a comfortable level. To further improve from Level 4, prioritize strengthening the microenvironmental balance between water accessibility and sunlight and shade.

Path 2 defines a five-level high-satisfaction cluster: These samples consistently fall into the highest category (support rate 8.608%) due to the presence of tranquility, a favorable waterfront environment, a moderately diverse commercial landscape, proximity to cultural heritage, and attractive leisure spaces. The result is a complex upward trajectory, emphasizing the desire to stay and emotional identification driven by vitality and culture, built on a foundation of tranquility. Strategic measures should be implemented through the meticulous development of waterfront accessibility and stay nodes, the selection of affordable small shops and community-oriented businesses, and pedestrian corridors connecting cultural attractions to strengthen this channel.

Combining these five paths creates a tiered governance sequence. The first tier is the Stop-Loss Package, addressing the conditions of Paths 1 and 4 by prioritizing noise, evacuation, slip resistance, odor, and pedestrian continuity. Public transit and pedestrian connections also enhance accessibility and affordability. The second tier is the Back-Up Package, addressing the key variables of Paths 3 and 5 by systematically improving nighttime lighting, rest areas, and greenery, ensuring 24/7 accessibility and easy access. The third tier is the Transition Package, aligning with Path 2. This package creates a complex, immer-

 https://doi.org/10.3390/buildings16010172

sive experience through waterfront nodes, cultural connections, and affordable, small-scale commercial facilities. This sequential approach will prioritize the exit of Levels 1 and 2, converge Level 3 toward Level 4, and transform Level 4 into Level 5 at key points, thereby maximizing overall satisfaction within a limited budget.

5. Conclusions

This study uses Tapgol Park in Jung-gu, Seoul, as a representative setting. This public space, frequently used by seniors, quantitatively identifies and refines pathways around residents' overall environmental satisfaction. The core question is to identify the environmental factors and their sequence of combinations that best enhance overall satisfaction and based on this, propose governance priorities within a limited budget. The research methodology employs an interpretable decision tree model, combining data preprocessing, stratified sampling and cross-validation, synthetic minority oversampling (SMOTE), hyperparameter search, and cost-complexity pruning (ccp_alpha). Evaluation is performed using multiple metrics, including Acc, F1 macro, BalancedAcc, QWK, and Acc@±1. A tree diagram and rule-based pathways are outputs for an evidence-based presentation.

The main conclusions of this study are as follows:

(1) The key to improving the management and satisfaction of Tapgol Park in Jung-gu, Seoul, is the quietness of the site.

(2) When the park is not quiet, users are most sensitive to the bottom-line factors, such as commercial supply, evacuation safety, transportation accessibility, price perception, barrier-free, and anti-slips. Any deficiency in any of them will reduce the overall satisfaction of the park to the lowest level.

(3) When the park meets the quiet standard, basic comfort factors such as smooth walking, all-day openness, sunlight, and no odor constitute the minimum condition set for entering the comfort zone. To further leap from comfort to optimal, water experience, waterfront accessibility, proximity to cultural resources, and moderate business and community-oriented leisure facilities are key bonus points.

(4) Through the high-confidence and high-coverage decision paths, we can identify that: (a) the low-satisfaction stop-loss combination emphasizes noise reduction, evacuation and accessibility, and transportation accessibility and price-friendliness; (b) the medium-satisfaction steady-state combination emphasizes maintenance details such as continuous walking, all-day openness, lighting, and deodorization; (c) the high-satisfaction transition combination emphasizes the superposition of water-friendly nodes, cultural connectivity, and small-scale commercial facilities on a quiet and comfortable base.

The data collection for this study focused on the summer period from June to August, which has certain seasonal limitations. Summer weather is suitable for outdoor activities, and the frequency, duration, and functional needs (such as cooling off and leisure) of older adults using outdoor spaces differ significantly from those in spring, autumn, and winter. This investigation may overestimate the actual intensity of outdoor space use and demand priorities, and it is difficult to capture the spatial use characteristics related to seasonal activities (such as winter fitness and spring outings). In addition, this study also has the risk of self-report bias and temporal instability because the sample is derived from a single scenario, and the questionnaire is cross-sectional. In the future, we should conduct replication studies in multiple scenarios and multiple time periods, introduce control experiments or pilot interventions to verify the causal validity of the path, introduce multi-source integration of spatial and pedestrian flow data, and incorporate fairness and accessibility of vulnerable groups into the evaluation framework so that the research results can better serve the refined governance of elderly-friendly cities.

https://doi.org/10.3390/buildings16010172

Author Contributions: Conceptualization, C.W., Y.C. and L.Z.; methodology, C.W., Y.C. and L.Z.; software, Y.C., J.L. and L.Z.; validation, Y.C., J.L. and L.Z.; formal analysis, C.W., Y.C. and L.Z.; investigation, C.W., F.Z., S.Y. and Y.W.; resources, C.W.; data curation, C.W.; writing—original draft preparation, C.W., Y.C. and L.Z.; writing—review and editing, C.W., Y.C., F.Z., L.Z., J.L., S.Y. and Y.W.; visualization, C.W., Y.C. and L.Z.; supervision, Y.C.; project administration, Y.C.; funding acquisition, Y.C. All authors have read and agreed to the published version of the manuscript.

Funding: This research was funded by the (1) Faculty Research Grants funded by Macau University of Science and Technology (FRG-MUST grant number: FRG-25-041-FA; FRG-25-067-FA); (2) Guangdong Provincial Department of Education's key scientific research platforms and projects for general universities in 2023: Guangdong, Hong Kong, and Macau Cultural Heritage Protection and Innovation Design Team (grant number: 2023WCXTD042); (3) Guangdong Provincial Philosophy and Social Sciences Planning 2025 Lingnan Cultural Project (grant number: GD25LN30). The funders had no role in study conceptualization, data curation, formal analysis, methodology, software, decision to publish, or preparation of the manuscript. There was no additional external funding received for this study.

Institutional Review Board Statement: Chunhong Wu, Fenrong Zhang, Shuai Yang and Yinqi Wang participated in the research and questionnaire distribution. The Institutional Review Board (IRB) of Hangzhou City University reviewed and approved the studies involving human subjects. The approval number is 2025-0308. It was conducted in full compliance with the principles of the Helsinki Declaration. As the study did not involve human clinical trials, animal experiments, or any unethical practices, and all participants were aged 18 years or older, no ethical concerns arose during the research process. The participant's legal guardian or next of kin has agreed to the written informed consent for this study prompted on the home page before filling out the questionnaire.

Informed Consent Statement: In this study, informed consent was obtained by clearly explaining the purpose, procedures, potential risks, and benefits of participation to all respondents. Participants were assured that their involvement was voluntary, and they could withdraw at any time without consequences. Confidentiality of the data was emphasized, and respondents were required to provide explicit agreement to participate before the study commenced.

Data Availability Statement: The original questionnaire involves personal privacy, and it is not convenient to disclose directly. The original questionnaire used and analyzed during the current study are available from Chunhong Wu (wuch0142@cau.ac.kr) upon reasonable request.

Acknowledgments: We would like to express our sincere gratitude to the students who helped distribute the questionnaires and the staff who assisted during the field survey.

Conflicts of Interest: The authors declare that they have no known competing financial interests or personal relationships that could have appeared to influence the work reported in this paper.

Appendix A. Questionnaire on the Elderly's Satisfaction with the Tapgol Park Environment

Please note that the information provided below is translated from the original questionnaire, which was written in Korean.

Part I: Informed Consent

This survey will serve as a baseline for understanding satisfaction with the park environment. This survey is targeted at seniors. We would appreciate it if you could take approximately 15 min to honestly and proactively share your thoughts on the park and surrounding facilities. If you agree to participate in the study, please complete the following questionnaire. Please note that you may withdraw at any time. All information related to this survey will be kept anonymous and will not be used for any purpose other than academic research.

(1) Agree with the survey [Single Choice]

☐ Agree

 https://doi.org/10.3390/buildings16010172

☐ Disagree (Terminate survey)
(2) Have you visited Tapgol Park? [Single Choice]
☐ Yes
☐ No (Terminate survey)
Thank you for your participation!

Table A1. Questionnaire on the elderly's satisfaction with the Tapgol Park environment. (All questions are required and must be filled in).

Dimensions	No.	Features	Feature Description	Strongly Disagree—Strongly Agree
Accessibility	Q1-1	Smooth and easy to navigate	The sidewalks have gentle slopes and steps.	1-2-3-4-5
	Q1-2	Barrier-free	The sidewalks are free of obstruction, allowing for smooth movement.	1-2-3-4-5
	Q1-3	Water-friendly environment	Access to waterfront areas (ponds, streams, etc.) is convenient.	1-2-3-4-5
	Q1-4	Convenient transportation	Access to various public transportation options (buses, subways, etc.) is also very convenient.	1-2-3-4-5
	Q1-5	Open all day	There are no time restrictions for entering the park.	1-2-3-4-5
	Q1-6	Satisfaction with accessibility (Approach satisfaction rating)	I am very satisfied with the overall road and transportation conditions in Tapgol Park, as well as its connectivity to other areas.	1-2-3-4-5
Convenience	Q2-1	Adequate rest facilities	There are ample facilities and space for seating and resting.	1-2-3-4-5
	Q2-2	Complete sunshade facilities	There are ample shade and space.	1-2-3-4-5
	Q2-3	Comprehensive public facilities	There are ample public amenities such as chairs and trash cans.	1-2-3-4-5
	Q2-4	Convenient and easy to find restrooms	Toilets are easily accessible and readily available nearby.	1-2-3-4-5
	Q2-5	Adequate welfare facilities	There are ample facilities for seniors around the park.	1-2-3-4-5
	Q2-6	Complete sports facilities	The nearby sports facilities allow me to engage in a variety of sports.	1-2-3-4-5
	Q2-7	Satisfaction with amenities (Satisfaction rating of amenities)	Overall, I am very satisfied with the basic amenities at Tapgol Park.	1-2-3-4-5
Interesting/Variety	Q3-1	Abundant shopping district	There are many shops and restaurants nearby that I can frequent.	1-2-3-4-5
	Q3-2	Abundant entertainment venues	There are also many entertainment venues nearby, such as karaoke bars and movie theaters.	1-2-3-4-5
	Q3-3	Proximity to cultural heritage sites	I often visit the park's ancient structures (e.g., ancient pagodas and stone monuments).	1-2-3-4-5
	Q3-4	Convenient food options	I often buy and eat at the carts along the roadside.	1-2-3-4-5
	Q3-5	Leisure facilities for interesting spaces	The area surrounding the park allows for a variety of recreational activities (e.g., playing chess, watching performances, etc.).	1-2-3-4-5
	Q3-6	Number of small shops	There are also many convenience stores and general stores nearby.	1-2-3-4-5
	Q3-7	Satisfaction with interesting	I think the overall facilities at Tapgol Park meet a wide range of needs and offer a wide variety of things to see.	1-2-3-4-5

https://doi.org/10.3390/buildings16010172

Table A1. *Cont.*

Dimensions	No.	Features	Feature Description	Strongly Disagree—Strongly Agree
Safety	Q4-1	No illegal parking	There's no illegal parking in pedestrian areas (sidewalks/around the park).	1-2-3-4-5
	Q4-2	Appropriate speed	I'm satisfied with the traffic volume and speed (standard speed of 30 km/h).	1-2-3-4-5
	Q4-3	Nighttime lighting	The roads and park are well lit at night.	1-2-3-4-5
	Q4-4	Non-slip road surface	The road surface is not slippery.	1-2-3-4-5
	Q4-5	Clear warning signs	Safety signs on the roads and in the park are clearly visible.	1-2-3-4-5
	Q4-6	Complete evacuation facilities	Emergency facilities (police station, emergency escape routes, etc.) are adequate.	1-2-3-4-5
	Q4-7	Satisfaction with safety	I think Tapgol Park is generally safe.	1-2-3-4-5
Comfort	Q5-1	Satisfied with greenery	The park has a rich and diverse flora.	1-2-3-4-5
	Q5-2	Satisfied with sunlight	The sunlight is pleasant.	1-2-3-4-5
	Q5-3	No noise	There was no unpleasant noise.	1-2-3-4-5
	Q5-4	Open view	Looking around, I felt there was minimal obstruction to my view.	1-2-3-4-5
	Q5-5	Good hygiene	I was pleased with the sanitation (little trash).	1-2-3-4-5
	Q5-6	Private and comfortable	I felt comfortable here, without being overly noticed.	1-2-3-4-5
	Q5-7	No odor	There were no unpleasant odors.	1-2-3-4-5
	Q5-8	Reasonable prices	I was satisfied with the prices.	1-2-3-4-5
	Q5-9	Satisfaction with comfort	I think Tapgol Park is generally comfortable.	1-2-3-4-5
Overall satisfaction	OS	Overall satisfaction	I am generally satisfied with the environment of Tapgol Park.	1-2-3-4-5

Source: Author's statistics.

Part II: Demographic characteristics

(1) What is your gender? [Single Choice]

☐ Male

☐ Female

(2) What is your marital status? [Single Choice]

☐ Single

☐ Married

☐ Other

(3) What is the current home you live in? [Single Choice]

☐ Owned (including your own, children's, or other family members' names)

☐ Rented

☐ Other

(4) What is your highest level of education? [Single Choice]

☐ High school graduate or below (or equivalent)

☐ High school graduate (or equivalent)

☐ Vocational college (or equivalent)

☐ University graduate (or equivalent)

(5) What was your occupation before retirement? [Single Choice]

☐ Company employees

☐ Managers

☐ Service industry workers

☐ Business owners

https://doi.org/10.3390/buildings16010172

☐ Other

(6) What is your age? [Single Choice]

☐ 65–70

☐ 71–80

☐ 80 or older

(7) What is your average monthly income? [Single Choice]

☐ Under 1 million KRW

☐ 1 million–2 million KRW

☐ 3 million–4 million KRW

☐ 4 million–5 million KRW

☐ 5 million or more

Appendix B. Decision Tree Paths Analysis Before Pruning

Figure A1 shows the optimal decision tree in this study.

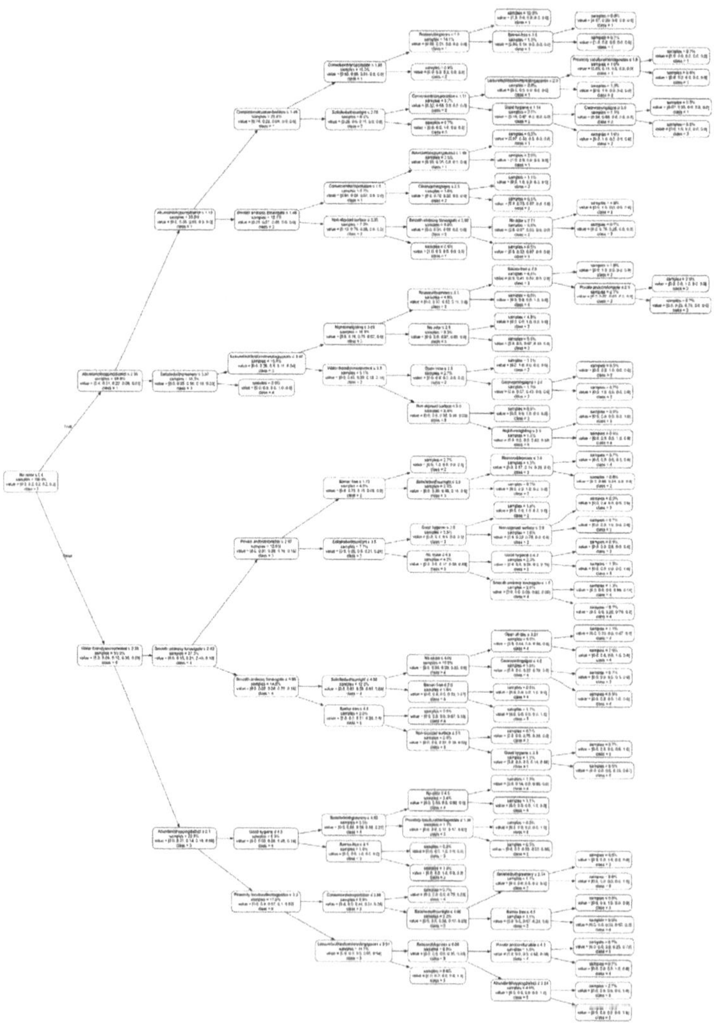

Figure A1. The optimal decision tree in this study. (Image source: drawn by the author).

 https://doi.org/10.3390/buildings16010172

Figures A2–A7 below shows the decision tree path analysis diagram before pruning.

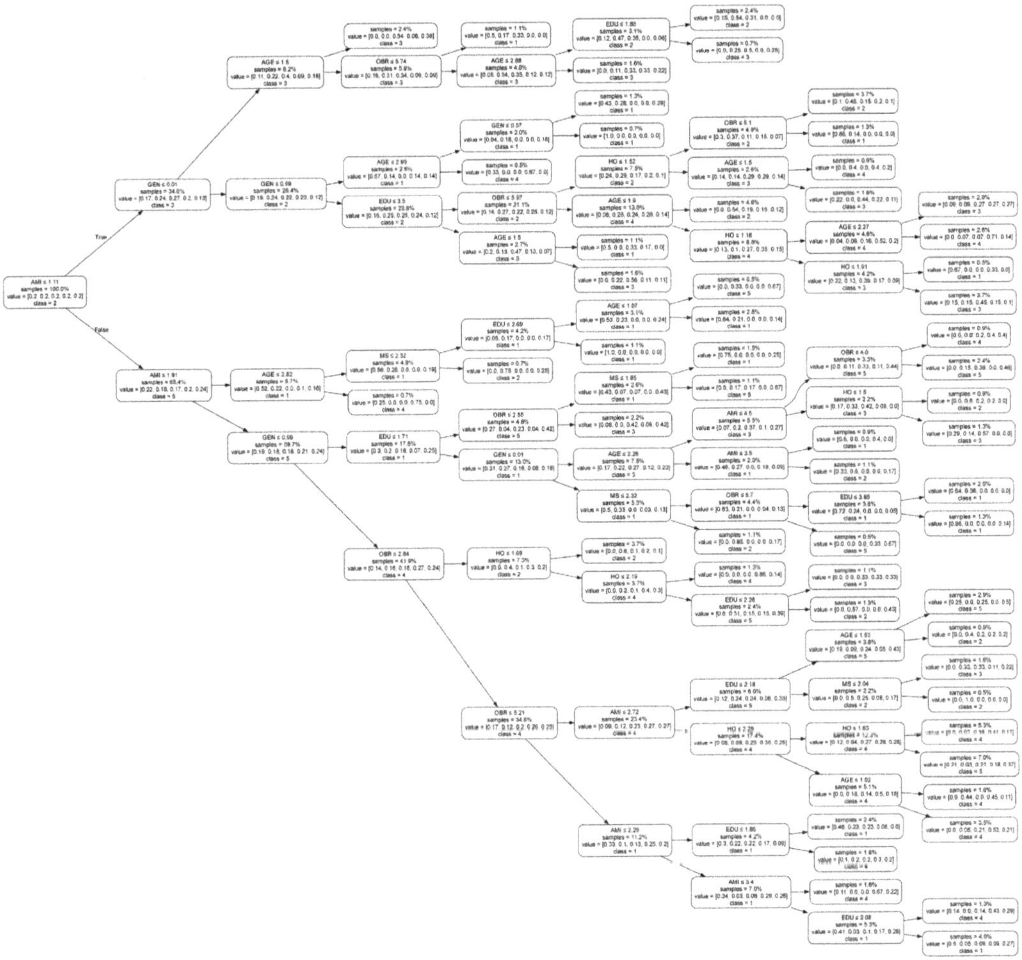

Figure A2. Group A Decision Tree. (Image source: drawn by the author).

https://doi.org/10.3390/buildings16010172

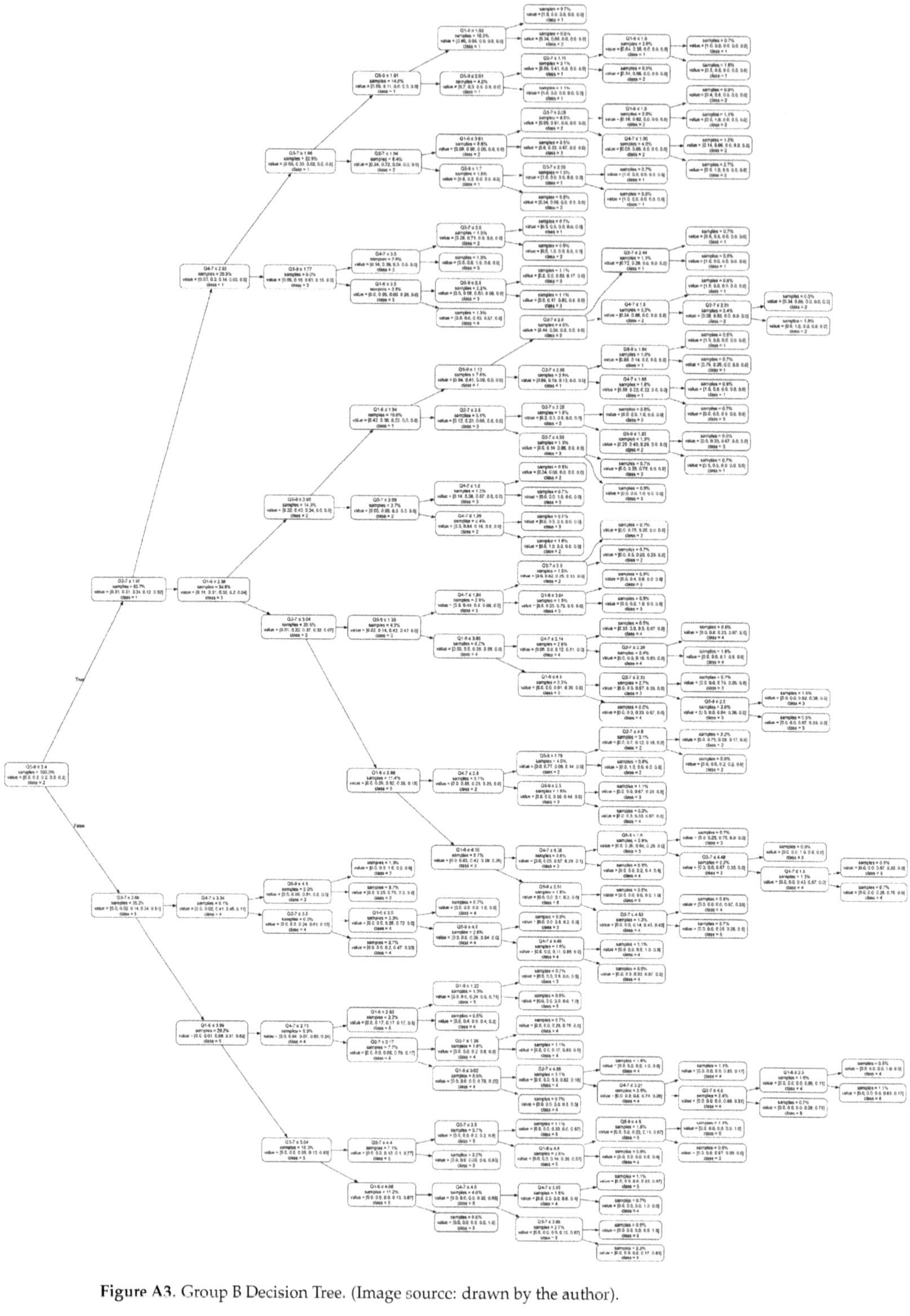

Figure A3. Group B Decision Tree. (Image source: drawn by the author).

https://doi.org/10.3390/buildings16010172

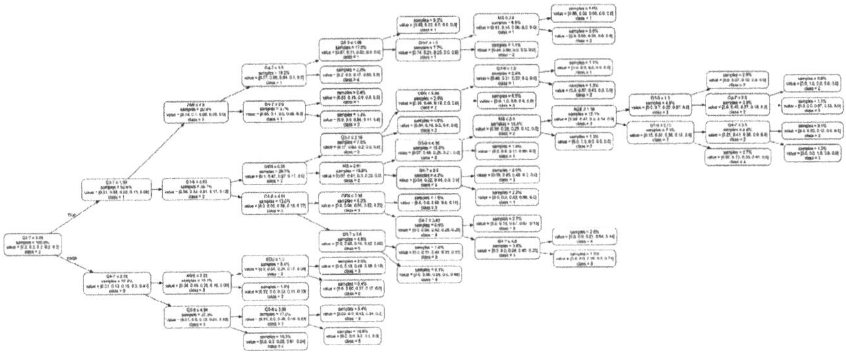

Figure A4. Group C Decision Tree. (Image source: drawn by the author).

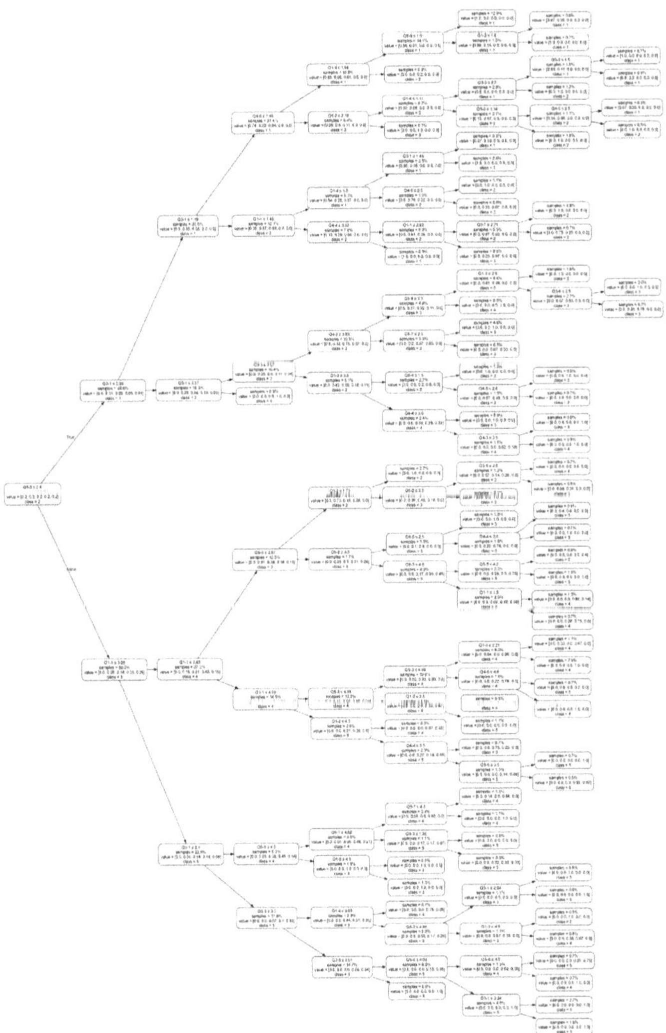

Figure A5. Group D Decision Tree. (Image source: drawn by the author).

https://doi.org/10.3390/buildings16010172

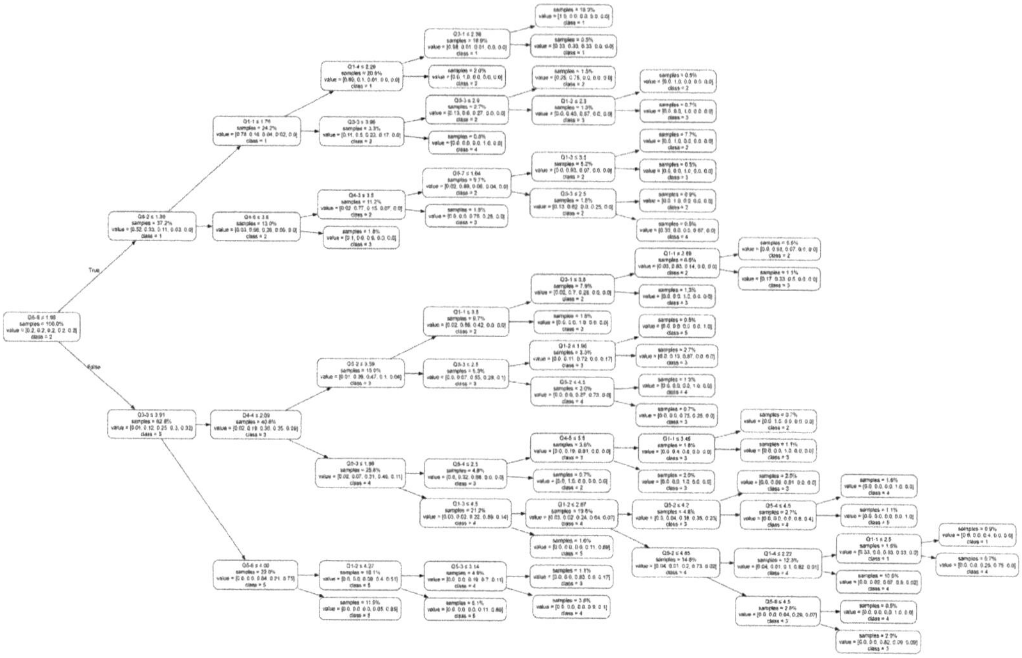

Figure A6. Group E Decision Tree. (Image source: drawn by the author).

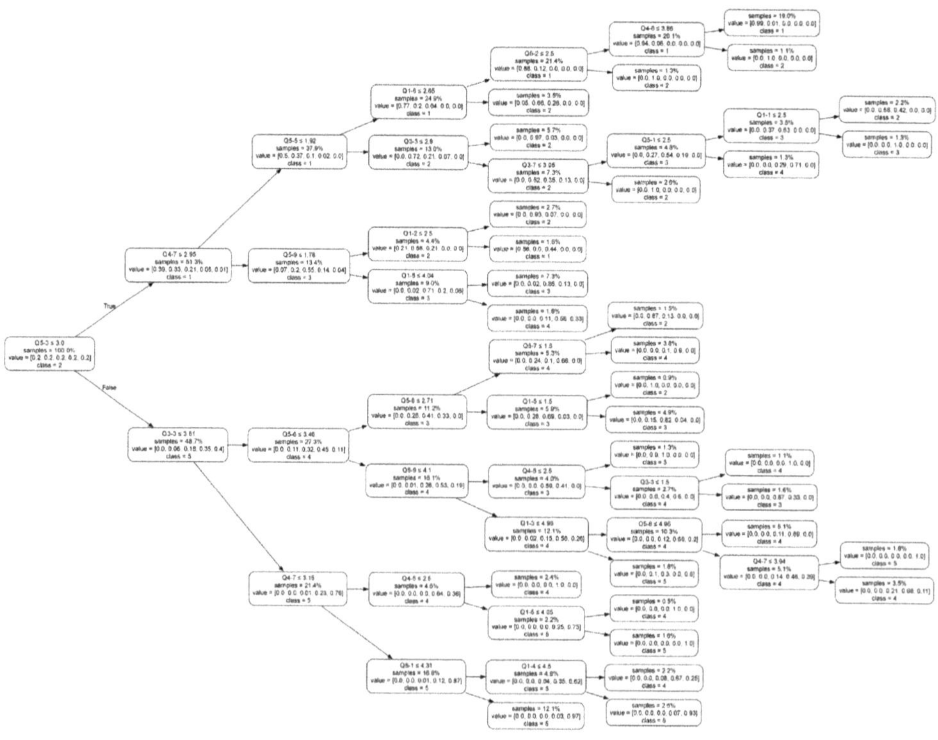

Figure A7. Group F Decision Tree. (Image source: drawn by the author).

Appendix C. Detailed Composition of Variable Sets Used in the Analysis

Table A2. Detailed Composition of Variable Sets used in the Analysis.

Variable Set	Included Variables
Respondent	Gender; Age; Marital Status; Housing Ownership; Education; Occupation Before Retirement; Average Monthly Income.
Major Category	Approach satisfaction rating; Satisfaction rating of amenities; Satisfaction with interesting; Satisfaction with safety; Satisfaction with comfort.
Subcategory	Smooth and easy to navigate; Barrier-free; Water-friendly environment; Convenient transportation; Open all day; Adequate rest facilities; Complete sunshade facilities; Complete public facilities; Convenient and easy to find restrooms; Adequate welfare facilities; Complete sports facilities; Abundant shopping district; Abundant entertainment venues; Proximity to cultural heritage sites; Convenient food options; Leisure facilities for interesting spaces; Number of small shops; No illegal parking; Appropriate speed; Nighttime lighting; Non-slip road surface; Clear warning signs; Complete evacuation facilities; Satisfied with greenery; Satisfied with sunlight; No noise; Open view; Good hygiene; Private and comfortable; No odor; Reasonable prices.

Source: Author's statistics.

References

1. Yung, E.H.K.; Conejos, S.; Chan, E.H.W. Social needs of the elderly and active aging in public open spaces in urban renewal. *Cities* **2016**, *52*, 114–122. [CrossRef]
2. Douglas, O.; Lennon, M.; Scott, M. Green space benefits for health and well-being: A life-course approach for urban planning, design and management. *Cities* **2017**, *66*, 53–62. [CrossRef]
3. Ajayi, A.O.; Amole, O.O. Open spaces and wellbeing: The impact of outdoor environments in promoting health. *Cities Health* **2022**, *6*, 1106–1121. [CrossRef]
4. Caronte-Veisz, A. Healthy Aging in Metropolitan Landscapes: Projects and Strategies to Promote the Quality of Life for Seniors. 2022. Available online: https://digital.casalini.it/9788822913005 (accessed on 25 September 2025).
5. Korea Has the Highest Old-Age Poverty Rate in the OECD. Available online: https://www.oecd.org/en/publications/oecd-economic-surveys-korea-2020_2ddc9180-en.html (accessed on 4 December 2025).
6. South Korea Now Officially 'Super-Aged' Society. Available online: https://www.koreaherald.com/article/10022641 (accessed on 25 September 2025).
7. Seoul Metropolitan Government: Statistics on the Status of Senior Citizens in Seoul. Available online: https://data.seoul.go.kr/dataList/10730/S/2/datasetView.do (accessed on 25 September 2025). (In Korean)
8. Statistics Korea: Elderly Poverty Rate. Available online: https://www.index.go.kr/unify/idx-info.do?idxCd=5057 (accessed on 25 September 2025). (In Korean)
9. Kang, J.; Park, J.; Cho, J. Inclusive Aging in Korea: Eradicating Senior Poverty. *Int. J. Environ. Res. Public Health* **2022**, *19*, 2121. [CrossRef]
10. Wang, S.; Yung, E.H.K.; Sun, Y. Effects of open space accessibility and quality on older adults' visit: Planning towards equal right to the city. *Cities* **2022**, *125*, 103611. [CrossRef]
11. Park, G.R.; Seo, B.K. Housing cost burden and material hardship among older adults: How do they influence psychological health? *Int. J. Geriatr. Psychiatry* **2022**, *37*, 1–9. [CrossRef]
12. Kim, Y.B.; Lee, S.H. Gender Differences in Correlates of Loneliness among Community-Dwelling Older Koreans. *Int. J. Environ. Res. Public Health* **2022**, *19*, 7334. [CrossRef] [PubMed]
13. Hwang, H.A.; Jeong, H.; Yim, H.W. Health-related quality of life of Korean older adults according to age, sex, and living arrangements: A cross-sectional study. *Front. Public Health* **2023**, *11*, 1281457. [CrossRef] [PubMed]
14. Seoul Welfare Foundation's Age-Friendly Seoul City 2.0 Report. 2025. Available online: https://welfare.seoul.kr/web/contents/archive1-1.do?&schM=view&page=1&viewCount=9&id=26095&schBdcode=&schGroupCode (accessed on 4 December 2025).
15. Jung, Y.J.; Chae, S.D. Analysis of usage behavior and preference in senior-friendly parks for the aging society. *J. People Plants Environ.* **2019**, *22*, 665–676. [CrossRef]
16. Han, S.M. Heritage conservation as urban regeneration policy in globalizing cities: Social exclusion and gentrification at the vicinity of the Jongmyo World Heritage site in Seoul, Korea. In Proceedings of the ICOMOS Scientific Symposium, Dublin, Ireland, 30 October 2010; p. 56. Available online: https://publ.icomos.org/publicomos/jlbSai?html=Bur&base=technica&ref=43605&file=2254.pdf&path=ICOMOS_Scientific_Symposium_DUBLIN_2010.pdf (accessed on 27 December 2025).
17. Seoul Creates Elderly Friendly Street Around Tapgol Park. Available online: https://english.seoul.go.kr/seoul-creates-elderly-friendly-street-around-tapgol-park (accessed on 4 December 2025).

https://doi.org/10.3390/buildings16010172

18. Tapgol Park 'Sacred Domain' Plan Proposal. Available online: https://archives.seoul.go.kr/item/1075943 (accessed on 12 December 2025).
19. Han, S.-M. A Study on the Place Identity of Tapgol Park—Focused on the Phenomena after Sacralization Project. *J. Korean Inst. Landsc. Archit.* **2016**, *44*, 25–36. [CrossRef]
20. Tapgol Park Becomes an Alcohol-Free Zone. Available online: https://app.jongno.go.kr/main/news/BBSMSTR_000000000271/257412 (accessed on 12 December 2025).
21. Debates on Social Equity in Low-Income Older Adults' Use of Public Open Spaces. Available online: https://www.news1.kr/society/incident-accident/6003258 (accessed on 12 December 2025).
22. Choung, E.-h. Space for the Elderly from the Perspective of Social Exclusion—Focusing on Tapgol Park and Rakhui Street in Jongno-gu. *J. Photo Geogr.* **2021**, *31*, 21–37. [CrossRef]
23. Kim, J.A.; Choi, S.M. An interpretative phenomenological analysis study on the experiences of the elderly living through the Corona Pandemic Crisis-Focused on the elderly male using Tapgol Park. *Korean J. Couns. Case Stud. Pract.* **2021**, *6*, 19–45. (In Korean) [CrossRef]
24. Tapgol Park Sees Many Elderly People Collect Free Food Every Day. Available online: https://v.daum.net/v/20231204144342359 (accessed on 12 December 2025).
25. Many Low-Income Seniors Gather at Tapgol Park Daily. Available online: https://www.donga.com/news/Opinion/article/all/20250831/132291151/2 (accessed on 12 December 2025).
26. Bonaccorsi, G.; Manzi, F.; Del Riccio, M.; Setola, N.; Naldi, E.; Milani, C.; Giorgetti, D.; Dellisanti, C.; Lorini, C. Impact of the Built Environment and the Neighborhood in Promoting the Physical Activity and the Healthy Aging in Older People: An Umbrella Review. *Int. J. Environ. Res. Public Health* **2020**, *17*, 6127. [CrossRef] [PubMed]
27. Mouratidis, K. Urban planning and quality of life: A review of pathways linking the built environment to subjective well-being. *Cities* **2021**, *115*, 103229. [CrossRef]
28. Fu, G.; Gai, Y.; Xiang, L.; Lin, L. Quantifying Older Adults' Spatial Perceptions of Outdoor Activity Areas for Embedded Retirement Facilities. *Buildings* **2025**, *15*, 271. [CrossRef]
29. Liu, F.; Zhou, H.; Xie, J.; Tang, Y.; Liu, S. Impact of Public Space in Primary and Secondary Schools Based on Natural Visibility Ratio. *Buildings* **2025**, *15*, 1472. [CrossRef]
30. Dennis, M.; Cook, P.A.; James, P.; Wheater, C.P.; Lindley, S.J. Relationships between health outcomes in older populations and urban green infrastructure size, quality and proximity. *BMC Public Health* **2020**, *20*, 626. [CrossRef]
31. Enssle, F.; Kabisch, N. Urban green spaces for the social interaction, health and well-being of older people—An integrated view of urban ecosystem services and socio-environmental justice. *Environ. Sci. Policy* **2020**, *109*, 36–44. [CrossRef]
32. Xian, Z.; Nakaya, T.; Liu, K.; Zhao, B.; Zhang, J.; Zhang, J.; Lin, Y. The effects of neighbourhood green spaces on mental health of disadvantaged groups: A systematic review. *Humanit. Soc. Sci. Commun.* **2024**, *11*, 488. [CrossRef]
33. Xiao, J.X.; Liao, J.; Zhao, B.; Long, Y.; Xu, X.; Liang, X.; Xia, T. The influence of community park characteristics on satisfaction in Guangzhou: Moderating and mediating effects analysis. *Heliyon* **2024**, *10*, e31043. [CrossRef]
34. Lak, A.; Khodakarim, S.; Myint, P.K.; Baradaran, H.R. The influencing factors of elder-friendly public open spaces promoting older adults' health in deprived urban neighborhoods: Partial Least Square Structural Equation Modeling approach. *Front. Public Health* **2023**, *11*, 1143289. [CrossRef]
35. Wen, F.; Peng, L.; Zhang, B.; Zhang, Y.; Zhang, Y. Satisfaction with Nighttime Outdoor Activity Environment Among Elderly Residents in Old Gated Communities. *Buildings* **2025**, *15*, 277. [CrossRef]
36. Jacobs, N.; Deforche, B.; Van Puyvelde, A.; Van Dyck, D.; Mertens, L.; Veitch, J.; Van Cauwenberg, J. Exploring emotions in relation to neighborhood environmental characteristics among older adults with low socioeconomic position: A qualitative study using walk-along interviews. *Cities* **2024**, *149*, 104950. [CrossRef]
37. Luo, J.; Ghim, G. A Study on the Elderly Friendly Design of Community Public Space for Aging in Place—Centered on around elderly people living in the central area of Shanghai. *J. Korea Intitute Spat. Des.* **2023**, *18*, 385–398. (In Korean) [CrossRef]
38. Zhang, C.; Loukaitou-Sideris, A. Chinese American older adults in urban and suburban public spaces: A comparative examination of two Los Angeles settings. *J. Urban Des.* **2025**, *30*, 203–222. [CrossRef]
39. Yang, J.; Shang, Y.; Cao, F.; Ying, H.; Luo, Y. Identify the direct and indirect impacts of the community built environment on the health of older adults. *Front. Public Health* **2025**, *13*, 1478337. [CrossRef] [PubMed]
40. Almahmood, M.A.M.; Scharnhorst, E.; Carstensen, T.A.; Jørgensen, G.; Schulze, O. Mapping the gendered city: Investigating the socio-cultural influence on the practice of walking and the meaning of walkscapes among young Saudi adults in Riyadh. *J. Urban Des.* **2017**, *22*, 229–248. [CrossRef]
41. Starkweather, S. Gender, perceptions of safety and strategic responses among Ohio University students. *Gend. Place Cult.* **2007**, *14*, 355–370. [CrossRef]
42. Li, J.; Tian, L.; Ouyang, W. Exploring the relationship between neighborhood-built environment and elderly health: A research based on heterogeneity of age and gender groups in Beijing. *Front. Public Health* **2022**, *10*, 882361. [CrossRef]

 https://doi.org/10.3390/buildings16010172

43. Navarrete-Hernandez, P.; Vetro, A.; Concha, P. Building safer public spaces: Exploring gender difference in the perception of safety in public space through urban design interventions. *Landsc. Urban Plan.* **2021**, *214*, 104180. [CrossRef]
44. Loukaitou-Sideris, A.; Levy-Storms, L.; Chen, L.; Brozen, M. Parks for an aging population: Needs and preferences of low-income seniors in Los Angeles. *J. Am. Plan. Assoc.* **2016**, *82*, 236–251. [CrossRef]
45. Zhang, Y.; Chen, G.; He, Y.; Jiang, X.; Xue, C. Social Interaction in Public Spaces and Well-Being among Elderly Women: Towards Age-Friendly Urban Environments. *Int. J. Environ. Res. Public Health* **2022**, *19*, 746. [CrossRef]
46. WHO. UN Decade of Healthy Ageing 2021–2030. 2021. Available online: https://unece.org/sites/default/files/2023-11/9_WHO_UN%20Decade%20of%20Healthy%20Ageing.pdf (accessed on 7 December 2025).
47. Shen, X.; Yin, F.; Jiao, C. Predictive Models of Life Satisfaction in Older People: A Machine Learning Approach. *Int. J. Environ. Res. Public Health* **2023**, *20*, 2445. [CrossRef]
48. Wang, L.; Sun, X.; Yan, J.; Xie, M.; Qin, H. Quantitative Assessment of Age-Friendly Design in Mountainous Urban Community Parks Based on Nonlinear Models: An Empirical Study in Chongqing, China. *Land* **2025**, *14*, 893. [CrossRef]
49. Wang, L.; Han, X.; He, J.; Jung, T. Measuring residents' perceptions of city streets to inform better street planning through deep learning and space syntax. *ISPRS J. Photogramm. Remote Sens.* **2022**, *190*, 215–230. [CrossRef]
50. Wang, R.; Lu, Y.; Zhang, J.; Liu, P.; Yao, Y.; Liu, Y. The relationship between visual enclosure for neighbourhood street walkability and elders' mental health in China: Using Street view images. *J. Transp. Health* **2019**, *13*, 90–102. [CrossRef]
51. Zhao, H.; Feng, G.; Zhao, W.; Wang, Y.; Chen, F. Analyzing Urban Parks for Older Adults' Accessibility in Summer Using Gradient Boosting Decision Trees: A Case Study from Tianjin, China. *Land* **2025**, *14*, 185. [CrossRef]
52. Cheng, L.; De Vos, J.; Zhao, P.; Yang, M.; Witlox, F. Examining non-linear built environment effects on elderly's walking: A random forest approach. *Transp. Res. Part D Transp. Environ.* **2020**, *88*, 102552. [CrossRef]
53. Yang, S.; Chen, Y.; Huang, Y.; Zheng, L.; Huang, Y. Investigating the Satisfaction of Residents in the Historic Center of Macau and the Characteristics of the Townscape: A Decision Tree Approach to Machine Learning. *Buildings* **2024**, *14*, 2925. [CrossRef]
54. The Protected Area of Tapgol Park. Available online: https://digital.khs.go.kr/heri/heriDetail.do?ctptNo=1331103540000&ctptUid=13898859677103300964 (accessed on 7 December 2025).
55. The Land Area of Seoul. Available online: https://en.wikipedia.org/wiki/Seoul (accessed on 7 December 2025).
56. The Land Area of Jongno-gu. Available online: https://www.jongno.go.kr/English.do?menuId=6101&menuNo=6101 (accessed on 7 December 2025).
57. Yoon, Y. An Interpretation of the Context Regarding Older People's Use of Outdoor Spaces: Focused on the Neighborhood Environment of Low-Income, Elderly Concentrated Residential Area. Ph.D. Thesis, Seoul National University Graduate School, Seoul, Republic of Korea, 2024. Available online: https://s-space.snu.ac.kr/handle/10371/215887 (accessed on 7 December 2025). (In Korean)
58. Gehl, J. "Three Types of Outdoor Activities," "Life Between Buildings," and "Outdoor Activities and the Quality of Outdoor Space". In *Life Between Buildings: Using Public Space*, 6th ed.; The City Reader; Routledge: London, UK, 2011; pp. 593–602.
59. UN-Habitat. SDG Indicator 11.7.1 Training Module: Public Space. United Nations Human Settlement Programme (UN-Habitat), Nairobi. 2018. Available online: https://data.unhabitat.org/documents/638d11fbfc464ca9a3289202076c128d (accessed on 27 December 2025).
60. Gagliardi, C.; Spazzafumo, L.; Marcellini, F.; Mollenkopf, H.; Ruoppila, I.; Tacken, M.; Szemann, Z. The outdoor mobility and leisure activities of older people in five European countries. *Ageing Soc.* **2007**, *27*, 683–700. [CrossRef]
61. Sugiyama, T.; Thompson, C.W. Outdoor environments, activity and the well-being of older people: Conceptualising environmental support. *Environ. Plan. A* **2007**, *39*, 1943–1960. [CrossRef]
62. Ryu, J.; Heo, J. Relationships between leisure activity types and well-being in older adults. *Leis. Stud.* **2018**, *37*, 331–342. [CrossRef]
63. Zhou, F.; Zhang, H.; Wang, H.Y.; Liu, L.F.; Zhang, X.G. Barriers and facilitators to older adult participation in intergenerational physical activity program: A systematic review. *Aging Clin. Exp. Res.* **2024**, *36*, 39. [CrossRef] [PubMed]
64. Guo, N.; Xia, F.; Yu, S. Enhancing Elderly Well-Being: Exploring Interactions between Neighborhood-Built Environment and Outdoor Activities in Old Urban Area. *Buildings* **2024**, *14*, 2845. [CrossRef]
65. Park, H.S.; Jung, Y.J.; Lee, K.H. Effects of neighborhood environment on elderly's walking route and outdoor activities. *J. Urban Des. Inst. Korea Urban Des.* **2017**, *18*, 17–35. [CrossRef]
66. Sugiyama, M.; Chau, H.W.; Abe, T.; Kato, Y.; Jamei, E.; Veeroja, P.; Mori, K.; Sugiyama, T. Third places for older adults' social engagement: A scoping review and research agenda. *Gerontologist* **2023**, *63*, 1149–1161. [CrossRef] [PubMed]
67. Schmidt, T.; Kerr, J.; Schipperijn, J. Associations between Neighborhood Open Space Features and Walking and Social Interaction in Older Adults—A Mixed Methods Study. *Geriatrics* **2019**, *4*, 41. [CrossRef]
68. Paudel, C.; Timperio, A.; Loh, V.; Deforche, B.; Salmon, J.; Veitch, J. Understanding the relative importance of micro-level design characteristics of walking paths in parks to promote walking among older adults. *Urban For. Urban Green.* **2023**, *89*, 128129. [CrossRef]
69. Sallis, J.F.; Cerin, E.; Conway, T.L.; Adams, M.A.; Frank, L.D.; Pratt, M.; Salvo, D.; Schipperijn, J.; Smith, G.; Cain, K.L.; et al. Physical activity in relation to urban environments in 14 cities worldwide: A cross-sectional study. *Lancet* **2016**, *387*, 2207–2217. [CrossRef]

70. Macintyre, S.; Macdonald, L.; Ellaway, A. Do poorer people have poorer access to local resources and facilities? The distribution of local resources by area deprivation in Glasgow, Scotland. *Soc. Sci. Med.* **2008**, *67*, 900–914. [CrossRef]
71. Kim, Y.S. Comparison of the decision tree, artificial neural network, and linear regression methods based on the number and types of independent variables and sample size. *Expert Syst. Appl.* **2008**, *34*, 1227–1234. [CrossRef]
72. Widodo, S.; Brawijaya, H.; Samudi, S. Stratified K-fold cross validation optimization on machine learning for prediction. *Sink. J. Dan Penelit. Tek. Inform.* **2022**, *6*, 2407–2414. [CrossRef]
73. Steinberg, D. CART: Classification and regression trees. In *The Top Ten Algorithms in Data Mining*; Chapman and Hall/CRC: Boca Raton, FL, USA, 2009; pp. 193–216.
74. Kumar, B.; Tamkute, P.; Saurabh, K.; Mishra, A.; Kumar, S.; Talesara, A. A Machine Learning Framework for Breast Cancer Detection and Classification. In *Robotics, Control and Computer Vision*; Muthusamy, H., Botzheim, J., Nayak, R., Eds.; Lecture Notes in Electrical Engineering; Springer: Singapore, 2023; Volume 1009. [CrossRef]
75. He, Y.; Xiao, L. Structured pruning for deep convolutional neural networks: A survey. *IEEE Trans. Pattern Anal. Mach. Intell.* **2023**, *46*, 2900–2919. [CrossRef] [PubMed]
76. Bademci, V. Cronbach's alpha is not a measure of unidimensionality or homogeneity. *J. Comput. Educ. Res.* **2014**, *2*, 19–27.
77. Lubiano, M.A.; Montenegro, M.; Pérez-Fernández, S.; Gil, M.Á. *Trends in Mathematical, Information and Data Sciences; Studies in Systems, Decision and Control*; Balakrishnan, N., Gil, M.Á., Martín, N., Morales, D., Pardo, M.d.C., Eds.; Springer: Cham, Switzerland, 2022; Volume 445, pp. 377–388. [CrossRef]

https://doi.org/10.3390/buildings16010172

buildings

MDPI

Article

Day–Night and Weekday–Weekend Heterogeneity in Built Environment Impacts on Public Space Vitality: A GWRF Analysis in Yuexiu District

Yingqian Yang [1,2,*], Xiuhong Lin [1,*], Xin Li [1], Qiufan Chen [1,3] and Xiaoli Sun [4]

[1] School of Urban Construction and Resource Exploration, Guangzhou Huali College, Guangzhou 511325, China
[2] Institute of Urban and Sustainable Development, City University of Macau, Macau 999078, China
[3] Faculty of Humanities and Arts, Macau University of Science and Technology, Macau 999078, China
[4] Department of Architecture, Qingdao City University, Qingdao 266000, China
* Correspondence: u23092120367@cityu.edu.mo (Y.Y.); u22092120351@cityu.edu.mo (X.L.)

Abstract

Existing studies on urban public space vitality predominantly focus on single temporal scales or macro-urban levels, lacking a systematic understanding of day–night and weekday–weekend differentiation patterns at the meso-scale. This study examines 149 public spaces in the Yuexiu District, Guangzhou, employing Baidu heatmap data and the geographically weighted random forest (GWRF) model to analyze built environment impacts across four temporal scenarios. The SHAP interaction analysis is incorporated to quantitatively evaluate factor interdependencies and their temporal variations. Findings reveal significant spatiotemporal heterogeneity. Building density shows greater night-time importance while residential density exhibits enhanced daytime importance, particularly on weekend. Weekday–weekend comparison demonstrates contrasting spatial reorganization patterns, with weekday showing divergence and weekend showing convergence in factor importance distributions. The factor interaction analysis highlights stable synergistic relationships between density and diversity, alongside temporal transitions in density–residential density interactions from competitive to synergistic during night-time. Low-vitality public spaces are concentrated in peripheral areas with high building density but insufficient commercial facilities and functional mix. These findings deepen our understanding of the spatiotemporal mechanisms underlying public space vitality generation and the interaction effects among built environment factors, thereby providing an empirical foundation for the formulation of temporally adaptive planning strategies.

Keywords: public space vitality; big data; daytime–night-time pattern; weekday–weekend pattern; GWRF; spatial heterogeneity; built environment

1. Introduction

Urban vitality reflects the quality of urban life and sustainable development [1,2]. Urban public spaces, as integral components of urban spatial structure, serve as primary venues where this vitality manifests most visibly through residents' diverse activities [1,3]. The theoretical understanding of public space vitality has evolved from early emphasis on physical design—as exemplified by Jane Jacobs's advocacy for diverse street life [4] and Jan Gehl's human-centered approach [5]—to contemporary recognition that public space vitality emerges from the dynamic interplay between the built environment and temporal patterns of human activities [6,7]. Their vitality characteristics are closely related to residents' well-being and socioeconomic equilibrium [7]. Understanding the spatiotemporal dynamics of public space vitality and its relationship with the built environment holds significant implications for urban planning and sustainable development.

Public spaces exhibit significant temporal differentiation in vitality levels, yet spatiotemporal coupling characteristics—the dynamic interaction between temporal usage patterns and spatial configurations—remain underexplored in existing research. Studies have revealed that community activities exhibit distinct temporal rhythms, with working residents' activities concentrated in morning and evening hours on weekdays, while facility operating schedules often fail to accommodate these patterns [8]. Research across multiple time scales has demonstrated that spatial segregation varies significantly throughout the day, week, and year, being stronger during evening hours, weekend, and winter months [9]. These temporal variations suggest that understanding public space vitality requires examining not only spatial patterns but also their evolution across different time periods.

The application of multi-source big data has enabled multi-temporal vitality research [2,10], providing new opportunities to examine spatiotemporal dynamics. However, existing studies exhibit limitations in temporal integration and spatial scale selection. Therefore, this study investigates the spatiotemporal heterogeneity of public space vitality and its built environment determinants at the district level, providing empirical evidence for developing differentiated planning strategies tailored to specific public space types and temporal contexts.

Consequently, this study focuses on the spatiotemporal differentiation characteristics of public space vitality in urban public spaces and proposes the following core research questions:

1. What spatial distribution patterns does public space vitality exhibit across four temporal scenarios (weekday/weekend × day/night)? What urban operational mechanisms underlie these spatiotemporal patterns?
2. How do built environment factors influence public space vitality across different temporal scenarios? What are the temporal variations and spatial heterogeneity in these influence mechanisms?
3. How can temporally adaptive and spatially differentiated planning strategies be formulated to enhance public space vitality based on the spatiotemporal coupling characteristics of built environment effects?

This study examines 149 public spaces in the Yuexiu District, Guangzhou, across four temporal scenarios (weekday daytime, weekday night-time, weekend daytime, and weekend night-time). The research integrates multi-source data, spatiotemporal analysis, and machine learning methods to investigate day–night vitality patterns and their influencing factors. The analytical framework consists of four interconnected components: First, mobile phone signaling data and multi-source-built environment data are acquired and processed. Second, vitality indicators and built environment factors are constructed and operationalized. Third, spatiotemporal patterns of public space vitality are analyzed through four-quadrant classification. Finally, the GWRF model is employed to reveal the spatiotemporal heterogeneity in how built environment factors influence vitality across different temporal scenarios.

2. Literature Review

2.1. Temporal Heterogeneity in Urban Public Space Vitality

Currently, there is substantial research on the vitality of urban public spaces [11]. Existing studies primarily focus on identifying the vitality characteristics of different

https://doi.org/10.3390/buildings16030523

types of public spaces and analyzing their influencing factors [7,12,13], while gradually expanding to comparative research across cities and across scales [14,15].

Scholars have examined vitality across various public space types, including streets, historic districts, parks, and waterfront spaces. Research has identified diverse influencing factors: street vitality is primarily driven by street width, permeability, and functional density [7,16]; historic district vitality varies with interface enclosure and accessibility [17]; park vitality exhibits significant spatial heterogeneity, highly dependent on adjacent residential communities and internal facilities [12]; and waterfront vitality demonstrates temporal rhythm influences, with spatial preferences varying across different vitality dimensions [13,18].

Research on temporal heterogeneity has progressed from single-period analysis to multi-temporal comparisons. Studies have examined day–night differences, revealing that daytime vitality is mainly driven by commercial density and thermal comfort, while night-time vitality depends more on the synergy between street width and shop transparency [19]. Weekday–weekend comparative studies have found that vitality exhibits similar clustering patterns across temporal periods [13], though the influences of vegetation, road density, and built environment factors show significant variations between weekdays and weekends in different spatial locations [20,21]. However, few studies have systematically integrated both day–night and weekday–weekend dimensions to explore the four temporal scenarios and their distinct mechanisms [22].

From a spatial scale perspective, existing research predominantly focuses on either macro-scale city-wide comparisons or micro-scale individual site analyses. Macro-scale studies have examined urban vitality across multiple cities [15] or within mega-cities [14], deconstructing vitality in dimensions such as density, accessibility, and diversity. Micro-scale research has concentrated on specific sites such as individual parks, plazas, or street segments [23], analyzing fine-grained spatial characteristics and user behavior. However, the meso-scale analysis at the district level remains under-explored [11].

2.2. Measurement and Modeling Methods for Urban Public Space Vitality

The scientific measurement of the vitality of urban public spaces is based on continuous innovations in data acquisition technologies and analytical modeling methods. From the perspective of data acquisition, research has evolved from traditional surveys to big data mining. Early studies relied primarily on field observations and questionnaire surveys to obtain spatial use data [24]. Although these methods could capture detailed behavioral information, they were limited in terms of spatial and temporal coverage. With the development of remote sensing technology, night-time light (NTL) data have been applied to quantify urban spatial vitality due to its wide coverage and strong temporal continuity [25]. The emergence of location-based big data has provided new possibilities for fine-grained measurement of public space vitality. Heat map data have been used to characterize the dynamic distribution of vitality in waterfront areas [13], while integration of mobile phone data and Points of Interest (POI) data has enabled more accurate measurement of the characteristics of diversity in street space [26]. Research on street spaces around metro stations has further integrated multi-source data, such as POI data and street view images, to construct three-dimensional evaluation frameworks based on deep learning [27].

Research methods have shifted from traditional regression to spatial regression, and from single- to multi-scale analysis. Multiple linear regression (MLR) models have been widely applied in early public space studies due to their simplicity [28–30]. However, traditional MLR assumes that the relationship between dependent and independent variables remains constant across space, neglecting the spatial heterogeneity and non-stationarity within public spaces. To overcome this limitation, geographically weighted regression

https://doi.org/10.3390/buildings16030523

(GWR) models have been introduced into public space research, revealing spatial variability in how built environment factors influence waterfront and street space vitality through localized regression [13,31]. Some studies have also proposed integrated frameworks that combine kernel density estimation (KDE), geographically and temporally weighted regression (GTWR) and the Herfindahl–Hirschman Index (HHI) to explore the spatiotemporal heterogeneity of public space vitality and its influencing factors [2]. However, the classical GWR assumes that all variables vary on the same spatial scale, overlooking the differences in spatial action scales between different factors that influence public space vitality, which may lead to estimation bias.

The application of geographically weighted random forest (GWRF) marks the entry of research into a new stage of multi-scale spatial heterogeneity analysis. GWRF effectively addresses the limitations of the classical GWR by allowing different bandwidths for each variable. A study in Shenzhen integrated NTL images and multi-source data into a multi-scale geographically weighted regression model (MGWR) to examine the impact of diversity indicators on urban vitality and identify the action scales of different variables [32]. In recent years, significant breakthroughs have been made in the application of GWRF at the micro-scale of public spaces. Research in the areas surrounding metro stations in Hangzhou employed GWRF models combined with SHAP interpretability methods, revealing the spatial heterogeneity and nonlinear relationships in how TOD characteristics influence station vitality. The study found that design elements (density of life and entertainment services), density elements (building density and height), and accessibility elements (density of public transport services) had the most significant impacts on vitality, with the GWRF model achieving an R^2 value (0.82) significantly higher than the traditional RF model (0.77) [33]. Similarly, research on park development potential in Shenzhen adopted the GWRF framework, integrating natural landscape elements such as vegetation coverage, slope, and elevation with economic activity elements, such as night-time light intensity, to achieve precise assessment of park site selection potential, with model accuracy (0.87) improving by 4% compared to the traditional RF model (0.83) [34].

In summary, while substantial progress has been made in the field of urban public space vitality, three key research gaps persist: First, temporal dimension analysis lacks integration. Although some studies have examined day–night or weekday–weekend differences separately, few have systematically integrated both dimensions to explore the four temporal scenarios (weekday daytime, weekday night-time, weekend daytime, and weekend night-time) and their distinct mechanisms. Second, spatial heterogeneity analysis in temporal contexts remains insufficient. While GWRF has demonstrated advantages in capturing multi-scale spatial heterogeneity, its application in analyzing how built environment impacts vary across different temporal scenarios is limited. Third, the meso-scale analysis at the district level is under-explored. Existing studies predominantly focus on either macro-scale city-wide comparisons or micro-scale individual site analyses, with limited attention to the meso-scale examination of multiple public spaces within a single administrative district. Moreover, few studies have systematically compared how vitality patterns and their temporal differentiation characteristics vary across different types of public spaces (e.g., parks, green spaces, plazas, and fitness facilities) within the same analytical framework.

3. Materials and Methods

3.1. Study Area

Guangzhou, as a national central city and core city of the Guangdong–Hong Kong–Macao Greater Bay Area, possesses a mature urban development system and comprehen-

https://doi.org/10.3390/buildings16030523

sive public space network. According to the "2021 Guangzhou National Economic and Social Development Statistical Bulletin" [35], Guangzhou's GDP reached CNY 2823.197 billion, ranking fourth nationally with an 8.1% year-on-year growth, providing a solid economic foundation for urban vitality development. Notably, Guangzhou's night-time economy ranks among the top in China, characterized by its commercialized nightlife landscape and cultural-tourism agglomeration zones [36,37], demonstrating balanced vitality between day and night periods.

Yuexiu District, one of Guangzhou's three traditional central urban areas, exhibits typical high-density urban characteristics. According to the Guangzhou Population Census Yearbook-2020 [38], Yuexiu District has a resident population density of 30,729 people/km², ranking first in the city. The district features mature built environments, highly mixed functions, and comprehensive public space systems, providing excellent research representativeness.

This study takes Yuexiu District's administrative boundary as the core area, extending outward by a 1000 m buffer zone, covering a total research area of 67.88 km² (Figure 1). Following existing literature's regional division methods and combining the "Yuexiu District Territorial Spatial Master Plan (2021–2035)" [39] with field surveys, urban public spaces are defined as publicly accessible urban spaces serving residents' daily exercise, leisure, and social needs, including sports and fitness facilities (sports grounds, fitness stations, and sports centers) and leisure and social spaces (parks, squares, and green spaces) [11,40].

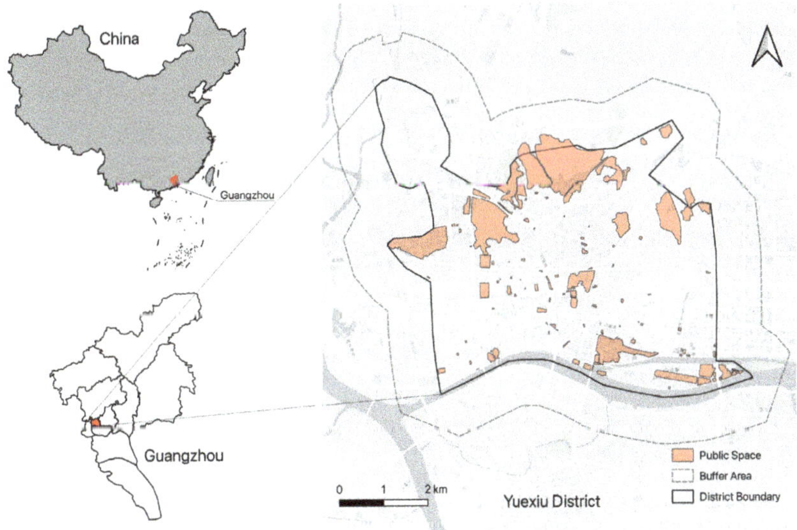

Figure 1. The location of the study area and public spaces.

3.2. Data Collection

This study employs a multi-source data integration approach to obtain spatiotemporal characteristic data of urban public spaces, primarily comprising Baidu Heat Map (BHM), Points of Interest (POI), building Area of Interest (AOI), and road network data. The BHM data are derived from Baidu Maps' location-based services (LBS), representing aggregated and anonymized mobile phone positioning data that visualize population density distributions. These data serve as the fundamental dataset to calculate the vitality of public spaces across day and night periods. Previous studies have demonstrated that the Baidu heat map data provide a reliable representation of spatiotemporal crowd activity patterns and have been effectively validated for urban spatial analysis applications [13,41].

https://doi.org/10.3390/buildings16030523

It is important to clarify the terminology used in this study. The term "heatmap" refers to a visualization technique that uses color gradients to represent data density or intensity, not actual temperature measurements. Throughout this paper, terms such as "heat values" are metaphorical expressions derived from the heatmap visualization method, representing population activity intensity rather than thermal comfort or temperature conditions. When discussing actual temperature-related factors (such as thermal comfort mentioned in the literature review), we explicitly use terms like "temperature" or "thermal comfort" to distinguish them from the metaphorical usage of "heat" in the context of population density visualization. For data characterizing urban public space environmental factors, POI data and road network data are sourced from OpenStreetMap, while building AOI vector data are obtained from Tencent Maps' first quarter 2024 dataset. To ensure data precision, all spatial data were verified and calibrated using high-resolution satellite imagery from 2024, supplemented by field survey data to validate functional attributes and usage patterns of public spaces, thus ensuring the reliability of research data.

Data collection was conducted on 15 October 2024 (Thursday) and 19 October 2024 (Saturday). The selection criteria for these dates were (1) suitable temperature conditions with similar weather patterns on both days, (2) absence of pandemic control measures or lockdown restrictions, and (3) non-holiday periods that could represent typical daily activities of Guangzhou residents over one weekday and one weekend day. Data sampling was performed hourly from 8:00 AM to 2:00 AM the following day, producing 17 data samples per day. The temporal periods were classified as daytime (8:00 AM to 6:00 PM) and night-time (6:00 PM to 2:00 AM the following day). The collected data were spatially visualized on the QGIS 3.40 (QGIS Development Team, 2024), and data points outside the defined study area were filtered to ensure spatial consistency with the scope of the investigation.

3.3. Methods

This section elaborates on the research methodology employed in this study. The overall research logic and analytical framework are first introduced to guide the investigation. The measurement procedures for quantifying public space vitality across different temporal scenarios are then described, followed by the selection and operationalization of built environment factors that potentially influence vitality. Finally, the geographically weighted random forest model is presented as the analytical method to examine the spatiotemporal heterogeneity of these influences.

3.3.1. Research Framework

Figure 2 illustrates the research framework, which integrates data acquisition, indicator construction, spatiotemporal pattern analysis, and GWRF modeling.

First, four types of spatial data sources are integrated: Baidu Heat Map (BHM), POI, road network, and building. The BHM data serve as the basis for quantifying public-space vitality across different temporal dimensions (weekday/weekend and daytime/night-time), whereas POI, road network, and building data are processed and categorized to construct an indicator system of factors influencing urban public-space vitality (see Table 1 for details).

The study adopts a dual-track analytical approach. The spatiotemporal track focuses on the distribution patterns of public-space vitality, including the identification of spatiotemporal patterns and the application of a four-quadrant classification method to recognize day–night vitality regimes. Concurrently, the GWRF model is employed to analyze the relationship between built-environment indicators and public-space vitality, thereby identifying key influencing factors and their spatial heterogeneity under different

https://doi.org/10.3390/buildings16030523

temporal contexts. By synthesizing findings from both tracks, the framework elucidates the temporal differentiation and spatial heterogeneity of public-space vitality.

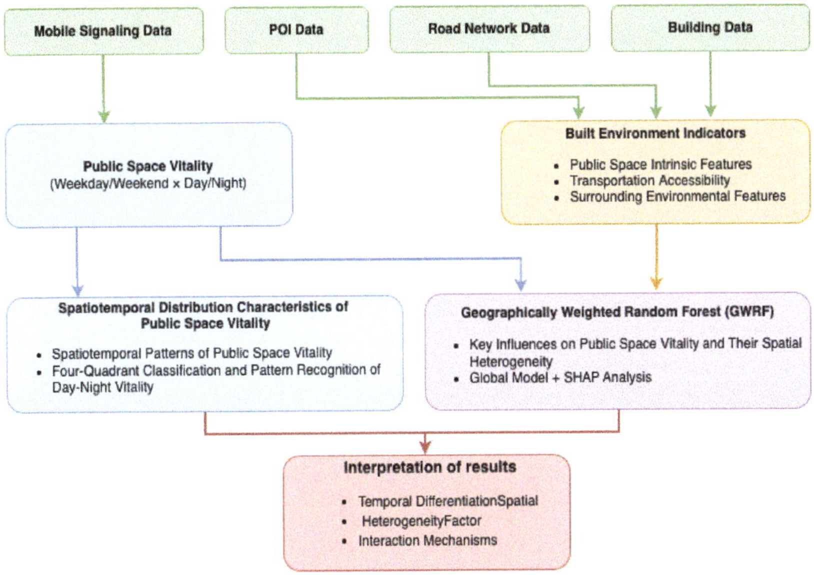

Figure 2. Research framework.

Table 1. Urban public space vitality impact factors indicator system.

Dimension	Variable	Calculation Method	Data Source
Public Space Intrinsic Features	Area	Public space unit area calculated using $A = \sum(\text{pixel area} \times \text{spatial resolution}^2)$	Remote sensing imagery
Transportation Accessibility	Road intersection density	$D_{\text{intersection}} = \frac{N_{\text{intersections}}}{A_{\text{service}}}$	OpenStreetMap
	Public transport stops density	$D_{\text{transit}} = \frac{N_{\text{stops}}}{A_{\text{service}}}$	OpenStreetMap
	Local integration	$I_{\text{local}} = \frac{1}{n}\sum_{i=1}^{n}\frac{1}{d_i}$	DepthMap
Surrounding Environmental Features	Total building density	$D_{\text{total}} = \frac{A_{\text{all\_buildings}}}{A_{\text{buffer}}}$ (1000 m buffer radius)	Tencent Map
	Residential density	$D_{\text{res}} = \frac{A_{\text{res\_buildings}}}{A_{\text{buffer}}}$ (1000 m buffer radius)	Tencent Map
	Residential proportion	$R_{\text{res}} = \frac{N_{\text{res\_buildings}}}{N_{\text{total\_buildings}}}$ (1000 m buffer radius)	Tencent Map
	POI density	$D_{\text{POI}} = \frac{N_{\text{POI}}}{A_{\text{service}}}$	OpenStreetMap
	Traffic density	$D_{\text{traffic}} = \frac{L_{\text{road}}}{A_{\text{service}}}$	OpenStreetMap
	Public service density	$D_{\text{services}} = \frac{N_{\text{services}}}{A_{\text{service}}}$	OpenStreetMap

A = Area; N = Number of facilities/points; D = Density; A_{service} = Service area; p_i = Proportion of category i; H = Shannon diversity index; $I_{\text{transport}}$ = Transport integration index.

3.3.2. Public Space Day–Night Vitality Measurement

The urban vitality values of this study are based on Baidu heatmap data, using the kernel density estimate (KDE) method to quantify spatial activity. The study area is divided

https://doi.org/10.3390/buildings16030523

into regular 30 × 30-m grids, and the vitality value for each grid cell is calculated using the following formula:

$$f(x) = \frac{1}{nh}\sum_{i=1}^{n} K\left(\frac{x - x_i}{h}\right) \quad (1)$$

where $f(x)$ is the density estimate at location x, n is the total number of data points, h is the bandwidth parameter, K is the Gaussian kernel function, and x_i is the location of the i-th data point. Vitality values reflect the frequency of population activities and the intensity of spatial usage within grid areas, providing comparable quantitative standards for spatial activity across different regions and periods. These values are dimensionless relative indices rather than absolute counts of individuals or devices, representing the aggregated intensity of human activities derived from mobile phone location-based services.

This study employs multi-source data fusion and spatial interpolation methods to process public space vitality data. First, the geographical information of 149 public spaces in the Yuexiu District was obtained from OpenStreetMap, with geometric centroids extracted and transformed to the UTM coordinate system (EPSG:32649) to ensure distance calculation accuracy. Second, the heatmap statistical data were loaded and aggregated by public space names to calculate average vitality values for daytime (8:00–18:00) and night-time (18:00–2:00) periods. Geographic and vitality data were merged through standardized name matching, with missing data estimated using the Inverse Distance Weighting (IDW) interpolation method, setting a search radius of 5 km, a maximum of 10 neighbors, and a power parameter of two. Finally, day–night vitality ratios were calculated to form a comprehensive dataset containing original data identifiers. This methodology effectively addresses data gaps, improves spatial coverage, and provides a reliable data foundation for subsequent public space vitality analysis.

3.3.3. Selection of Vitality Influencing Factors

Based on a review of the literature [7,13,17], the environmental characteristics influencing public space vitality can be categorized into three dimensions: intrinsic characteristics of the parcel, surrounding environmental characteristics, and transportation network attributes. Considering indicator applicability and data availability, this study selects 10 factors across these three dimensions to establish an indicator system for analyzing urban public space vitality (Table 1).

3.3.4. Geographically Weighted Random Forests

Geographically weighted random forest (GWRF) is an analytical method that integrates geographically weighted regression (GWR) with random forest (RF). This approach introduces a geographical weight matrix to dynamically adjust the contribution of neighboring spatial units during the decision tree construction process in random forest, effectively addressing the limitations of traditional global models that ignore spatial heterogeneity. Leveraging the ensemble learning mechanism of random forest, it automatically identifies complex nonlinear interactions among variables through collaborative exploration of high-dimensional feature space by multiple decision trees, quantifying the influence of different factors on vitality. This study employs the GWRF model to analyze the correlation between urban environmental characteristics and urban public space vitality, interpreting spatial heterogeneity in the data and determining how the importance of different variables varies across geographical space.

This study constructs a comprehensive indicator system comprising 10 independent variables: diversity score, local integration, total building density, residential density, residential proportion, POI density, traffic density, intersection density, public service density, and public space area. Using daytime and night-time vitality values as dependent variables, separate

https://doi.org/10.3390/buildings16030523

GWRF models were established for comparative diurnal analysis. The model employs a Gaussian kernel function to calculate spatial weights $w_{ij} = \exp(-0.5 \times (d_{ij}/h)^2)$, where bandwidth h is set as the median of all pairwise distances. For each target location, local random forest models are trained based on spatial weights (*n_estimators* = 100), extracting feature importance and calculating local R^2 values, ultimately obtaining global R^2 and RMSE to evaluate model performance. Through comparison of feature importance, local R^2 distribution, and prediction accuracy between daytime and night-time models, key factors influencing park vitality and their spatiotemporal variation characteristics are identified, with correlation analysis employed to validate the rationality of inter-variable relationships.

4. Results

4.1. Spatiotemporal Distribution Characteristics of Public Space Vitality

This subsection comprises two components: Section 4.1.1 presents the spatial distribution and statistical characteristics of vitality values across four temporal scenarios (weekday/weekend × daytime/night-time), revealing the overall spatiotemporal patterns. Section 4.1.2 applies a four-quadrant classification method to categorize public spaces based on their day–night vitality relationships, thereby identifying distinct temporal response patterns and examining how these patterns vary across different public space types.

4.1.1. Spatiotemporal Patterns of Public Space Vitality

Considering that the vitality of public spaces is not only reflected within their boundaries but also manifested through pedestrian activities in surrounding areas, which indicate the attractiveness and radiating effects of these spaces, this study establishes a 200-m buffer zone around each public space as the statistical unit. This distance corresponds to a walkable range (approximately 2–3 min walking time) and can encompass a sufficient number of 30 × 30 m heat map grid cells, ensuring the stability and representativeness of the statistical results.

For the 200 m buffer zone of each public space, all heat values from the 30 × 30 m grid points falling within the buffer area are extracted to form a heat value set $\{v_1, v_2, \ldots, v_N\}$, where v_i represents the heat value of the i-th grid point, and N denotes the total number of grid points within the buffer zone. Three statistical indicators are calculated:

$$\text{mean} = \frac{1}{N}\sum_{i=1}^{N} v_i. \tag{2}$$

The mean heat value (Equation (1)) eliminates the impact of area differences among public spaces, facilitating cross-comparisons between public spaces of different scales. The sum of heat values ($\sum_{i=1}^{N} v_i$) comprehensively considers both vitality intensity and spatial extent, better reflecting the attractiveness of public spaces. The count of heat points (N) indicates the data coverage within each buffer zone.

The mean heat values of public spaces in the Yuexiu District during daytime and night-time periods, calculated from the Baidu heat map data, were imported into QGIS for visualization and analysis (Figure 3). The map illustrates the spatial distribution patterns of public space vitality across four spatiotemporal scenarios: weekday vs. weekend, and daytime vs. night-time within the Yuexiu District. The visualization clearly reveals significant spatiotemporal heterogeneity in public space vitality. To facilitate visual comparison across different temporal scenarios, a unified color scale is applied to all four maps, with specific vitality value ranges for each period described in the text.

In general, the levels of vitality of the public space on weekend are higher than on weekday. The mean heat value of daytime vitality on weekend is 47.51, exceeding the weekday value of 40.33; the mean night-time value is 33.10, surpassing the weekday value

https://doi.org/10.3390/buildings16030523

of 26.49. These values represent dimensionless indices reflecting relative population activity density. From a day–night perspective, daytime vitality (mean: 40.33–47.51) is generally higher than night-time vitality (mean: 26.49–33.10), though the spatial distribution patterns differ significantly. During daytime hours (8:00–18:00), vitality values range from 2 to 194. During night-time hours (18:00–02:00), the overall vitality level decreases (range: 2.7–128), though the spatial distribution pattern shows notable differences compared to daytime.

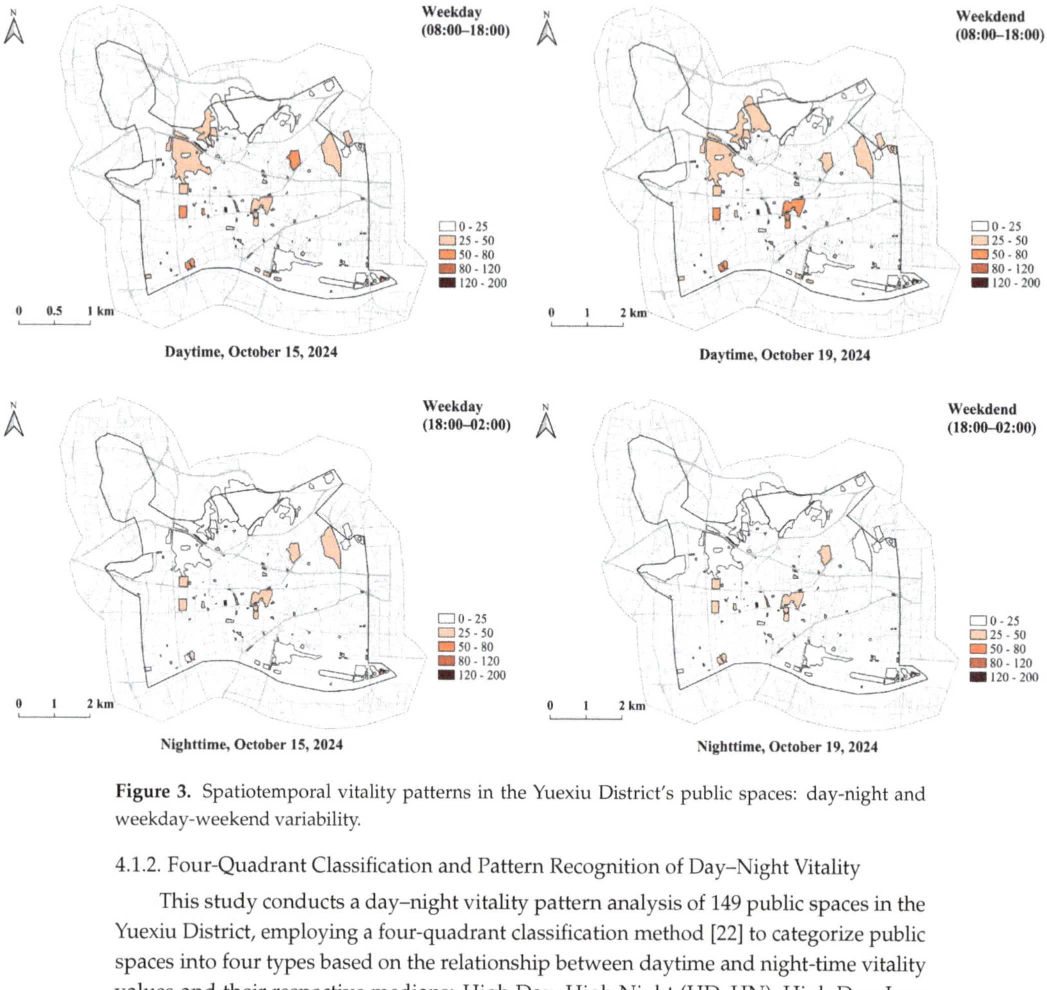

Figure 3. Spatiotemporal vitality patterns in the Yuexiu District's public spaces: day-night and weekday-weekend variability.

4.1.2. Four-Quadrant Classification and Pattern Recognition of Day–Night Vitality

This study conducts a day–night vitality pattern analysis of 149 public spaces in the Yuexiu District, employing a four-quadrant classification method [22] to categorize public spaces into four types based on the relationship between daytime and night-time vitality values and their respective medians: High Day–High Night (HD–HN), High Day–Low Night (HD–LN), Low Day–High Night (LD–HN), and Low Day–Low Night (LD–LN). During the data preprocessing stage, the Inverse Distance Weighting (IDW) interpolation method is applied to spatially complete the missing vitality values. Among the 149 public spaces, only three (2.0%) required IDW interpolation due to the absence of overlapping heatmap grid cells, while the remaining 146 (98.0%) obtained direct measurements from heatmap data. The IDW parameters (power = 2, max neighbors = 10, max distance = 5000 m) ensured spatial accuracy, with interpolated values falling within reasonable ranges consistent with the surrounding areas. This method fully considers the spatial distribution characteristics of heat values and significantly improves computational accuracy compared to simple statistical imputation methods. Through a multi-level missing value processing mechanism, all public spaces are ensured to be accurately classified. Based on this classification framework, this study conducts a systematic comparative analysis of vitality pattern distribution characteristics between weekday (15 October 2024) and weekend (19 October 2024). The classification results are spatially visualized using QGIS software, with different colors representing the spatial distribution patterns of the four vitality types across the Yuexiu District (Figure 4).

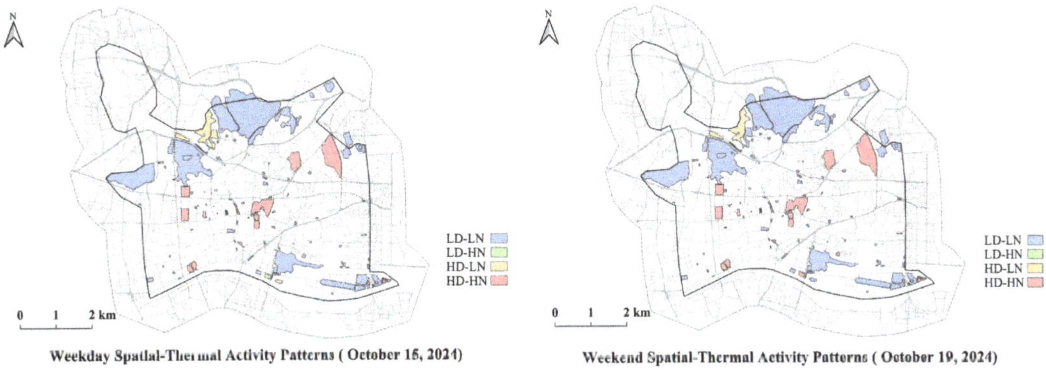

Figure 4. Four-quadrant classification of day–night vitality patterns in the Yuexiu District public spaces.

Spatial distributions reveal distinct patterns. LD–LN spaces predominantly concentrate in the northern and eastern peripheral areas of the Yuexiu District, indicating consistently low vitality throughout both day and night. HD–HN spaces scatter across central and southern districts, representing areas with sustained high activity levels. LD–HN and HD–LN spaces show limited distribution, suggesting that most public spaces do not experience dramatic day–night vitality transitions.

Public spaces in the Yuexiu District were extracted from OpenStreetMap (OSM) data and classified into six categories based on their functional attributes. A total of 149 public spaces were identified, distributed as follows: green spaces (n = 66, 44.3%), parks (n = 62, 41.6%), plazas (n = 7, 4.7%), fitness and sports facilities (n = 7, 4.7%), playgrounds (n = 3, 2.0%), and leisure venues (n = 3, 2.0%) (Figure 5a). Among these, green spaces and parks constitute the predominant public space types, accounting for 85.9% of the total distribution.

The comparative analysis between weekday and weekend reveals variations in the temporal stability of vitality patterns across different types of public spaces (Figure 5b). Green spaces demonstrate high temporal stability (change rate of 6.1%, defined as the proportion of samples whose vitality patterns changed between weekday and weekend), with the HD–HN pattern predominating on both weekday and weekend (56.1% and 57.6%, respectively). However, the persistent polarization phenomenon (39.4% LD–LN on weekday, 42.4% on weekend) indicates significant spatial heterogeneity and accessibility differences within the green space system. Parks also show consistent pattern distributions between weekday and weekend (change rate of 11.3%), characterized by the coexistence of two dominant patterns: LD–LN (approximately 50%) and HD–HN (approximately 35–40%), reflecting functional differentiation across different periods and locations. In contrast, Fitness and Squares exhibit more pronounced pattern changes with the arrival of weekend (change rates of 28.6% and 14.3%, respectively), demonstrating sensitivity to leisure time shifts. Playgrounds and Leisure facilities (each n = 3, with limited sample sizes introducing certain specificity) are predominantly characterized by LD–LN and LD–HN patterns during both time periods, with change rates of 0%.

https://doi.org/10.3390/buildings16030523

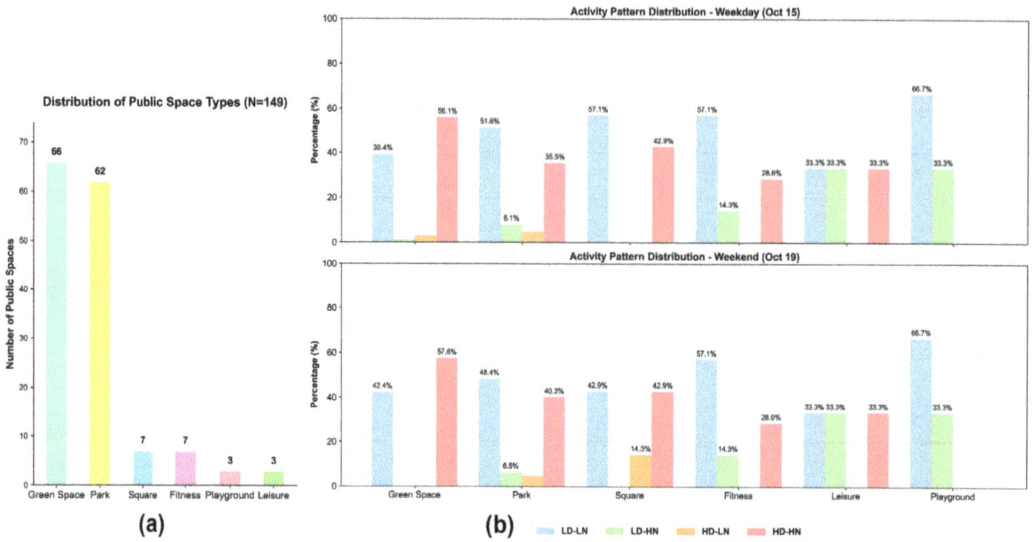

Figure 5. Analysis of day–night vitality patterns across public space types: (**a**) proportion of four-quadrant classifications by public space type and (**b**) comparison of four-quadrant classification across different types of public spaces: weekday vs. weekend.

4.2. GWRF-Based Analysis of Built Environment Effects and Spatiotemporal Patterns

This subsection comprises four components that sequentially address model validation, variable relationships, factor importance, and spatial heterogeneity. Section 4.2.1 presents the GWRF model construction process and performance evaluation results. Section 4.2.2 examines the correlation patterns among variables and their temporal variations. Section 4.2.3 identifies key influencing factors and analyzes their importance across day–night and weekday–weekend scenarios. Section 4.2.4 reveals the spatial heterogeneity of built environment effects, examining how factor importance varies across different locations within the district.

4.2.1. GWRF Model Construction and Performance Evaluation

Model Construction Workflow. The GWRF model construction follows a standard spatial machine learning workflow comprising four key steps. First, standardization was applied to 10 feature variables across 149 public space samples using Z-score transformation to achieve a mean of zero and standard deviation of one. Second, a Gaussian kernel function was adopted to calculate the spatial weight matrix, formulated as follows:

$$w_i = \exp\left(-0.5 \times \left(\frac{d_i}{h}\right)^2\right) \tag{3}$$

where d_i represents the Euclidean distance between sample points and h denotes the bandwidth parameter. This function ensures that neighboring sample points receive higher weights, embodying the spatial correlation principle of Tobler's First Law of Geography. Third, cross-validation was employed to determine the optimal spatial bandwidth of 2.76 km, balancing localized modeling precision with adequate sample size. Finally, a weighted random forest model was constructed for each spatial location with the following parameters: 100 trees, maximum depth of 10 layers, and minimum leaf node sample size of five.

https://doi.org/10.3390/buildings16030523

Model Validation and Performance. To improve the model's reliability and generalizability, commonly adopted validation practices in the geographically weighted modeling literature were followed [42,43]. Our validation approach comprises three components:

(1) goodness-of-fit metrics including global and local R^2, RMSE, and spatial stability assessment through Coefficient of Variation (CV) of local R^2 values; (2) spatial autocorrelation diagnostics using Moran's I tests at both pre-modeling and post-modeling stages to confirm adequate capture of spatial dependencies; (3) residual analysis examining spatial distribution patterns and statistical properties to detect systematic bias.

The GWRF model demonstrated superior predictive performance across all temporal periods, with global R^2 values exceeding 0.88 (Tables 2 and 3). Model performance varied systematically by time: weekday night-time models (R^2 = 0.9172) outperformed daytime models, while weekend models showed the reverse pattern. This reflects distinct spatial structures—night-time weekday vitality exhibits regular patterns linked to residential areas, whereas weekend daytime involves more complex, heterogeneous recreational and commercial activities, resulting in higher prediction errors.

Table 2. Daytime model performance comparison: weekday vs. weekend.

Evaluation Metrics	Weekday (Oct 15)	Weekend (Oct 19)	Difference	Change (%)
Model Performance				
Global R^2	0.9096 *	0.9060 *	−0.0036	−0.40%
RMSE	3.79	5.36	+1.57	+41.42%
Local R^2 Minimum	0.8769	0.8717	−0.0052	−0.59%
Local R^2 Mean	0.9014	0.8981	−0.0033	−0.37%
Local R^2 Median	0.8967	0.8959	−0.0008	−0.09%
Local R^2 Maximum	0.9336	0.9283	−0.0053	−0.57%
CV of Local R^2 (%)	1.52	1.53	+0.01	+0.66%
Spatial Autocorrelation				
Raw Vitality Moran's I	0.1457 **	0.1681 **	+0.0224	+15.38%
Residual Moran's I	−0.0315 (ns)	−0.0130 (ns)	—	—
Spatial Autocorrelation Reduction (%)	78.4	92.3	+13.9	—

* $p < 0.001$ for model significance; ** $p < 0.05$ for spatial autocorrelation; ns = not significant ($p > 0.05$). CV (Coefficient of Variation) measures the spatial stability of local R^2 values. CV < 2% indicates very high stability, and 2–3% indicates good stability. Spatial Autocorrelation Reduction = (Raw Moran's I − | Residual Moran's I |)/Raw Moran's I × 100%, indicating the percentage of spatial clustering eliminated by the GWRF model.

Table 3. Night-time model performance comparison: weekday vs. weekend.

Evaluation Metrics	Weekday (Oct 15)	Weekend (Oct 19)	Difference	Change (%)
Model Performance				
Global R^2	0.9172 *	0.8880 *	−0.0292	−3.18%
RMSE	2.09	3.79	+1.70	+81.34%
Local R^2 Minimum	0.8850	0.8357	−0.0493	−5.57%
Local R^2 Mean	0.9119	0.8810	−0.0309	−3.39%
Local R^2 Median	0.9093	0.8786	−0.0307	−3.38%
Local R^2 Maximum	0.9395	0.9214	−0.0181	−1.93%
CV of Local R^2 (%)	1.41	2.36	+0.95	+67.38%
Spatial Autocorrelation				
Raw Vitality Moran's I	0.1478 **	0.1444 **	−0.0034	−2.30%
Residual Moran's I	−0.0248 (ns)	0.0159 (ns)	—	—
Spatial Autocorrelation Reduction (%)	83.2	89.0	+5.8	—

* $p < 0.001$ for model significance; ** $p < 0.05$ for spatial autocorrelation; ns = not significant ($p > 0.05$). CV (Coefficient of Variation) measures the spatial stability of local R^2 values. CV < 2% indicates very high stability, and 2–3% indicates good stability. Spatial Autocorrelation Reduction = (Raw Moran's I − | Residual Moran's I |)/Raw Moran's I × 100%, indicating the percentage of spatial clustering eliminated by the GWRF model.

Temporal stability analysis revealed contrasting patterns. Daytime models maintained consistent performance across weekday and weekend (R^2 difference only 0.36%), while night-time models exhibited substantial variation (2.92 percentage points), with weekday

https://doi.org/10.3390/buildings16030523

performance considerably exceeding weekend. This confirms that daytime vitality follows stable spatial determinants regardless of day type, while night-time patterns are more temporally variable.

Diagnostic Analysis. Pre-modeling analysis revealed significant positive spatial autocorrelation in raw vitality data (Moran's I = 0.1444–0.1681, all $p < 0.05$), justifying geographically weighted approaches. Post-modeling diagnostics demonstrated successful elimination of spatial autocorrelation, with all residual Moran's I values becoming non-significant ($p > 0.05$) and near zero, representing 78.4–92.3% reduction compared to the original data (Figure 6). The reduction rate is calculated as: Reduction (%) = $(|I_{raw}| - |I_{residual}|) / |I_{raw}| \times 100\%$, where I_{raw} is the Moran's I of original data and $I_{residual}$ is the Moran's I of model residuals.

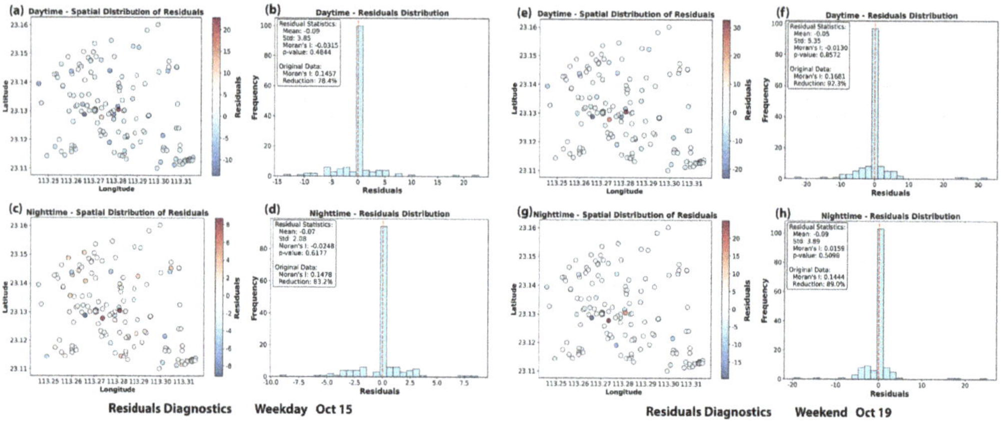

Figure 6. Spatial diagnostics of GWRF model residuals: (**a–d**) weekday models; (**e–h**) weekend models; (**a**,**c**,**e**,**g**) spatial distribution maps; and (**b**,**d**,**f**,**h**) residual histograms. Red dashed lines in residual histograms mark the zero baseline for reference.

Local R^2 distributions indicated consistent spatial stability (CV: 1.41–2.36%), with weekday models showing slightly tighter distributions than weekend models (Figure 7). Visual inspection of residuals revealed random spatial patterns without clustering, near-normal distributions centered at zero, and strong linear relationships in predicted vs. actual scatter plots (Figure 8). These diagnostics collectively confirm that GWRF achieves high predictive accuracy while satisfying fundamental spatial modeling assumptions, ensuring reliable interpretation of spatially varying relationships between urban form and vitality patterns.

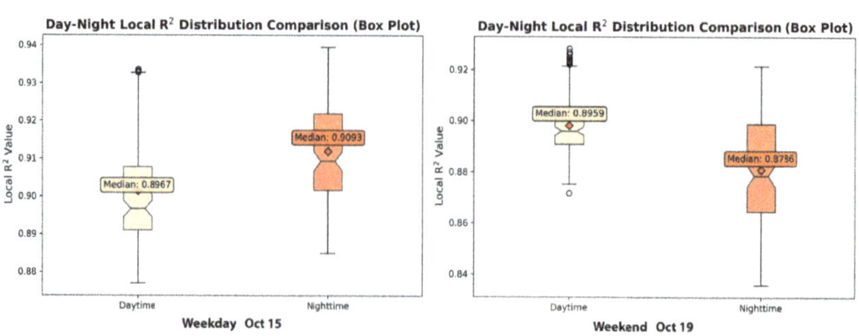

Figure 7. Box Plot Comparison of local R^2 distribution between daytime and night-time models.

https://doi.org/10.3390/buildings16030523

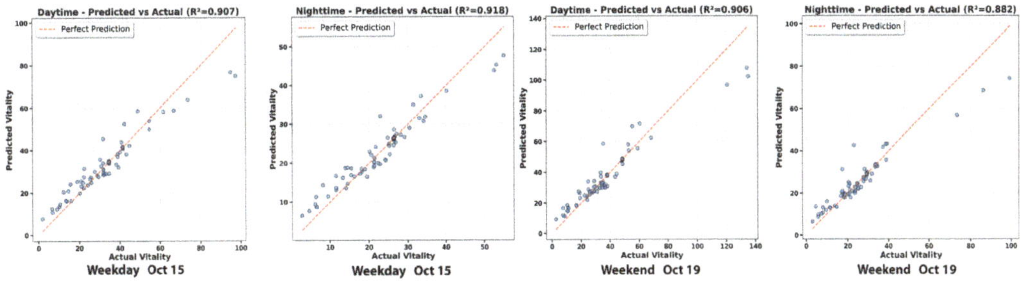

Figure 8. Predicted vs. actual vitality scatter plots across four temporal scenarios.

4.2.2. Temporal Variation Characteristics of Variable Correlations

The correlation analysis between variables revealed fundamental association patterns and the temporal heterogeneity characteristics of influencing factors (Figure 9). The results showed that correlations between various factors and vitality showed systematic differences in weekday–weekend and day–night cycles. Some variables showed strong correlations between them, while others showed relatively independent characteristics.

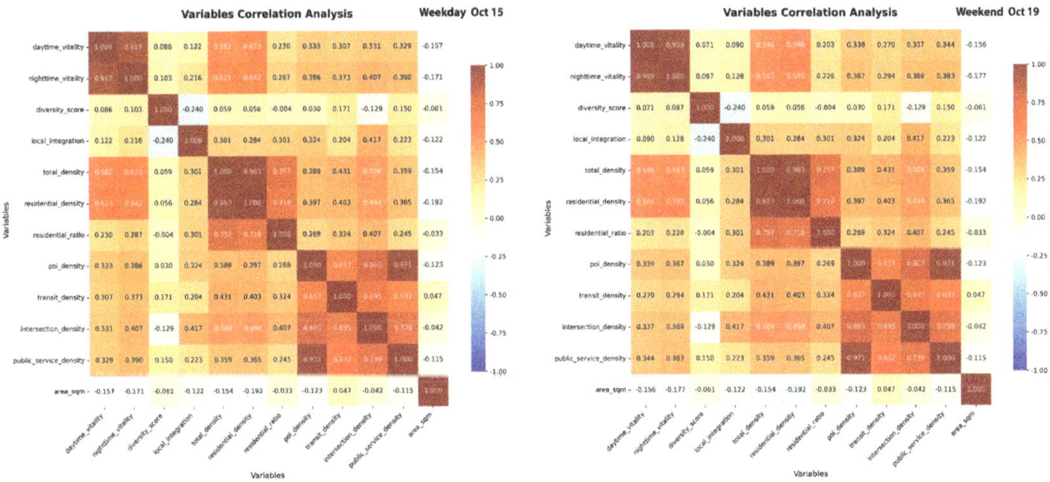

Figure 9. Correlation analysis of influencing factors.

Correlations between daytime and nightlife show weekday–weekend differences. Weekend periods showed a significantly higher correlation between day and night vitality (r = 0.959) compared to weekday (r = 0.917), indicating greater continuity and consistency in weekday vitality patterns. On the other hand, weekday showed a more pronounced diurnal differentiation, reflecting the structural influence of commuting patterns and functional zoning on day–night vitality dynamics.

The association between density variables and vitality showed temporal variations. During the week, the correlations of total building density with daytime vitality (r = 0.582) and night-time vitality (r = 0.621) were higher than weekend vitality (r = 0.546, night r = 0.563), indicating a greater dependence of weekday activities on the density of the urban built environment. The residential density showed similar patterns of association, with weekly daytime correlations (r = 0.623) exceeding weekend (r = 0.586). In particular, the correlation between the residence ratio and nightlife was higher during weekend

https://doi.org/10.3390/buildings16030523

(r = 0.287) than on weekday (r = 0.226), reflecting the greater dependence of weekend night activities on residential functions.

Some variables showed relatively independent characteristics. Diversity points generally show weak correlations with other variables (absolute values 0.24), indicating that functional mixtures, as a relatively independent dimension of urban forms, are not strongly constrained by density or other factors that influence vitality. The density of sections showed similar low correlations with most variables, representing the independent role of the network connectivity of the streets. Local integration, as a space syntax indicator, maintained moderate correlations with density variables (r = 0.20–0.42), reflecting the relative independence of the accessibility dimension. Public space areas showed weak negative correlations with vitality (r = −0.15 to −0.19), but weak correlations with other influencing factors, reflecting the particularity of the spatial scale factor.

4.2.3. Difference in Impact of Factors on Day–Night Vitality

The GWRF model revealed the differential importance of factors influencing public space vitality between daytime and night-time across different temporal contexts (Figure 10).

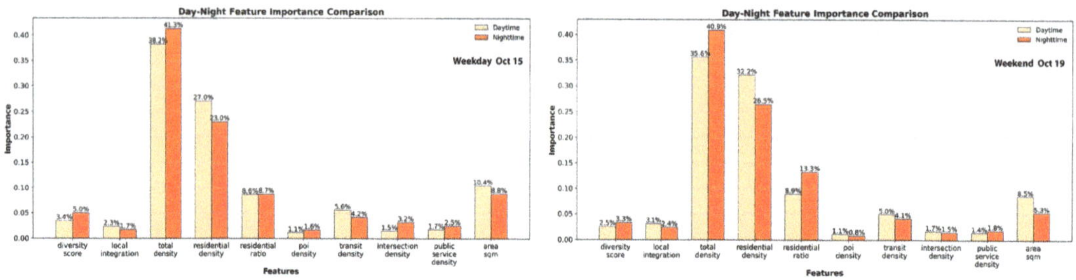

Figure 10. Feature importance comparison between daytime and night-time models: weekday vs. weekend.

On weekday (Oct 15), total density (total building density) emerged as the dominant factor for both periods, showing higher importance during night-time (41.3%) compared to daytime (38.2%). Residential density ranked second, with greater influence during daytime (27.0%) than night-time (23.0%). The residential ratio maintained similar importance across both periods (8.6% daytime vs. 8.7% night-time). Notably, the diversity score showed higher importance during night-time (5.0%) compared to daytime (3.4%), suggesting an enhanced role of functional diversity in night-time activities. Public space area demonstrated relatively high importance (10.4% daytime, 8.8% night-time), while other factors such as POI density, transit density, and public service density showed limited contributions (all below 6%). The greater daytime influence of residential density (27.0% vs. 23.0% night-time) suggests that daytime public space usage is more closely tied to local residential populations engaging in routine activities near their homes. In contrast, the higher night-time importance of total density (41.3% vs. 38.2% daytime) indicates that evening activities are less constrained by residential proximity and more attracted to areas with higher overall building density. This pattern is further supported by the increased night-time importance of POI density (1.6% vs. 1.1% daytime), suggesting that night-time activities show stronger associations with areas offering diverse amenities and services.

Weekend patterns (Oct 19) exhibited distinct shifts in factor importance. While total density remained the most critical factor, its daytime importance decreased to 35.6% (compared to 38.2% on weekday), though night-time importance remained stable at 40.9%. Residential density showed more pronounced day–night difference on weekend (32.2% daytime vs. 26.5% night-time), with enhanced daytime importance compared to week-

https://doi.org/10.3390/buildings16030523

day. Most notably, the residential ratio demonstrated substantially increased night-time importance (13.3%) compared to both weekend daytime (8.9%) and weekday levels (approximately 8.7%), indicating the heightened role of residential composition in weekend evening activities. Diversity score showed relatively low importance on weekend (2.5% daytime, 3.3% night-time), markedly lower than weekday levels. Public space area exhibited reduced weekend importance (8.5% daytime, 5.3% night-time) compared to weekday, while other factors, including POI, transit, intersection, and public service densities, maintained consistently low importance (all below 5%) across both temporal contexts. The enhanced daytime importance of residential density on weekend (32.2% vs. 27.0% on weekday) reflects increased leisure-time usage of public spaces by residential populations, with the small gap between total density (35.6%) and residential density (32.2%) indicating that weekend daytime activities are strongly tied to residential areas. The substantially elevated night-time importance of residential ratio (13.3%) suggests that weekend evening activities show stronger associations with residential community characteristics, indicating more neighborhood-oriented social gatherings compared to weekday patterns.

4.2.4. Spatial Heterogeneity of Built Environment Effects

Beyond the temporal differentiation of factor importance discussed above, the GWRF model reveals spatial heterogeneity in how built environment factors influence public space vitality across different locations within the district. Figures 11 and 12 visualize the spatial distributions of local importance coefficients, with black dots representing public space locations. The coefficient of variation (CV) in Tables 4 and 5 quantifies the degree of spatial heterogeneity for each factor.

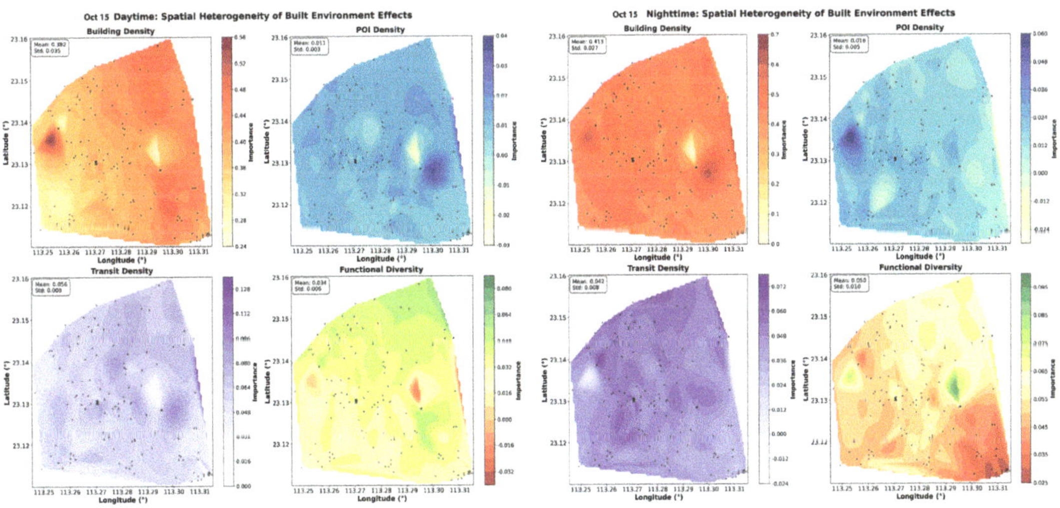

Figure 11. Spatial heterogeneity of built environment effects on weekday: local importance coefficients of four built environment factors.

As shown in Figure 11 and Table 4, weekday spatial heterogeneity exhibits distinct daytime and night-time patterns. During daytime, POI density exhibits the highest spatial variation (CV = 0.231). Building density shows moderate spatial heterogeneity (CV = 0.092) with importance coefficients averaging 0.382, while transit density (CV = 0.136) and functional diversity (CV = 0.189) display moderate heterogeneity. At night-time, building density effects increase in overall importance (mean: 0.413 vs. daytime 0.382) while becoming more spatially homogeneous (Std: 0.027 vs. daytime 0.055), with CV decreasing

https://doi.org/10.3390/buildings16030523

to 0.065. Meanwhile, POI density maintains high spatial variation (CV = 0.278), transit density shows CV = 0.184, and functional diversity exhibits CV = 0.198.

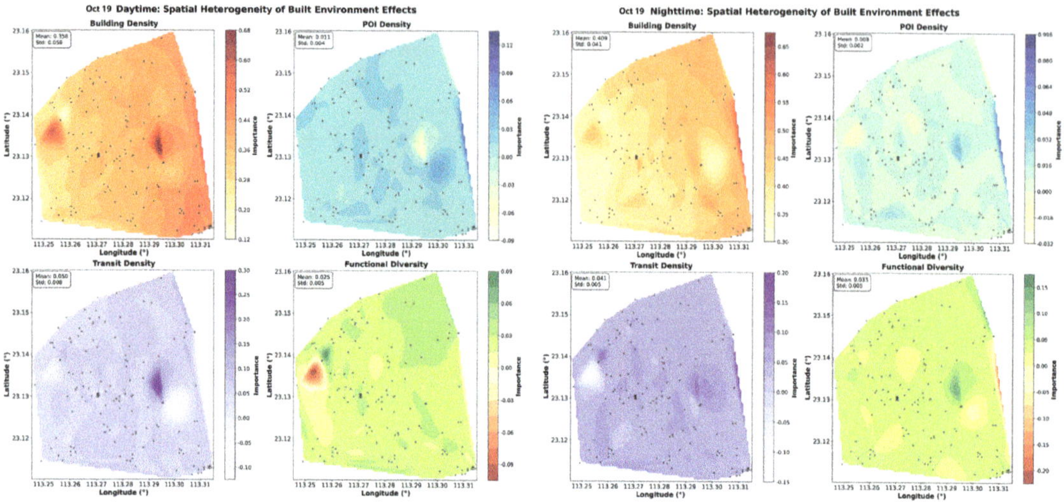

Figure 12. Spatial heterogeneity of built environment effects on weekend: local importance coefficients of four built environment factors.

Table 4. Spatial heterogeneity statistics of built environment factors on weekday.

Factor	Period	Mean	SD	Range	CV
Building Density	Day	0.383	0.035	0.278–0.482	0.092
	Night	0.413	0.027	0.356–0.462	0.065
POI Density	Day	0.011	0.003	0.006–0.017	0.231
	Night	0.016	0.005	0.008–0.029	0.278
Transit Density	Day	0.056	0.008	0.038–0.072	0.136
	Night	0.042	0.008	0.024–0.057	0.184
Functional Diversity	Day	0.034	0.007	0.019–0.058	0.189
	Night	0.050	0.010	0.027–0.068	0.198

CV = Coefficient of Variation, measuring the spatial variability of local importance values across different locations.

Table 5. Spatial heterogeneity statistics of built environment factors on weekend.

Factor	Period	Mean	SD	Range	CV
Building Density	Day	0.356	0.059	0.245–0.476	0.166
	Night	0.409	0.041	0.330–0.484	0.101
POI Density	Day	0.011	0.004	0.004–0.024	0.328
	Night	0.008	0.002	0.004–0.017	0.292
Transit Density	Day	0.050	0.008	0.028–0.069	0.154
	Night	0.041	0.005	0.030–0.055	0.129
Functional Diversity	Day	0.025	0.005	0.015–0.040	0.192
	Night	0.033	0.005	0.025–0.056	0.158

CV = Coefficient of Variation, measuring the spatial variability of local importance values across different locations.

These day-to-night transitions reveal three key patterns. Building density effects become more homogeneous (CV decreases from 0.092 to 0.065). POI density heterogeneity intensifies (CV increases from 0.231 to 0.278), and spatial disparities in transit density and functional diversity widen (CV increases from 0.136 to 0.184 and from 0.189 to 0.198, respectively).

As shown in Figure 12 and Table 5, weekend spatial heterogeneity reveals intensified patterns. During daytime, POI density exhibits the most pronounced heterogeneity (CV = 0.328). Building density shows substantial spatial variation (CV = 0.166), while

https://doi.org/10.3390/buildings16030523

transit density (CV = 0.154) and functional diversity (CV = 0.192) also display clear spatial differentiation. At night-time, building density importance increases (mean: 0.409 vs. daytime 0.356), while spatial heterogeneity decreases (CV = 0.101). POI density exhibits CV = 0.292, transit density shows CV = 0.129, and functional diversity displays CV = 0.158.

The day-to-night transition reveals a convergence pattern: building density (CV decreasing from 0.166 to 0.101), POI density (CV decreasing from 0.328 to 0.292), transit density (CV decreasing from 0.154 to 0.129), and functional diversity (CV decreasing from 0.192 to 0.158) all exhibit reduced spatial variation.

The weekday–weekend comparison reveals contrasting spatial reorganization patterns. Weekdays show divergence, with night-time heterogeneity increasing for POI density, transit density, and functional diversity, indicating spatial concentration of activities. Weekends demonstrate convergence, with all factors exhibiting reduced spatial variation at night, suggesting more dispersed activity patterns. This contrast reflects the shift from routine-bound weekday behaviors anchored to residential-work commutes to more spatially flexible weekend recreational activities.

4.3. Factor Interaction Patterns

The above analysis elucidates the individual effects of built environment factors and their spatial heterogeneity; however, it has yet to address the following critical questions: how do these factors interact with each other? Do synergistic or competitive relationships exist among different factors, and how do these interaction patterns vary across temporal contexts?

To explore factor interactions, SHAP (SHapley Additive exPlanations) interaction analysis was applied to the global GWRF model to quantify how the effect of one built environment factor depends on the value of another [44]. SHAP interaction values measure synergistic effects (positive values, where combined effects exceed the sum of individual effects) vs. competitive or substitution effects (negative values, where one factor's high value reduces another's marginal benefit). Figures 13 and 14 visualize the top factor interactions for weekday and weekend, respectively, revealing distinct temporal patterns in how built environment factors combine to influence public space vitality.

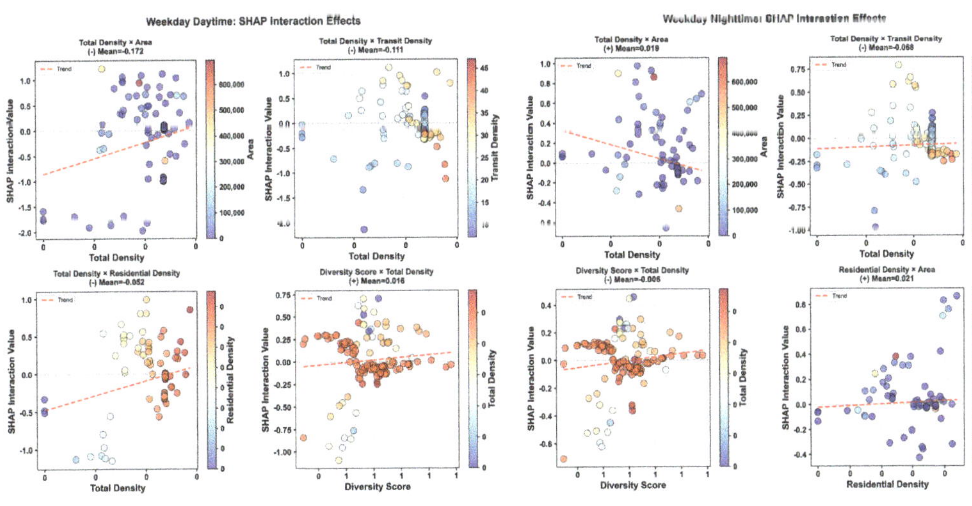

Figure 13. SHAP interaction effects on weekday: daytime (**left**) and night-time (**right**).

https://doi.org/10.3390/buildings16030523

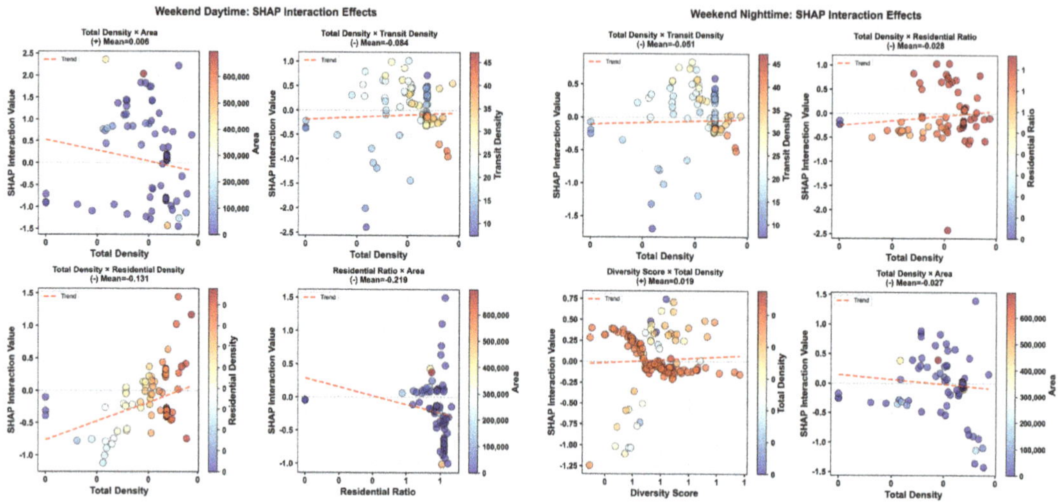

Figure 14. SHAP interaction effects on weekend: daytime (**left**) and night-time (**right**).

Weekday Interaction Patterns (Figure 13). On weekday daytime, the strongest interaction is total density × area (interaction strength: 0.489), exhibiting a competitive relationship (mean interaction value: −0.172). This negative interaction suggests that in high-density areas, a larger public space provides diminishing marginal benefits, while in low-density areas, a public space becomes more critical for vitality generation. Similarly, total density × transit density shows strong competitive interaction (0.325, −0.111), indicating spatial substitution, where high building density can partially compensate for lower transit accessibility in generating vitality. In contrast, diversity score × total density demonstrates synergistic interaction (0.213, 0.016), where functional diversity and building density reinforce each other—areas with both high density and diverse land uses achieve vitality levels exceeding the sum of individual effects. Residential density × residential ratio also shows synergistic effects (0.102, 0.010), suggesting that residential concentration works best when accompanied by high proportions of residential buildings. During weekday night-time, interaction patterns shift notably: total density × area transforms from competitive to synergistic (0.169, 0.019), indicating that larger public spaces in high-density areas become more valuable for evening activities. Total density × residential density also becomes synergistic (0.086, 0.047), reflecting the importance of residential support for night-time vitality in dense urban areas.

Weekend Interaction Patterns (Figure 14). Weekend daytime shows total density × area as the strongest interaction (0.618, 0.006), with moderate synergistic effects, indicating that public space area and building density work better together during leisure periods. However, total density × transit density remains competitive (0.328, −0.084), suggesting that weekend visitors rely less on transit density when building density is high, possibly due to automobile-based travel. Total density × residential density exhibits strong competitive interaction (0.348, −0.131), contrasting sharply with weekday night-time patterns. This suggests that on weekend, high building density areas dominated by commercial functions can attract visitors even without strong residential support. Diversity score × total density maintains moderate synergistic effects (0.279, 0.007), though weaker than on weekday. Local Integration × residential density shows strong synergistic interaction (0.116, 0.011), indicating that well-connected street networks enhance the vitality contribution of residential areas during weekend leisure activities. Weekend night-time demonstrates a return to density-centric patterns: total density × transit density remains competitive

https://doi.org/10.3390/buildings16030523

(0.262, −0.051), while diversity score × total density becomes strongly synergistic (0.233, 0.019), and total density × residential density shifts to synergistic (0.148, 0.098), mirroring weekday night-time patterns where residential support becomes critical.

Temporal Transitions in Factor Interactions. The total density × area interaction transitions from competitive (weekday daytime: −0.172) to synergistic (weekday night-time: 0.019, weekend daytime: 0.006), then back to competitive (weekend night-time: −0.027), demonstrating that the value of public space area in high-density areas varies with temporal activity patterns. Total density × residential density shows the most dramatic shift: strongly competitive on weekday daytime (−0.052) and weekend daytime (−0.131), but strongly synergistic during both night-time periods (weekday: 0.047, weekend: 0.098). This pattern reveals that residential density becomes a critical complement to building density, specifically during evening hours, when residents engage in proximate leisure activities. Diversity score × total density maintains consistently positive interactions across all periods, though with varying strength, confirming that functional diversity and building density form a stable synergistic relationship regardless of temporal context.

5. Discussion

The results presented in Section 4 reveal significant spatiotemporal heterogeneity in both public space vitality patterns and built environment effects. This section provides a mechanistic interpretation of these findings by examining two interconnected dimensions: the temporal differentiation mechanisms that drive day–night and weekday–weekend variations in factor importance, and the geographic variations in built environment factor effects across different locations and temporal scenarios. Together, these analyses elucidate the underlying processes through which temporal rhythms and spatial contexts interact to shape public space vitality.

5.1. Temporal Heterogeneous Effects of Built Environment on Vitality

The four-quadrant classification findings reveal distinct temporal mechanisms underlying built environment effects on public space vitality. Total building density, as the most critical factor, exhibits enhanced night-time importance across both weekday and weekend, consistent with existing studies [13,28,29]. This pattern reflects the spatiotemporal rhythm of urban population mobility. Daytime cross-regional commuting flows mean vitality depends on both local density and inter-regional mobility, while night-time reduced mobility confines activities to residential neighborhoods, making building density the dominant driver. Weekday–weekend comparison reveals attenuated effects on weekend, indicating reduced dependence on the immediate built environment when residents have greater temporal freedom.

Residential density exhibits opposite temporal patterns, showing greater daytime importance, particularly on weekend. This stems from two mechanisms: non-employed populations (elderly, children) constitute the main daytime users with limited mobility ranges, and high-density residential areas provide well-developed supporting facilities [45,46]. Specifically, elderly populations and children exhibit distinct spatiotemporal constraints that amplify residential density effects during daytime. In the Yuexiu District, the elderly population (aged 60+) accounts for 30.1% of the total population as of late 2024, ranking second among all districts in Guangzhou [47]. This substantial elderly demographic engages primarily in proximity-based activities due to reduced physical mobility and a preference for familiar environments. Research shows elderly residents prefer shady trees, peaceful settings, and accessible walking paths in neighborhood parks [48], with activity patterns concentrated in paved open spaces for social gatherings and pathways for walking exercises [49]. Their outdoor mobility is highly sensitive to environmental conditions, favoring cooler morning/evening hours and areas with greenery and pedestrian-friendly infrastructure [50]. Children's activities are similarly constrained by parental supervision and school schedules, concentrating public space usage in neighborhood parks during after-school hours. High residential density areas provide larger user bases and support age-appropriate facilities that reinforce proximity-based usage. This demographic structure explains why residential density shows particularly strong daytime importance (27.0% on weekday, 32.2% on weekend), as these non-employed populations constitute the primary daytime users whose activity rarely extends beyond immediate residential neighborhoods. Beyond these daytime patterns, weekend intensification reflects the shift from workplace- to residence-centered activities. Residential proportion shows dramatic night-time enhancement on weekend, indicating that residential functions provide stable population bases for weekend nocturnal vitality.

Among accessibility factors, local integration and intersection density exhibit low overall importance but show notable temporal differentiation [41,46,51]. Local integration demonstrates contrasting weekday–weekend patterns. On weekday, daytime importance (2.3%) slightly exceeds night-time (1.7%), while weekend shows substantially elevated daytime importance (3.1%) compared to weekday. This weekend enhancement reflects greater freedom of movement as residents actively explore distant neighborhoods for leisure activities, with destination choices becoming more flexible and showing stronger preferences for areas with high transportation network connectivity [7]. During weekday daytime, most individuals remain in relatively fixed workplaces, rendering public space users primarily lunch-break crowds and non-working populations whose activity ranges prioritize proximity. Regardless of weekday or weekend, night-time importance remains relatively low (1.7% and 2.4%, respectively), as night-time activities are confined to spaces near residences within walking distance, reducing dependence on broader transportation networks.

Intersection density exhibits pronounced weekday–weekend contrasts. Weekdays show distinctive diurnal variation, with night-time importance (3.2%) more than doubling daytime importance (1.5%). This pattern may stem from proximate activity patterns, as individuals tend to select nearby leisure venues after work, where dense intersections offer more destination options within walking range. In contrast, weekend maintains temporal stability (1.7% daytime, 1.5% night-time) with consistently low importance. With more ample time for cross-regional travel on weekend, people rely more on macro-level transportation accessibility rather than micro-level road intersection density, underscoring the context-dependent role of intersection density in proximate vs. cross-regional mobility.

5.2. Geographic Variations in the Influence of Built Environment Factors

The four-quadrant classification identified LD–LN zones (consistently low vitality) concentrated in the northern periphery and southeastern corner. The spatial heterogeneity maps (Figures 11 and 12) reveal different built environment patterns underlying this persistence. Northern LD–LN areas exhibit moderate building density importance but low importance for POI density, transit density, and functional diversity, indicating these areas have residential populations but lack commercial facilities, transit access, and functional mix necessary for vitality generation. Southeastern zones exhibit low importance across all four factors, suggesting both insufficient residential population and land uses incompatible with public space vitality.

Weekday–weekend comparisons reveal contrasting spatial patterns [20,52]. On weekday, night-time shows increased heterogeneity for POI, transit, and functional diversity, with importance concentrating in central areas while peripheral LD–LN zones remain low. Building density shows the opposite trend, becoming more spatially uniform at night. Weekends show reduced heterogeneity for all factors, with more uniform distributions.

https://doi.org/10.3390/buildings16030523

However, this spatial homogenization does not equalize vitality—LD–LN zones remain low because they possess lower absolute quantities of POIs, transit services, and functional diversity, regardless of how importance patterns distribute spatially.

These contrasting weekday–weekend patterns are fundamentally driven by socioeconomic and cultural factors. From a time-space perspective [53], these divergent patterns reflect how institutional, economic, and cultural forces shape distinct spatiotemporal regimes. On weekdays, institutional constraints—fixed schedules, centralized employment, and transit-dependent commuting—concentrate night-time activities around residential and commercial hubs, amplifying spatial heterogeneity [20,54]. Limited evening time and job-housing imbalance further reinforce this concentration by constraining mobility to proximity-based activities. In contrast, weekend free individuals from temporal-spatial rigidities, enabling leisure-oriented mobility toward diverse destinations—particularly natural hills and coastal areas. This expanded spatial accessibility, reinforced by cultural preferences for family-oriented outdoor recreation [46], results in more dispersed vitality patterns [53,54].

Beyond these temporal variations in spatial patterns, the GWRF model also reveals differential stability characteristics among built environment factors [55]. Building density maintains stable spatial patterns across temporal periods, while POI density and functional diversity vary spatially across different periods. This explains why some areas maintain stable vitality while others fluctuate temporally.

5.3. Factor Interaction Mechanisms

The SHAP interaction analysis reveals a fundamental characteristic of urban vitality generation. Built environment factors do not operate independently but interact through synergistic or competitive mechanisms that vary across temporal contexts [33]. These complex nonlinear interaction relationships and their significant temporal dependencies provide more refined and dynamic planning guidance for enhancing urban public space vitality. The following differentiated, temporally adaptive planning and design recommendations are proposed based on major interaction patterns:

Promoting "Density + Diversity" Mixed-Use Development. Diversity score and total density maintain stable positive synergistic interactions across all temporal periods. In urban renewal and redevelopment, priority should be given to mixed-use configurations (residential + commercial + cultural + recreational) to achieve multiplicative effects. This approach is particularly effective in areas with high vitality fluctuations, reducing disparities between different time periods.

Strengthening Residential Support for Night-time Vitality. Total density × residential density shifts from strong competition during daytime to strong synergy during both night-time periods, indicating that night time activities are dominated by residents' proximate leisure behavior, making residential density an indispensable complement to building density. Planning interventions should prioritize residential function enhancement in areas targeting night-time vitality improvement.

Context-Sensitive Public Space Area Planning. Total density × area exhibits a clear temporal cycle: competitive during weekday daytime, synergistic during weekday night-time and weekend daytime, then competitive again during weekend night-time. This demonstrates that the marginal benefit of large public spaces in high-density environments is highly dependent on activity types. High-density areas should not blindly pursue oversized plazas or green spaces; instead, priority should be given to improving space utilization efficiency and maximizing vitality capacity during leisure periods.

Transit-Density Substitution in Planning Strategies. The competitive interaction between total density and transit density during weekday daytime reveals substitution

https://doi.org/10.3390/buildings16030523

relationships. In high-density areas, building density can partially compensate for lower transit accessibility. However, in peripheral areas with high vitality fluctuations, addressing transit density deficiencies should be prioritized to enhance overall accessibility and vitality generation capacity.

6. Conclusions and Limitations

6.1. Conclusions

This study investigated the spatiotemporal heterogeneity of public space vitality and its built environment determinants across 149 public spaces in the Yuexiu District, Guangzhou, using the geographically weighted random forest model to examine four temporal scenarios.

This research demonstrates that public space vitality emerges from spatiotemporal coupling between built environment configurations and temporal activity patterns, rather than being a static spatial attribute. Methodologically, the GWRF model effectively captures spatial non-stationarity and nonlinear relationships, revealing localized mechanisms that conventional global models fail to detect.

The findings reveal significant spatiotemporal differentiation: Public space vitality varies substantially between day and night, weekday and weekend, while built environment impacts exhibit both temporal variation and spatial non-stationarity, with different factors dominating in different temporal contexts and spatial locations. The four-quadrant classification identified four distinct vitality patterns, with LD–LN spaces (consistently low vitality) concentrated in peripheral areas and HD–HN spaces (sustained high vitality) distributed across central districts. Weekday–weekend comparison reveals contrasting spatial reorganization patterns. Weekdays show divergence with night-time heterogeneity increasing for POI density, transit density, and functional diversity, while weekend demonstrates convergence with all factors exhibiting reduced spatial variation at night, reflecting the shift from routine-bound weekday behavior to spatially flexible weekend recreational activities.

These spatiotemporal patterns have direct implications for urban planning practice. First, recognizing diurnal differentiation patterns, planning interventions should provide temporally appropriate facility configurations: prioritizing night-time service facilities and lighting in high-density residential areas while supporting daytime activities for non-working populations. Second, addressing weekday–weekend differences requires multi-scale accessibility improvements: enhancing regional transportation connectivity for weekend leisure travel while refining community-scale pedestrian networks for weekday evening activities. Third, spatial heterogeneity analysis identifies priority intervention zones where high building density coexists with low functional mix. In these areas, supplementing diverse commercial and public service facilities is recommended to optimize community life circle configurations and enhance residents' daily convenience.

Furthermore, the SHAP interaction analysis reveals that built environment factors do not operate independently but interact through synergistic or competitive mechanisms that vary across temporal contexts. Diversity score and total density maintain stable synergistic interactions across all periods, while total density and residential density exhibit dramatic temporal transitions from competitive during daytime to synergistic during night-time, demonstrating that the same socioeconomic forces driving temporal activity patterns also shape factor interaction mechanisms at different periods. This finding necessitates a shift from factor-by-factor optimization to interaction-aware planning strategies that maximize synergistic effects while minimizing competitive trade-offs across different temporal scenarios.

https://doi.org/10.3390/buildings16030523

6.2. Limitations

The Baidu location-based services (LBS) data serve as a reliable proxy for human activity and interaction by capturing the dynamic distribution of the population across urban space; however, they cannot fully represent all age and social groups—particularly the elderly—or encompass all types of activity.

This study focuses primarily on quantifiable attributes of the built-environment that shape the diurnal characteristics of the vitality of public spaces, thereby overlooking the critical roles of qualitative factors such as aesthetic quality, perceived safety, and social atmosphere. These elements also exert substantial influence on vitality, which warrants deeper exploration in future research.

The temporal scope of this study is limited to data collected from only two dates (one weekday and one weekend in October 2024), which constrains the representation of seasonal variations in public space vitality. Different seasons may exhibit distinct activity patterns due to variations in weather conditions, daylight hours, and seasonal events, which are not captured in this analysis.

The findings were based of a single case study in Guangzhou's Yuexiu District, and their generalizability to cities with differing morphological, cultural, and socioeconomic profiles remains to be validated. Future investigations could enhance theoretical and methodological transferability through multi-source data fusion, cross-seasonal comparisons, and multi-city validation.

Author Contributions: Conceptualization: Y.Y., X.L. (Xiuhong Lin) and X.S.; data curation: Y.Y.; formal analysis: Y.Y.; funding acquisition: X.L. (Xiuhong Lin) and X.L. (Xin Li); methodology: Y.Y.; software: Y.Y. and X.L. (Xin Li); visualization: Y.Y. and Q.C.; writing—original draft: Y.Y. and X.L. (Xiuhong Lin); writing—review and editing: Y.Y. and X.L. (Xiuhong Lin). All authors have read and agreed to the published version of the manuscript.

Funding: This research was funded by the Guangdong Province Philosophy and Social Science Planning Project (Grant No. GD23XSH16); the project "Construction of BIM Course Faculty Team under the Background of Industry-Education Integration"; and the 2021 Guangdong Provincial First-Class Undergraduate Major Construction Point Project in Engineering Management.

Data Availability Statement: The data presented in this study are available in this article. Further inquiries can be directed to the corresponding author.

Conflicts of Interest: The authors declare no conflicts of interest.

References

1. Huang, B.; Zhou, Y.; Li, Z.; Song, Y.; Cai, J.; Tu, W. Evaluating and characterizing urban vibrancy using spatial big data: Shanghai as a case study. *Environ. Plan. B Urban Anal. City Sci.* **2020**, *47*, 1543–1559. [CrossRef]
2. Wu, C.; Ye, X.; Ren, F.; Du, Q. Check-in behaviour and spatio-temporal vibrancy: An exploratory analysis in Shenzhen, China. *Cities* **2018**, *77*, 104–116. [CrossRef]
3. Lopes, M.N.; Camanho, A.S. Public Green Space Use and Consequences on Urban Vitality: An Assessment of European Cities. *Soc. Indic. Res.* **2013**, *113*, 751–767. [CrossRef]
4. Jacobs, J. *The Death and Life of Great American Cities*; Random House: New York, NY, USA, 1961.
5. Gehl, J. *Life Between Buildings: Using Public Space*; Van Nostrand Reinhold: New York, NY, USA, 1987.
6. Montgomery, J. Making a City: Urbanity, Vitality and Urban Design. *J. Urban Des.* **1998**, *3*, 93–116. [CrossRef]
7. Li, X.; Li, Y.; Jia, T.; Zhou, L.; Hijazi, I.H. The Six Dimensions of Built Environment on Urban Vitality: Fusion Evidence from Multi-Source Data. *Cities* **2022**, *121*, 103482. [CrossRef]
8. Ta, N.; Zeng, Y.; Zhu, Q.; Wu, J. Understanding community rhythm from a spatiotemporal behavior perspective. *Hum. Geogr.* **2023**, *38*, 34–43. [CrossRef]
9. Silm, S.; Ahas, R. Ethnic Differences in Activity Spaces: A Study of Out-of-Home Nonemployment Activities with Mobile Phone Data. *Ann. Assoc. Am. Geogr.* **2014**, *104*, 542–559. [CrossRef]

https://doi.org/10.3390/buildings16030523

10. Niu, H.; Silva, E.A. Crowdsourced Data Mining for Urban Activity: Review of Data Sources, Applications, and Methods. *J. Urban Plan. Dev.* **2020**, *146*, 04020007. [CrossRef]
11. Li, X.; Kozlowski, M.; Salih, S.A.; Ismail, S.B. Evaluating the Vitality of Urban Public Spaces: Perspectives on Crowd Activity and Built Environment. *Archnet-IJAR Int. J. Archit. Res.* **2025**, *19*, 562–583. [CrossRef]
12. Qin, L.; Zong, W.; Peng, K.; Zhang, R. Assessing Spatial Heterogeneity in Urban Park Vitality for a Sustainable Built Environment: A Case Study of Changsha. *Land* **2024**, *13*, 480. [CrossRef]
13. Fan, Z.; Duan, J.; Luo, M.; Zhan, H.; Liu, M.; Peng, W. How Did Built Environment Affect Urban Vitality in Urban Waterfronts? A Case Study in Nanjing Reach of Yangtze River. *ISPRS Int. J. Geo-Inf.* **2021**, *10*, 611. [CrossRef]
14. Xia, C.; Yeh, A.G.O.; Zhang, A. Analyzing spatial relationships between urban land use intensity and urban vitality at street block level: A case study of five Chinese megacities. *Landsc. Urban Plan.* **2020**, *193*, 103669. [CrossRef]
15. Zeng, C.; Song, Y.; He, Q.; Shen, F. Spatially explicit assessment on urban vitality: Case studies in Chicago and Wuhan. *Sustain. Cities Soc.* **2018**, *40*, 296–306. [CrossRef]
16. Zheng, Y.; Ye, R.; Hong, X.; Tao, Y.; Li, Z. What Factors Revitalize the Street Vitality of Old Cities? A Case Study in Nanjing, China. *ISPRS Int. J. Geo-Inf.* **2024**, *13*, 282. [CrossRef]
17. Yu, B.; Sun, J.; Wang, Z.; Jin, S. Influencing Factors of Street Vitality in Historic Districts Based on Multisource Data: Evidence from China. *ISPRS Int. J. Geo-Inf.* **2024**, *13*, 277. [CrossRef]
18. Liang, S.; Lu, M. Computer vision framework for site-scale multidimensional vitality assessment: Lakeside waterfront spaces as a testing ground. *Habitat Int.* **2025**, *166*, 103603. [CrossRef]
19. Zhang, J.; Li, X.; Lian, H.; Li, H.; Zhang, J. Day–Night Synergy between Built Environment and Thermal Comfort and Its Impact on Pedestrian Street Vitality: Beijing–Chengdu Comparison. *Buildings* **2025**, *15*, 2118. [CrossRef]
20. Chen, Y.; Yu, B.; Shu, B.; Yang, L.; Wang, R. Exploring the Spatiotemporal Patterns and Correlates of Urban Vitality: Temporal and Spatial Heterogeneity. *Sustain. Cities Soc.* **2023**, *91*, 104440. [CrossRef]
21. Liu, S.; Zhu, Z.; Gao, Y.; Wang, S.; Zhou, Y. Exploring Factors behind Weekday and Weekend Variations in Public Space Vitality in Traditional Villages, Using Wi-Fi Sensing Method. *ISPRS Int. J. Geo-Inf.* **2025**, *14*, 386. [CrossRef]
22. Zhuo, B.; Wang, Z.; Chen, T.; Tian, Z. Urban Vitality Characters at Day and Night in Qingdao. *Planners* **2021**, *37*, 94–100.
23. Xu, X.; Xu, X.; Guan, P.; Ren, Y.; Wang, W.; Xu, N. The Cause and Evolution of Urban Street Vitality under the Time Dimension: Nine Cases of Streets in Nanjing City, China. *Sustainability* **2018**, *10*, 2797. [CrossRef]
24. Wang, Y.; Dewancker, B.J.; Qi, Q. Citizens' preferences and attitudes towards urban waterfront spaces: A case study of Qiantang riverside development. *Environ. Sci. Pollut. Res.* **2020**, *27*, 45787–45801. [CrossRef]
25. Xu, G.; Su, J.; Xia, C.; Li, X.; Xiao, R. Spatial mismatches between nighttime light intensity and building morphology in Shanghai, China. *Sustain. Cities Soc.* **2022**, *81*, 103851. [CrossRef]
26. Yue, Y.; Zhuang, Y.; Yeh, A.G.O.; Xie, J.Y.; Ma, C.L.; Li, Q.Q. Measurements of POI-based mixed use and their relationships with neighbourhood vibrancy. *Int. J. Geogr. Inf. Sci.* **2017**, *31*, 658–675. [CrossRef]
27. Guo, Z.; Luo, K.; Yan, Z.; Hu, A.; Wang, C.; Mao, Y.; Niu, S. Assessment of the street space quality in the metro station areas at different spatial scales and its impact on the urban vitality. *Front. Archit. Res.* **2024**, *13*, 1270–1287. [CrossRef]
28. Ye, Y.; Li, D.; Liu, X. How block density and typology affect urban vitality: An exploratory analysis in Shenzhen, China. *Urban Geogr.* **2018**, *39*, 631–652. [CrossRef]
29. Mouratidis, K.; Poortinga, W. Built environment, urban vitality and social cohesion: Do vibrant neighborhoods foster strong communities? *Landsc. Urban Plan.* **2020**, *204*, 103951. [CrossRef]
30. Tu, W.; Zhu, T.; Xia, J.; Zhou, Y.; Lai, Y.; Jiang, J.; Li, Q. Portraying the spatial dynamics of urban vibrancy using multisource urban big data. *Comput. Environ. Urban Syst.* **2020**, *80*, 101428. [CrossRef]
31. Zhang, A.; Li, W.; Wu, J.; Lin, J.; Chu, J.; Xia, C. How can the urban landscape affect urban vitality at the street block level? A case study of 15 metropolises in China. *Environ. Plan. B Urban Anal. City Sci.* **2021**, *48*, 1245–1262. [CrossRef]
32. Zhang, J.; Liu, X.; Tan, X.; Jia, T.; Senousi, A.M.; Huang, J.; Yin, L.; Zhang, F. Nighttime Vitality and Its Relationship to Urban Diversity: An Exploratory Analysis in Shenzhen, China. *IEEE J. Sel. Top. Appl. Earth Obs. Remote Sens.* **2021**, *15*, 309–322. [CrossRef]
33. Xu, Y.; Mao, W.; Hu, S. Unveiling the Pulse of Urban Metro Stations: A TOD-driven Approach to Measuring Vibrancy Using Geographically Weighted Random Forest. *Int. J. Digit. Earth* **2025**, *18*, 2524054. [CrossRef]
34. Li, H.; He, L. Park Development, Potential Measurement, and Site Selection Study Based on Interpretable Machine Learning—a Case Study of Shenzhen City, China. *ISPRS Int. J. Geo-Inf.* **2025**, *14*, 184. [CrossRef]
35. Guangzhou Municipal Bureau of Statistics and National Bureau of Statistics Guangzhou Survey Team. *Guangzhou Statistical Bulletin on National Economic and Social Development 2021*; Statistical Bulletin, Guangzhou Municipal Bureau of Statistics: Guangzhou, China, 2022. (In Chinese)
36. Liu, Y.; Zhang, Y.; Zhang, X.; Han, F.; Zhao, Y. A Geographical Perspective on the Formation of Urban Nightlife Landscape. *Humanit. Soc. Sci. Commun.* **2023**, *10*, 483. [CrossRef]

https://doi.org/10.3390/buildings16030523

37. Ren, W.; Lu, P.; Wang, Y.; Hong, X.C. Understanding the Association between Urban Noise and Nighttime Light in China. *Sci. Rep.* **2024**, *14*, 31472. [CrossRef]
38. Guangzhou Municipal Bureau of Statistics. *Guangzhou Population Census Yearbook-2020*; Guangzhou Municipal Bureau of Statistics: Guangzhou, China, 2022.
39. Guangzhou Yuexiu District People's Government. *Notice on the Issuance of Yuexiu District Territorial Spatial Master Plan (2021–2035)*; Government Document No. Yuefu [2025], 9, 2025; Guangzhou Yuexiu District People's Government: Guangzhou, China, 2025. (In Chinese)
40. Francis, J.; Giles-Corti, B.; Wood, L.; Knuiman, M. Creating Sense of Community: The Role of Public Space. *J. Environ. Psychol.* **2012**, *32*, 401–409. [CrossRef]
41. Dong, Q.; Cai, J.; Chen, S.; He, P.; Chen, X. Spatiotemporal Analysis of Urban Green Spatial Vitality and the Corresponding Influencing Factors: A Case Study of Chengdu, China. *Land* **2022**, *11*, 1820. [CrossRef]
42. Khan, S.N.; Li, D.; Maimaitijiang, M. A Geographically Weighted Random Forest Approach to Predict Corn Yield in the US Corn Belt. *Remote Sens.* **2022**, *14*, 2843. [CrossRef]
43. Georganos, S.; Grippa, T.; Niang Gadiaga, A.; Linard, C.; Lennert, M.; Vanhuysse, S.; Mboga, N.; Wolff, E.; Kalogirou, S. Geographical Random Forests: A Spatial Extension of the Random Forest Algorithm to Address Spatial Heterogeneity in Remote Sensing and Population Modelling. *Geocarto Int.* **2021**, *36*, 121–136. [CrossRef]
44. Gao, F.; Deng, X.; Liao, S.; Liu, Y.; Chen, W.; Li, G. Comparing Visitation Patterns in Distinct Types of Parks Using Geospatial Data and Interpretable Machine Learning. *Urban For. Urban Green.* **2025**, *112*, 128986. [CrossRef]
45. Wang, F.; Zhao, M.X.; Meng, Q.L. Analysis of the vitality measurement and correlation factors of urban waterfront space. *IOP Conf. Ser. Earth Environ. Sci.* **2020**, *612*, 012013. [CrossRef]
46. Zhu, J.; Lu, H.; Zheng, T.; Rong, Y.; Wang, C.; Zhang, W.; Yan, Y.; Tang, L. Vitality of urban parks and its influencing factors from the perspective of recreational service supply, demand, and spatial links. *Int. J. Environ. Res. Public Health* **2020**, *17*, 1615. [CrossRef]
47. Guangzhou Civil Affairs Bureau. Statistical Data on Elderly Population and Aging Affairs in Guangzhou City for 2024; Technical Report; Guangzhou Civil Affairs Bureau: 2025. Released on 20 May 2025. Official Announcement with Attached PDF Containing Detailed Statistics. Available online: http://mzj.gz.gov.cn/gk/zdlyxxgk/ylfwxx/content/post_10271932.html (accessed on 9 January 2026).
48. Veitch, J.; Ball, K.; Rivera, E.; Loh, V.; Deforche, B.; Best, K.; Timperio, A. What Entices Older Adults to Parks? Identification of Park Features That Encourage Park Visitation, Physical Activity, and Social Interaction. *Landsc. Urban Plan.* **2022**, *217*, 104254. [CrossRef]
49. Zhai, Y.; Li, D.; Wu, C.; Wu, H. Spatial Distribution, Activity Zone Preference, and Activity Intensity of Senior Park Users in a Metropolitan Area. *Urban For. Urban Green.* **2023**, *79*, 127761. [CrossRef]
50. Močnik, Š.; Moogoor, A.; Yuen, B. Exploring Facilitators and Barriers of Older Adults' Outdoor Mobility: A Walk-along Study in Singapore. *J. Transp. Health* **2022**, *26*, 101386. [CrossRef]
51. Yang, Y.; Ma, Y.; Jiao, H. Exploring the Correlation between Block Vitality and Block Environment Based on Multisource Big Data: Taking Wuhan City as an Example. *Land* **2021**, *10*, 984. [CrossRef]
52. Park, S.; Lee, D. Exploring the Spatiotemporal Heterogeneities in Urban Vitality Through Scalable Proxies from Mobile Data. *Land* **2024**, *13*, 1772. [CrossRef]
53. Huang, Y.; Du, J.; Chen, M.; Lin, Y.; Huang, S.; Cai, Y. Evaluating the Spatial-Temporal Impact of Urban Nature on Urban Vitality in Vancouver: A Social Media and GPS Data Approach. *Land Use Policy* **2026**, *160*, 107824. [CrossRef]
54. Li, Z.; Zhao, G. Revealing the Spatio-Temporal Heterogeneity of the Association between the Built Environment and Urban Vitality in Shenzhen. *ISPRS Int. J. Geo-Inf.* **2023**, *12*, 433. [CrossRef]
55. Tang, S.; Ta, N. How the Built Environment Affects the Spatiotemporal Pattern of Urban Vitality: A Comparison among Different Urban Functional Areas. *Comput. Urban Sci.* **2022**, *2*, 39. [CrossRef] [PubMed]

https://doi.org/10.3390/buildings16030523

buildings

MDPI

Article

Identification and Configuration Optimization of Key Campus Landscape Features Using Augmentation-Based Machine Learning and Configuration Analysis

Xiaowen Zhuang [1,†], Yi Cai [2,†], Zhenpeng Tang [2,†], Zheng Ding [1,*] and Christopher Gan [3]

1 College of Landscape Architecture and Art, Fujian Agriculture and Forestry University, Fuzhou 350002, China; 12319097012@fafu.edu.cn
2 College of Economics and Management, Fujian Agriculture and Forestry University, Fuzhou 350002, China; 2221410003@fafu.edu.cn (Y.C.); zhenpt@fzu.edu.cn (Z.T.)
3 Department of Financial and Business Systems, Faculty of Agribusiness and Commerce, Lincoln University, Christchurch 7647, New Zealand; christopher.gan@lincoln.ac.nz
* Correspondence: 000q151039@fafu.edu.cn
† These authors contributed equally to this work.

Abstract: A university campus is a composite built environment integrating research, daily life, culture, and ecological green space. Its landscape elements shape environmental perception and overall spatial quality. This study assesses spatial quality by identifying key features and optimizing their joint effects across three perceptions: safety, comfort, and belonging. Using a Chinese campus, we captured street-view images, applied semantic segmentation to quantify elements (grass, trees, buildings, roads, sidewalks), and used explainable machine learning with data augmentation to identify the features most relevant to these perceptions. This study then employed fuzzy-set Qualitative Comparative Analysis (fsQCA) to reveal configuration pathways that enhance spatial quality. Results show that data augmentation mitigates class imbalance and improves prediction accuracy. Key features include sky, river, bridge, people, grass, and sidewalks, and path analysis indicates that greater sky openness and higher densities of people, roads, sidewalks, and grass, together with fewer buildings, cars, and bare earth, enhance safety, comfort, and belonging. This study delivers globally transferable design rules and a replicable, policy-ready workflow that enables evidence-based campus upgrades across diverse regions.

Keywords: campus buildings; landscape features; data augmentation; combined effects

1. Introduction

In a world increasingly focused on sustainable development and environmental quality, higher education institutions play a vital and growing role. University campuses are not only centers for teaching and research, they also serve as living laboratories for green infrastructure and eco-friendly design. Campus landscape planning is now a key component of the contemporary university environment. As campuses confront climate change and the urban heat island effect, their spatial form strongly shapes microscale adaptation and mitigation, making it an essential pillar of sustainability.

Prior studies show that campus green and blue infrastructure is critical. Natural elements, including vegetation, green spaces, and water bodies, regulate the thermal environment, mitigate urban heat islands, and enhance water cycles and ecosystem services [1,2]. The campus landscape, as a complex system, also supports environmental regulation and biodiversity conservation. Its structural design and functional optimization are crucial for maintaining ecological balance and achieving higher environmental quality.

Beyond their ecological functions, campus landscapes strongly shape users' psychological experiences. Studies show that building scale, landscape diversity, and amenities in public spaces directly affect students' mental health and restorative capacity [3]. At the microscale, trees, water, and lawns reduce stress, enhance well-being, and provide restorative experiences [4,5]. Greenness and access to green space are closely linked to emotional health and depression, underscoring the role of landscape design in mental health. Consequently, high-quality landscape structure and natural settings foster pleasant experiences that support study, leisure, and social interaction, which in turn improve overall well-being [6,7]. However, despite extensive research on the ecological and psychological benefits of campus landscapes, little is known about how specific combinations of landscape features jointly shape perceptions of safety, comfort, and belonging. This constitutes the critical research gap that the present study addresses.

To address the limited research on optimizing combinations of campus landscape features, this study focuses on a comprehensive university in China. We combine semantic image segmentation, explainable machine learning with data augmentation, and fuzzy-set Qualitative Comparative Analysis (fsQCA) to examine how campus environments shape perception. First, the Mask2Former model identifies landscape elements such as plants, grass, and rivers. A survey then provides three perception scores: safety, comfort, and sense of belonging. Second, to mitigate class imbalance between high and low score samples, we use a Conditional Tabular Generative Adversarial Network (CTGAN) for data augmentation and train interpretable models to improve prediction of the perception indices. Finally, guided by SHAP-based feature rankings, we apply fsQCA to uncover configuration pathways through which landscape attributes enhance the three perceptions and to offer practical guidance for improving campus spatial quality.

The novelty of this study lies first in its theoretical contribution. By integrating Prospect Refuge Theory and Thermal Comfort Theory, we move beyond purely ecological or aesthetic indicators and evaluate campus spatial quality across three psychological dimensions: perceived safety, environmental comfort, and sense of belonging. This holistic evaluation clarifies how the environment shapes students' psychological experiences and expands the theoretical framework for campus landscape research.

Second, methodologically, this study combines semantic image segmentation, explainable machine learning with data augmentation, and fuzzy-set qualitative comparative analysis (fsQCA). Segmentation yields high-quality landscape features for empirical analysis. The augmented and interpretable models address the imbalance between high and low perception scores and markedly improve prediction of campus spatial quality. Building on machine learning identification of key features, fsQCA examines configuration pathways that enhance spatial quality. Together, these three methods reinforce one another to achieve the research goal.

Third, regarding findings, this study moves beyond prior work that explained restorative quality with a single landscape feature. We show that greater sky openness, higher crowd density, and moderate reductions in building and bare-earth densities jointly increase perceived campus safety. Grounded in established theory and a systematic configuration approach, the study optimizes campus spatial quality and provides new evidence.

Overall, this study contributes a unified three-tier methodology that links semantic segmentation, interpretable machine learning with data augmentation, and fsQCA into an end-to-end replicable pipeline. It connects perception data to feature attribution and then to configuration rules, moving from prediction to prescription and providing generalizable, implementation-ready guidance for campus design.

The remainder of this paper is organized as follows. Section 2 reviews the relevant literature and outlines the theoretical basis. Section 3 describes the street view image data and the environmental perception scores, and provides a brief introduction to the models used. Section 4 reports the empirical results. Section 5 discusses the findings. Section 6 concludes the study.

2. Literature Review and Theory

2.1. Literature Review

2.1.1. Ecological and Environmental Functions of Campus Landscapes

Campus landscapes serve both aesthetic and cultural roles while also functioning as core ecosystems. Open lawns, water features, and vegetation provide venues for interaction among students and staff, which can foster a sense of belonging and social cohesion [8,9]. Natural settings such as lawns and ponds reduce stress and support recovery through "soft fascination," restoring directed attention and sustaining focus. Campus studies consistently document these restorative effects in students [10,11].

For stress relief, natural environments offer non-threatening stimuli that reduce sympathetic activation and lower cortisol levels, thereby easing anxiety and stress. Multiple mechanisms have been proposed. Hartig et al. (2003) [12] reviewed how nature regulates physiological stress indicators such as heart rate, blood pressure, and cortisol. Kaplan (1995) [13] argued that the joint action of affective soothing and physiological recovery is the key pathway for stress reduction.

Green spaces deliver core ecosystem services: they improve air quality, reduce particulate matter, regulate the microclimate to temper urban heat islands, and manage stormwater through lawns, trees, and rain gardens [14,15]. On campuses, trees contribute to cleaner air, better thermal comfort, and carbon sequestration [16]. Optimizing green-area configuration strengthens these benefits and improves spatial efficiency [17]. Together, these functions elevate environmental quality and support the physical and mental health of students and staff, underscoring the central role of natural elements in campus design.

2.1.2. Campus Landscapes and Psychological Perception

Campus landscapes strongly shape students' psychological perceptions and behaviors. Environmental psychology shows that they regulate mood and can influence social interaction and academic performance [6,18]. Perceived restorativeness is a key mediator that links campus green features to restorative experiences [19,20]. Multiple studies support this pathway. Gao et al. (2025) and Felsten (2009) [21,22] found that spatial restorative qualities affect students' emotions and health through perceived restorativeness, and emphasized that subjective experience is central to recovery.

Sensory dimensions have a direct and stable positive impact on the restorative experience. Mechanistically, Zhang et al. (2024) [23] used structural equation modeling to show that sensory dimensions are not only strongly associated with restoration but also act through the mediating paths of "capacity for solitude" and "perceived restorativeness." In parallel, Foellmer et al. (2021) [24] reported that students' subjective perceptions of green space, including their sense of identity and emotional attachment, are closely tied to their physical and mental well-being. At the feature level, spaces with richer water and vegetation, as well as those with cultural symbolism, more effectively elevate positive emotions and reduce negative affect [25,26]. At the behavioral level, Diao et al. (2024) [27] found a significant positive link between students' subjective ratings of landscape quality and mental health. Although the role of objective indicators remains uncertain, higher visitation frequency clearly strengthens psychological well-being. Hence, as a key setting of daily academic life, campus landscapes shape students' perceptions and behaviors through multiple pathways.

2.1.3. Research Methods and Emerging Trends in Campus Landscape Research

In recent years, driven by digital technologies, a more profound commitment to sustainability, and user-centered design, campus landscape research has shifted from traditional approaches to interdisciplinary integration. Early studies relied on field mapping, questionnaires, and spatial statistics to quantify greening rate, vegetation structure, and spatial accessibility, and to relate these metrics to students' behaviors and psychological perceptions [28]. With the adoption of GIS and advanced spatial analytics, researchers can more directly examine the distribution, functions, and configurations of campus green space, providing crucial tools to reveal links between landscape attributes and user experience [29].

Driven by these advances, researchers have begun using remote sensing to conduct multi-scale spatiotemporal analyses that track changes in campus and urban green space. For example, Ma et al. (2022) [30] showed that GIS and spatial statistics can effectively reveal the spatiotemporal evolution of Beijing's urban green space from 1990 to 2019. This methodological toolkit applies not only to city-scale studies but also provides powerful tools for dynamic analysis, scenario simulation, and forward prediction of campus landscapes.

With the deepening of campus informatization, big data and machine learning are being incorporated into landscape performance assessment. For example, Zhuang et al. (2025) [31] utilized image semantic segmentation and explainable machine learning to assess the restorative quality of university outdoor spaces, employing XGBoost with SHAP to identify key landscape features and optimization options. Chai et al. (2025) [32] proposed a machine-learning-based planning framework that demonstrates how to integrate large datasets with complexity analysis to systematically enhance campus landscape design.

In summary, campus landscape research is shifting from traditional qualitative descriptions and spatial statistics toward fine-grained, data-driven evaluations powered by big data and artificial intelligence. This transition broadens the research perspective and offers new tools and approaches for the scientific design and management of campus landscapes.

2.2. *Theoretical Framework*

In campus landscape research, students' psychological and behavioral responses are shaped by environmental features [20]. As a core setting for study and daily life, the campus not only delivers knowledge but also shapes users' emotions and social identity [33]. Ecological or aesthetic indicators alone cannot capture the full value of campus space.

The conceptual framework in Figure 1 reveals how campus environmental features influence students' psychological, behavioral, and social responses through three perception dimensions. Sense of security (based on Prospect-Refuge Theory), as the foundational layer, creates the necessary conditions by reducing threats and enhancing spatial safety. Environmental comfort (based on Thermal Comfort Theory), as the intermediary layer, transforms physical and sensory inputs into emotional enhancement and sustained engagement. Sense of belonging (based on Social Identity Theory), as the higher-level layer, strengthens identity recognition and maintenance intentions through repeated exposure and place attachment. The arrows in the figure have been annotated to explain the key mechanisms: sense of security provides the foundation for comfort and belonging, comfort moderates the efficiency of the transition from security to belonging, and belonging generates protective behaviors and long-term resilience. Together, these three dimensions form the path from environmental input to student outcomes, ultimately promoting long-term academic achievement and social well-being.

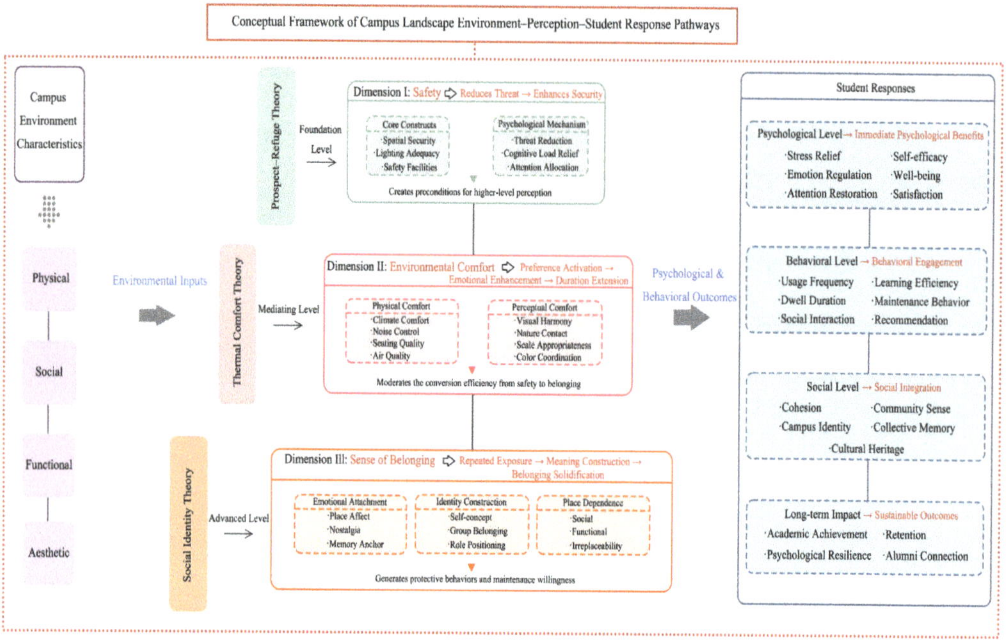

Figure 1. Conceptual framework of this work.

First, a sense of safety is the fundamental precondition for students' normal study and life on campus. The Prospect–Refuge Theory suggests that spaces offering both clear outward views and moderate shelter can reduce uncertainty and perceived threats, thereby improving safety experiences [34]. Recent empirical studies have further validated this. Peng et al. (2024) [35] developed a model of day-and-night sense of security in Chinese universities, showing that lighting, visibility, and landscape configuration significantly influence students' sense of security. Zhu et al. (2023) [36] identified five key environmental factors that affect the sense of security during nighttime walking.

Second, environmental comfort reflects students' direct experience of the physical and sensory setting. Thermal Comfort Theory provides the foundation, noting that comfort arises from a combination of factors. Beyond temperature and humidity, it also depends on factors such as light, ventilation, noise, and visual landscape quality [37]. In campus contexts, natural elements such as lawns, water, and trees enhance comfort by improving the microclimate, easing heat island effects, and creating a pleasant sensory environment [38]. Specifically, Zhang et al. (2024) [39] assessed the thermal comfort of students in semi-outdoor corridors of universities in hot-humid regions, while Elshabshiri et al. (2025) [40] evaluated the microclimate differences between various urban areas on university campuses.

Ultimately, a sense of belonging reflects the emotional bond and social identification that form between students and the campus environment. Social Identity Theory provides the foundation: individuals construct their identity through interactions with settings and groups, which in turn fosters a sense of belonging [41]. Recent studies have shown that Chang et al. (2025) [42] used structural equation modeling to demonstrate that landscape perception and place attachment can enhance school belonging and promote psychological well-being.

3. Materials and Methods

At the data level, this study uses Baidu Maps street-view images and the Mask2Former semantic segmentation model to extract spatial features of landscape elements, including grass, trees, roads, buildings, and water. The overall technical workflow is shown in Figure 2. At the perception level, we conducted a questionnaire survey to collect students' subjective ratings on three core dimensions: perceived safety, environmental comfort, and sense of belonging (Step 1).

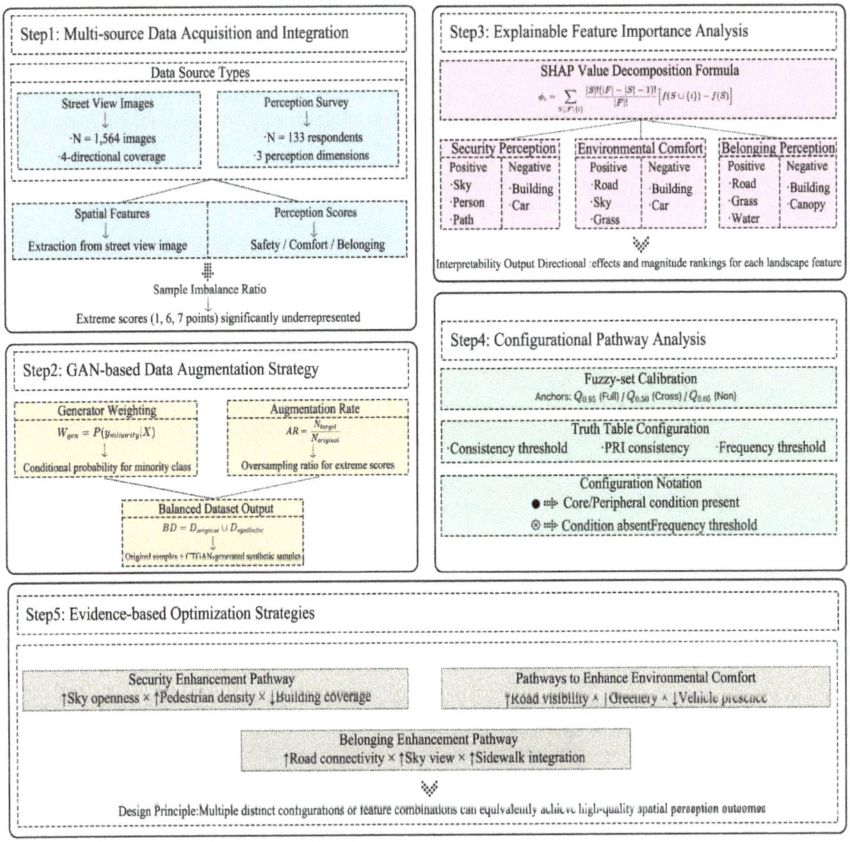

Figure 2. Technical workflow. **Note:** In Figure 2, ↑ denotes the increasing of characteristic values, and ↓ denotes the decreasing of characteristic values.

To address the imbalance between high and low scores in the perception indices, we apply a CTGAN model to augment minority extreme-score samples (Step 2). We then combine explainable machine learning to identify key landscape elements that significantly affect safety, comfort, and belonging (Step 3). On this basis, we use fsQCA to explore, from a configurational perspective, how different elements interact to form effective pathways for campus spatial quality (Steps 4 and 5). In Step 4 of the fuzzy-set Qualitative Comparative Analysis (fsQCA), we conduct variable calibration, truth-table configuration, and the identification of ancillary conditions. Variable calibration is based on the 5th, 50th, and 95th percentiles of each variable.

The 5th, 50th, and 95th percentiles are pragmatic, theory-informed anchors for fsQCA calibration. The 5th percentile marks full nonmembership, reserving a score of 0 for clearly "out" cases and limiting misclassification at the low end. The 95th percentile marks full membership, assigning 1 only to unequivocally "in" cases at the top of the distribution. The 50th percentile serves as the crossover (0.5), the point of maximum ambiguity. Together, these anchors translate interval data into consistent fuzzy scores while remaining open to theoretical justification.

3.1. Data Collection and Feature Quantification

- Campus street-view acquisition

Fujian Normal University (FNU), founded in 1907, is a key provincial university and an early national pilot under the "Program for Training Talents to Meet Special National Needs." This study focuses on its Qishan Campus, and it can be seen at https://www.openstreetmap.org/relation/12658062 (accessed on 23 October 2025). As shown in Figure 3, the campus is located in the University New District of Shangjie Town, Minhou County, Fuzhou, covering approximately 2800 mu (≈187 hectares). It serves as the main campus and the primary base for teaching and research. The setting is framed by hills and water, with a well-organized internal layout and emblematic nodes such as the Youxuan Library, Xingyu Lake, and Baochen Square.

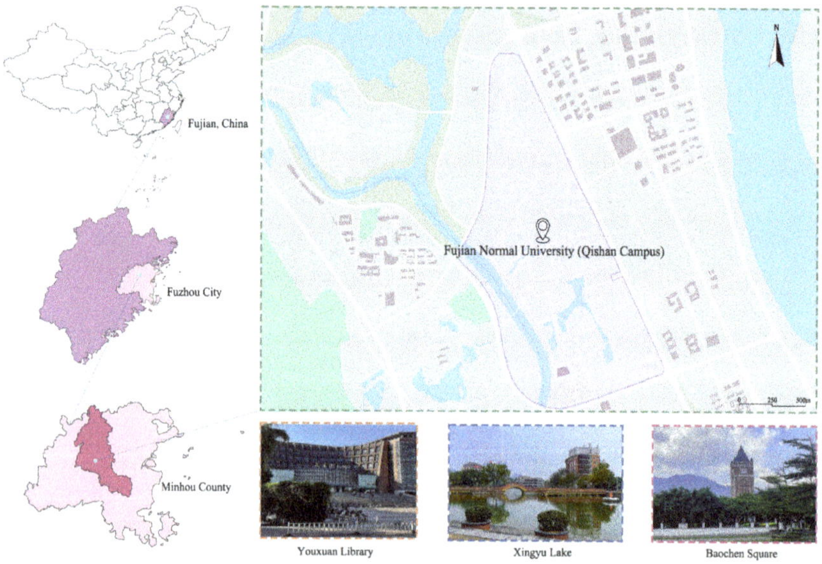

Figure 3. Geographical location of the study area.

We first obtained campus road data from OpenStreetMap (OSM) and, using QGIS, generated 391 sampling points at 50 m intervals, recording their latitude–longitude coordinates. For each point, with a fixed pitch angle of 0°, we captured street-view images facing 0°, 90°, 180°, and 270° azimuth angles. Each location produced four images, resulting in a total of 1564 images. This dataset provides near-complete coverage of the campus and supplies the data basis for subsequent semantic segmentation and environmental perception modeling [43].

- Quantification of environmental perception indicators

To evaluate street-view images on three dimensions (perceived safety, environmental comfort, and sense of belonging), we followed the rule "use validated scales when available; otherwise adapt classic items": (1) Safety: drawing on environmental-psychology research on public-space safety, we integrated lighting, visibility, escape potential, and maintenance, and translated them into items that can be judged directly from images [44,45]. (2) Environmental comfort: based on an outdoor thermal and psychological adaptation framework, we contextualized visual cues for thermal comfort, visual comfort, and ventilation/noise [37,46]. (3) Sense of belonging: using a validated place-attachment scale, we selected representative items from "place identity" and "place dependence," adapted to the campus context [47].

Table 1 shows example semanticized items for the evaluation. All items used a 7-point Likert scale (1 = "strongly disagree," 7 = "strongly agree"). Using the FNU Qishan Campus image library, we applied stratified random sampling by road class and spatial zone to select 200 representative images. A total of 133 enrolled students completed online ratings. Note that our focus is subjective visual perception under image conditions, not actual use behaviors. This study assessed the reliability and validity of the questionnaire. The overall Cronbach's α was 0.995, indicating excellent internal consistency. The KMO measure of sampling adequacy was 0.992, well above the 0.90 threshold, showing that the data are highly suitable for factor analysis. Taken together, the questionnaire demonstrates strong reliability and construct validity, providing a sound statistical basis for subsequent analyses.

Table 1. Example of semantic judgment items for street view image perception evaluation.

Indicators Used	Example of Semantic Judgment Items (Based on Street View Images)
Sense of security	The street environment, including its lighting and visibility, makes me feel safe and aware of my surroundings.
Sense of comfort	The overall environment, including openness, cleanliness, greenery, and air circulation, makes me feel comfortable and at ease.
Sense of belonging	The campus environment, with its quiet atmosphere, harmonious design, and strong connection to campus life, enhances my overall sense of belonging.

3.2. Data Processing Workflow

- Workflow for semantic information extraction from street-view images

With broad coverage and rich visual detail, street-view images have become an important data source for environmental perception and spatial analysis. Figure 4 presents the Transformer workflow and the resulting semantic segmentation outputs. In this study, we utilize the Mask2Former model to extract semantic features from images. The model is based on a Transformer architecture, which fuses a convolutional backbone with a set-prediction mechanism to jointly optimize semantic and panoptic segmentation tasks [48]. With this approach, key campus elements such as vegetation, pedestrian paths, sky openness, and building contours are identified at the pixel level, providing a solid foundation for subsequent perception modeling.

For Mask2Former on ADE20K, train semantic and panoptic models separately under the same configuration. Use a Swin B backbone pretrained on ImageNet 1K, AdamW with β values 0.9 and 0.999, a base learning rate of 0.0001, weight decay 0.02, and a linear warm up of 1500 steps followed by cosine decay to step 180,000 at an effective batch size of 32. Apply moderate augmentation that includes random resize with scale from 0.8 to 1.6, horizontal flip, random crop 640 $\times$ 640, photometric jitter, and Gaussian blur with probability 0.10. Use drop path 0.2, gradient clipping at 1.0, and focal loss with γ equal to 2 for mask classification together with the standard task losses. During inference, keep aspect ratio and resize the long edge to 1024 pixels; for large scenes, use sliding windows of 640 $\times$ 640 with a stride of 426 pixels, and optionally apply multi scale testing at 0.75, 1.0, and 1.25 for semantic segmentation. Set 150 queries for semantic runs with Swin B and 250 queries for panoptic runs to stabilize instance grouping without excessive memory use.

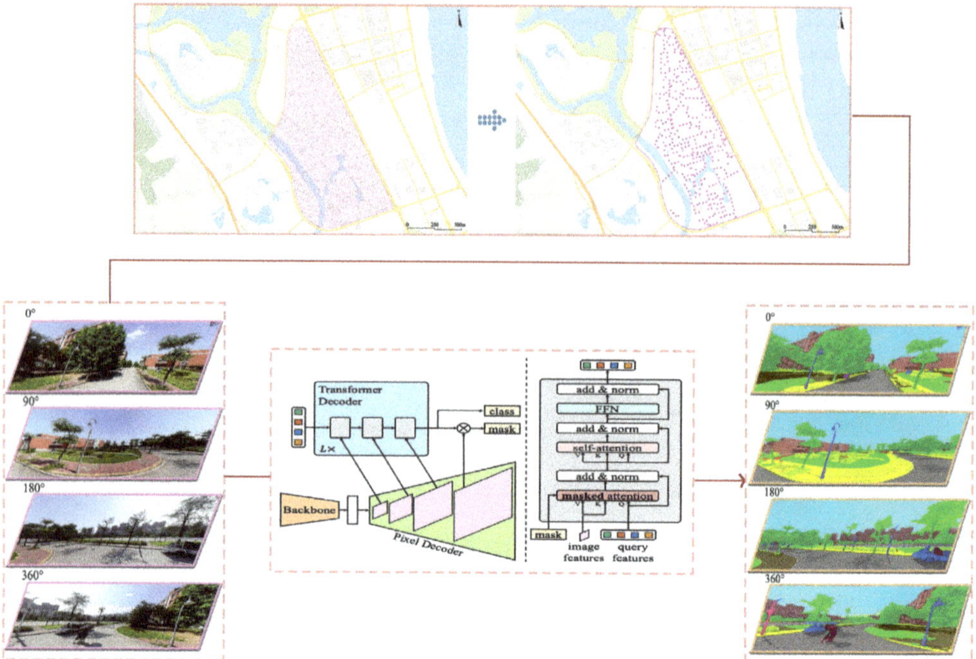

Figure 4. Workflow of the Transformer model and semantic segmentation results. **Note:** Workflow of the Transformer model from [48].

- Application of CTGAN to imbalanced data

In recent years, generative adversarial networks (GANs) have gained importance in handling tabular data and mitigating class imbalance. Xu et al. (2019) [49] introduced CTGAN, which combines conditional generation with mode-specific normalization. It supports both continuous and categorical variables and produces high-quality, diverse samples under multimodal and imbalanced distributions. Subsequent work extended this line of research: Engelmann & Lessmann (2021) [50] proposed a conditional Wasserstein GAN to stabilize oversampling; Zhao et al. (2024) [51] developed CTAB-GAN by adding a classification loss, information-preserving constraints, and improved conditional vectors, which enhanced statistical fidelity and downstream predictions; CTAB-GAN+ further broadened the training scheme to address more complex imbalance patterns [51]. CTGAN models multi-peaked continuous features through mode-specific normalization, and its conditional generator and sampling strategy strengthen learning from minority classes in categorical features. The generator and discriminator are fully connected networks, enabling a unified treatment of both continuous and discrete attributes [52]. In summary, CTGAN-based approaches enhance data quality and mitigate distributional imbalance, providing a robust methodological foundation for modeling and interpreting campus environmental perception data in this study.

- Fuzzy-Set Qualitative Comparative Analysis (fsQCA)

The core idea of fsQCA is that outcomes arise from configurations of multiple causal conditions. Research should therefore examine the complex causal links between different condition sets and the outcome. In recent years, QCA has seen growing use in management studies. For example, Li et al. (2025) [53] employed fsQCA to investigate how modern agricultural industry systems influence farm household incomes, including their effects on income structure and crowding-out across various dimensions. Wen et al. (2025) [54] applied fsQCA to university public spaces. They found that an environment–facility mix, a static rest pattern, a dynamic activity pattern, and an interaction-driven pattern all foster substantial psychological restoration.

From the perspective of variable adaptability, QCA includes three types: crisp-set (csQCA), multi-value (mvQCA), and fuzzy-set (fsQCA). Compared with the other two, fsQCA offers greater analytical flexibility. It handles categorical variables while also accommodating continuous change and partial membership. By transforming fuzzy-set data into a truth table, fsQCA preserves the strengths of traditional QCA in dealing with qualitative evidence, limited diversity, and configuration simplification, while combining the advantages of qualitative inquiry and quantitative analysis.

4. Empirical Results

4.1. Modeling Results of Augmentation-Based Explainable Machine Learning

Using features derived from semantic segmentation as inputs and students' perceived safety, comfort, and belonging as targets, we built machine learning models. We first evaluated eXtreme Gradient Boosting (XGBoost), Random Forest (RF), and Gradient Boosting Decision Tree (GBDT) on the test set, as shown in Table 2. For each perceptual dimension, we selected the best single algorithm. On that basis, for scores of 1, 6, and 7 whose counts were below the sample mean, we applied a Conditional Generative Adversarial Network (CTGAN) to augment minority cases, then retrained with the chosen algorithm. Model performance was assessed using mean squared error (MSE), mean absolute error (MAE), and mean absolute percentage error (MAPE). The error indicators in Table 2 show that the CTGAN-enhanced models, paired with XGBoost, RF, and GBDT for safety, comfort, and belonging, respectively, outperformed their single-algorithm baselines.

Table 2. The performance of enhanced decision tree algorithms.

	Sense of Security				**Sense of Comfort**				**Sense of Belonging**			
	XGBoost	RF	GBDT	**CTGAN-XGBoost**	XGBoost	RF	GBDT	**CTGAN-XGBoost**	XGBoost	RF	GBDT	**CTGAN-XGBoost**
MSE	1.33	1.69	1.51	**1.21**	1.34	1.21	1.19	**0.88**	1.56	1.59	1.58	**1.00**
MAE	0.93	0.98	0.95	**0.83**	0.95	0.89	0.89	**0.72**	1.02	1.03	1.02	**0.76**
MAPE	35.08	37.98	36.06	**33.25**	35.72	34.24	34.10	**28.29**	35.66	36.33	35.65	**29.56**

Note: The bold entries indicate the best-performing model.

4.2. Key Semantic Features Influencing Campus Environmental Perception

Figure 5 maps the spatial distribution of six key semantic features. Building clusters in core teaching zones and functional precincts, forming a high-density pattern that reflects spatial concentration. Grass is widespread across green areas and ecological buffers, forming the campus's natural base. The path shows a mesh-like layout that supports walkability and spatial permeability. The road functions as the backbone network, covering the campus and enhancing circulation and spatial organization. Sky appears in open areas and between buildings, enhancing visual openness. Car points are dense along roads and parking areas, indicating traffic activity. Together, these patterns shape campus functions and perceptual experience, offering clear guidance for optimizing landscape structure and environmental quality.

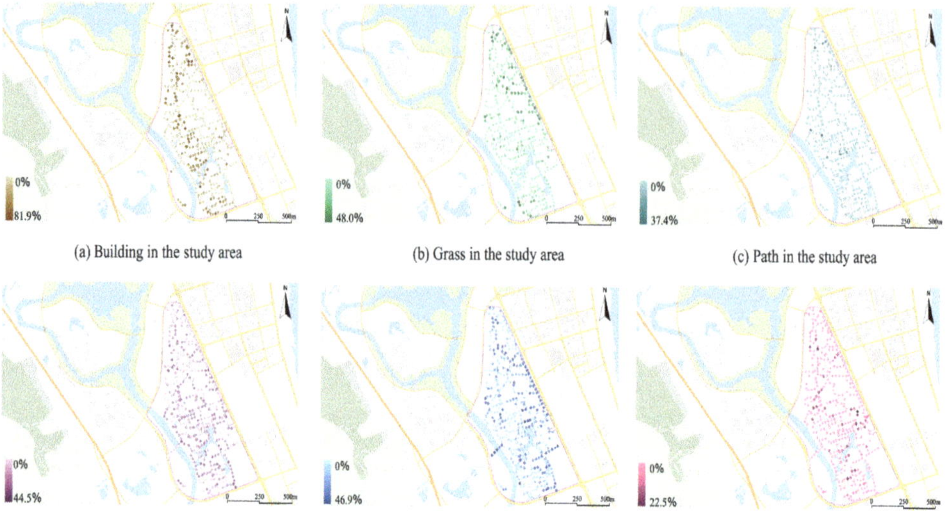

(a) Building in the study area (b) Grass in the study area (c) Path in the study area

(d) Road in the study area (e) Sky in the study area (f) Car in the study area

Figure 5. Spatial distribution of six key semantic features in the study area. **Note:** The legend in the top right corner of Figure 5 represents a compass. The legend in the bottom right corner indicates that one unit corresponds to a distance of 250 m on the map. The legend in the bottom left corner uses a gradient of colors to show the distribution of specific landscape semantic feature values across all coordinate points in this study.

Figure 6 presents the feature-importance results for safety based on the combined CTGAN–XGBoost and SHAP models. Each row displays the SHAP value distribution for a single feature, with most features exhibiting few extremes and many values close to zero. The top six image semantic features affecting campus safety, in order, are River, Sky, Car, Person, Hill, and Path, followed by Plant, Building, and Earth. The SHAP distributions indicate that River contributes mainly to adverse effects on perceived safety. In contrast, higher SHAP values for Sky, Person, Hill, and Path correspond to positive effects on safety. For cars and buildings, Figure 6 indicates a negative effect on perceived safety. Figures 7 and 8 further show that these two landscape features have a significant negative impact on comfort and sense of belonging. In summary, for campus safety, natural features such as rivers and broad sky show stronger negative associations, possibly due to the uncertainty of water and the unease that very open skies may evoke. In contrast, human-presence features (e.g., persons and paths) tend to enhance safety, suggesting that visible, controllable activity benefits students' sense of security.

Figure 7 reports feature-importance results for environmental comfort using the combined CTGAN–GBDT and SHAP models. Ranked by importance, the top six image semantic features are Boat, River, Sky, Road, Bridge, and Building, followed by Car, Stairs, and Ball. The SHAP distributions show that for Boat, River, Sky, Road, and Bridge, high SHAP values cluster on the right, indicating mainly positive effects on campus comfort. For Ball, high SHAP values cluster on the left, indicating primarily adverse effects on comfort. Overall, for environmental comfort, the positive effects of Boat, River, Sky, Road, and Bridge imply that water reflections, coherent road layouts, and bridge design create open and pleasant settings that raise comfort. The adverse effects of buildings and cars suggest that dense building clusters and heavy traffic can lead to crowding and noise, thereby reducing overall comfort.

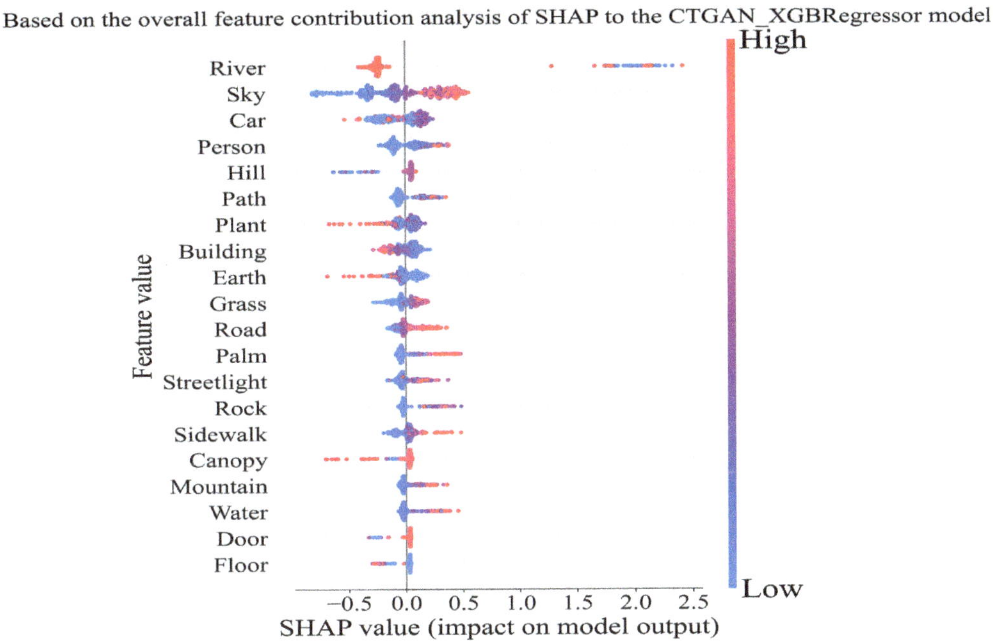

Figure 6. Feature-importance results for safety based on the combined CTGAN-XGBoost and SHAP models.

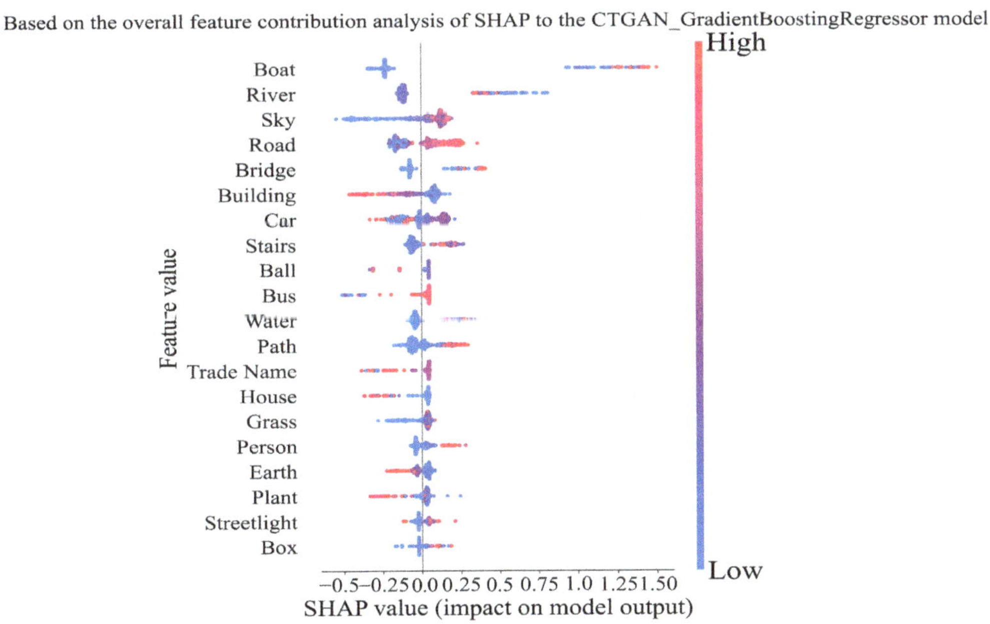

Figure 7. Feature-importance modeling results for environmental comfort based on the combined CTGAN-GBDT and SHAP models.

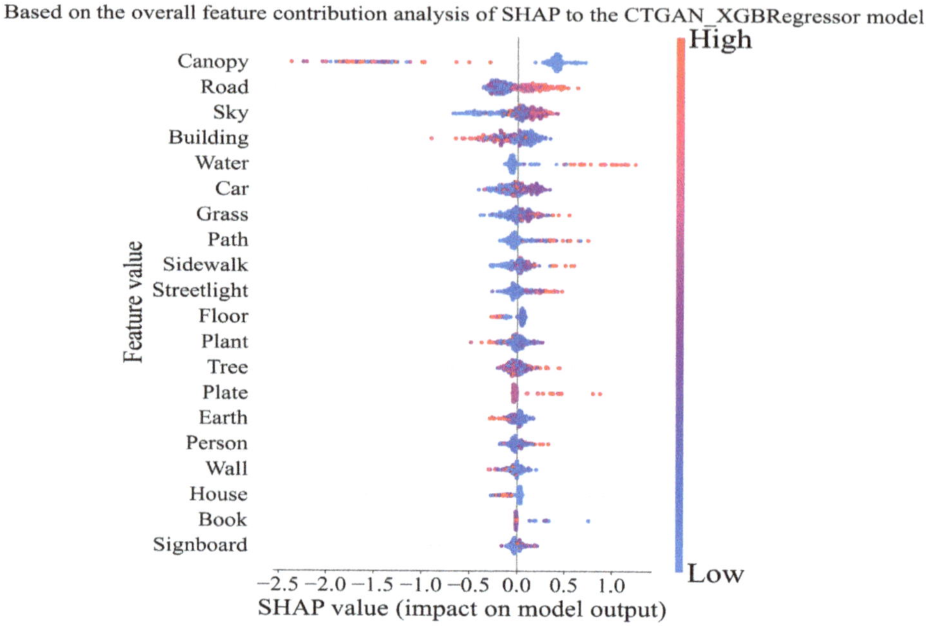

Figure 8. Feature-importance modeling results for sense of belonging based on the combined CTGAN-XGBoost and SHAP models.

Figure 8 presents feature-importance results for sense of belonging using the combined CTGAN–XGBoost and SHAP models. Ranked by importance, the top six image semantic features are Canopy, Road, Sky, Building, Water, and Car, followed by Grass, Path, and Sidewalk. The SHAP distributions show that for Road, Water, Grass, Path, and Sidewalk, high SHAP values cluster on the right, indicating strong positive effects on belonging. For Canopy and Sky, high SHAP values cluster on the left, indicating pronounced adverse impact on belonging. In conclusion, for a sense of belonging, natural elements foster affinity and attachment, mainly grass and water, which symbolize open and free spaces. By contrast, Canopy, Sky, and Building show negative links that may stem from excessive enclosure and an isolating built context. Such occlusion can induce a feeling of isolation and weaken one's sense of belonging.

4.3. Configuration Analysis for Enhancing Campus Environmental Perception

We further employ fuzzy-set Qualitative Comparative Analysis (fsQCA) to examine the configuration pathways by which campus landscape features enhance safety, comfort, and a sense of belonging. The recommended number of condition variables is 6–8 [55,56]. Before analysis, we calibrate fuzzy-set membership for both conditions and outcomes [57]. For safety, we initially selected River, Sky, Car, Person, Hill, and Path based on feature importance. The values at the 0.05, 0.25, and 0.50 percentiles for River, Hill, and Path are identical, making variable calibration and subsequent fsQCA analysis unfeasible. This is due to the rarity of these landscape features on campus, leading to their exclusion. Based on the previously mentioned feature importance, we have replaced them with Plant, Building, and Earth. For comfort, the six final features are Sky, Road, Building, Car, Grass, and Person. For belonging, the six final features are Road, Sky, Building, Car, Grass, and Sidewalk. Table 3 reports the full membership, crossover, and full nonmembership anchors for each set of features and the outcome.

Table 3. Calibration of condition variables and outcome variables in different environmental perception dimensions.

	Variable Name	Full Membership Point	Calibration Crossover Point	Full Non-Membership Point
Outcome variable	Sense of security	6	4	2
	Sense of comfort	6	3	2
	Sense of belonging	6	4	5
Condition variable	Sky	0.16	14.26	35.47
	Plant	0	1.10	11.88
	Person	0	0.0046	1.17
	Car	0	3.34	11.01
	Grass	0	6.12	28.02
	Sidewalk	0	3.53	23.81
	Building	0.16	7.00	47.72
	Earth	0	0.13	17.10
	Road	0	17.64	32.97

Second, we present the necessity analysis for single landscape features, as shown in Table 4. All individual features have consistency below 0.9, indicating that no single feature is a necessary condition for campus safety, comfort, or sense of belonging. This suggests that changes in one feature have limited explanatory power for environmental perception, and that multiple features must act together through configuration and synergy to have a significant impact.

Table 4. Results of necessity tests for individual landscape features.

Condition Variable	High Level of Sense of Security	
	Overall Consistency	Overall Coverage
Sky	0.77	0.63
Car	0.56	0.61
Person	0.54	0.62
Plant	0.55	0.55
Building	0.55	0.50
Earth	0.49	0.52
Condition Variable	High Level of Sense of Comfort	
	Overall Consistency	Overall Coverage
Sky	0.68	0.74
Road	0.68	0.74
Building	0.53	0.63
Car	0.52	0.74
Grass	0.58	0.69
Person	0.49	0.75
Condition Variable	High Level of Sense of Belonging	
	Overall Consistency	Overall Coverage
Road	0.66	0.67
Sky	0.69	0.69
Building	0.53	0.59
Car	0.49	0.64
Grass	0.59	0.65
Sidewalk	0.57	0.66

Based on the studies in [53,54,58], the consistency threshold is initially set to 0.75, the PRI consistency to 0.5, and the frequency threshold to 3. Given 391 sampling points (1564 images) and many rare condition combinations, a consistency threshold of 0.75 with a PRI of 0.50 balances substantive sufficiency against overlap, reducing spurious solutions un-der moderately overlapping sets. A frequency cutoff of 3 filters out configurations backed by only one or two cases, which suits this sample size and limited diversity, keeping the final solutions stable and interpretable. The decision to adjust the threshold boundaries was based on the tuning results of the intermediate solutions. Subsequent results show that the valid paths are consistently 2–3, so there is no need to increase the thresholds further to reduce the number of paths.

Panel A of Table 5 reports the fsQCA configuration results for campus safety. In the notation, the presence of a peripheral condition is marked "•", its absence is marked "⊗", the presence of a core condition is marked "•", and the absence of a core condition is marked "⊗". Interpretation of the optimization pathways should be aligned with the effect directions reported in Section 4.2.

Table 5. Configuration analysis results for high-level environmental perception.

Condition Variable	Panel A. High Level of Campus Sense of Security	
	Configuration A1	Configuration A2
Sky	•	•
Car		⊗
Person	•	•
Plant		
Building	•	
Earth		•
Consistency	0.81	0.82
PRI	0.57	0.55
Coverage	0.38	0.35
Unique coverage	0.05	0.04
Overall PRI	0.53	
Overall consistency	0.78	
Overall coverage	0.50	
Condition Variable	**Panel B. High Level of Campus Sense of Comfort**	
	Configuration B1	**Configuration B2**
Sky		•
Road	•	
Building		
Car		•
Grass	•	•
Person	•	•
Consistency	0.90	0.91
PRI	0.72	0.72
Coverage	0.25	0.21
Unique coverage	0.01	0.01
Overall PRI	0.69	
Overall consistency	0.85	
Overall coverage	0.41	

Table 5. *Cont.*

Condition Variable	Panel C. High Level of Campus Sense of Belonging		
	Configuration C1	Configuration C2	Configuration C3
Road	•		•
Sky	•	•	
Building			
Car			
Grass			•
Sidewalk		•	•
Consistency	0.75	0.81	0.86
PRI	0.56	0.64	0.65
Coverage	0.51	0.42	0.29
Unique coverage	0.02	0.03	0.01
Overall PRI	0.50		
Overall consistency	0.66		
Overall coverage	0.83		

Note: • denotes the presence of a peripheral condition, and ⊗ indicates its absence.

Panel A of Table 5 indicates that Configuration A1 achieves a high level of campus security with high sky density, high crowd density, and low building density as peripheral conditions. Configuration A2 attains the same outcome with high sky density, high crowd density, and low earth density as peripheral conditions, while low car density is not required (some traffic is permitted). Thus, there are two routes to a high level of campus safety: (i) high sky density + high crowd density + low building density; or (ii) high sky density + high crowd density + low earth density, even when traffic is present. The overall consistency is 0.78, indicating strong sufficiency in relation to the standard criteria. Overall coverage is 0.50, meaning that about half of the cases are explained, and the overall PRI is 0.53, indicating moderate discriminating power. The above results suggest that in areas with high building density, enhanced vertical surveillance and crowd management should be prioritized, while in areas primarily for ground activities, focus should be on spatial planning and optimizing traffic flow.

Panel B of Table 5 shows two configurations for a high level of campus comfort. Configuration B1 achieves the outcome with high road density, high pedestrian flow, and low building density as peripheral conditions. Configuration B2 achieves the outcome with high sky density, low vehicle density, high grass density, and high pedestrian flow as peripheral conditions. The two configurations have an overall consistency of 0.85, overall coverage of 0.41, and overall PRI of 0.69. Their shared peripheral condition is high pedestrian flow. Under low building density, B1 attains high comfort by increasing the share of roads. Under low vehicle density, B2 attains high comfort by increasing pedestrian density, sky openness, and grass coverage. Based on the above results, two differentiated practical approaches can be adopted to improve campus comfort: prioritize optimizing network accessibility in areas with sparse buildings, and focus on open space design and green space permeability in areas with traffic restrictions. Both approaches should ensure sufficient support for pedestrian activity.

Panel C of Table 5 identifies three configurations that contribute to a high level of campus belonging. Configuration C1 reaches the outcome with high road density and high sky openness as peripheral conditions. Configuration C2 reaches the outcome with high sky openness and high sidewalk density as peripheral conditions. Configuration C3 reaches the outcome with high road density, high grass density, and high sidewalk density as peripheral conditions. The overall consistency is 0.66, indicating solid sufficiency, and the overall coverage is 0.83, demonstrating broad case coverage. In summary, C1 and C2 share high sky openness, C1 and C3 share high road density, and C2 and C3 share high sidewalk density. This suggests that broader sky views and a dense network of walking paths strengthen students' sense of belonging to the campus. Therefore, campus planning should focus on enhancing the visual permeability of open spaces and the continuity of pedestrian networks. For example, controlling building heights along main roads and walkways and increasing green visibility can help create a campus environment with a stronger sense of belonging.

The results of this study contribute to campus landscape optimization. For safety management, differentiated strategies should be adopted based on building density and ground activity. In high-density areas, vertical surveillance and crowd guidance should be strengthened, while spatial layout and traffic flow should be optimized in open activity areas. To improve comfort, network accessibility should be prioritized in areas with sparse buildings, and open space design and green permeability in traffic-restricted areas, ensuring a pedestrian-friendly environment. For fostering a sense of belonging, building heights along major roads and walkways should be controlled, while enhancing sky visibility, green experience, and pedestrian network connectivity.

5. Discussion

In the discussion, and in line with prior work, we first employ a multiple linear regression model to examine how the landscape elements in Table 5 influence environmental perception. This step demonstrates that numerous regressions can reveal single-feature effects but cannot provide guidance on optimizing campus spatial quality from the perspective of combined effects across multiple features.

The regression results in Table 6 align closely with the SHAP distributions in Figures 5–7. For campus safety (Table 6, Panel A), the estimated coefficients are: Sky = 0.0354, Person = 0.1215, Plant = −0.0263, Building = −0.0109, and Earth = −0.0344. All are significant at the 5% level or higher. The model's R^2 and F-statistic indicate a good fit. However, Table 6 only reveals the effects of single landscape features on environmental perception; it does not show how multiple features act in combination to enhance campus spatial quality.

Table 6. Multiple linear regression analysis results.

Panel A	**Sky**	**Car**	**Person**	**Plant**	**Building**	**Earth**	R^2	**F Statistics**
Sense of security	0.0354 *** (0.00)	0.0005 (0.96)	0.1215 ** (0.02)	−0.0263 *** (0.00)	−0.0109 *** (0.00)	−0.0344 *** (0.00)	0.16	51.19
Panel B	**Sky**	**Road**	**Building**	**Car**	**Grass**	**Person**	R^2	**F Statistics**
Sense of comfort	0.02 *** (0.00)	0.0223 *** (0.00)	−0.0106 *** (0.00)	0.0041 (0.63)	0.0072 ** (0.05)	0.0639 (0.19)	0.12	36.28
Panel C	**Road**	**Sky**	**Building**	**Car**	**Grass**	**Sidewalk**	R^2	**F Statistics**
Sense of belonging	0.0323 *** (0.00)	−0.0154 *** (0.00)	−0.0091 *** (0.00)	−0.0241 *** (0.01)	0.0239 *** (0.00)	0.0374 *** (0.00)	0.12	36.49

Note: ***, ** denote statistical significance at the 1%, 5% levels, respectively.

In the discussion, we also compare our work with prior studies to highlight its advantages. Table 7, Panel A, shows that explainable machine learning methods, such as tree models with SHAP, can identify key landscape elements and reveal threshold effects; however, they have their limits. A central issue is the accuracy of feature identification. Although [31,42,59] employed explainable machine learning, their predictions were weaker because, as Table 2 indicates, they did not utilize CTGAN to address class imbalance in perception scores, which may have biased their conclusions. In contrast, this study effectively mitigates model bias caused by uneven sample distribution by introducing CTGAN to generate synthetic data. This significantly improves the identification accuracy of key landscape features and the overall predictive stability of the model.

Table 7. Comparisons with previous research.

Panel A. Application of Explainable Machine Learning Models in Campus Landscape Research			
Author (Year)	**Research Object/Dependent Variable**	**Key Points of Explainable Machine Learning**	**Limitations**
Zhuang et al. (2025) [31]	Psychological restoration, emotional uplift, and social interaction	SHAP identifies key features & thresholds; regression checks hidden significant factors	Explainable machine learning identifies key landscape features and thresholds, but its results depend on feature quality, lack causal interpretability, and have limited generalizability across contexts.
Chang et al. (2025) [42]	Psychological well-being via psychosocial pathways	SHAP reveals threshold & interaction patterns among predictors	
Chen et al. (2025) [59]	Attention restoration quality	SHAP ranks soundscape & visual metrics; finds interaction/threshold effects	
Panel B. Application of Multiple Linear Regression Model in Campus Landscape Research			
Author (Year)	**Research Object/Dependent Variable**	**Independent Variables (Land-scape/Environmental Features)**	**Limitations**
Kim & Huh (2023) [60]	Students' Mental Health/Psychological Well-Being	Symbolism, visual beauty, frequency of use, and other factors.	Multiple linear regression quantifies relationships between landscape features and perceptions, but it relies on subjective data, struggles with nonlinearities and interactions, and faces issues with multicollinearity and limited explanatory power.
Qin et al. (2025) [61]	aesthetics, security, depression, vitality	Buildings, vegetation, grassland, walls, sky openness, etc.	
Fu (2023) [62]	Campus landscape design quality/aesthetics-ecological balance	Ecological Diversity Index, Space Utilization Efficiency, Landscape Aesthetics Indicators.	

Table 7, Panel B, summarizes the use of multiple linear regression (MLR) in campus landscape studies [60–62]. MLR can quantify links between landscape elements and mental health or satisfaction. However, it only captures single-feature effects on environmental perception and cannot uncover the combined effects of multiple features needed to improve perception, which restricts its usefulness. Compared to multiple linear regression, the configurational approach used in this study identifies multiple equivalent pathways leading to high perception results. This explains why similar senses of belonging can emerge under different conditions (e.g., high sky openness with high walkway density or high road density with high greenery). This not only expands the understanding of landscape synergistic effects but also provides a theoretical basis for differentiated planning strategies.

The results of this study contribute to campus landscape optimization. For safety management, differentiated strategies should be adopted based on regional building density and ground activity characteristics. In high-density built areas, vertical surveillance and crowd guidance should be enhanced, while spatial layout and traffic flow should be optimized in open ground activity areas. To improve comfort, network accessibility should be prioritized in areas with sparse buildings, while open space design and green space permeability should be emphasized in traffic-restricted areas, ensuring a pedestrian-friendly environment. For fostering a sense of belonging, attention should be given to controlling building heights along major roads and walkways, enhancing sky visibility and green views, and establishing a coherent pedestrian network.

6. Conclusions

This study investigated the relationship between campus landscape features and environmental perception, specifically examining safety, comfort, and a sense of belonging. By combining semantic image segmentation, augmentation-based explainable machine learning, and fuzzy-set qualitative comparative analysis, we identified key features that shape students' perceptions and uncovered effective configuration pathways for landscape optimization. The main conclusions are:

(1) The sense of security is achieved through the combination of high sky openness and high pedestrian traffic, supported by either low building density or low hardscape density. This pathway remains valid even with a certain level of traffic flow.

(2) The formation of comfort follows two distinct pathways: the first is a combination of high road visibility, high pedestrian traffic, and low building density; the second results from the combined effects of high sky openness, low traffic volume, abundant greenery, and active pedestrian activity.

(3) The formation of comfort follows two distinct pathways, while the creation of a sense of belonging can be achieved in three ways: the combination of high road density and open sky views, the pairing of sky openness with a clear walkway system, or the synergy of high road density, ample greenery, and well-defined walkways.

This configurational framework, identifying multiple equifinal pathways to positive perceptual outcomes, is replicable and provides a transferable model for evidence-based planning in a wide range of campus and urban environments.

Based on our findings, we propose the following design guidelines. To enhance safety, create open, obvious spaces, moderately control building density, and provide paths with visible pedestrian flow to improve sightlines and a sense of control. For comfort, incorporate natural elements such as green spaces and water, and refine the visual and functional design of roads and bridges to enhance the microclimate, mitigate heat island effects, and provide pleasant areas for rest. For belonging, pair broad sky views with a continuous, accessible walkway system to strengthen legibility and connectivity, encourage community interaction, and deepen students' identification with the campus.

This study has three limitations. First, the case site is a regular university in China, so the findings may not generalize to other regions or educational settings; future research should include cross-cultural comparisons to test their applicability. Second, some high-ranking SHAP features (e.g., River, Boat, Hill) were excluded from fsQCA because many observations had zero values in the segmentation outputs, which failed the percentile calibration rule. Future work should introduce continuous variables with fewer zeros—such as tree canopy, green space area, light intensity, and road networks—to analyze configuration pathways for optimizing campus spatial quality. Third, we relied on static street-view images, which do not capture temporal and seasonal variation in perception; future studies can use multi-temporal imagery to explore differences across seasons and times of day.

Author Contributions: Conceptualization, X.Z.; Methodology, X.Z., Y.C. and Z.T.; Software, Z.D. and C.G.; Writing—original draft preparation, X.Z.; Writing—review and editing, Y.C. and Z.D.;

Supervision, Visualization, Z.D. and C.G.; Funding acquisition, Z.T. and Z.D. All authors have read and agreed to the published version of the manuscript.

Funding: This research work was supported by the National Natural Science Foundation of China under Grant No. 72341030, and the Special Funding Project of the China Agricultural and Forestry University Design Art Alliance under Grant No. 111900050.

Data Availability Statement: The original contributions presented in this study are included in the article. Further inquiries can be directed to the corresponding author.

Conflicts of Interest: The authors declare no conflicts of interest.

References

1. Demuzere, M.; Orru, K.; Heidrich, O.; Olazabal, E.; Geneletti, D.; Orru, H.; Bhave, A.G.; Mittal, N.; Feliu, E.; Faehnle, M. Mitigating and Adapting to Climate Change: Multifunctional and Multi-Scale Assessment of Green Urban Infrastructure. *J. Environ. Manag.* **2014**, *146*, 107–115. [CrossRef] [PubMed]
2. Kumar, P.; Debele, S.E.; Khalili, S.; Halios, C.H.; Sahani, J.; Aghamohammadi, N.; Andrade, M.d.F.; Athanassiadou, M.; Bhui, K.; Calvillo, N.; et al. Urban Heat Mitigation by Green and Blue Infrastructure: Drivers, Effectiveness, and Future Needs. *Innovation* **2024**, *5*, 100588. [CrossRef]
3. Guo, W.; Wen, H.; Liu, X. Research on the Psychologically Restorative Effects of Campus Common Spaces from the Perspective of Health. *Front. Public Health* **2023**, *11*, 1131180. [CrossRef] [PubMed]
4. Liu, H.; Wang, Y.; He, Q.; Wang, X.; Cui, C.; Gong, Y. Assessing the Restorative Effects of Campus Greenness on Student Depression: A Comparative Study Across Three Distinct University Campus Types in Macau. *BMC Public Health* **2025**, *25*, 907.
5. Qi, F.; Parra, A.O.; Block-Lerner, J.; McManus, J. Psychological Impacts of Urban Environmental Settings: A Micro-Scale Study on a University Campus. *Urban Sci.* **2024**, *8*, 73. [CrossRef]
6. Gulwadi, G.B.; Mishchenko, E.D.; Hallowell, G.; Alves, S.; Kennedy, M. The Restorative Potential of a University Campus: Objective Greenness and Student Perceptions in Turkey and the United States. *Landsc. Urban Plan.* **2019**, *187*, 36–46. [CrossRef]
7. Carrus, G.; Scopelliti, M.; Lafortezza, R.; Colangelo, G.; Ferrini, F.; Salbitano, F.; Agrimi, M.; Portoghesi, L.; Semenzato, P.; Sanesi, G. Go Greener, Feel Better? The Positive Effects of Biodiversity on the Well-Being of Individuals Visiting Urban and Peri-Urban Green Areas. *Landsc. Urban Plan.* **2015**, *134*, 221–228. [CrossRef]
8. Speake, M. Campus Landscape and Social Cohesion: The Role of Open Spaces. *Urban Des. Int.* **2013**, *18*, 201–210.
9. Sun, S. The Role of Campus Green Spaces in Enhancing Social Interaction and Well-Being. *J. Urban Green Spaces* **2025**, *24*, 112–125.
10. Lee, J.; Park, B.; Lee, C. Effects of Green Space on Recovery from Mental Fatigue. *Environ. Behav.* **2008**, *40*, 517–540.
11. Berman, M.G.; Jonides, J.; Kaplan, S. The Cognitive Benefits of Interacting with Nature. *Psychol. Sci.* **2008**, *19*, 1207–1212. [CrossRef] [PubMed]
12. Hartig, T.; Mang, M.; Evans, G.W. Restorative Effects of Natural Environment Experiences. *Environ. Psychol.* **2003**, *23*, 3–14. [CrossRef]
13. Kaplan, R. The Restorative Benefits of Nature: Toward an Integrative Framework. *J. Environ. Psychol.* **1995**, *15*, 169–182. [CrossRef]
14. Li, J.; Wang, X.; Zhang, Y. Urban Green Space and Microclimate: Impacts on Thermal Comfort and Urban Heat Island. *Landsc. Urban Plan.* **2019**, *184*, 109–118.
15. Wang, Y.; Huang, S.; Li, X. The Role of Green Infrastructure in Urban Water Management: A Case Study of Rainwater Gardens. *Ecol. Eng.* **2020**, *156*, 103–112.
16. Wang, J.; Zhang, Z.; Li, X. The Ecological Value of Campus Trees in Urban Areas: A Comprehensive Evaluation. *Urban For. Urban Green.* **2019**, *44*, 126–137.
17. Li, L.; Zhang, Y.; Zhao, Q. Optimization of Campus Green Space Layout for Enhancing Ecological Services and Space Utilization. *Sustainability* **2021**, *13*, 5141.
18. McFarland, A.L.; Waliczek, T.M.; Zajicek, J.M. The relationship between student use of campus green spaces and perceptions of quality of life. *HortTechnology* **2008**, *18*, 232–238. [CrossRef]
19. Malekinezhad, F.; Courtney, P.; bin Lamit, H.; Vigani, M. Investigating the mental health impacts of university campus green space through perceived sensory dimensions and the mediation effects of perceived restorativeness on restoration experience. *Front. Public Health* **2020**, *8*, 578241. [CrossRef]
20. Hipp, J.A.; Gulwadi, G.B.; Alves, S.; Sequeira, S. The relationship between perceived greenness and perceived restorativeness of university campuses and student-reported quality of life. *Environ. Behav.* **2016**, *48*, 1292–1308. [CrossRef]
21. Gao, W.; Tang, B.M.; Liu, B. Effects of Landscape Characteristic Perception of Campus on College Students' Mental Restoration. *Behav. Sci.* **2025**, *15*, 470. [CrossRef]

22. Felsten, G. Where to take a study break on the college campus: An attention restoration theory perspective. *J. Environ. Psychol.* **2009**, *29*, 160–167. [CrossRef]
23. Zhang, J.; Jin, J.; Liang, Y. The Impact of Green Space on University Students' Mental Health: The Mediating Roles of Solitude Competence and Perceptual Restoration. *Sustainability* **2024**, *16*, 707. [CrossRef]
24. Foellmer, J.; Kistemann, T.; Anthonj, C. Academic greenspace and well-being-Can campus landscape be therapeutic? Evidence from a German university. *Wellbeing Space Soc.* **2021**, *2*, 100003. [CrossRef]
25. Zhang, P.; He, Q.; Chen, Z.; Li, X.; Ma, J. An empirical study on the promotion of students' physiological and psychological recovery in green space on campuses in the postepidemic era. *Int. J. Environ. Res. Public Health* **2022**, *20*, 151. [CrossRef]
26. Xu, Y.; Wang, T.; Wang, J.; Tian, H.; Zhang, R.; Chen, Y.; Chen, H. Campus landscape types and pro-social behavioral mediators in the psychological recovery of college students. *Front. Psychol.* **2024**, *15*, 1341990. [CrossRef]
27. Diao, W.; Li, S.; Zhao, B.; Zhang, F. Estimation of Mediators in the Associations between Campus Green Spaces and Students' Anxiety: A Case Study in Nanjing. *Front. Psychol.* **2024**, *15*, 1396548. [CrossRef] [PubMed]
28. Li, L.; Zhang, Y.; Zhao, Q. Campus open spaces and students' psychological well-being: A meta-analysis. *Buildings* **2025**, *15*, 2835. [CrossRef]
29. Chen, W.Y.; Wang, D.; Zhang, W. Social functional mapping of urban green space using remote sensing and GIS. *J. Photogramm. Remote Sens.* **2019**, *149*, 436–452.
30. Ma, Y.; Zheng, X.; Liu, M.; Liu, D.; Ai, G.; Chen, X. Spatio-temporal evolution characteristics analysis and optimization prediction of urban green infrastructure: A case study of Beijing, China. *Sci. Rep.* **2022**, *12*, 10702. [CrossRef]
31. Zhuang, X.; Tang, Z.; Lin, S.; Ding, Z. Prediction and Optimization of the Restoration Quality of University Outdoor Spaces: A Data-Driven Study Using Image Semantic Segmentation and Explainable Machine Learning. *Buildings* **2025**, *15*, 2936. [CrossRef]
32. Chai, X.; Liu, Y.; Niu, J.; Miao, Y. Landscape Complex Planning in the Big Data Era. *Int. J. Agric. Environ. Inf. Syst.* **2025**, *16*, 1–19. [CrossRef]
33. Weaver, B.J. Mission and Place: Strengthening Learning and Community through Campus Design. *Rev. High. Educ.* **2006**, *29*, 545–546. [CrossRef]
34. Dosen, A.S.; Ostwald, M.J. Evidence for prospect-refuge theory: A meta-analysis of the findings of environmental preference research. *City Territ. Archit.* **2016**, *3*, 4. [CrossRef]
35. Peng, Y.-L.; Li, Y.; Cheng, W.-Y.; Wang, K. Evaluation and Optimization of Sense of Security during the Day and Night in Campus Public Spaces Based on Physical Environment and Psychological Perception. *Sustainability* **2024**, *16*, 1256. [CrossRef]
36. Zhu, M.; Teng, R.; Wang, C.; Wang, Y.; He, J.; Yu, F. Key Environmental Factors Affecting Perceptions of Security of Night-Time Walking in Neighbourhood Streets: A Discussion Based on Fear Heat Maps. *J. Transp. Health* **2023**, *32*, 101636. [CrossRef]
37. Nikolopoulou, M.; Steemers, K. Thermal comfort and psychological adaptation as a guide for designing urban spaces. *Energy Build.* **2003**, *35*, 95–101. [CrossRef]
38. Ambarwati, N.; Faida, L.R.W.; Marhaento, H. The Effects of Green Open Spaces on Microclimate and Thermal Comfort in Three Integrated Campus in Yogyakarta, Indonesia. *Geoplan. J. Geomat. Plan.* **2023**, *10*, 37–44. [CrossRef]
39. Zhang, Q.; Li, Y.; Lin, C. A Study on Outdoor Thermal Comfort of College Students in the Outdoor Corridors of Teaching Buildings in Hot and Humid Regions. *Buildings* **2024**, *14*, 2756. [CrossRef]
40. Elshabshiri, A.; Aly, M.; Alharbat, R.; Abdalla, M.J.; Alsyouf, L.; Ghanim, A.; Yahia, M.W. Enhancing the microclimate of outdoor campus spaces in hot humid climates: The example of the University of Sharjah. *Comput. Urban Sci.* **2025**, *5*, 35. [CrossRef]
41. Liu, X.; Zhao, J. Chinese Migrant Adolescents' Perceived Discrimination and Psychological Well-Being: The Moderating Roles of Group Identity and the Type of School. *PLoS ONE* **2016**, *11*, e0146559. [CrossRef]
42. Chang, Y.; Yang, Y.; Cai, X.; Zhou, L.; Li, J.; Liu, S. How Campus Landscapes Influence Mental Well-Being through Place Attachment and Perceived Social Acceptance: Insights from SEM and Explainable Machine Learning. *Land* **2025**, *14*, 1712. [CrossRef]
43. Biljecki, F.; Ito, K. Street view imagery in urban analytics and GIS: A review. *Landsc. Urban Plan.* **2021**, *215*, 104217. [CrossRef]
44. Blöbaum, A.; Hunecke, M. Perceived danger in urban public space: The impacts of physical features and personal factors. *Environ. Behav.* **2005**, *37*, 465–486. [CrossRef]
45. Nasar, J.L.; Fisher, B. "Hot spots" of fear and crime: A multi-method investigation. *J. Environ. Psychol.* **1993**, *13*, 187–206. [CrossRef]
46. Vázquez-Torres, C.E.; Beizaee, A.; Bienvenido-Huertas, D. The Impact of Human Occupancy in Thermal Performance of a Historic Religious Building in Sub-Humid Temperate Climate. *Energy Build.* **2022**, *259*, 111912. [CrossRef]
47. Williams, D.R.; Vaske, J.J. The measurement of place attachment: Validity and generalizability of a psychometric approach. *For. Sci.* **2003**, *49*, 830–840. [CrossRef]
48. Cheng, B.; Misra, I.; Schwing, A.G.; Kirillov, A.; Girdhar, R. Masked-attention mask transformer for universal image segmentation. In Proceedings of the IEEE/CVF Conference on Computer Vision and Pattern Recognition (CVPR), New Orleans, LA, USA, 18–24 June 2022; pp. 1290–1299.

49. Xu, L.; Skoularidou, M.; Cuesta-Infante, A.; Veeramachaneni, K. Modeling tabular data using conditional gan. In *Advances in Neural Information Processing Systems*; MIT Press: Cambridge, MA, USA, 2019; Volume 32.
50. Engelmann, J.; Lessmann, S. Conditional Wasserstein GAN-based oversampling of tabular data for imbalanced learning. *Expert Syst. Appl.* **2021**, *174*, 114582. [CrossRef]
51. Zhao, Z.; Kunar, A.; Birke, R.; Van der Scheer, H.; Chen, L.Y. CTAB-GAN+: Enhancing Tabular Data Synthesis. *Front. Big Data* **2024**, *6*, 1296508. [CrossRef]
52. Lázaro, C.; Angulo, C. Iterative application of UMAP-based algorithms for fully synthetic healthcare tabular data generation. *Algorithms* **2024**, *17*, 591. [CrossRef]
53. Li, X.; Zhu, X.; Cao, H.; Huang, W. Research on the Paths of the Modern Agricultural Industrial System Promoting Income Increases and Prosperity for Farmers Based on the fsQCA Method. *Sustainability* **2025**, *17*, 2799. [CrossRef]
54. Wen, H.; Lin, H.; Liu, X.; Guo, W.; He, B.-J. Analysis of the characteristics of university common spaces that affect university students' psychological restoration. *Sci. Rep.* **2025**, *15*, 14146. [CrossRef]
55. Guo, X.; Zheng, D.; Huang, D.; Gu, J. Research on Influencing Factors of Digital Transformation of Construction Enterprises Based on SEM and fsQCA Methods. *Buildings* **2025**, *15*, 3302. [CrossRef]
56. Zeng, H.; Wang, C.; Chen, J.; Tang, D.; Xu, A. Pathways to tourism industry ecologization: A technology-organization-environment configuration framework. *Ecol. Indic.* **2023**, *156*, 111119. [CrossRef]
57. Wang, X.; Zhang, T.; Luo, S.; Abedin, M.Z. Pathways to improve energy efficiency under carbon emission constraints in iron and steel industry: Using EBM, NCA and QCA approaches. *J. Environ. Manag.* **2023**, *348*, 119206. [CrossRef]
58. Kumar, S.; Sahoo, S.; Lim, W.M.; Kraus, S.; Bamel, U. Fuzzy-set qualitative comparative analysis (fsQCA) in business and management research: A contemporary overview. *Technol. Forecast. Soc. Change* **2022**, *178*, 121599. [CrossRef]
59. Chen, S.; Chen, Z.; Hong, J.; Zhuang, X.; Su, C.; Ding, Z. Exploring the Relationship Between Audio–Visual Perception in Fuzhou Universities and College Students' Attention Restoration Quality Using Machine Learning. *Front. Psychol.* **2025**, *16*, 1572426. [CrossRef]
60. Kim, Y.; Huh, Y. Analysis of Satisfaction and Characteristics of Using University Campus Gardens for Enhancing the Mental Well-Being of University Students. *Int. J. Environ. Res. Public Health* **2023**, *20*, 1977. [CrossRef]
61. Qin, J.; Zhang, Y.; Li, X.; Wang, Y. Enhancing Student-Centered Walking Environments on University Campuses: A Streetscape Perception Perspective. *PLoS ONE* **2025**, *20*, e0321028.
62. Fu, W. Enhancing University Campus Landscape Design Through Regression Analysis Integrating Ecological and Aesthetic Indicators. *Soft Comput.* **2023**, *27*, 11945–11958. [CrossRef]

buildings

MDPI

Article

Optimizing University Campus Functional Zones Using Landscape Feature Recognition and Enhanced Decision Tree Algorithms: A Study on Spatial Response Differences Between Students and Visitors

Xiaowen Zhuang [1,†], Yi Cai [2,†], Zhenpeng Tang [2,†], Zheng Ding [1,*] and Christopher Gan [3]

1 College of Landscape Architecture and Art, Fujian Agriculture and Forestry University, Fuzhou 350002, China; 12319097012@fafu.edu.cn
2 College of Economics and Management, Fujian Agriculture and Forestry University, Fuzhou 350002, China; 2221410003@fafu.edu.cn (Y.C.); zhenpt@fzu.edu.cn (Z.T.)
3 Department of Financial and Business Systems, Faculty of Agribusiness and Commerce, Lincoln University, Christchurch 7647, New Zealand; christopher.gan@lincoln.ac.nz
* Correspondence: 000q151039@fafu.edu.cn
† These authors contributed equally to this work.

Abstract: As universities become increasingly open, campuses are no longer only places for study and daily life for students and faculty, but also essential spaces for public visits and cultural identity. Traditional perception evaluation methods that rely on manual surveys are limited by sample size and subjective bias, making it challenging to reveal differences in experiences between groups (students/visitors) and the complex relationships between spatial elements and perceptions. This study uses a comprehensive open university in China as a case study to address this. It proposes a research framework that combines street-view image semantic segmentation, perception survey scores, and interpretable machine learning with sample augmentation. First, full-sample modeling is used to identify key image semantic features influencing perception indicators (nature, culture, aesthetics), and then to compare how students and visitors differ in their perceptions and preferences across campus spaces. To overcome the imbalance in survey data caused by group–space interactions, the study applies the CTGAN method, which expands minority samples through conditional generation while preserving distribution authenticity, thereby improving the robustness and interpretability of the model. Based on this, attribution analysis with an interpretable decision tree algorithm further quantifies semantic features' contribution, direction, and thresholds to perceptions, uncovering heterogeneity in perception mechanisms across groups. The results provide methodological support for perception evaluation of campus functional zones and offer data-driven, human-centered references for campus planning and design optimization.

Keywords: campus space; campus functional zones; image semantic segmentation; student–visitor differences; SHAP

1. Introduction

University campuses are central places for knowledge transmission, talent development, and symbolic spaces that reflect cultural traditions and social interaction [1]. Within them, campus functional zones serve not just to connect buildings, but also to support diverse activities such as learning, leisure, and communication [2]. Studies have shown that green infrastructure and multifunctional nodes can help promote mental health and social interaction [3,4]. At the same time, with their cultural atmosphere and landscape appeal, universities are increasingly becoming "semi-public" landscape resources within cities. In the context of cultural and tourism integration, they attract visitors and serve as essential carriers of urban image and cultural communication [5–9].

Existing research has focused chiefly on students, showing that the visibility, layout, accessibility, and landscape design of campus functional zones significantly affect their frequency of use and comfort [10–12]. Shade, quietness, and scenic views are also key factors for staying in these spaces [11,13]. Even under unfavorable weather conditions, students still prefer outdoor areas of campus functional zones [10]. Green spaces and natural elements have improved mental health and satisfaction, while short-term exposure can enhance resilience and creativity [14]. At the same time, students generally regard campus functional zones as essential for rest and social interaction, positively influencing their well-being and sense of belonging [15,16]. With increasing campus openness, research has gradually expanded to include tourism perspectives. Connell (1996) highlighted the need to balance the interests of universities, students, and visitors in using campus spaces [17]. Cheng et al. (2020) noted that landmark buildings are the core attractions for visitors [18]. McManus et al. (2021) showed how visitors reconstruct spaces' physical and symbolic meaning through social media [19]. Almeida and Silveira (2022) further pointed out that different types of visitors show varied preferences for campus spaces, including prospective students, residents, and highly educated family groups with children [20].

Although previous studies have shown that campus spaces play an essential role in environmental experience and tourism, there are still three main limitations. First, most research has focused on a single group, lacking systematic comparisons between students, faculty, and visitors. Second, while traditional surveys and interviews can capture subjective experiences, they struggle to reveal the complex and non-linear relationships between spatial features and perception. They are also often affected by data imbalance (such as too few visitor samples or extreme ratings), which weakens the detection of differences and thresholds and may lead to biased results. Third, existing studies have paid insufficient attention to the discrepancies between campus functional zones. Evaluations often stay at the overall or green-space level, without detailed comparisons across other types of spaces.

The research questions addressed in this paper are: How can suggestions for campus space optimization be made from the perspective of multi-group perception? How can the issue of imbalanced high and low score samples in traditional questionnaire surveys be resolved? How can the functional areas of the campus be effectively divided based on the results of explainable machine learning?

To address these limitations, this study developed an integrated framework combining landscape feature recognition, interpretable machine learning, and spatial clustering, using a Chinese comprehensive open university as a case study. First, perception surveys were conducted among students and tourists to collect subjective ratings across three dimensions: natural, cultural, and aesthetic aspects. Second, the study employed the Mask2Former model to extract campus landscape features and used the CTGAN algorithm to augment data for samples with extremely high and low scores. An "enhanced decision tree–SHAP" model was then developed to identify key landscape features influencing student and tourist perceptions, while analyzing the threshold effects of these features. Finally, based on the spatial distribution of these influential features, the campus was divided into different functional zones. By integrating driving mechanisms with spatial effects, the study proposes a "group–context" interaction model, from which optimization strategies for functional zones are derived, providing evidence-based guidance for inclusive design in open campuses. The innovation of this research lies in:

First, this study employs the Mask2Former model to conduct semantic segmentation on campus street view images, obtaining objective measurements of natural elements, architectural forms, and facility layouts. These segmentation results are then combined with questionnaire survey scores from both students and visitors. Previous campus landscape research typically focused solely on students, overlooking visitor perceptions at open universities. However, studying visitor perceptions helps optimize campus spaces from an external perspective and enhances institutional reputation. By integrating image semantic segmentation with multi-group perceptions, this study compares how students and visitors differ in their sensitivity, preferences, and thresholds regarding campus spatial characteristics. This analysis identifies synergies between "teaching-daily use" and "visiting-short-term experience" functions. The research provides a valuable complement to existing studies by optimizing campus landscapes through a multi-group perception approach.

Second, this study introduces methodological innovations compared to previous research. Earlier studies overlooked a critical issue: sample imbalance in participants' perceptual ratings. For instance, on a 1–7 scale, participants rarely assign extreme scores of 1 or 7, resulting in limited samples at these endpoints. However, the underlying factors driving these extreme ratings contain valuable insights and warrant increased representation in the dataset. This research innovatively employs Conditional Tabular Generative Adversarial Networks (CTGAN) to augment extreme high and low scoring samples, combined with SHAP-enabled interpretable decision trees and ensemble tree models to analyze the importance and thresholds of semantic features. This approach offers two key advantages: First, it enables the model to better capture the critical factors driving very high or very low perceptual scores, providing more precise guidance for subsequent landscape optimization. Second, regarding model performance, integrating CTGAN with SHAP-based interpretable tree models achieves lower error rates than single interpretable models while maintaining explainability of perceptual differences between students and visitors.

Finally, this study offers a novel perspective on research content by projecting key landscape elements that influence student and visitor groups onto spatial maps and comparing them with street view images to identify governance units, including green-blue ecological corridors, academic and cultural axes with structural nodes, aesthetic enhancement zones along interfaces, gateway and landmark display areas, and open activity or hard-surfaced plazas. These findings directly translate group perceptual differences into functional zone planning strategies. Compared to existing studies that typically rely on overall evaluations or focus on single groups without connecting features.

Based on the above analysis, this study achieved fine-grained recognition of visual elements and multidimensional analysis of perception differences in its methods. In content, it expanded the pathways for campus landscape optimization from both student and visitor perspectives. Together, these contributions provide a practical technical framework and reference for refined design and functional zone optimization in open university spaces.

This paper is organized into six chapters: Chapter 2 reviews the theoretical foundations; Chapter 3 develops the integrated methodological framework; Chapter 4 presents the empirical impact analysis; Chapter 5 discusses the research findings; and Chapter 6 concludes with a summary and future directions.

2. Literature Review and Theory

2.1. Literature Review

2.1.1. Analysis of Spatial Perception Differences Among Multiple Campus User Groups

As a composite public space, the campus must meet the diverse needs of university members and the wider public [21]. Environmental psychology shows that group perception differences stem from functional needs, cultural background, and psychological expectations [22]. In this context, place attachment has become essential for understanding such differences. Lewicka (2011) noted that it includes three dimensions: cognitive, emotional, and behavioral [23]. Scannell and Gifford (2010) proposed a tripartite framework that emphasizes the interaction of "person–psychological process–place," showing that students tend to form functional attachment through daily use, while visitors more often build emotional connections through cultural symbols [24]. Williams and Vaske (2003) further divided place attachment into "place identity" and "place dependence," offering valuable tools for analyzing group perception differences [25].

From the student perspective, as frequent campus users, their spatial perceptions are strongly utility-oriented, mainly focusing on academic support, social interaction, and restorative experiences [26,27]. Studies show that students prefer areas with seating, shade, and opportunities for socializing, and they often use informal spaces when formal ones are lacking [21]. At the same time, campus green spaces have been proven to reduce stress and restore attention, with students favoring natural, quiet, and socially supportive environments [28]. These preferences highlight students' core needs for stress relief and social support within the campus.

In contrast, visitors' perceptions are more oriented toward cultural experience, focusing mainly on cultural identity, aesthetic appreciation, and symbolic meaning [29]. Their evaluations place greater emphasis on architectural style, historical and cultural significance, and landscape aesthetics [30–32]. In addition, visitors' place attachment often depends on the length of stay, the diversity of activities, and the depth of exploration. Their spatial perceptions rely more on short-term sensory and cultural impressions than long-term functional needs.

The familiarity effect and social identity are key explanatory paths for group differences. Through long-term study and life on campus, students are more familiar with the environment and thus more likely to develop positive emotions and attachment. At the same time, visitors rely mainly on sensory impressions to make judgments [33,34]. At the same time, social identity strongly shapes spatial perception: students, as insiders, experience a sense of belonging and identification, whereas visitors, viewing from an "outsider" perspective, place greater emphasis on aesthetics and cultural experience [35].

In summary, the spatial perception differences between students and visitors reflect a division between functional and cultural needs, highlighting the key roles of place attachment and identity in the perception process [33–35]. This differentiated analysis provides empirical support for campus landscape design and offers theoretical reference for universities in balancing the needs of multiple groups and optimizing campus spaces.

2.1.2. Advances in Intelligent Recognition of Landscape Features in Campus Space Research

Intelligent recognition of landscape features has developed rapidly with advances in deep learning and computer vision [36,37]. As an essential approach in landscape research, urban green space (UGS) analysis shows that accurate measurement is central to assessing landscape quality and provides methodological insights for campus studies [38]. Breakthroughs in semantic segmentation have enabled image recognition to be widely applied in green space identification and spatial pattern measurement [39,40], overcoming the limitations of traditional methods in calculating greening rates and recognizing patterns [41]. Applying these techniques to campus research makes it possible to precisely quantify landscape elements and provide technical support for the optimization of open spaces.

Studies have combined semantic segmentation with various analytical methods at the application level. He et al. (2024) used semantic segmentation to quantify campus landscape elements and applied variance analysis to identify key factors and combinations that help relieve short-term stress [42]. Gao et al. (2025) employed Segmenter with PLS-

SEM and Bootstrapping to verify the moderating role of landscape elements, providing support for healing-oriented campus design [43]. Qin et al. (2025) used DeepLabV3+ to extract semantic features from street-view images and developed a modeling framework for walking space perception [44]. Chen et al. (2025) applied HRNet to extract indicators such as vegetation, water bodies, paving, building enclosure, and openness, revealing the non-linear mechanisms linking audiovisual perception and attention restoration [45].

In summary, intelligent recognition of landscape features has already shown its applicability and potential in campus space research. Combining semantic segmentation with multiple analytical methods makes it possible to quantify and analyze complex campus landscape elements and uncover their effects on students' psychological and behavioral responses. This provides technical and theoretical support for campus landscape optimization and health-oriented design.

2.1.3. Applications of Interpretable Machine Learning in Spatial Evaluation

With the growing use of machine learning in urban space analysis, the "black box" problem has increasingly become a barrier to planning and design practice. Interpretable machine learning addresses this by improving model transparency and directly revealing the complex relationships between space and human behavior, offering new approaches for spatial evaluation and optimization [46,47]. Current research mainly relies on two frameworks: SHAP (Shapley Additive exPlanations) and LIME (Local Interpretable Model-agnostic Explanations) [48].

Interpretable machine learning has shown strong potential in spatial perception evaluation to reshape our understanding of human–environment relationships. Xiao et al. (2024) combined population data from Shenzhen with a Random Forest–SHAP model, revealing the non-linear interaction effects between community park visitation and built environment factors [49]. Chen et al. (2024) integrated hotspot analysis with SHAP to uncover how informal green space features influence residents' complaint emotions [50]. Yang et al. (2024) explained how the built environment affects residents' health [51]. In transport and urban form research, Wagner et al. (2022) analyzed 3.5 million commuting records to show that the development of Berlin's sub-centers reduced commuting distances, confirming the "15 min city" hypothesis and identifying a carbon emission density threshold effect [52]. In spatial prediction and optimization, Ruan et al. (2025) combined street-view imagery with a deep vision model–SHAP Random Forest, finding that non-linear building and facility density interactions significantly affect non-motorized travel vitality [53]. Eshraghi et al. (2025) proposed an AI-based optimization framework that integrates SHAP and CFX to evaluate and improve skyline openness in open spaces, balancing design flexibility with computational efficiency [54].

Overall, interpretable machine learning has become an essential tool for spatial evaluation. It can reveal the non-linear and interaction effects of environmental features on perception and behavior while providing optimization strategies for planning and design [46,48]. In campus studies, this approach is equally applicable, as it can uncover perception differences and underlying mechanisms between students and visitors, offering scientific support for optimizing open spaces and fostering cultural identity [45,53].

2.2. *Indicator Construction and Theoretical Framework*

Integrating architectural and environmental design theories has provided a new analytical framework for campus space research, encouraging interdisciplinary exploration of how spatial configuration and environmental quality influence user experience and behavior [55,56]. Within this context, campus open spaces serve as meaningful learning, communication, and rest venues. Their ecological attributes and spatial organization affect environmental comfort and functional efficiency, and shape social interaction and cultural atmosphere. Based on this, the present study builds a comprehensive evaluation model from three dimensions—natural perception, artistic perception, and aesthetic perception—to reveal the environmental benefits and optimization mechanisms of campus functional zones, while offering references for planning renewal and sustainable governance. Figure 1 illustrates the connections between the three indicators and the different user groups in this study. These indicators have cross-group applicability: students, as long-term users, focus more on support for learning, communication, and belonging, whereas visitors form perceptions mainly through natural settings, cultural imagery, and aesthetic impressions. Thus, this framework can explain students' everyday experiences and uncover visitors' immediate perceptions, providing theoretical support for optimizing campus spaces to meet the needs of multiple groups.

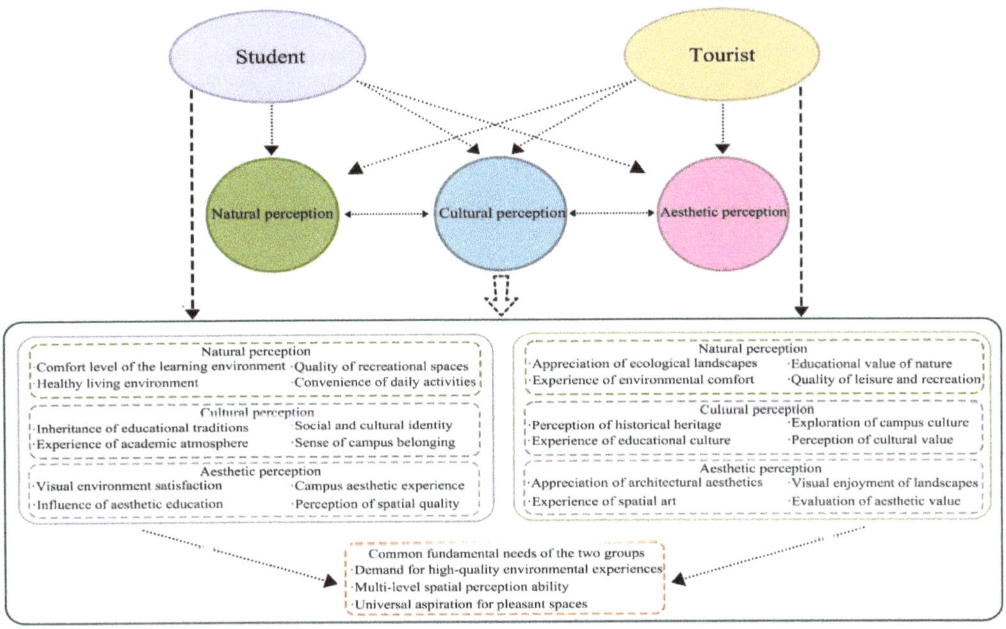

Figure 1. Relationships among the three indicators and between different groups in this study.

Natural perception is mainly grounded in Attention Restoration Theory (ART) and Stress Recovery Theory (SRT). ART highlights how natural environments support attention recovery through features such as "being away," "soft fascination," "extent/coherence," and "compatibility," while SRT emphasizes that positive responses to natural cues can help reduce both physiological and psychological stress [57–59]. The Perceived Restorativeness Scale (PRS) has been widely used to assess environmental restorative capacity [60]. Research has shown that green spaces and exposure to nature significantly improve health and well-being [61]. For students, nature-rich spaces enhance feelings of restoration and improve learning efficiency [62]. Higher landscape quality improves visitors' experience and satisfaction; even short-term exposure can ease rumination and emotional stress [63,64].

Cultural perception is rooted in place attachment theory, which emphasizes the emotional–cognitive bonds between people and places, including "place identity" and "place dependence" [24,25]. Within the cultural landscape framework, landscapes are seen as products of social–natural interaction, carrying collective memory and supporting the reproduction of identity [65]. For students, cultural perception relates to their sense of belonging in study and daily life and recognition and inheritance of campus historical symbols [23]. For visitors, it relies more on cultural imagery and narrative experiences to build symbolic understanding and cultural connections [66]. Aesthetic perception is mainly derived from Kaplan's environmental preference model, whose four core dimensions—coherence, complexity, legibility, and mystery—have been shown to predict landscape preference and visual evaluation [57,67] reliably. At the level of buildings and streetscapes, factors such as façade composition, material colors, and the degree of spatial enclosure significantly influence aesthetic judgments [30,68]. At the urban scale, aesthetic perception has been widely applied in studies of public spaces such as squares and parks [69]. For students, the aesthetic quality of learning and activity spaces directly affects their comfort and satisfaction with the environment [70]. For visitors, novelty in architectural form and substantial visual impact play a greater role, shaping their overall impression of the campus and their aesthetic memories [71].

As shown in Figure 2, the three dimensions—natural perception, cultural perception, and aesthetic perception—together form the core framework of this study. They capture both students' long-term experiences and visitors' immediate perceptions, thus providing theoretical support for campus space optimization and sustainable governance from a multi-group perspective.

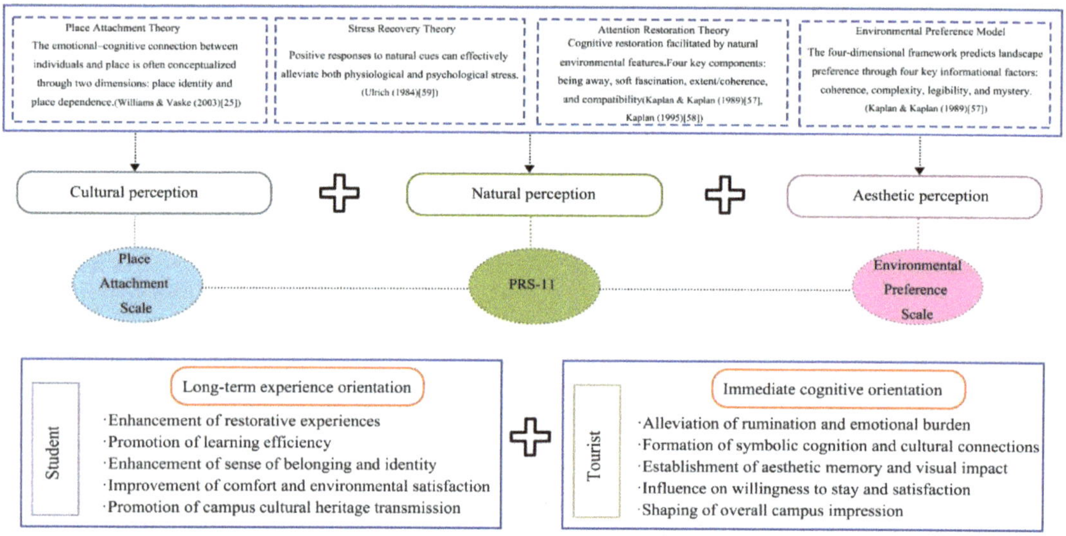

Figure 2. The theoretical framework of the three indicators [25,57–59].

3. Materials and Methods

This study used campus street-view images retrieved from Baidu Maps and applied a Mask2Former-based semantic segmentation method to extract landscape features. At the same time, a questionnaire survey was conducted to collect student and visitor scores on the three indicators of natural perception, cultural perception, and aesthetic perception. Next, a CTGAN–interpretable decision tree model was employed to identify the key landscape features that significantly affect each type of perception, and to analyze their influence patterns and threshold effects. Based on the actual distribution of the critical landscape features identified earlier, the campus was divided into functional zones for heterogeneity analysis. Figure 3 illustrates the technical framework of this study.

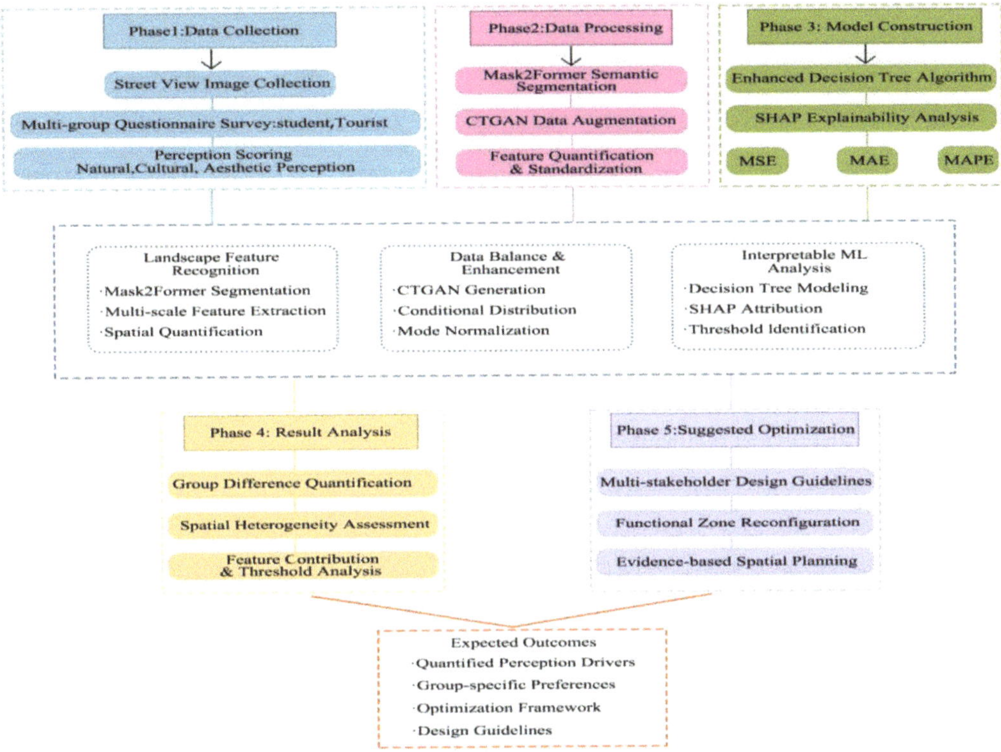

Figure 3. Technical framework diagram.

3.1. Data Collection and Feature Quantification

- Campus Street-View Acquisition

Fuzhou University, founded in 1958, is a national "Double First-Class" university and a key institution under the "211 Project." The study site is the Qishan Campus, which is located in the University Town of Fuzhou, and it can be seen at https://www.openstreetmap.org/search?lat=26.06143&lon=119.19346&zoom=16#map=16/26.06143/119.19345 (accessed on 20 September 2025). The campus has a planned total floor area of 174,000 square meters and is surrounded by mountains and water, with a favorable natural environment. Figure 4 shows the geographical location of the study site. Representative campus buildings include Fuyou Pavilion, Jinjiang Building, and the library.

In this study, the road network was first extracted from OpenStreetMap (OSM), and 378 sampling points were generated in QGIS at 50 m intervals, with their latitude and longitude recorded. At each point, the pitch angle was set to 0°, and images were captured in four directions: 0°, 90°, 180°, and 270°. This produced four images per point and a total of 1516 street-view photos. The dataset provided near 360° panoramic coverage of the sampling locations, offering a systematic data foundation for semantic segmentation and perception modeling.

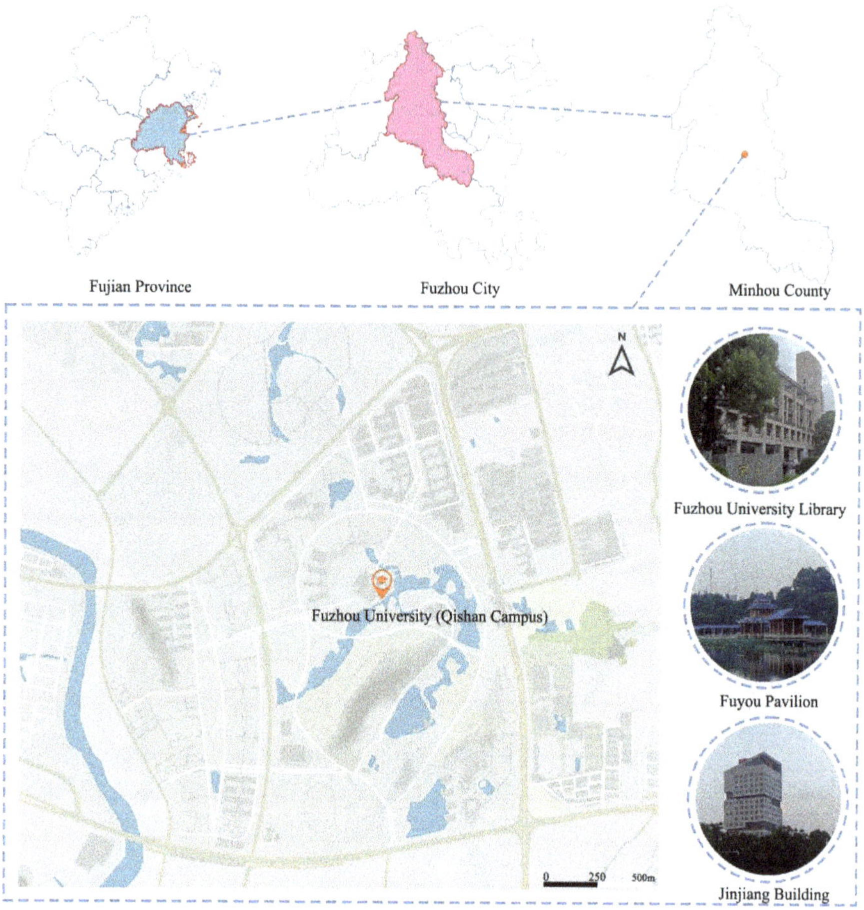

Figure 4. Geographical location of the study area.

- Quantification of perception indicators

This study aimed to assess street-view images in terms of three dimensions: natural perception, cultural perception, and aesthetic perception. To do so, we adapted three classic scales: the Perceived Restorativeness Scale (PRS-11) [60], the Place Attachment Scale [25], and the Environmental Preference Scale [57,67]. The questionnaire design selected 3–4 representative items from each scale and semantically rephrased to fit the evaluation context based on street-view images. All items were rated on a 7-point Likert scale, where 1 indicated "strongly disagree" and 7 indicated "strongly agree," quantifying participants' subjective perceptions of the images. Using a stratified random sampling approach, 200 representative images were selected from the street-view image database of Fuzhou University's Qishan Campus. 235 participants completed the evaluation task, including 123 university students (recruited through on-campus channels to ensure reliable identity) and 112 tourists (recruited via online platforms, with screening questions at the beginning of the survey to exclude current students). Since tourists' perceptions of the campus are primarily shaped by short-term visits and immediate impressions, online image-based ratings were an effective way to simulate their perception patterns. Each image received approximately 50 independent evaluations, and Table 1 presents the complete questionnaire for street-view image perception evaluation. Perception indicators were quantified through an online rating experiment involving university students and the general public. Student participants were recruited through campus channels to ensure reliable identity verification, while visitor participants were recruited via public platforms. Screening questions at the start of the survey (e.g., whether the participant was a current student, or had visited or planned to visit the campus) were used to distinguish the groups. It should be emphasized that this study focuses on visual perception differences under street-view image conditions, rather than actual usage behavior. Since visitors' understanding of campuses is often based on short visits and immediate impressions, image-based online ratings can effectively simulate their real perception process. It should be emphasized that this study focuses on visual perception differences under street-view image conditions, rather than actual usage behavior. Since visitors' understanding of campuses is often based on short visits and immediate impressions, image-based online ratings can effectively simulate their real perception process.

Table 1. Questionnaire for street-view image perception evaluation.

Section	Scales Used	Item Code	Question	Scale
Background Information		Q1	What is your identity?	Student/Visitor
		Q2	What is your gender?	Male/Female/Other
		Q3	What is your age?	≤18/19–25/26–35/36–45/≥46
		Q4	Have you ever visited Fuzhou University's Qishan Campus?	Never/Once/2–3 times/Frequently
		Q5	What is your current residence?	Fuzhou/Other cities in Fujian/Other provinces/Abroad
Natural Perception	PRS-11	N1	Viewing this scene makes me feel relaxed, both physically and mentally.	1 (Strongly Disagree)–7 (Strongly Agree)
		N2	This environment seems helpful for relieving stress.	1 (Strongly Disagree)–7 (Strongly Agree)
		N3	The landscape in the image appears to help me restore my attention and energy.	1 (Strongly Disagree)–7 (Strongly Agree)
		N4	This scene conveys a sense of natural comfort.	1 (Strongly Disagree)–7 (Strongly Agree)
Cultural perception	Place Attachment Scale	C1	This scene helps me feel the campus's unique cultural atmosphere.	1 (Strongly Disagree)–7 (Strongly Agree)
		C2	I think this environment represents the school's history and cultural features well.	1 (Strongly Disagree)–7 (Strongly Agree)
		C3	When I look at this image, I feel an emotional connection to the campus.	1 (Strongly Disagree)–7 (Strongly Agree)
		C4	This scene seems to help visitors feel more connected to the school.	1 (Strongly Disagree)–7 (Strongly Agree)

Table 1. *Cont.*

Section	Scales Used	Item Code	Question	Scale
Aesthetic Perception	Environmental Preference Scale	A1	I think this scene is visually appealing.	1 (Strongly Disagree)–7 (Strongly Agree)
		A2	The environmental elements in the image work well together and look harmonious.	1 (Strongly Disagree)–7 (Strongly Agree)
		A3	This landscape has some complexity that makes me want to explore it further.	1 (Strongly Disagree)–7 (Strongly Agree)
		A4	This scene looks neat, clear, and easy to understand.	1 (Strongly Disagree)–7 (Strongly Agree)

3.2. Data Processing Procedure

- Semantic feature extraction process of street-view images

Street view images are widely used in environmental perception and spatial analysis research because they provide comprehensive spatial coverage and detailed visual information. This study uses the Mask2Former model to extract semantic features [72]. Figure 5 shows how this model works. The model is built on a Transformer structure and combines convolutional backbone networks with set prediction methods to achieve unified optimization for panoptic segmentation. This approach can accurately identify typical elements in campus scenes—such as vegetation, roads and walkways, open sky areas, and building outlines—at the pixel level. The model also uses the ADE20K dataset to extract semantic features from street view images. Figure 6 illustrates the process of collecting street view images and semantic segmentation, while Table 2 presents all feature names identified through semantic segmentation.

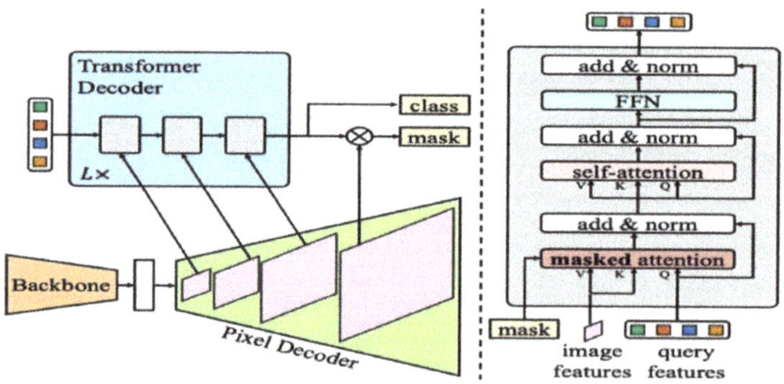

Figure 5. Mask2Former model workflow. Note: This figure is adapted from reference [72].

For Mask2Former training on the ADE20K dataset [72], semantic and panoptic segmentation are trained separately with identical settings, using an ImageNet-22K pre-trained backbone. Data augmentation includes random scale jittering (0.5–2.0), horizontal flip, 512 × 512 cropping, and color perturbation to match the dataset's resolution and scene variety. A polynomial learning rate schedule ensures stable convergence (batch size 16, 160k–180k iterations), with AdamW optimizer, linear warm-up, and weight decay set to 0.05. In inference, images are resized so the long side is 1024 px, with optional multi-scale testing for semantic segmentation. The number of queries depends on task complexity and backbone (200 for Swin-L panoptic, otherwise 100) to balance accuracy and stability.

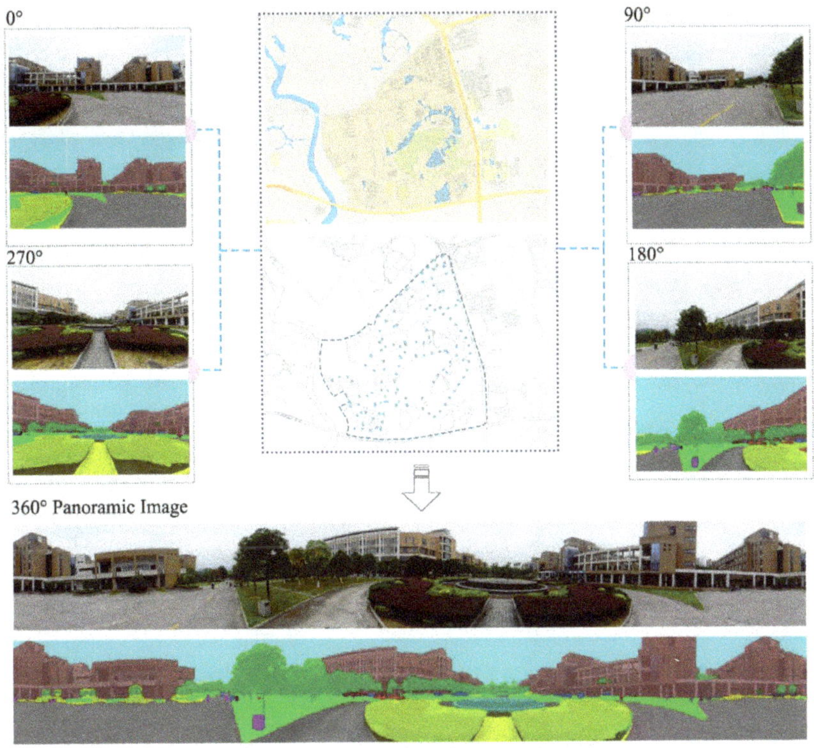

Figure 6. The process of collecting street view images and semantic segmentation.

Table 2. Complete image semantic segmentation features name.

Tree	Road	Earth	Sky	Grass	Car	Plant	Pole	Palm	Path
Building	Signboard	Rock	Dirt Track	Bridge	Water	Column	Bicycle	Wall	Person
Minibike	Van	Mountain	Sidewalk	Fence	Streetlight	Railing	Stairs	Ashcan	Canopy
Bench	Truck	Table	Awning	Bus	Booth	House	Base	Tower	Box
Chair	Bag	Stairway	Banister	Pot	Flag	Sculpture	Poster	Clock	Grandstand
Fountain	Field	Hill	Vase	Light	Traffic Light	Tank	Sand	Book	Trade Name
Flower	Step	Floor	Mirror	Door	Airplane	Boat	River	Ball	Seat
Blanket	Tent	Sea	Land	Basket	Bulletin Board	Escalator	Curtain	Food	Sconce
Hovel	Plaything	Ceiling	Apparel	Tray	Bottle				

- Application of CTGAN in imbalanced date

In recent years, Generative Adversarial Networks (GANs) have gained increasing attention for their use in tabular data modeling and imbalanced learning. Xu et al. (2019) first introduced CTGAN, which uses conditional generation mechanisms and mode normalization to solve problems like mixed-type variables (continuous and discrete), multimodal distributions, and class imbalance. This approach achieves high-quality and diverse tabular data synthesis [73]. Building on this work, Engelmann & Lessmann (2021) [74] introduced a conditional Wasserstein GAN for more stable oversampling of imbalanced tabular data. Later, Zhao et al. (2021) [75] proposed CTAB-GAN, which further improved the statistical similarity of synthetic data and model performance by adding classification loss, information loss, and an improved conditional vector structure. The latest research, CTAB-GAN+, extends the training mechanism to handle complex imbalanced distributions better. These methods together show that CTGAN-based generative modeling has significant advantages in reducing class imbalance and improving data quality. This provides strong methodological support for modeling and interpreting campus perception data in this study.

3.3. SHAP-Based Interpretable Decision Tree Modeling

To evaluate how landscape features affect student and visitor perception scores, this study uses three ensemble methods based on decision trees: Random Forest (RF), Gradient Boosting Decision Trees (GBDT), and AdaBoost [76–78]. The data uses semantic features from street view images and perception scores as output variables. The data is divided into training, validation, and test sets. CTGAN is used for conditional data augmentation during the training phase and is based only on the training set to address uneven distribution and minor sample problems. Model performance is evaluated using MSE, MAE, and validation set EVS, and the best model is selected for prediction and interpretation. To improve interpretability, SHAP (Python "shap" v0.47.2, TreeExplainer) calculates sample-level SHAP values on the test set to show local contributions. Global importance rankings are formed by averaging SHAP values across samples. To enhance robustness, 50 random resampling iterations are performed, and the average importance is reported.

4. Empirical Results

4.1. Building Optimal Decision Tree Models for Campus Functional Areas

To model and predict the restorative effects of campus outdoor spaces, this study uses three ensemble algorithms based on decision trees: Random Forest (RF), Gradient Boosting Decision Trees (GBDT), and AdaBoost. The approach uses semantic features from street view images as input variables and perception indicator scores as output variables. The data is divided into training, validation, and test sets. Before modeling, CTGAN is introduced for conditional data augmentation based only on the training set to address uneven distribution and minor sample problems and improve generalization ability. Model performance is evaluated using Mean Squared Error (MSE), Mean Absolute Error (MAE), and Mean Absolute Percentage Error (MAPE). The best-performing model on the validation set is selected as the final solution. The overfitting prevention strategies for the three algorithms are: RF limits maximum tree depth, sets minimum samples per leaf node, and uses random feature subsets for node splitting; GBDT controls model complexity through small learning rates, limited tree depth and number, and subsampling; AdaBoost achieves regularization by constraining the depth of base learners (such as shallow trees) and setting learning rates and the number of weak learners. Combining results from six tasks, the CTGAN + tree model achieves better error performance compared to the baseline in most cases. Tables 3 and 4 show the final optimal models for each task in the two groups of students and visitors, along with their primary hyperparameters and indicators.

Table 3. The performance of decision tree algorithms was trained on complete student samples on the validation set.

Student Full-Sample Perceived Naturalness Index					**Student Full-Sample Cultural Perception Index**					**Student Full-Sample Aesthetic Perception Index**				
	RF	GBDT	AdaBoost	CTGAN-GBDT		RF	GBDT	AdaBoost	CTGAN-AdaBoost		RF	GBDT	AdaBoost	CTGAN-RF
MSE	0.93	0.93	0.91	0.75	MSE	0.98	0.99	0.97	0.62	MSE	1.29	1.32	1.31	0.95
MAE	0.78	0.77	0.77	0.63	MAE	0.81	0.82	0.81	0.50	MAE	0.92	0.94	0.93	0.72
MAPE	25.41	25.036	25.29	22.87	MAPE	32.04	32.02	32.01	20.40	MAPE	31.66	31.95	32.01	28.03

Table 4. Performance of decision tree algorithms trained on complete visitor samples on the validation set.

Tourist Full-Sample Perceived Naturalness Index					Tourist Full-Sample Cultural Perception Index					Tourist Full-Sample Aesthetic Perception Index				
	RF	GBDT	AdaBoost	CTGAN-GBDT		RF	GB DT	AdaBoost	CTGAN-AdaBoost		RF	GB DT	AdaBoost	CTGAN-RF
MSE	2.00	2.00	1.99	1.65	MSE	0.92	0.93	0.92	0.60	MSE	1.32	1.30	1.29	0.87
MAE	1.18	1.19	1.17	1.06	MAE	0.82	0.82	0.81	0.58	MAE	0.97	0.95	0.95	0.69
MAPE	45.13	45.38	45.40	40.15	MAPE	31.30	31.39	30.72	23.23	MAPE	31.84	31.19	31.26	23.14

4.2. Analysis of the Working Mechanism of Semantic Features in Campus Spaces

4.2.1. Visual Analysis of Key Semantic Features Within the Study Area

Figure 7 shows the spatial distribution of semantic features that significantly affect perception indicators across the entire study area. Combining student and visitor perception evaluations, 8 core image semantic features were selected after removing duplicates, highlighting representative landscape space distribution. Among these, the building has the highest density in core teaching and functional areas, with an impact weight of 52.91. Road (weight 39.32) shows a network-like distribution that strengthens spatial connectivity. Sky (weight 38.08) is found in open spaces and gaps between buildings, enhancing the sense of visual openness. Grass features rank fourth with a weight value of 31.98, serving as an important component of natural landscapes. Plant (weight 31.92) is widely distributed in green areas and building transition spaces, adding natural qualities. Earth (weight 22) is concentrated in open ground areas. Fence (weight 13) has a relatively small influence but is key in defining spatial boundaries. Path (weight 9.43) is distributed throughout the walking network, creating intimate experiences. These diverse semantic features' distribution density and combination patterns directly affect users' perception experiences, providing data support for campus optimization design. These diverse semantic features' distribution density and combination patterns directly affect users' perception experiences, providing data support for campus optimization design.

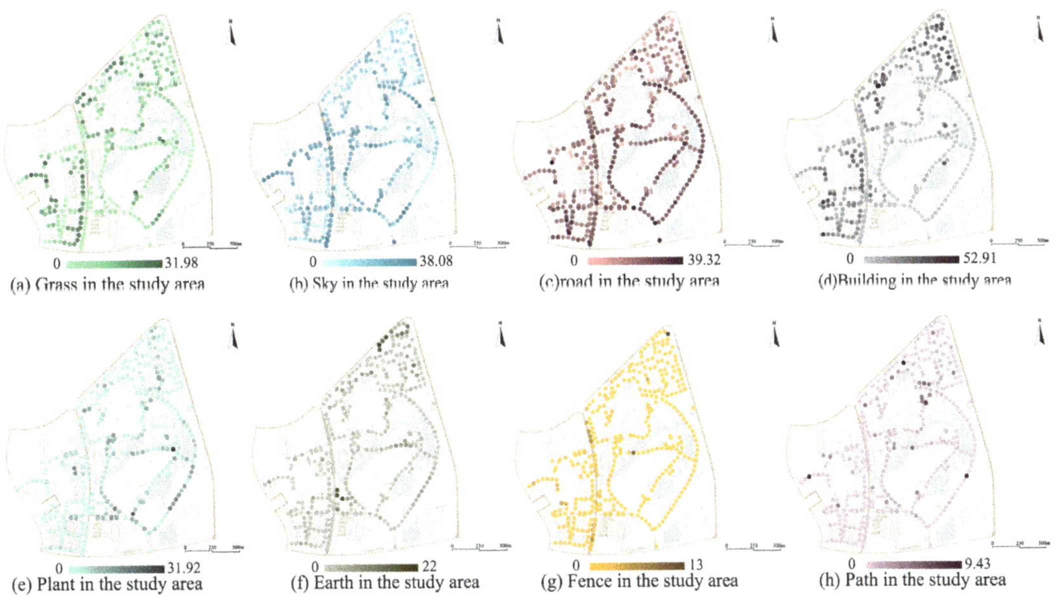

Figure 7. Spatial distribution and selection results of semantic features significantly affect the study area's perception indicators across all samples.

4.2.2. Analysis of Semantic Features Affecting Students' and Visitors' Natural Perception Indicators

Figure 8 shows the SHAP analysis results of the critical semantic features that influence natural perception for students and visitors. The different sectors in the figure reflect the contribution and distribution of each feature. Features with higher importance have more polarized SHAP value distributions, while less important features are concentrated around zero, suggesting limited influence. On the left are the important features affecting students' natural perception, and on the right are those affecting visitors. Overall, both students and visitors rely strongly on natural elements such as grass and sky, which generally have a positive effect. For students, besides grass, features like buildings, fences, and awnings also play a role, and in many cases high values of these reduce natural perception. Visitors, on the other hand, depend more on straightforward natural elements such as grass, sky, paths, and sand, but they are also more sensitive to facilities like chairs and steps, which can lead to negative experiences if they appear too frequently. In general, visitors' natural perception relies more on direct natural elements, but their sensitivity to man-made facilities is higher, and excessive presence of such features can easily turn into a negative experience.

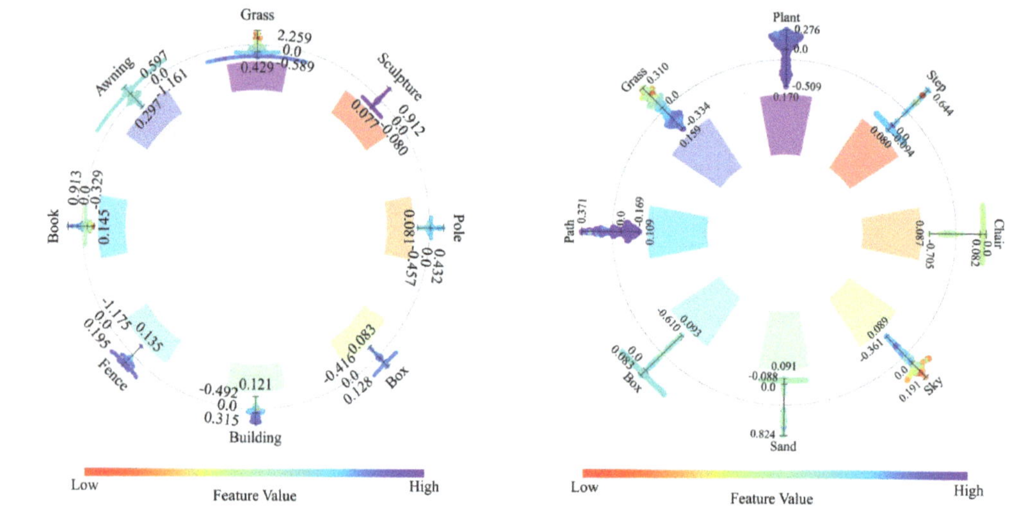

Important semantic features influencing students' natural perception indicators

Important semantic features influencing tourists' natural perception indicators

Figure 8. SHAP analysis results of important semantic features influencing natural perception of students and visitors.

Figure 9 shows the influence patterns of the top eight important semantic features affecting students' natural perception. The results indicate that grass significantly enhances natural perception once it exceeds about 9.23, making it the most important positive factor. Buildings have a slight positive effect when below 11.47 but become negative when this threshold is exceeded. Fences are unfavorable below 0.33 but turn positive after 26.83. Awnings and poles show fluctuating effects, shifting from positive to negative within certain ranges, though their overall influence is limited. Books, boxes, and sculptures have weaker effects, showing only minor influence in specific situations.

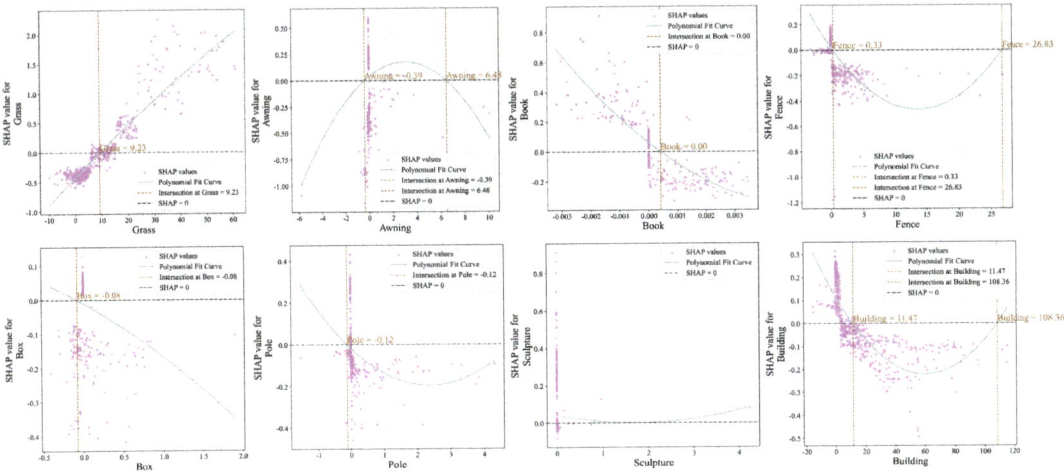

Figure 9. Influence patterns of the top eight important semantic features affecting students' natural perception.

Figure 10 shows the top eight key semantic features influencing visitors' natural perception and threshold effects. The results indicate that grass improves natural perception above 8.45, making it the strongest positive factor. Sky also continues to have a positive impact once it exceeds 11.88. Paths contribute positively when above about 0.30, while plants enhance perception significantly within the range of 3.39 to 93.51, but their impact weakens when values are too low or too high. In contrast, boxes become negative once they exceed 0.02. Sand is positive when below −0.01, but its effect weakens as the value increases. Chairs shift from negative to positive around zero, despite limited influence. Steps contribute positively when above 0.02.

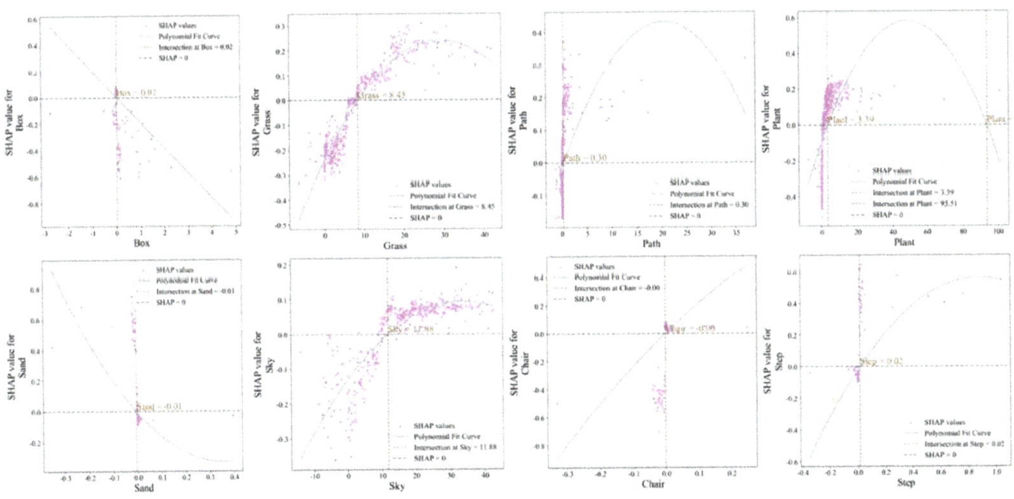

Figure 10. The top eight key semantic features influencing visitors' natural perception and their threshold effects.

4.2.3. Analysis of Semantic Features Influencing Cultural Perception of Students and Visitors

Figure 11 shows the SHAP analysis results of important semantic features influencing cultural perception for students and visitors. Both groups are commonly influenced by symbolic spatial elements such as buildings, which form the foundation of the cultural atmosphere. For students, elements like bridges, fountains, houses, and trade names are the main drivers, with symbolic or memory-related landscapes playing a stronger role in enhancing cultural perception. For visitors, cultural perception relies more on directly recognizable site elements such as stairways, roads, trees, and tents, which shape the cultural atmosphere through spatial use and visual symbols. At the same time, visitors are more sensitive to features like poles, boxes, and earth, which tend to weaken cultural perception when their values are high. Students' cultural perception is generally rooted in landscape symbols connected to campus history and daily life, while direct environmental elements and experiential facilities influence visitors' cultural perception.

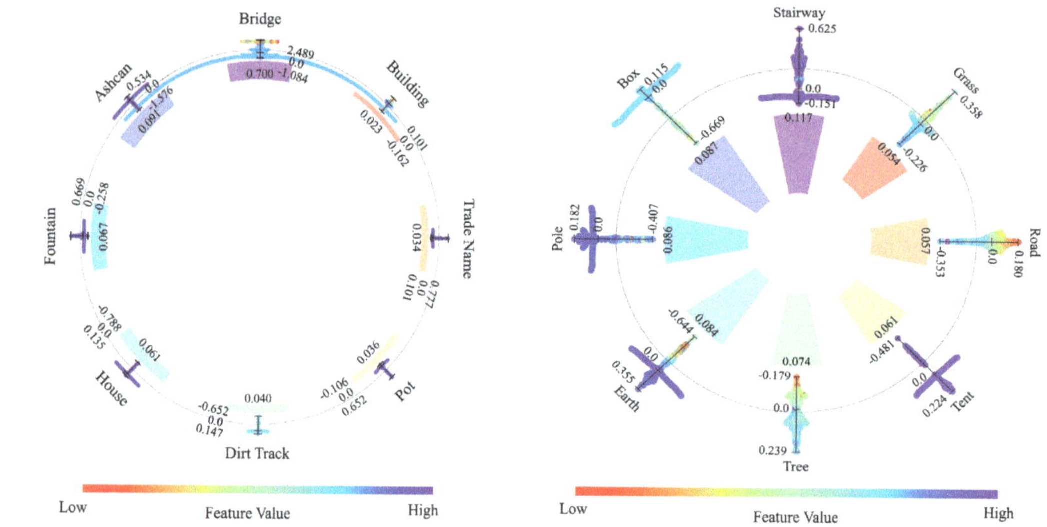

Important semantic features influencing students' cultural perception indicators   Important semantic features influencing tourists' cultural perception indicators

Figure 11. SHAP analysis results of important semantic features influencing the cultural perception of students and visitors.

Figure 12 shows the influence patterns of the top eight important semantic features affecting students' cultural perception. The results indicate that bridges significantly enhance cultural perception once they exceed about 5.16, making them the most representative positive factor. Houses contribute clearly within the range of 0–3.62, but their effect weakens when values are too low or too high. Fountains generally have a positive impact, especially when values are greater than 0, where the effect remains steady. Ashcans show a positive impact within the range of 0.04–7.36, suggesting that their moderate presence helps the cultural atmosphere. Buildings gradually enhance perception when above 7.27, though their influence is limited. Trade names have a positive contribution within the range of 0–1.37. In contrast, dirt tracks and pots show weaker effects; in some ranges, they even have negative impacts.

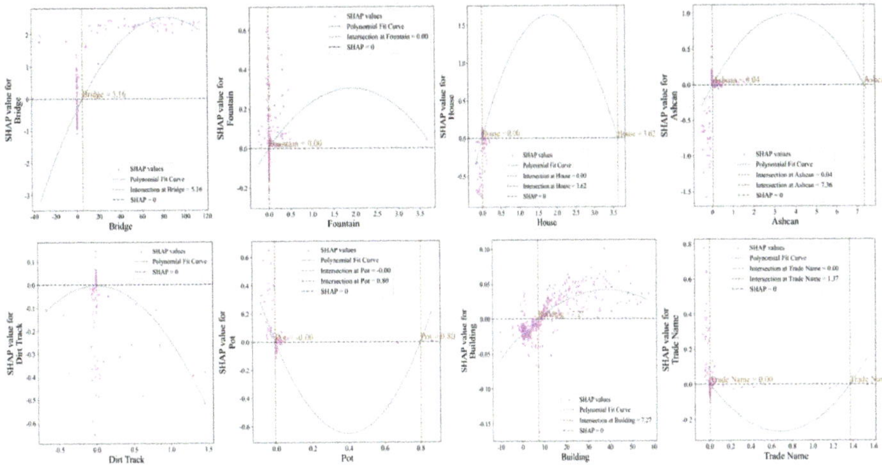

Figure 12. Influence patterns of the top eight important semantic features affecting students' cultural perception.

Figure 13 shows the top eight important semantic features influencing visitors' cultural perception and threshold effects. The results indicate that stairways have an overall positive impact once they exceed about -0.04, making them an essential factor in strengthening cultural perception. Roads continue to influence positively when above 16.45, and grass also shows positive contributions when greater than 7.13. In contrast, boxes, poles, and earth gradually become negative factors at higher values, weakening the cultural experience. Trees have little effect below 23.71 but show a clear negative impact when exceeding this threshold. Similarly, tents shift to an adverse effect once they are above 0.04.

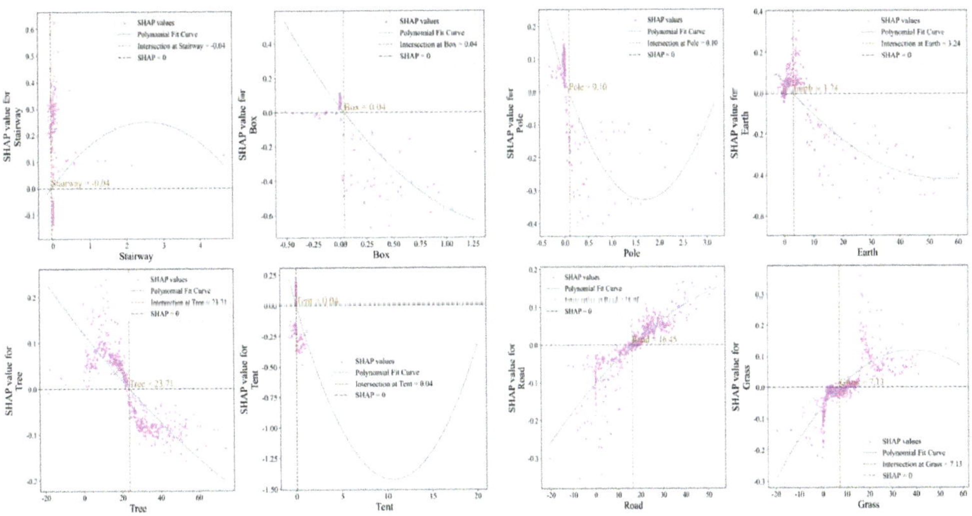

Figure 13. The top eight important semantic features influencing visitors' cultural perception and their threshold effects.

4.2.4. Analysis of Semantic Features Influencing Aesthetic Perception of Students and Visitors

Figure 14 shows the SHAP analysis results of important semantic features influencing the aesthetic perception of students and visitors. For both groups, grass plays a shared role, as natural greenery generally enhances the beauty of the environment. Students rely more on sculptures, sand, and fences, which carry symbolic meaning or help create atmosphere. On the other hand, visitors place greater emphasis on features like grandstands, fields, and hills, while being more sensitive to facilities such as tents, poles, and earth, which can reduce aesthetic experience when overly present. Overall, visitors' aesthetic perception depends more on spatial settings and scene elements, while students are more likely to be influenced by cultural symbols and landscape details.

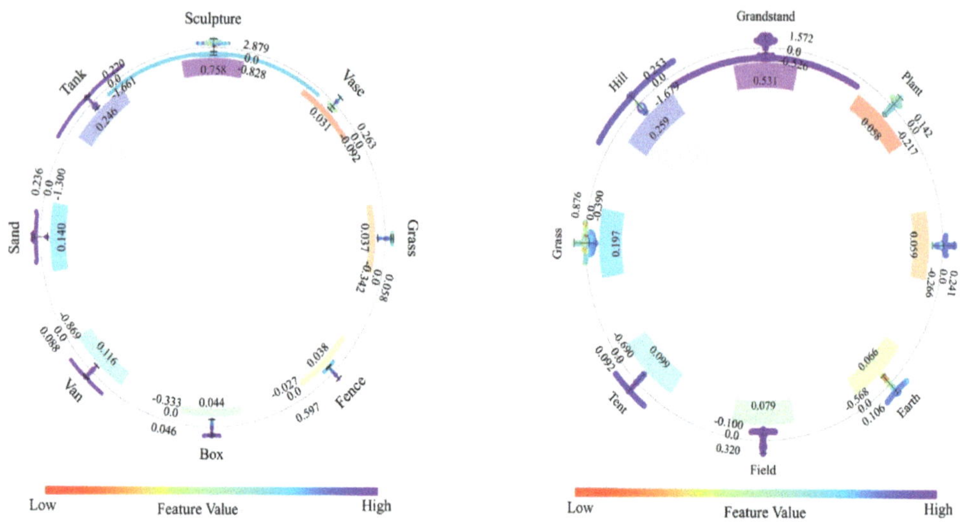

Important semantic features influencing students' aesthetic perception indicators

Important semantic features influencing tourists' aesthetic perception indicators

Figure 14. SHAP analysis results of important semantic features influencing the aesthetic perception of students and visitors.

Figure 15 shows the top eight semantic features influencing students' aesthetic perception and threshold effects. The results indicate that sculptures make a steady positive contribution once they exceed 0.01, making them the core factor for aesthetic perception. Sand significantly enhances perception when its value is greater than 0, while tanks have a positive effect in the range of 0–0.01. Vans have a positive impact between 0.01 and 6.49, but their effect weakens when the value is too high. Boxes contribute positively within the range of 0–0.05. Grass shows a positive impact within 6.59–40.16. In contrast, vases and fences have limited overall influence and show weak adverse effects in specific ranges.

Figure 16 shows the top eight semantic features influencing visitors' aesthetic perception and threshold effects. The results indicate that grandstands significantly enhance aesthetic perception once they exceed 13.18. Grass provides a steady positive impact within the range of 5.30–42.74, while plants contribute positively between 1.07 and 22.93. Fields also bring a positive influence when above 0.01. In contrast, hills and tents generally show adverse effects. Earth turns negative when below 3.68, and poles display a weaker adverse impact within the range of 0.13–2.69.

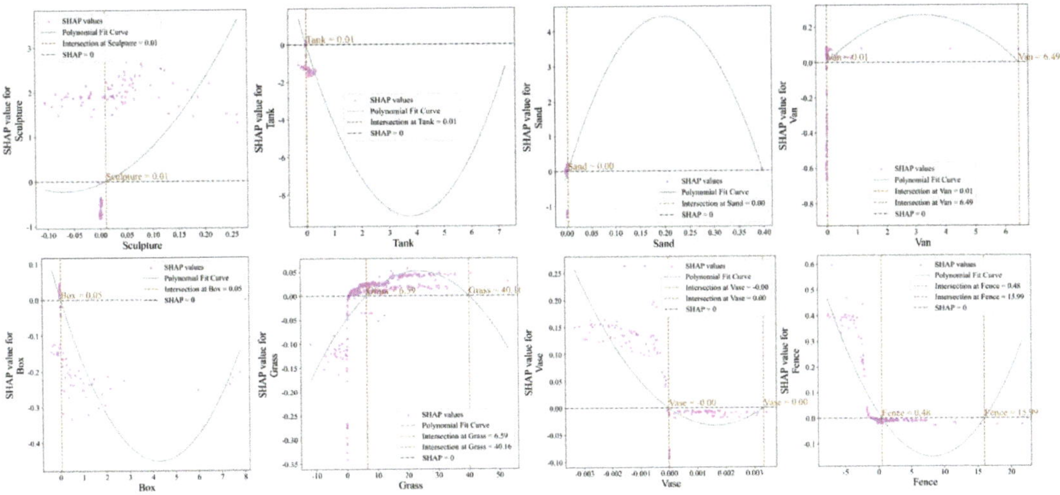

Figure 15. The top eight semantic features influence students' aesthetic perception and threshold effects.

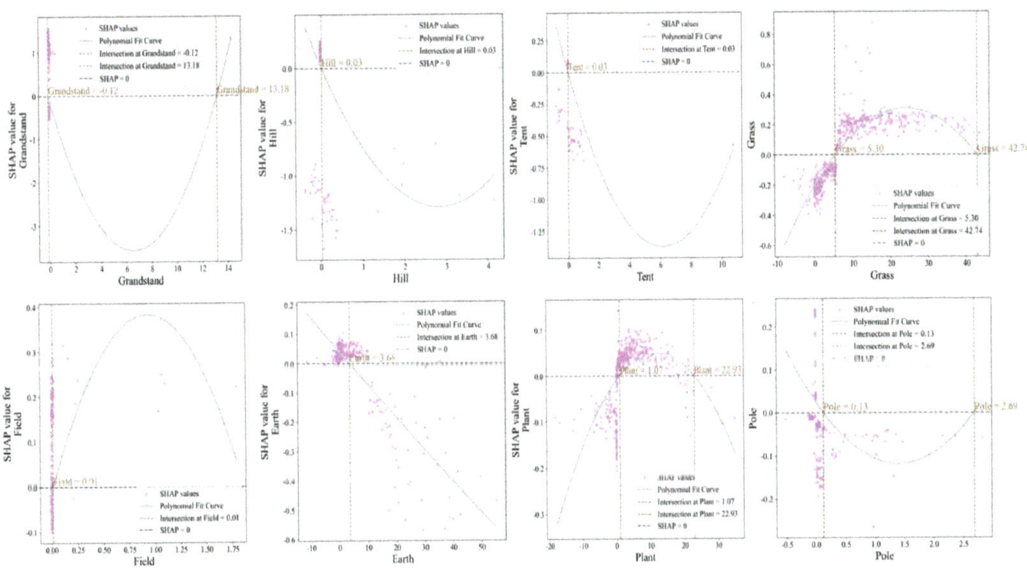

Figure 16. The top eight semantic features influence visitors' aesthetic perception and threshold effects.

4.3. Spatial Distribution of Key Semantic Features Influencing Students and Visitors, with Street View Comparison

Figure 17 shows the spatial distribution of the key semantic features influencing students and visitors and the corresponding real-world street views. The six semantic features presented here come directly from the earlier SHAP results. We divided the task into six sub-items ("student/visitor × natural/cultural/aesthetic"). Based on the final model determined from the training set (with CTGAN augmentation), we calculated sample-level SHAP values on the test set, using the mean of their absolute values as the measure of global contribution, and applied 50 rounds of random resampling for robustness. For each sub-item, only the top-ranked semantic feature was kept for spatial projection and real-world examples. This sampling method ensures a clear one-to-one comparison between the two groups and the three perception dimensions while guaranteeing that the visualized features are the key factors with the most substantial marginal impact on perception. In this way, the results can directly support later spatial interventions and management decisions.

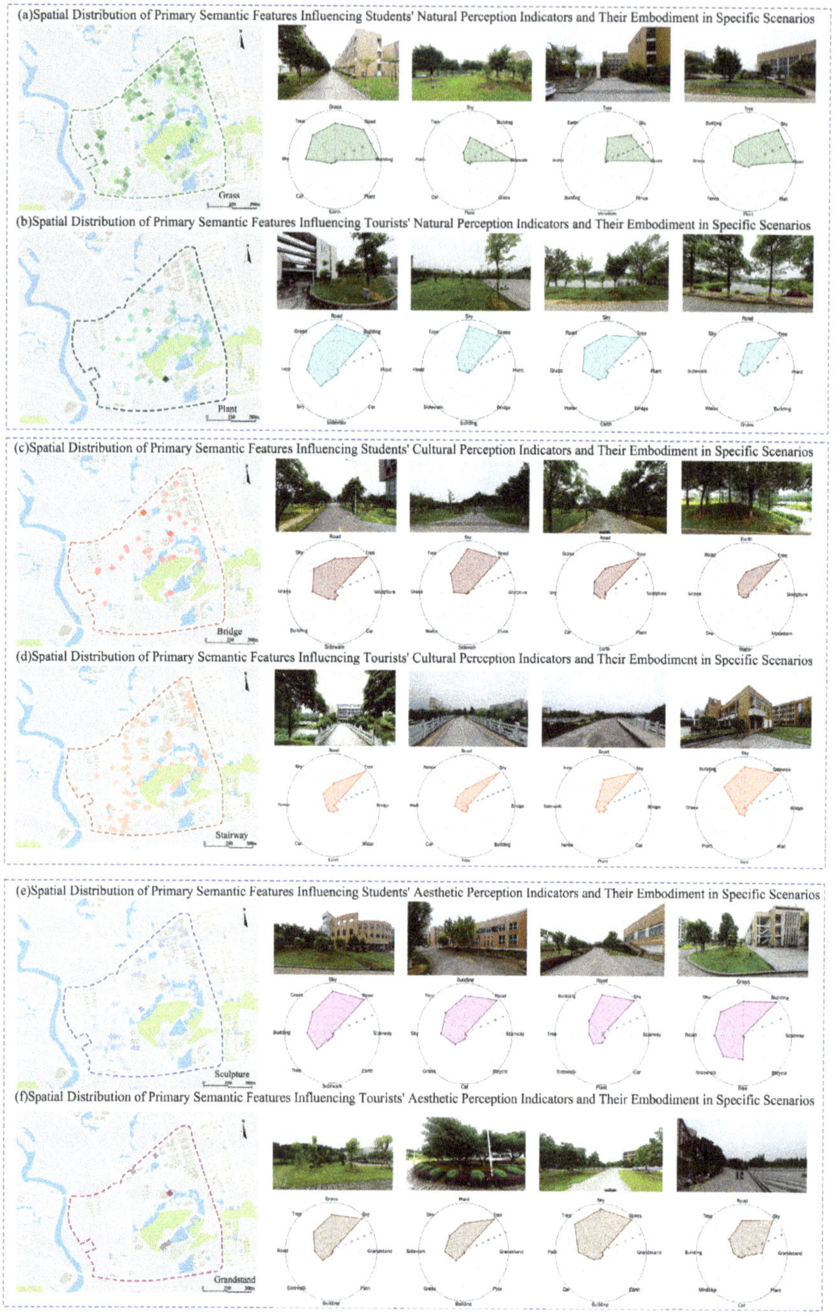

Figure 17. Spatial distribution of key semantic features influencing students and visitors, with corresponding real-world street views.

This figure presents, in six panels (a–f), the spatial patterns of the key semantic features influencing students' and visitors' perception and their corresponding forms in typical street views. On the left are the distributions of high-impact locations, the top right shows the related street views, and the bottom right presents the semantic radar charts. The key semantic features for each of the six sub-items are: Students—Natural perception: Grass; Visitors—Natural perception: Plant; Students—Cultural perception: Bridge; Visitors—Cultural perception: Stairway; Students—Aesthetic perception: Sculpture; Visitors—Aesthetic perception: Grounds.

Regarding overall patterns, for natural perception, grass for students forms hotspots along green areas and waterfront strips, closely linked with commuting corridors and the edges of teaching complexes. In the radar chart, grass carries the highest weight, followed by trees and sky, while buildings and roads are lower, showing that continuous greenery and open views shape the core image. For visitors, plants cluster at gateways, forest edges, and openings near water, with street views often featuring single tall trees, groups of shrubs, or ornamental flowerbeds. The radar chart shows plants rise clearly, followed by trees and sky, reflecting a "visually recognizable" scenic attribute.

For cultural perception, bridges for students connect key walking routes along rivers or corridors, with the radar chart showing bridges together with roads, sidewalks, and buildings as dominant, representing a linear cultural clue of "structure–access–order." For visitors, stairways appear more often in entrance forecourts and at elevation changes. In the radar chart, stairways rise together with buildings and roads, highlighting ceremony and guidance in the visiting experience. For aesthetic perception, sculptures for students mainly appear in front of buildings and in transition zones between buildings and greenery. The street views show a balance of "sculpture–greenery–facade" as the core of the aesthetic setting, with the radar chart giving higher weights to sculptures and buildings, while grass and trees provide a backdrop. For visitors, grounds and points should be opened, and hard squares and forecourts should be cleaned along principal axes. In the radar chart, grounds are dominant, with buildings and roads framing the space and supporting composition, creating areas suitable for viewing, photography, and activities with strong carrying capacity.

In summary, high-value areas for students tend to extend along continuous corridors and linear interfaces, while visitors depend more on symbolic nodes.

4.4. Campus Open Space Functional Zoning Based on Key Semantic Features of Students and Visitors

In this section, the study identifies the top-ranking landscape features that influence students' and visitors' sense of nature, culture, and aesthetics: Grass, Bridge, Sculpture, Plant, Stairway, and Grandstand. Furthermore, coordinate points containing these landscape features are filtered out and, combined with the actual campus conditions, the study area can be categorized into five types of spatial units:

- Green–blue ecological corridor zone

This zone, formed by continuous grass, trees, sky, and waterfront background, creates the green–blue base of the campus. It stretches in strips along commuting axes and riverbanks, with small resting spots appearing at corners and openings. In its semantic composition, grass and trees are the largest share, while sky and water provide an open backdrop, and buildings and roads remain at lower levels. This area supports students' daily walking and commuting, offering visitors stable scenic frames at key nodes.

- Academic and cultural axis with structural nodes

Bridges, together with roads and sidewalks, link building complexes and plazas to form an organizational pattern of "linear framework with rhythmic nodes." Typical nodes appear at river crossings, corridor turns, and building entrances, serving the functions of stopping, recognition, and guidance. Regarding semantics, bridges work with roads, sidewalks, and buildings as dominant features, making them the leading carriers through which the campus structure is perceived and understood.

- Interface aesthetic enhancement zone

Located between building forecourts and greenery, this zone centers on the combination of sculpture, building, and landscaping. The forecourt remains open with clear sightlines, while the sculptures' scale matches the building facades, and grass and trees provide the underlying layers. Regarding semantic composition, sculptures and buildings are relatively high, with grass and trees at medium levels. This area directly shapes students' daily aesthetic and resting experiences and determines the completeness of visitors' viewing and photo-taking.

- Gateway and landmark display zone

Centered on the main gate, landmark buildings, and distinctive sculptures, this zone is characterized by ceremonial sightlines and a broad hardscape foreground (grounds). A vast open space is kept at the front, with plants and trimmed hedges framing the sides, while buildings and roads in the background organize the entry and exit flow. In terms of semantics, grounds and dominates, buildings and roads define boundaries and rhythm, and plants enhance recognition. This area provides the first impression for visitors and represents the image of the campus.

- Open activity and hardscape plaza zone.

This zone is centered on open forecourts and sports or gathering spaces dominated by grounds. Its geometric boundaries and entrance rhythms are clear, allowing for circulation, gathering, and activity use. Regarding semantic composition, grounds are the main element, while buildings and roads define the boundaries and access framework. Plants and trees locally soften the scale and provide shade. This area is highly adaptable to different contexts and serves as a primary setting for various activities and photo opportunities.

5. Discussion

For the top eight features identified by the enhanced explainable decision tree algorithm, the Section 5 first analyzes the impact coefficients using a multiple regression model. Table 5 shows that for the natural perception of the student group, the influences of Grass (0.03), Box (−0.24), Building (−0.02), Fence (−0.06), and Awning (−0.04) are notable. For the natural perception of the tourist group, the effects of Plant (0.02), Step (2.20), Chair (−1.81), Sky (0.03), Sand (6.53), Box (−0.33), Path (0.03), and Grass (0.03) are all significant. Regarding the cultural perception of the student group, the impacts of Bridge (0.04), Building (0.004), Pot (1.01), House (−0.13), and Ashcan (−0.26) are evident. In the case of the tourist group's cultural perception, the effects of Stairway (0.46), Grass (0.01), Road (0.008), Tree (−0.01), Earth (−0.009), Pole (−0.13), and Box (−0.16) are prominent. For the student group's aesthetic perception, the influences of Grass (0.009), Fence (−0.02), Box (−0.28), Van (−0.09), and Tank (−0.38) stand out. In terms of the tourist group's aesthetic perception, the effects of Grandstand (0.02), Earth (−0.04), Tent (−0.12), Grass (0.01), and Hill (−0.12) are notable. The R^2 and F−statistic values for the six equations in Table 5 demonstrate a satisfactory fit, highlighting the key features' influence on the psychological perceptions of both groups.

Table 5. Empirical results of multivariable regression.

	Grass	**Sculpture**	**Pole**	**Box**	**Building**	**Fence**	**Book**	**Awning**	R^2	**F**
Natural perception (Students)	0.03 *** (0.00)	0.12 (0.58)	−0.09 (0.23)	−0.24 *** (0.01)	−0.02 *** (0.00)	−0.06 *** (0.00)	−183.67 (0.34)	−0.04 *** (0.00)	0.14	32.07
	Plant	Step	Chair	Sky	Sand	Box	Path	Grass	R^2	F
Natural perception (Vistors)	0.02 *** (0.00)	2.20 * (0.08)	−1.81 * (0.08)	0.03 *** (0.00)	6.53 ** (0.05)	−0.33 *** (0.01)	0.03 ** (0.05)	0.03 *** (0.00)	0.08	16.98
	Bridge	Building	Trade Name	Pot	Dirt Track	House	Fountain	Ashcan	R^2	F
Cultural perception (Students)	0.04 *** (0.01)	0.004 ** (0.03)	−0.12 (0.33)	1.01 ** (0.04)	−0.21 (0.29)	−0.13 *** (0.01)	0.07 (0.40)	−0.26 *** (0.01)	0.01	3.44
	Stairway	Grass	Road	Tent	Tree	Earth	Pole	Box	R^2	F
Cultural perception (Vistors)	0.46 *** (0.00)	0.01 *** (0.00)	0.008 *** (0.01)	−0.01 (0.87)	−0.01 *** (0.00)	−0.009 ** (0.05)	−0.13 ** (0.05)	−0.16 ** (0.05)	0.06	13.12
	Sculpture	Vase	Grass	Fence	Box	Van	Sand	Tank	R^2	F
Aestheticperception (Students)	−0.08 (0.72)	−86.25 (0.20)	0.009 *** (0.01)	−0.02 ** (0.05)	−0.28 *** (0.000)	−0.09 *** (0.00)	−0.18 (0.95)	−0.38 ** (0.04)	0.01	3.66
	Grandstand	Plant	Pole	Earth	Field	Tent	Grass	Hill	R^2	F
Aestheticperception (Vistors)	0.02 *** (0.003)	−0.0042 (0.43)	−0.11 (0.20)	−0.04 *** (0.00)	0.08 (0.60)	−0.12 *** (0.00)	0.01 *** (0.00)	−0.12 *** (0.00)	0.06	12.52

Note: ***, **, and * denote statistical significance at the 1%, 5%, and 10% levels, respectively.

Regarding research methods, many international studies on campus outdoor spaces have used post-occupancy evaluation (POE) and questionnaires to describe place quality and user experience, or applied space syntax to identify social and accessibility hotspots [12,79]. These studies provide a solid framework for campus environment evaluation, but they are relatively limited in revealing the non-linear effects and threshold influences of specific spatial elements on perception. This study develops an integrated approach of "street view semantic segmentation—interpretable learning—spatial classification." First, Mask2Former is used for pixel-level segmentation of street views to quantify semantic elements in a unified way. Then, random forest, gradient boosting trees, and AdaBoost are applied as model backbones, with SHAP introduced to explain feature contributions, directions, and potential thresholds. To address the problem of sample imbalance caused by "population × space" interactions, CTGAN is used for conditional data augmentation during training. Finally, the key semantic features the model identifies are projected onto space and matched with street views, translating statistical evidence directly into governable units such as "corridors, nodes, interfaces, and forecourts." This approach has three main advantages: (1) it compares both students and visitors within the same segmentation framework for cross-space analysis; (2) it uses SHAP to clearly identify the positive and negative effects of features and their possible thresholds, avoiding the "black box" problem; and (3) it connects statistical findings with functional zoning and figure–scene interpretation, turning results into practical governance tools. In addition, international evidence on campus green spaces and college students' mental health provides interdisciplinary support for focusing on green–blue elements in natural and aesthetic perception dimensions, strengthening the rationality and broader relevance of the proposed strategies.

Regarding research findings, existing campus studies mainly focus on overall quality evaluation and often consider a single group, usually students [80,81]. Supported by pixel-level semantics and interpretable modeling, this study identifies a stable correspondence among "semantics, space, and groups." Natural perception is mainly shaped by continuous green–blue corridors (grass/plant plus sky/water); cultural perception is tied to structural frameworks such as bridges and roads (bridge/stairway plus road/sidewalk/building); and aesthetic perception centers on transitions between buildings and greenery as well as open forecourts (sculpture/grounds and building). More importantly, the study reveals apparent group differences: students favor linear continuity along commuting corridors and the edges of teaching areas, while visitors are drawn to symbolic nodes such as gateways, bridges, and forecourts. Based on SHAP evidence, we identified six key semantic features (grass, plant, bridge, stairway, sculpture, grounds) and mapped them onto the spatial distribution of "key elements—typical locations—scene forms." From this, we propose layered interventions and five categories of functional zones: green–blue ecological corridors, academic and cultural axes with structural nodes, interface aesthetic enhancement zones, gateway and landmark display zones, and open activity and hardscape plazas. Thus, this study addresses students and visitors, explains nonlinear and threshold-driven mechanisms, and directly translates statistical results into actionable strategies for path optimization, node design, and interface improvement.

6. Conclusions

Using the Qishan Campus of Fuzhou University as a case study, this research builds an integrated framework of "street view image semantic segmentation (Mask2Former)—enhanced interpretable machine learning (combining CTGAN and decision tree algorithms with SHAP)—functional zoning." The study systematically describes the chain of "elements—perception—space types" in campus open spaces by linking questionnaire-based measures of natural, cultural, and aesthetic perception. On the data side, the approach combines multi-view street view collection with semantic feature quantification. On the modeling side, it introduces enhanced interpretable decision trees, using SHAP to reveal directions and thresholds of influence. This provides a practical path for translating from element-level analysis to spatial governance. The main findings and corresponding directions for landscape improvement are as follows:

- At the full-sample level, the three types of perception show stable and interpretable semantic differences between students and visitors, which can form a set of improvement suggestions. For natural perception, both groups benefit from the grass and the sky. However, when the proportions of buildings, fences, and awnings become too high, the natural experience is weakened for students. Visitors rely more on the direct combination of grass, sky, paths, and sand, and they are more sensitive to facilities such as chairs and steps, which become negative when overused. Therefore, along commuting and waterfront axes, continuous green–blue corridors should be strengthened, visual obstructions cleared, and hard edges softened with hedges, open railings, and rain shelters of controlled scale. At visitor nodes, planting design should focus on "small but refined" compositions, keeping facilities to a minimum. For cultural perception, the most potent positive effect for students comes from the linear structure guided by bridges (with an apparent threshold effect), and this can be further reinforced by adding fountains, houses, or trade names to enhance recognition. For visitors, entrances and elevation-change nodes are enhanced by stairways, higher levels of roads, and medium-to-high grass, while high proportions of boxes, poles, and bare earth, as well as poorly placed trees or tents, reduce the experience. The corresponding design direction is to strengthen accessibility and recognition at bridges and corridor nodes, create a "stairway–road" ceremonial sequence at gateways, unify signage and lighting, and reduce visual clutter by merging or burying boxes and poles, and minimizing exposed earth and temporary structures. For aesthetic perception, both groups generally benefit from grass. For students, even small amounts of sculpture can significantly improve the visual experience, and this can be complemented with limited sand elements. Visitors, in contrast, see clear improvement from grandstands, fields, and moderate levels of plants, while hills, tents, low-threshold earth, and specific ranges of poles tend to disrupt the scene and movement. The recommended approach is to implement coordinated "sculpture–building–greenery–paving" compositions in front of teaching buildings and at building–landscape transition zones, keep grounds clean and continuous along principal axes, use plants as a secondary compositional layer, and strictly control the number and placement of hills, tents, earth, and poles.

- In the functional zoning of campus open spaces, the green–blue ecological corridor zone is structured by continuous grass, trees, sky, and waterfront backgrounds. It supports students' commuting and walking, while providing visitors with stable green–blue scenic frames. Overall, it shows a pattern of "linear continuity with point stops." The design recommendations are strengthening the continuity of street trees and lawns, adding small-scale resting spots under trees, softening the edges of buildings and fences, limiting motor vehicle disturbance, and improving basic nighttime lighting and anti-slip measures. The academic and cultural axis with structural nodes links building clusters and plazas through bridges, roads, and sidewalks, forming "stop, view, and guide" points at bridge locations, corridor turns, and building forecourts. The recommendations are to enhance bridge recognition and accessibility, unify paving and railing systems, improve signage and barrier-free connections, and regulate stalls, shared bicycles, and temporary parking. The interface aesthetic enhancement zone lies between building forecourts and green transition areas, with coordinated compositions of "sculpture–building–greenery" as the core, ensuring openness and clear sightlines. Suggested improvements include integrating micro-renovation of facades, greenery, and paving, removing foreground obstructions and illegal parking, keeping edges tidy, and using even, glare-controlled lighting. The gateway and landmark display zone centers on campus entrances, landmark forecourts, and distinctive sculptures. It keeps safe space in front for photos and gathering, framed on both sides by plants and decorative hedges, while buildings and roads in the background organize circulation and order. Recommendations are strengthening ceremonial sightlines, providing designated photo stops and separate pedestrian and vehicle flows, improving barrier-free access and layered signage, and refining nighttime lighting and event-related traffic control. The open activity and hardscape plaza zone is based on open forecourts and sports or gathering spaces dominated by grounds. Buildings and roads define precise geometry and entrance rhythms, supporting quick transitions between daily use and events. Recommended strategies include using durable, non-slip, and well-drained paving, reserving points for power, internet, and shading structures, providing movable seating and temporary performance interfaces, and softening boundaries with plants and tree rows while establishing a "quick in, quick out" recovery mechanism after events.

In summary, the optimization of campus open spaces should follow the principles of "enhancing nature, guiding structure, adapting to perception, and coordinating management." Along commuting routes, the continuity of green–blue spaces should be strengthened, while bridges and corridors should be used to build cultural guidance sequences. Layered interventions should be applied for different groups—for example, softening hard edges along student pathways and simplifying facility layouts at visitor nodes. Ultimately, this forms an integrated path of "quantifying ecological base—responding to group perceptions—targeted spatial governance," providing a data-driven model for open space optimization in similar universities.

This study still has several limitations. First, its generalizability is limited. The research is based on a single campus sample, which makes it challenging to represent different climates and built environments. Second, relying on static street view images has significant limitations in terms of the time dimension. This "snapshot" approach fails to capture seasonal changes in the campus scenery, daily time variations, and the dynamic perception patterns of students and visitors. Third, the image recognition method has semantic biases, as the general segmentation model can easily confuse or miss similar categories such as plant, grass, and tree and small-scale facilities. Fourth, the measurement and modeling assumptions are relatively strong. The perception data are cross-sectional self-reports, and the RF/GBDT/AdaBoost + SHAP + CTGAN approach mainly reveals correlations and thresholds, without explicitly controlling for spatial autocorrelation or causal direction. Fifth, the functional zoning is simplified. The zoning is primarily based on the top SHAP features and the "semantic–group" logic, which does not fully capture the coordination of multiple elements or the uncertainty of boundaries.

Future research improvements: First, future research should conduct multi-campus comparative studies across diverse geographical locations and institutional settings to enhance the generalizability of findings. Second, future research should integrate multi-temporal street view collection, behavioral trajectory tracking, and environmental sensor monitoring to develop a dynamic perception assessment framework that captures how temporal changes influence spatial perception. Third, improve recognition accuracy through detailed scene labeling and model fine-tuning, refine the semantic ontology, and introduce uncertainty quantification. Fourth, the methodological framework can be strengthened by incorporating spatial econometrics, causal/non-linear models, robustness, and sensitivity analyses. Fifth, empirical validation is carried out by combining field observations and intervention assessments to test the effectiveness and transferability of zoning and thresholds, and external cross-validation is conducted.

Author Contributions: Conceptualization, X.Z.; Methodology, X.Z., Y.C. and Z.T.; Software, Z.D. and C.G.; Writing—original draft preparation, X.Z.; Writing—review and editing, Y.C. and Z.D.; Supervision, Visualization, Z.D. and C.G.; Funding acquisition, Z.T. and Z.D. All authors have read and agreed to the published version of the manuscript.

Funding: This research work was supported by the National Natural Science Foundation of China under Grant No. 72341030, and the Special Funding Project of the China Agricultural and Forestry University Design Art Alliance under Grant No. 111900050.

Institutional Review Board Statement: The study was conducted according to the guidelines of the Declaration of Helsinki, and approved by the Ethics Committee of Fujian Agriculture and Forestry University (protocol code FAFUIRB-2025-04023 on 2 April 2025).

Informed Consent Statement: Informed consent was obtained from all participants involved in the study.

Data Availability Statement: The raw data supporting the conclusions of this article will be made available by the authors on request.

Conflicts of Interest: The authors declare no conflicts of interest.

References

1. Hajrasouliha, A. Campus score: Measuring university campus qualities. *Landsc. Urban Plan.* **2017**, *158*, 166–176. [CrossRef]
2. Söderström, O.; Paasche, T.; Klauser, F. Smart cities as corporate storytelling. In *The Routledge Companion to Smart Cities*; Willis, K.S., Aurigi, A., Eds.; Routledge: London, UK, 2020; pp. 283–300.
3. Van den Berg, A.E.; Jorgensen, A.; Wilson, E.R. Evaluating restoration in urban green spaces: Does setting type make a difference? *Landsc. Urban Plan.* **2014**, *127*, 173–181. [CrossRef]
4. Turk, Y.A.; Sen, B.; Ozyavuz, A. Students exploration on campus legibility. *Procedia Soc. Behav. Sci.* **2015**, *197*, 339–347. [CrossRef]
5. Madanipour, A. *Public and Private Spaces of the City*; Routledge: London, UK, 2003.
6. Peschardt, K.K.; Stigsdotter, U.K. Associations between park characteristics and perceived restorativeness of small public urban green spaces. *Landsc. Urban Plan.* **2013**, *112*, 26–39. [CrossRef]

7. Chapman, M.P. American Places. In *Search of the Twenty-First Century Campus*; Greenwood Publishing Group: Westport, CT, USA, 2006.
8. Ujang, N.; Zakariya, K. The notion of place, place meaning and identity in urban regeneration. *Procedia Soc. Behav. Sci.* **2015**, *170*, 709–717. [CrossRef]
9. Toth, E. Teaching contested university histories through campus tours. *Transformations* **2017**, *27*, 104–112.
10. Alnusairat, S.; Ayyad, Y.; Al-Shatnawi, Z. Towards meaningful university space: Perceptions of the quality of open spaces for students. *Buildings* **2021**, *11*, 556. [CrossRef]
11. Abid, N.; Haque, M. Exploring and assessing user perception and preferences for open spaces in a university campus: A case study of IIT Roorkee, India. *New Des. Ideas* **2024**, *8*, 412–432. [CrossRef]
12. El-Darwish, I.I. Enhancing outdoor campus design by utilizing space syntax theory for social interaction locations. *Ain Shams Eng. J.* **2022**, *13*, 101524. [CrossRef]
13. Tao, Y.; Zhao, F.; Xue, M.; Jiang, B.; Lau, S.S.; Zhang, L. Factors influencing seating preferences in semi-outdoor learning spaces at tropical universities. *Buildings* **2023**, *13*, 982. [CrossRef]
14. Xu, Z.; Zhong, Y.; Han, L.; Shang, Z.; Xu, F. Catalyzing college students' well-being and creativity with campus outdoor spaces: A field study in China. *Think. Skills Creat.* **2025**, *56*, 101744. [CrossRef]
15. Tudorie, C.A.M.; Vallés-Planells, M.; Gielen, E.; Arroyo, R.; Galiana, F. Towards a greener university: Perceptions of landscape services in campus open space. *Sustainability* **2020**, *12*, 6047. [CrossRef]
16. Tourinho, A.C.C.; Barbosa, S.A.; Göçer, Ö.; Alberto, K.C. Post-occupancy evaluation of outdoor spaces on the Federal University of Juiz de Fora, Brazil campus. *Archnet-IJAR Int. J. Archit. Res.* **2021**, *15*, 617–633. [CrossRef]
17. Connell, J. A study of tourism on university campus sites. *Tour. Manag.* **1996**, *17*, 541–544. [CrossRef]
18. Cheng, D.; Gao, C.; Shao, T.; Iqbal, J. A landscape study of Sichuan University (Wangjiang Campus) from the perspective of campus tourism. *Land* **2020**, *9*, 499. [CrossRef]
19. McManus, P.; Connell, J.; Ding, X. Chinese tourists at the University of Sydney: Constraints to co-creating campus tourism? *Curr. Issues Tour.* **2021**, *24*, 3508–3518. [CrossRef]
20. Almeida, I.; Silveira, L. Who is visiting universities? General considerations on the demand characteristics of campus-based tourism. *J. Tour. Herit. Res.* **2022**, *5*, 179–198.
21. Agheyisi, J.E.; Ebinum, G.O. Students' perception and use of open spaces in a university campus. *IFE Res. Publ. Geogr.* **2019**, *17*, 1–13.
22. Stedman, R.C. Is It Really Just a Social Construction? The Contribution of the Physical Environment to Sense of Place. *Soc. Nat. Resour.* **2003**, *16*, 671–685. [CrossRef]
23. Lewicka, M. Place attachment: How far have we come in the last 40 years? *J. Environ. Psychol.* **2011**, *31*, 207–230. [CrossRef]
24. Scannell, L.; Gifford, R. Defining place attachment: A tripartite organizing framework. *J. Environ. Psychol.* **2010**, *30*, 1–10. [CrossRef]
25. Williams, D.R.; Vaske, J.J. The measurement of place attachment: Validity and generalizability of a psychometric approach. *For. Sci.* **2003**, *49*, 830–840. [CrossRef]
26. Wang, S.; Han, C. The influence of learning styles on perception and preference of learning spaces in the university campus. *Buildings* **2021**, *11*, 572. [CrossRef]
27. Zhang, J.; Li, Y. The impact of campus outdoor space features on students' emotions based on the emotion map. *Int. J. Environ. Res. Public Health* **2023**, *20*, 4277. [CrossRef]
28. Xu, Y.; Wang, T.; Wang, J.; Tian, H.; Zhang, R.; Chen, Y.; Chen, H. Campus landscape types and pro-social behavioral mediators in the psychological recovery of college students. *Front. Psychol.* **2024**, *15*, 1341990. [CrossRef]
29. Poria, Y.; Reichel, A.; Biran, A. Heritage site management: Motivations and expectations. *Ann. Tour. Res.* **2006**, *33*, 162–178. [CrossRef]
30. Nasar, J.L. Urban design aesthetics: The evaluative qualities of building exteriors. *Environ. Behav.* **1994**, *26*, 377–401. [CrossRef]
31. Poria, Y. Clarifying Heritage Tourism: Distinguishing Heritage Tourists from Tourists in Heritage Places. Ph.D. Thesis, University of Surrey, Guildford, UK, 2001.
32. Daniel, T.C. Whither scenic beauty? Visual landscape quality assessment in the 21st century. *Landsc. Urban Plan.* **2001**, *54*, 267–281. [CrossRef]
33. Meneghetti, C.; Muffato, V.; Toffalini, E.; Altoè, G. The contribution of visuo-spatial factors in representing a familiar environment: The case of undergraduate students at a university campus. *J. Environ. Psychol.* **2017**, *54*, 160–168. [CrossRef]
34. Peng, J.; Strijker, D.; Wu, Q. Place identity: How far have we come in exploring its meanings? *Front. Psychol.* **2020**, *11*, 294. [CrossRef] [PubMed]
35. Lewicka, M. Place attachment, place identity, and place memory: Restoring the forgotten city past. *J. Environ. Psychol.* **2008**, *28*, 209–231. [CrossRef]

36. Limei, N.; Dongfan, W.; Bo, Z. Landscape image recognition and analysis based on a deep learning algorithm. *J. Intell. Fuzzy Syst.* **2024**, JIFS-239654. [CrossRef]
37. Malik, K.; Robertson, C. Landscape similarity analysis using texture encoded deep-learning features on unclassified remote sensing imagery. *Remote Sens.* **2021**, *13*, 492. [CrossRef]
38. Kabisch, N.; Qureshi, S.; Haase, D. Human–environment interactions in urban green spaces—A systematic review of contemporary issues and prospects for future research. *Environ. Impact Assess. Rev.* **2015**, *50*, 25–34. [CrossRef]
39. Li, X.; Zhang, C.; Li, W.; Ricard, R.; Meng, Q.; Zhang, W. Assessing street-level urban greenery using Google Street View and a modified green view index. *Urban For. Urban Green.* **2015**, *14*, 675–685. [CrossRef]
40. Zhu, X.X.; Tuia, D.; Mou, L.; Xia, G.S.; Zhang, L.; Xu, F.; Fraundorfer, F. Deep learning in remote sensing: A comprehensive review and list of resources. *IEEE Geosci. Remote Sens. Mag.* **2017**, *5*, 8–36. [CrossRef]
41. Helbich, M.; Yao, Y.; Liu, Y.; Zhang, J.; Liu, P.; Wang, R. Using deep learning to examine street view green and blue spaces and their associations with geriatric depression in Beijing, China. *Environ. Int.* **2019**, *126*, 107–117. [CrossRef] [PubMed]
42. He, H.; Zhang, T.; Zhang, Q.; Rong, S.; Jia, Y.; Dong, F. Exploring the impact of campus landscape visual elements combination on short-term stress relief among college students: A case from China. *Buildings* **2024**, *14*, 1340. [CrossRef]
43. Gao, W.; Tang, B.M.; Liu, B. Effects of landscape characteristic perception of campus on college students' mental restoration. *Behav. Sci.* **2025**, *15*, 470. [CrossRef] [PubMed]
44. Qin, Y.; Wu, X.; Yu, T.; Jiang, S. Enhancing student-centered walking environments on university campuses through street view imagery and machine learning. *PLoS ONE* **2025**, *20*, e0321028. [CrossRef]
45. Chen, S.; Chen, Z.; Hong, J.; Zhuang, X.; Su, C.; Ding, Z. Exploring the relationship between audio-visual perception in Fuzhou universities and college students' attention restoration quality using machine learning. *Front. Psychol.* **2025**, *16*, 1572426. [CrossRef]
46. Ribeiro, M.T.; Singh, S.; Guestrin, C. "Why should I trust you?": Explaining the predictions of any classifier. In Proceedings of the 22nd ACM SIGKDD International Conference on Knowledge Discovery and Data Mining (KDD '16), San Francisco, CA, USA, 13–17 August 2016; ACM: New York, NY, USA, 2016; pp. 1135–1144.
47. Gebru, T.; Krause, J.; Wang, Y.; Chen, D.; Deng, J.; Aiden, E.L.; Fei-Fei, L. Using deep learning and Google Street View to estimate the demographic makeup of neighborhoods across the United States. *Proc. Natl. Acad. Sci. USA* **2017**, *114*, 13108–13113. [CrossRef]
48. Lundberg, S.M.; Lee, S.-I. A unified approach to interpreting model predictions. In *Advances in Neural Information Processing Systems*; Curran Associates, Inc.: Red Hook, NY, USA, 2017; Volume 30, pp. 4765–4774.
49. Xiao, Z.; Zhang, C.; Li, Y.; Chen, Y. Community park visits determined by the interactions between built environment attributes: An explainable machine learning method. *Appl. Geogr.* **2024**, *172*, 103423. [CrossRef]
50. Chen, Z.; Yang, H.; Ye, P.; Zhuang, X.; Zhang, R.; Xie, Y.; Ding, Z. How does the perception of informal green spaces in urban villages influence residents' complaint sentiments? A machine learning analysis of Fuzhou City, China. *Ecol. Indic.* **2024**, *166*, 112376. [CrossRef]
51. Yang, W.; Fei, J.; Li, Y.; Chen, H.; Liu, Y. Unraveling nonlinear and interaction effects of multilevel built environment features on outdoor jogging with explainable machine learning. *Cities* **2024**, *147*, 104813. [CrossRef]
52. Wagner, F.; Milojevic-Dupont, N.; Franken, L.; Zekar, A.; Thies, B.; Koch, N.; Creutzig, F. Using explainable machine learning to understand how urban form shapes sustainable mobility. *Transp. Res. Part D Transp. Environ.* **2022**, *111*, 103442. [CrossRef]
53. Ruan, Y.; Zhang, X.; Wang, S.; Chen, X.; Chen, Q. Street view-enabled explainable machine learning for spatial optimization of non-motorized transportation-oriented urban design. *Land* **2025**, *14*, 1347. [CrossRef]
54. Eshraghi, P.; Dehnavi, A.N.; Mirdamadi, M.; Talami, R.; Zomorodian, Z.S. An AI-driven framework for rapid and localized optimizations of urban open spaces. *arXiv* **2025**, arXiv:2501.08019. [CrossRef]
55. Rafiei, S.; Gifford, R. The meaning of the built environment: A comprehensive model based on users traversing their university campus. *J. Environ. Psychol.* **2023**, *87*, 101975. [CrossRef]
56. Higuera-Trujillo, J.L.; Llinares, C.; Macagno, E. The cognitive-emotional design and study of architectural space: A scoping review of neuroarchitecture and its precursor approaches. *Sensors* **2021**, *21*, 2193. [CrossRef]
57. Kaplan, R.; Kaplan, S. *The Experience of Nature: A Psychological Perspective*; Cambridge University Press: Cambridge, UK, 1989.
58. Kaplan, S. The restorative benefits of nature: Toward an integrative framework. *J. Environ. Psychol.* **1995**, *15*, 169–182. [CrossRef]
59. Ulrich, R.S. View through a window may influence recovery from surgery. *Science* **1984**, *224*, 420–421. [CrossRef]
60. Hartig, T.; Korpela, K.; Evans, G.W.; Gärling, T. A measure of restorative quality in environments. *Scand. Hous. Plan. Res.* **1997**, *14*, 175–194. [CrossRef]
61. Twohig-Bennett, C.; Jones, A. The health benefits of the great outdoors: A systematic review and meta-analysis of greenspace exposure and health outcomes. *Environ. Res.* **2018**, *166*, 628–637. [CrossRef]
62. Felsten, G. Where to take a study break on the college campus: An attention restoration theory perspective. *J. Environ. Psychol.* **2009**, *29*, 160–167. [CrossRef]

63. Kyle, G.; Graefe, A.; Manning, R.; Bacon, J. Effects of place attachment on users' perceptions of social and environmental conditions in a natural setting. *J. Environ. Psychol.* **2004**, *24*, 213–225. [CrossRef]
64. Bratman, G.N.; Hamilton, J.P.; Hahn, K.S.; Daily, G.C.; Gross, J.J. Nature experience reduces rumination and subgenual prefrontal cortex activation. *Proc. Natl. Acad. Sci. USA* **2015**, *112*, 8567–8572. [CrossRef] [PubMed]
65. Ashworth, G.J.; Tunbridge, J.E. *A Geography of Heritage: Power, Culture, and Economy*; Arnold: London, UK, 2000.
66. Poria, Y.; Butler, R.; Airey, D. The core of heritage tourism. *Ann. Tour. Res.* **2003**, *30*, 238–254. [CrossRef]
67. Stamps, A.E., III. Mystery, complexity, legibility and coherence: A meta-analysis. *J. Environ. Psychol.* **2004**, *24*, 1–16. [CrossRef]
68. Malewczyk, M.; Taraszkiewicz, A.; Czyż, P. Preferences of the facade composition in the context of its regularity and irregularity. *Buildings* **2022**, *12*, 169. [CrossRef]
69. Nasar, J.L. The evaluative image of the city. *J. Am. Plan. Assoc.* **1990**, *56*, 41–53. [CrossRef]
70. Hipp, J.A.; Gulwadi, G.B.; Alves, S.; Sequeira, S. The relationship between perceived greenness and perceived restorativeness of university campuses and student-reported quality of life. *Environ. Behav.* **2016**, *48*, 1292–1308. [CrossRef]
71. Baloglu, S.; McCleary, K.W. A model of destination image formation. *Ann. Tour. Res.* **1999**, *26*, 868–897. [CrossRef]
72. Cheng, B.; Misra, I.; Schwing, A.G.; Kirillov, A.; Girdhar, R. Masked-attention mask transformer for universal image segmentation. In Proceedings of the IEEE/CVF Conference on Computer Vision and Pattern Recognition (CVPR), New Orleans, LA, USA, 19–24 June 2022; pp. 1290–1299.
73. Xu, L.; Skoularidou, M.; Cuesta-Infante, A.; Veeramachaneni, K. Modeling tabular data using conditional GAN. In *Advances in Neural Information Processing Systems 32 (NeurIPS 2019)*; Curran Associates, Inc.: Red Hook, NY, USA, 2019; pp. 7333–7343.
74. Engelmann, J.; Lessmann, S. Conditional Wasserstein GAN-based oversampling of tabular data for imbalanced learning. *Expert Syst. Appl.* **2021**, *174*, 114582. [CrossRef]
75. Zhao, Z.; Kunar, A.; Birke, R.; Chen, L.Y. CTAB-GAN: Effective table data synthesizing. In Proceedings of the 13th Asian Conference on Machine Learning (ACML 2021), Virtual, 17–19 November 2021; PMLR: Lawrence, KS, USA, 2021; Volume 157, pp. 97–112. Available online: https://proceedings.mlr.press/v157/zhao21a (accessed on 3 October 2025).
76. Breiman, L. Random forests. *Mach. Learn.* **2001**, *45*, 5–32. [CrossRef]
77. Friedman, J.H. Greedy function approximation: A gradient boosting machine. *Ann. Stat.* **2001**, *29*, 1189–1232. [CrossRef]
78. Freund, Y.; Schapire, R.E. A decision-theoretic generalization of on-line learning and an application to boosting. *J. Comput. Syst. Sci.* **1997**, *55*, 119–139. [CrossRef]
79. Aydin, D.; Ter, U. Outdoor space quality: Case study of a university campus plaza. *Archnet-IJAR Int. J. Archit. Res.* **2008**, *2*, 189–203.
80. Liu, W.; Sun, N.; Guo, J.; Zheng, Z. Campus green spaces, academic achievement and mental health of college students. *Int. J. Environ. Res. Public Health* **2022**, *19*, 8618. [CrossRef] [PubMed]
81. Działek, J.; Homiński, B.; Miśkowiec, M.; Świgost-Kapocsi, A.; Gwosdz, K. The assessment of the quality of campus public spaces as key parts of the learning landscape: Experience from a crowdsensing study on the Third Campus of Jagiellonian University, Krakow, Poland. *Urban Des. Int.* **2024**, *29*, 77–92. [CrossRef]

buildings

MDPI

Article

YOLOv11-Safe: An Explainable AI Framework for Data-Driven Building Safety Evaluation and Design Optimization in University Campuses

Jing Hou [1], Yanfeng Hu [2,*], Bingchun Jiang [1], Zhoulin Chang [1], Mingjie Cao [1] and Beili Wang [1]

1 School of Mechanical and Electrical Engineering, Guangdong University of Science and Technology, No. 1, Wenhui Road, Fulong Village, Shipai Town, Dongguan 523070, China; houjing@gdust.edu.cn (J.H.)
2 Faculty of Design and Architecture, University Putra Malaysia, Selangor 43400, Malaysia
* Correspondence: huyanfeng1996@gmail.com

Abstract: Campus buildings often present hidden safety risks such as falls and wheelchair instabilities, which are closely related to architectural layout, material selection, and accessibility design. This study develops YOLOv11-Safe, an attention-enhanced and geometry-aware framework that functions as both a detection model and a spatial diagnostic tool for building safety assessment. The framework integrates a modified SimAM attention mechanism and a normalized Wasserstein distance (NWD) loss to improve detection accuracy in complex indoor environments, trained on a dataset of 1000 annotated images covering fall and wheelchair accident scenarios. To interpret spatial risk patterns, a Random Forest classifier combined with SHAP analysis was applied to quantify the contribution of five architectural–behavioral variables: body–ground contact ratio (BGCR), accessibility index (AI), event duration (D), body posture angle (PA), and spatial density (SD). Results show that BGCR and AI dominate the risk-level prediction, while D, PA, and SD refine boundary conditions. Scene-based verification further demonstrated that the framework accurately localized unsafe features—such as uneven drainage edges and discontinuous handrails—and translated them into actionable design feedback. The proposed approach thus links deep-learning detection with interpretable spatial analysis, offering a quantitative foundation for evidence-based architectural safety optimization in university campuses.

Keywords: campus building safety; deep learning; spatial risk assessment; architectural safety improvement; SHAP explainability

1. Introduction

In the era of digitalization and intelligent systems, public safety remains a pressing concern across diverse built environments. In university campuses, behavior-related safety risks are often aggravated by architectural and spatial design factors, such as poor lighting, uneven flooring, or congested circulation zones. Conventional surveillance systems can record these events but lack the analytical capacity to identify where and why such risks frequently occur, highlighting the need for adaptive monitoring methods that link behavioral detection with spatial safety improvement. With recent advances in artificial intelligence—particularly deep learning and computer vision—new opportunities have emerged to use intelligent models not only for event recognition but also as diagnostic tools for spatial risk evaluation and evidence-based design enhancement.

In recent years, abnormal behavior detection has become a major focus of deep learning research. Most existing studies concentrate on outdoor scenarios or public transportation

https://doi.org/10.3390/buildings15224125

hubs, emphasizing object recognition and risk warning [1,2]. Commonly used methods include the YOLO series (YOLOv4/v5/v7/v8), CNNs, and LSTMs, which have demonstrated promising real-time performance and accuracy [3,4]. In the educational domain, some scholars have applied such algorithms to classroom engagement monitoring [5], student emotion and attendance detection [6], and classroom state recognition [7]. Improvements in system responsiveness have also been achieved through head pose recognition and multimodal sensing [8,9]. In addition, for dormitory safety and contraband detection, Jahid et al. [10] proposed the DormGuardNet framework, which optimized YOLO specifically for student dormitory environments and constructed the dedicated PISD dataset to address challenges such as high concealment and low inspection efficiency.

In the architecture, engineering, and construction (AEC) domain, activity recognition techniques have similarly evolved from early sensor-based machine learning to advanced computer-vision approaches, achieving higher accuracy and broader applicability. Zhang et al. combined LSTM algorithms with the MediaPipe framework to recognize workers' actions, providing an effective solution for improving construction management and ensuring on-site safety [11]. Onososen et al. analyzed facial and ocular features of construction workers to identify fatigue symptoms and develop preventive strategies to reduce accidents [12]. Wang et al. employed machine learning models with points-of-interest (POI) and thermal environmental features as inputs to generate high-resolution insights into behavioral drivers [13]. Zhao et al. [14] improved the equipment pose estimation accuracy using enhanced AlphaPose and YOLOv5-FastPose models, while Feng et al. [15] proposed a camera-marking network to estimate complex equipment postures and reduce uncertainty. Liang et al. [16] implemented real-time 2D–3D pose estimation for construction robots through deep convolutional networks without additional markers or sensors, and Luo et al. [17] predicted equipment posture and potential safety hazards using historical monitoring data and neural networks. In terms of human posture assessment, Ray et al. [18] applied deep learning and vision based methods to monitor workers' body positions in real time and evaluate ergonomic compliance, while Paudel et al. [19] integrated CMU OpenPose with ergonomic assessment tools to automatically identify risky postures and enhance workplace safety. These studies collectively demonstrate that human activity recognition serves as an effective foundation for diagnosing spatial risks and informing safety-oriented design and management strategies in the built environment.

In the field of architectural safety and human factors, previous studies have emphasized that accident prevention in built environments depends not only on hazard detection but also on the spatial and ergonomic design of circulation areas [20]. Recent studies highlight that safety in built environments is inherently multidimensional, integrating physical, perceptual, and environmental dimensions of indoor quality [21–23]. Research on human–environment interaction shows how floor materials, illumination, and accessibility features directly influence postural stability and fall likelihood [24–27]. Moreover, Atlas [28] demonstrated that poor architectural decisions—such as uneven level transitions, discontinuous handrails, excessive thresholds, and inadequate lighting—are recurrent design failures contributing to slip, trip, and fall accidents. He further noted that "the lack of perception by the human brain to detect a change in elevation or a change in surface" is a key cause of falls in unfamiliar environments, underscoring the inseparability of human factors and spatial design in safety management. This aligns with findings from Wittek et al. [29], who emphasized that inadequate visual cues and spatial ambiguity reduce occupants' sense of control and increase behavioral risk in complex indoor settings. Similarly, ergonomic risk assessment frameworks, such as REBA and RULA, have been applied to evaluate posture-related hazards in construction and public interiors [30]. Within campus contexts, indoor safety audits reveal that glare, narrow corridor design, and slippery floor materials significantly increase accident risk [31,32]. These studies collectively suggest that effective safety management requires integrating behavioral monitoring with architectural diagnostics and environmental quality indicators.

For broader human activity recognition tasks, such as abnormal behavior detection, posture change, and collective behavior analysis, numerous studies have introduced diverse deep models, including C3D [33], LRCN [34,35], and deep belief networks [36]. Most of these models have achieved recognition accuracies exceeding 90% and are gradually evolving toward intelligent surveillance, real-time response, and behavior prediction. Nevertheless, these approaches remain limited when applied to university campuses, which are inherently more complex and dynamic. Existing models often address specific scenarios but struggle with the fragmented, sporadic, and time-sensitive nature of crises that occur in higher-education environments. University settings, therefore, present additional challenges: they are characterized by open and heterogeneous spaces, infrequent crisis events, and the urgent need for rapid emergency responses. Addressing these issues requires solutions that not only ensure accurate and efficient detection but also support lightweight deployment and provide transparent reasoning.

The present study addresses these gaps by introducing YOLOv11-Safe, an intelligent monitoring and spatial diagnostic framework specifically designed for campus safety. The main contributions are: (i) an improved YOLOv11 detector enhanced with attention and geometry-aware loss functions for robust crisis detection, (ii) a risk-level prediction mechanism based on Random Forest and SHAP analysis to identify spatial zones with higher structural or behavioral risk and guide modification strategies, and (iii) a dedicated dataset of crisis scenarios enabling systematic evaluation against conventional baselines. Experimental results demonstrate that the framework achieves superior detection performance while remaining lightweight and interpretable. While the proposed B-SAFE system is conceptually outlined to demonstrate potential integration into campus safety infrastructure, its implementation remains a prototype-level framework that requires further empirical validation through long-term deployment and user studies.

2. Materials and Methods

The deep learning framework consists of three stages: data preprocessing, model comparison and improvement, and risk-level prediction, as shown in Figure 1. First, a dedicated dataset of building-related safety events, including falls and wheelchair accidents that frequently occur in interior campus environments such as corridors, staircases, and ramps, was constructed and preprocessed with extensive data augmentation strategies. These events were selected because they directly reflect the relationship between human behavior and architectural safety performance, providing a foundation for evidence-based design evaluation. Second, a comparative analysis of mainstream detectors was conducted, leading to the development of the improved YOLOv11-Safe framework, which integrates a modified SimAM attention mechanism and a normalized Wasserstein distance (NWD) loss function to enhance localization accuracy and robustness in indoor environments. Finally, a risk-level prediction stage was implemented using a Random Forest model with SHAP-based interpretability to quantify how spatial attributes—such as floor material, lighting, and slope—affect the likelihood and severity of safety incidents. Together, these stages establish a unified framework linking deep learning-based behavior detection with architectural safety assessment and spatial retrofit decision support.

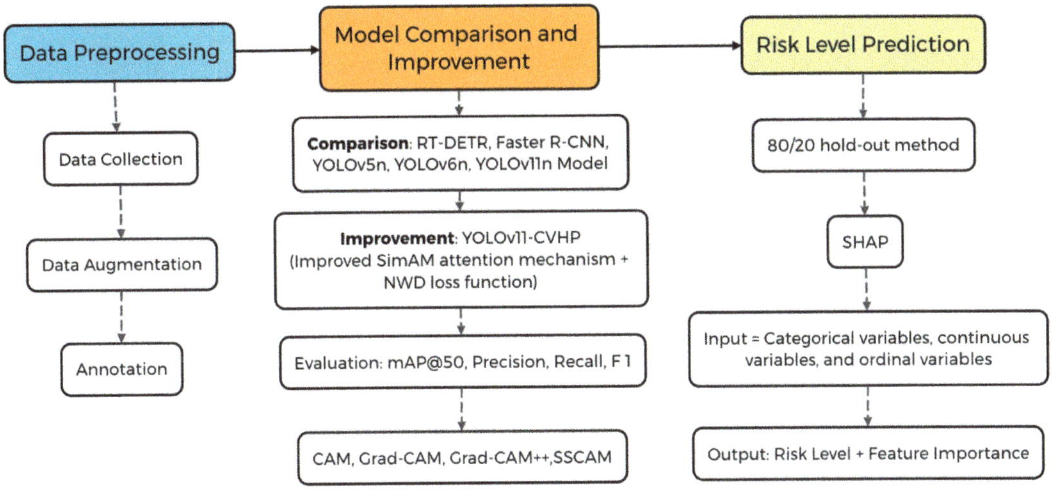

Figure 1. Overview of the deep learning framework.

2.1. Data Preprocessing

A dedicated dataset of 1000 annotated images was constructed to represent two common safety-critical categories—falling and wheelchair instability—each comprising 500 samples. Although the dataset size is moderate, diversity and representativeness were prioritized through multi-source collection and extensive augmentation. Data were obtained from two sources: (i) Real-world campus recordings, collected in collaboration with university security and medical departments across three different semesters (2023–2024) to capture seasonal lighting variations and occupancy patterns. Recordings covered daytime and nighttime conditions under varied illumination and crowd densities, using Hikvision DS-2CD2021G1-IDW and Dahua IPC-HFW3249T cameras (1080 p, 25 fps). (ii) Open-access safety video repositories and emergency drill datasets, including curated subsets collected from public behavior recognition datasets for safety and emergency response at https://blog.csdn.net/guyuealian/article/details/130184256 (accessed on 10 August 2025). These publicly available resources compile representative fall and accident scenarios captured in controlled environments, which were further screened by our research team to ensure contextual alignment with university campus conditions. All videos and images were anonymized and reprocessed to remove personal identifiers, following institutional ethical standards. All external data were screened to ensure contextual relevance to campus environments and anonymized according to institutional privacy guidelines (no identifiable faces or personal data). All data collection and processing procedures complied with institutional research ethics guidelines and the relevant national data protection regulations of the People's Republic of China, including the Personal Information Protection Law (PIPL) and the Cybersecurity Law. All recordings were fully anonymized before analysis: automatic face and body blurring were applied using the OpenCV and MediaPipe libraries, followed by manual inspection to ensure that no personally identifiable information was visible. Metadata such as timestamps, camera IDs, and location details were removed. All videos were collected either from publicly available safety datasets or controlled campus recordings conducted in non-sensitive, low-risk environments (e.g., corridors, staircases, dormitories, and open walkways) without capturing identifiable individuals. The resulting dataset contains no personal information and is used solely for academic research purposes. The dataset was fully anonymized prior to analysis, and no identifiable facial or personal information was stored, displayed, or used in training. The study followed institutional privacy protection principles consistent with national data security standards in China. In addition, this study focuses on two representative categories—falls and wheelchair instability—chosen due to their frequency in campus environments and the ethical feasibility of data collection under controlled conditions. These categories serve as proxies for broader safety-critical behaviors that share similar spatial and biomechanical features.

To enhance model robustness, multiple augmentation strategies were applied: random cropping, $\pm 90^{\circ}$ rotation, horizontal flipping, Gaussian blur, and background normalization. For cluttered indoor scenes, GrabCut-based foreground extraction was applied to highlight motion-active regions. This ensured coverage of diverse conditions, including occlusion, background clutter, low illumination, and multi-person overlaps.

Annotation was performed by a team of three trained security researchers, each with at least two years of surveillance analysis experience. Every image was independently annotated by two annotators and then reviewed in a third consensus round. Bounding boxes followed YOLOv11 format standards and were verified using an IoU ≥ 0.5 criterion for inter-annotator consistency. Cases with disagreement (IoU < 0.5 or >10 px centroid displacement) were reviewed jointly, and consensus was reached through majority voting. Across all scenes, the mean Cohen's Kappa coefficient was 0.91 (SD = 0.04), with per-scene values ranging from 0.87 (dormitory corridors) to 0.94 (staircases), indicating high annotation reliability. To further assess labeling consistency across different contexts, Cohen's Kappa coefficients were computed separately by category (fall vs. wheelchair), period (daytime vs. nighttime), and scene type. The overall mean κ remained 0.91 (SD = 0.04). Stratified analysis showed $\kappa = 0.92$ for fall incidents and $\kappa = 0.89$ for wheelchair instability, indicating consistent annotation quality across event types. Temporal analysis yielded $\kappa = 0.90$ for daytime and $\kappa = 0.91$ for nighttime samples, demonstrating negligible illumination-related bias. Scene-level values ranged from 0.87 in dormitory corridors (where partial occlusions were common) to 0.94 in staircases, confirming robust inter-annotator agreement across spatial conditions. Disputed samples (IoU < 0.5 or centroid displacement > 10 px) accounted for approximately 3.2% of all annotated instances and were resolved through a third-round consensus review.

To prevent data leakage, all frames belonging to the same event sequence (same camera, same timestamp cluster within ± 2 s) were grouped as a single unit and assigned to one split only. This ensured that no temporally adjacent frames or correlated video segments appeared across training, validation, and test subsets. The final dataset split was 70% training, 20% validation, and 10% testing, with 20% of the training subset reserved internally for hyperparameter tuning. Although the test subset represents approximately 10% of the total data (around 100 images), it includes samples from all seven spatial scenes and both behavioral categories, ensuring adequate coverage of lighting, crowding, and viewpoint variations.

The YOLOv11-Safe framework was trained entirely from scratch, as the newly introduced components—SimAM attention layers and the NWD loss function—do not have any pretrained weights available. Following the standard practice of YOLO-based detectors, the stochastic gradient descent (SGD) optimizer was employed by YOLOv11-Safe with an initial learning rate of 1×10^{-2}, momentum = 0.9, and weight decay = 1×10^{-4}. The batch size was set to 16, and training images were resized to 640×640. A cosine-annealing learning-rate scheduler was applied to enhance convergence stability.

All experiments were conducted using an NVIDIA RTX 4090 GPU (24 GB VRAM) and an Intel Core i9-13900K CPU on a workstation with 64 GB RAM. The models were implemented in Python 3.10 and PyTorch 2.1.0 under Ubuntu 22.04 LTS, with CUDA 12.1 and cuDNN 8.9 acceleration.

The training process was executed for a maximum of 200 epochs, with early stopping triggered after 50 consecutive epochs without improvement in the validation F1 score. To ensure reproducibility and robustness, all random processes (data shuffling, weight initialization, and augmentation order) were controlled by fixed random seeds. Each full training configuration was repeated five times with independent seeds, and the final model was selected based on the highest mean validation F1 score across runs. For the ablation experiments in Results section, each configuration was trained and evaluated three times, and results are reported as mean ± standard deviation across runs to reflect statistical consistency under limited data conditions.

2.2. Model Comparison and Improvement

To establish a performance baseline, several mainstream detectors were evaluated, including RT-DETR [37], Faster R-CNN [38], and the YOLO family. While YOLOv11 offers competitive accuracy–efficiency trade-offs, challenges remain in (i) focusing on critical body regions under diverse postures, (ii) discriminating behaviors under complex backgrounds, and (iii) ensuring robustness under low-quality inputs.

To address the challenges, this study introduces the YOLOv11-Safe framework, a task-driven extension of YOLOv11 tailored for safety-critical behavior detection in educational environments, as shown in Figure 2. Unlike generic object detectors, YOLOv11-Safe strategically integrates two core improvements—an enhanced SimAM attention mechanism and a Normalized Wasserstein Distance (NWD) loss function [39]—to jointly improve localization accuracy, robustness under occlusion, and interpretability of predictions. The integration of these modules is not an arbitrary stacking of components but rather a functionally coordinated strategy designed specifically to meet the challenges of heterogeneous crisis events: (i) significant pose variability across different categories, (ii) interference from complex campus backgrounds (crowds, occlusion, illumination changes), and (iii) degraded input quality due to camera placement or motion blur.

At the backbone level, the framework preserves the standard YOLOv11 feature extraction pipeline to maintain efficiency on large-scale surveillance data. To mitigate the difficulty of detecting subtle local features under significant posture variations, an improved SimAM module is introduced at the P5 layer. By fusing global statistics (mean, variance of feature maps) with local contextual cues (regional pooling and variance estimation), the modified SimAM dynamically balances long-range context with fine-grained local signals. Additionally, a position-aware weighting scheme is employed to emphasize central body regions, which typically carry higher semantic salience. This design enables the network to focus on critical body parts more effectively (e.g., head, arms, torso tilt) under conditions of overlapping individuals or visual noise.

For bounding box regression, the traditional IoU-based losses are replaced by the NWD loss function, which models predicted and ground-truth boxes as spatial probability distributions. NWD jointly considers the center displacement and scale discrepancy between boxes and normalizes the Wasserstein distance with a constant factor to achieve scale-invariant optimization. Compared with IoU and its variants, NWD provides smoother gradients and superior convergence stability, particularly in cases of incomplete boundaries, small object detection, and blurred inputs. This ensures reliable localization even under adverse imaging conditions.

Furthermore, the framework explicitly accounts for event diversity across scales by optimizing detection at three hierarchical branches: P3 (small objects), P4 (medium objects), and P5 (large objects). A LargeObject-SimAM variant is deployed at P5, enabling enhanced attention allocation to large-scale, whole-body movements. This multi-branch optimization ensures robustness across both fine-scale subtle motions and coarse-scale full-body dynamics.

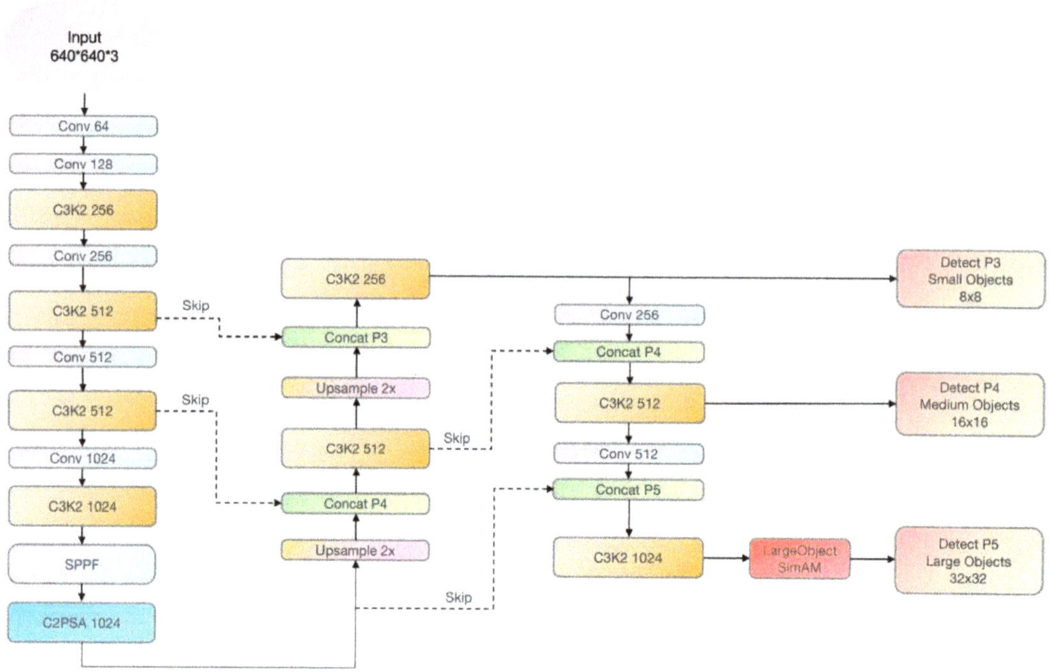

Figure 2. YOLO-Safe framework. Overall architecture of YOLOv11-Safe with multi-scale detection heads. The network takes an RGB image of 640 × 640 as input and produces predictions at three pyramid levels: P3 (stride = 8; 80 × 80) for small objects, P4 (stride = 16; 40 × 40) for medium objects, and P5 (stride = 32; 20 × 20) for large objects. The Backbone extracts hierarchical features (Conv/C2f/SPPF), the Neck fuses them via up/down-sampling and lateral connections, and the Heads output class/confidence/box offsets for each scale. Channel widths are labeled on blocks (e.g., 256/512/1024). Arrows indicate the data flow; dashed arrows denote skip connections. Backbone blocks (Conv, C2f, SPPF): pastel-blue/teal tones. Neck fusion (Upsample/Concat/Conv): pastel-yellow/orange tones. Detection heads (Detect P3/P4/P5): pastel-red/pink. SimAM attention modules: light-red rounded blocks (inserted before P3/P4/P5 heads). NWD loss: not a layer; shown as a red tag near the P5 head to indicate it replaces IoU-based regression during training. Scale and tensor sizes (example with 640 × 640 input): Backbone outputs → C3@P3: 80 × 80 × 256; C4@P4: 40 × 40 × 512; C5@P5: 20 × 20 × 1024. Neck fusions → upsample (×2) and concat to form P3: 80 × 80 × 256, P4: 40 × 40 × 512, P5: 20 × 20 × 1024 (exact channels shown on blocks). Heads → Detect P3 (small), Detect P4 (medium), Detect P5 (large).

Specifically, to improve the robustness of YOLOv11 in detecting abnormal behaviors, we propose a modified SimAM attention module, as shown in Figure 3. Unlike the original SimAM, which relies solely on global statistics, our version integrates global and local statistical information and introduces a position-aware weighting scheme to better capture feature variations across different object scales. For recognizing posture-related abnormalities, an improved SimAM attention mechanism was integrated into the backbone. The module was inserted after the C3 block at the P5 feature layer (stride = 32), where semantic abstraction is high but local spatial cues are partially lost. This placement enables refined attention to human body regions before multi-scale fusion in the PANet head.

Unlike the original SimAM, which relies solely on global statistics, the improved version jointly computes global and local attention weights and fuses them using a position-aware spatial mask. Global statistics capture overall semantic distribution, while local pooling preserves intra-object consistency for large-scale movements. A fusion coefficient $\alpha = 0.7$ balances these two branches, empirically determined through ablation among {0.3, 0.5, 0.7, 0.9}. To maintain convergence stability, α is fixed during training rather than optimized as a learnable parameter, as dynamic α values led to gradient variance and slower convergence. The improved SimAM introduces no trainable parameters and adds only +0.01 GFLOPs (~0.15%) to total computation. The core computation process can be summarized as Supplementary Code S1.

The workflow of the module is as follows:

(i) Global statistics: For an input feature map $F \in \mathbb{R}^{H\times W\times C}$, the global mean μ_g and variance σ_g^2 are computed across spatial dimensions. These are used to construct a global attention weight A_g, which captures holistic feature distribution:

$$A_g = \frac{F - \mu_g}{\sigma_g + \epsilon} \tag{1}$$

(ii) Local statistics: A local branch applies a 3 × 33\times 33 × 3 average pooling to obtain regional means μ_l, followed by local variance estimation σ_l^2. This preserves intra-object consistency for large-scale targets and prevents uniform weighting across heterogeneous body regions. The local attention weight is defined as:

$$A_l = \frac{F - \mu_l}{\sigma_l + \epsilon} \tag{2}$$

(iii) Position-aware weighting: A spatial distance-based mask $M(x, y)$ is generated to emphasize central regions of the target:

$$M(x,y) = exp\left(-\frac{(x-x_c)^2 + (y-y_c)^2}{2\sigma_p^2}\right) \tag{3}$$

where (x_c, y_c) denotes the target center and σp\sigma_pσp controls the spread. This design leverages the empirical prior that central regions often contain semantically salient cues.

(iv) Fusion and normalization: Global and local attentions are fused using adaptive weighting:

$$A = \alpha A_g + (1-\alpha)A_l, \ \ \alpha = 0.7 \tag{4}$$

And multiplied with the positional mask:

$$A' = \sigma(A \cdot M) \tag{5}$$

where $\sigma(\cdot)$ denotes the Sigmoid activation. The refined attention map A' is then applied elementwise to the input feature map, yielding the enhanced representation.

Compared with the original design, our modified SimAM introduces (i) Local statistics computation (configurable pooling window, default kernel size = 3) for improved consistency in large objects. (ii) Position-aware weighting to highlight central body regions, enhancing robustness in large-scale human action scenarios. (iii) Adaptive global–local fusion (fixed ratio, extendable to learnable) for balanced context modeling. (iv) Configurable local window size (local_sizelocal_sizelocal_size), enabling adaptation to objects of varying scales.

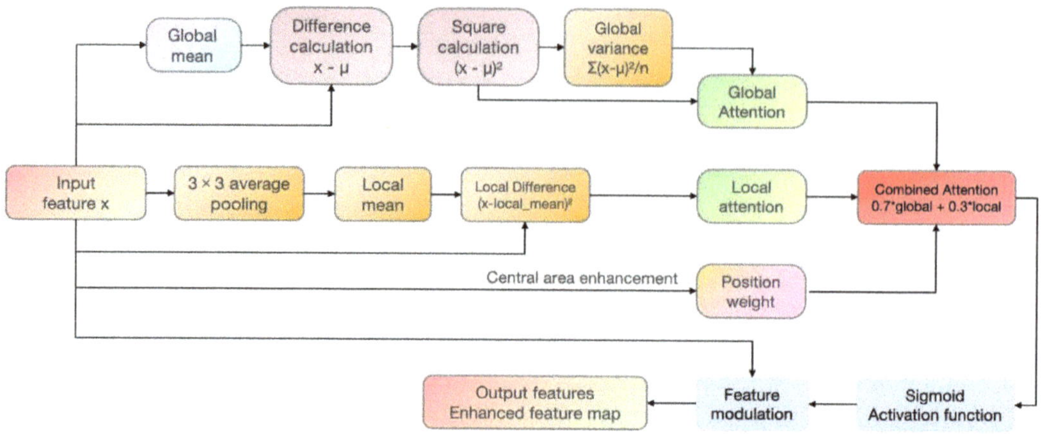

Figure 3. Improved SimAM framework. The architecture of the SimAM (Simple Attention Module) applied within the YOLOv11-Safe framework. The diagram illustrates the computation flow from the input feature map x to the enhanced output features through global and local attention integration. The process begins with global mean and local mean calculations obtained via a 3 × 3 average pooling operation. The local difference (x − local mean) and its squared form $(x - \mu)^2$ are used to compute global variance $E[(x - \mu)^2/\varphi]$ and local variance components, which together generate global and local attention weights. These weights are then combined into a position-weighted attention map to enhance central spatial regions (central-area enhancement). The enhanced feature map is further modulated through a sigmoid activation to produce the final output. Blue blocks–input and output feature tensors. Orange/yellow blocks–arithmetic or pooling operations (mean, difference, variance). Green blocks–attention weight generation (global/local). Red/magenta block–combined attention (global + local). Purple blocks–activation and feature modulation modules. Scale: All operations are performed element-wise over the spatial resolution of the input feature map (e.g., C × H × W), where attention is computed per pixel location without learnable parameters.

By explicitly modeling local differences and spatial priors, the improved SimAM produces more precise attention maps. It enhances the model's ability to locate critical body regions (head, limbs, torso) under occlusion and background noise, thereby improving both detection accuracy and interpretability.

To overcome the limitations of IoU-based losses in capturing geometric relations between bounding boxes, we introduce the Normalized Wasserstein Distance (NWD) loss, as shown in Figure 4. Unlike IoU, which only measures overlapping areas, NWD treats predicted and ground-truth boxes as two-dimensional Gaussian distributions and evaluates their similarity through a Wasserstein-based distance metric.

(i) Bounding box decomposition. For a predicted box $b_p = (x_p, y_p, w_p, h_p)$ and a ground-truth box $b_g = (x_g, y_g, w_g, h_g)$, we compute width and height as:

$$w = x_{\text{right}} - x_{\text{left}} + \epsilon, \qquad h = y_{\text{bottom}} - y_{\text{top}} + \epsilon \tag{6}$$

where a small constant ϵ is added during computation to avoid division by zero.

(ii) Center distance. The box centers are defined as:

$$c_p = (x_p + 0.5w_p, y_p + 0.5h_p), c_g = (x_g + 0.5w_g, y_g + 0.5h_g) \tag{7}$$

The squared Euclidean distance between centers is:

$$D_c = \sqrt{(x_p - x_g)^2 + (y_p - y_g)^2} \tag{8}$$

(iii) Width–height discrepancy. The scale difference is defined as:

$$D_s = \sqrt{(w_p - w_g)^2 + (h_p - h_g)^2} \tag{9}$$

(iv) Wasserstein distance and normalization. The total distance is then formulated as a first-order approximation of the closed-form 2D Gaussian Wasserstein distance under isotropic covariance assumption $\sum = diag(w^2/12, h^2/12)$:

$$D_w(b_p, b_g) = D_c + D_S \tag{10}$$

This formulation decomposes the true Wasserstein cost into translational and scale components, preserving geometric interpretability while maintaining real-time efficiency. The simplification assumes diagonal covariance and neglects rotation and cross-covariance terms, which are discussed as a limitation in Section 4.

To ensure scale-invariant gradient magnitudes, the Wasserstein distance is normalized by a constant factor C:

$$L_{NWD} = \frac{D_w}{C} \tag{11}$$

The normalization constant $C = 12.8$ corresponds to the expected diagonal standard deviation of a 640×640 input image $\left(\sqrt{640^2/12}\right)$, ensuring balanced gradient magnitudes across pyramid scales (P3–P5). Empirically, varying C within [8.0, 16.0] produced negligible effects (<0.3%) on training stability and mAP performance, confirming robustness across scales. This theoretical and empirical consistency supports the use of a fixed constant in all experiments.

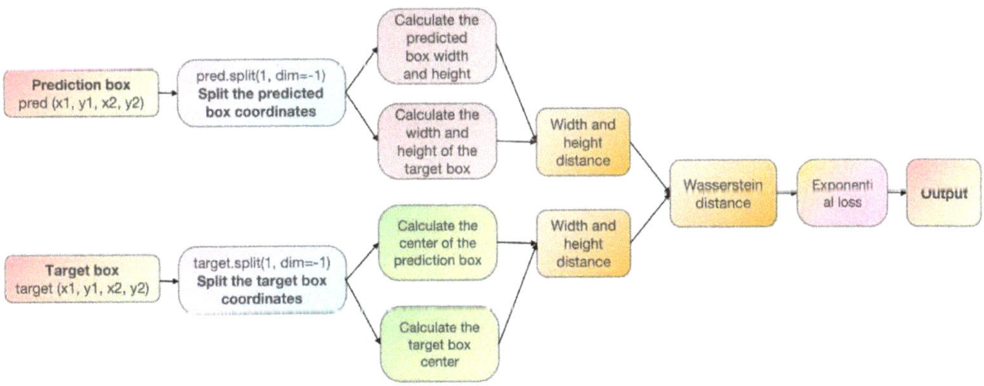

Figure 4. Normalized Wasserstein Distance (NWD) Loss framework. Computational flow of the Normalized Wasserstein Distance (NWD) loss function used for bounding-box regression in the YOLOv11-Safe framework. The process begins by splitting the prediction box (pred(x_1, y_1, x_2, y_2)) and target box (target(x_1, y_1, x_2, y_2)) coordinates into their respective widths and heights. The centers of both boxes are calculated, and the width/height differences and center-point distances are used to compute the Wasserstein distance between the predicted and target Gaussian representations. This distance is then passed through an exponential function to obtain the final NWD loss output, providing smoother gradients and improved localization stability compared with IoU-based metrics. Blue-green blocks–input data (prediction and target boxes). Orange/yellow blocks–arithmetic operations (width, height, center computation). Purple blocks–distance calculations (width/height and center). Pink block–Wasserstein-distance computation. Light-red block–final exponential loss output. Scale: All computations are performed element-wise per bounding-box pair. Each box is modeled as a 2-D isotropic Gaussian with covariance $\Sigma = diag(w^2/12, h^2/12)$, following the simplified formulation used for real-time detection.

Compared to IoU-based losses, NWD offers the following advantages: First, it captures both positional and scale discrepancies through probabilistic distance rather than area overlap. Second, the normalization constant C ensures stable optimization across varying object sizes. Third, the Wasserstein metric provides smoother and more interpretable gradients, improving convergence stability and localization accuracy under occlusion, scale variation, and blurred inputs. Consequently, NWD enhances the geometric fidelity and training stability of YOLOv11-Safe, enabling more reliable detection of safety-critical events under complex campus environments.

2.3. Model Evaluation Metrics

To ensure fair and statistically reliable comparisons, all models—including Faster R-CNN, RT-DETR, YOLOv5n, YOLOv6n, YOLOv10n, YOLOv11n, and YOLOv11-Safe—were evaluated using a consistent set of performance metrics following the COCO evaluation protocol. These metrics include Precision, Recall, F1 score, and mean Average Precision at 50% IoU (mAP@50). For baseline detectors (Faster R-CNN to YOLOv11n), training was conducted using their publicly available COCO-pretrained weights to ensure initialization consistency. In contrast, YOLOv11-Safe was trained entirely from scratch due to the inclusion of newly designed components—SimAM attention and NWD loss—which lack pretrained parameters. The model was trained using the stochastic gradient descent (SGD) optimizer with an initial learning rate of 1×10^{-2}, momentum = 0.9, and weight decay = 1×10^{-4}. The batch size was set to 16, and input images were resized to 640×640.

(i) *Precision*: Measures the proportion of samples predicted to be positive that are positive.

$$Precision = \frac{True\ Positives\ (TP)}{True\ Positives\ (TP) + False\ Positives\ (FP)} \tag{12}$$

(ii) Recall: measures the proportion of correctly identified samples among all positive samples.

$$Recall = \frac{True\ Positives\ (TP)}{True\ Positives\ (TP) + False\ Negatives\ (FN)} \tag{13}$$

(iii) F1 score: strikes a balance between Precision and Recall.

$$\text{F1 Score} = 2 \cdot \frac{Precision \times Recall}{Precision + Recall} \tag{14}$$

(iv) Mean Average Precision (mAP@50): The average detection accuracy of all categories is evaluated under an IoU threshold of 0.5.

$$\text{mAP@50} = \frac{1}{N} \sum_{i=1}^{N} AP_i \tag{15}$$

(v) Deployment efficiency indicators. Considering that campus security systems are mostly deployed on edge servers or resource-constrained monitoring devices, this paper further evaluates the operational efficiency of the model in actual deployment using FLOPs, parameter count (Params), and inference speed (FPS). FLOPs (Floating Point Operations) measure the amount of computation required for a single inference process; parameter count (Params) indicates the model storage size and determines its deployability on lightweight devices; FPS (Frames Per Second) directly reflects the model's inference speed in real-time monitoring video streams.

2.4. Danger Level Definition and Random Forest Classification

To classify building-related safety incidents into interpretable risk levels and link behavioral patterns to architectural factors, this study employed a Random Forest (RF) classifier combined with SHAP (SHapley Additive exPlanations) analysis. SHAP, grounded in cooperative game theory, quantifies the marginal contribution of each input variable to the model's output, providing post hoc interpretability for complex learning models [40].

Rather than relying on subjective scoring, a quantitative evaluation framework was developed to reflect practical safety-management requirements in campus indoor environments. Five measurable indicators were selected according to established building-safety assessment standards and the physical characteristics of fall and wheelchair-instability events: event duration (D), spatial density (SD), posture angle (PA), body–ground contact ratio (BGCR), and accessibility index (AI). All indicators were normalized to [0, 1] for comparability across heterogeneous dimensions:

(i) Event Duration (D): The total duration (in seconds) during which a fall or wheelchair-instability event persists within a single sequence is defined as:

$$D = \frac{(t_{end} - t_{start})}{f} \tag{16}$$

where f denotes the frame rate. Longer durations typically correspond to more severe or unresolved incidents. Durations are normalized by the maximum observed event length D_{max}.

(ii) Spatial Density (SD): SD measures *spatial occupation* rather than direct risk probability:

$$SD = \frac{\sum_{i=1}^{n} A_i}{A_f} \tag{17}$$

where A_i is the i-th bounding-box area and A_f the total frame area. High SD values indicate crowding, which may constrain evacuation but does not necessarily imply danger; low SD values may still coincide with isolated hazards. SD is therefore interpreted jointly with posture-based cues in subsequent risk evaluation and normalized via Min–Max scaling.

(iii) Posture Angle (PA): PA quantifies the deviation of the body's principal axis from the vertical direction, computed from the key points of the detected human bounding box:

$$PA = arccos\left(\frac{|y_{top} - y_{bottom}|}{\sqrt{(x_{top} - x_{bottom})^2 + (y_{top} - y_{bottom})^2}}\right) \tag{18}$$

where (x_{top}, y_{top}) and (x_{bottom}, y_{bottom}) denote the coordinates of the upper and lower body points, respectively. Smaller PA values indicate a greater body tilt and higher likelihood of balance loss.

(iv) Body–Ground Contact Ratio (BGCR): BGCR reflects the proportion of body area in contact with the floor and thus indicates potential impact severity. To approximate the floor plane from monocular 2D images, a hybrid method combining edge-based contour extraction and RANSAC plane fitting was adopted. Temporal averaging over five consecutive frames mitigated short-term occlusions, and low-confidence detections (IoU < 0.5 with the bounding-box base) were discarded.

A small-scale validation (120 frames) yielded an average vertical-position error of ±3–4 pixels (≈1–1.5% of frame height). Although monocular estimation is inherently approximate, this accuracy is adequate because BGCR functions as a relative indicator of body-to-floor proximity rather than an absolute geometric measurement. The revised text also acknowledges potential bias under extreme viewpoints or heavy occlusion and notes that future work will explore depth-assisted or multi-view calibration to further reduce this error.

(v) Accessibility Index (AI): AI evaluates the presence of safety-support features such as ramps, handrails, and unobstructed exits that facilitate movement and reduce fall risk. To ensure applicability in non-BIM environments, AI is derived through vision-based detection confidence using auxiliary YOLOv11-Safe classes ("ramp", "handrail", "exit"). Each detected element contributes its confidence score $p \in [0, 1]$, and the scene-level AI is defined as the mean confidence across all detected elements.

In cases where no auxiliary elements are detected (e.g., due to occlusion or limited field of view), the missing values are approximated using the average detection confidence of scenes with the same spatial type (corridor, staircase, dormitory). This imputation allows the RF model to maintain completeness while acknowledging potential bias in small or heterogeneous samples.

To mitigate such effects, the conclusion section explicitly notes that future work will adopt probabilistic or multi-imputation modeling to quantify uncertainty and examine the sensitivity of SHAP feature importance to AI estimation. When BIM or floor-plan data are available, they can optionally enhance calibration accuracy but are not required for model deployment.

(vi) Composite Risk Scoring and Discretization. The five normalized indicators were integrated into a composite score through an equal-weight linear combination:

$$R = \sum_{i=1}^{5} w_i \cdot s_i \quad w_i = 0.2 \tag{19}$$

Equal weighting was adopted as a neutral initialization to prevent dominance by any single metric and to maintain interpretability in the absence of empirical priors.

A sensitivity test was performed by perturbing each w_i by ± 0.1 while keeping $\sum w_i = 1$ on the 200-sample validation subset. The resulting variation in accuracy and mAP was below $\pm 1.5\%$, confirming robustness to moderate weight changes. For label generation, both equal-width and quantile-based binning strategies were evaluated. Equal-width binning achieved higher cross-validation consistency (+7.9% Top-1 accuracy) and clearer interpretability for architectural-risk communication and was therefore retained to define four danger levels (L1–L4).

This composite score served as the input label for the Random Forest (RF) classifier, enabling prediction of discrete risk levels (L1–L4) corresponding to distinct spatial safety conditions within campus buildings. SHAP analysis further decomposed each prediction into feature-level contributions, identifying dominant risk factors in alignment with expert judgment.

Each indicator directly corresponds to a measurable physical or behavioral dimension of the built environment, thereby linking human-centered observations with actionable design feedback. For instance, long event duration (*D*) indicates insufficient floor friction or lighting conditions that delay recovery after imbalance; high spatial density (*SD*) often suggests circulation bottlenecks or furniture layout constraints; steep posture angles (*PA*) and large body–ground contact ratios (BGCR) reveal fall-prone zones likely associated with inadequate handrail positioning or step geometry; and low accessibility index (AI) values highlight missing or malfunctioning ramps and exits.

By mapping these quantified indicators into four ordered risk levels (L1–L4) through equal-width binning (Table 1), the system translates behavioral observations into spatially interpretable diagnostics. Levels L1–L2 denote manageable conditions where design meets minimum safety standards, but routine maintenance or signage enhancement may still be required. Levels L3–L4 indicate spatial zones requiring design-level interventions, such as surface material replacement, slope correction, installation of additional handrails, or reconfiguration of narrow corridors. This hierarchical mapping establishes a bridge between machine-learning outputs and evidence-based architectural decision-making.

Table 1. Risk Levels Standards.

AvgScore Interval	Risk Level	Spatial Safety Condition	Dominant Indicators (Behavior + Space)	Recommended Design/Management Strategy
0–0.55	L1/low	Safe and well-maintained area; events short and self-recoverable	D < 5 s, SD < 0.25, PA > 70°, BGCR < 0.2, AI > 0.8	Very low risk: short duration (<5 s)/crowd size: only 1–3 people/very low frequency/stable posture/body far from ground
0.56–0.72	L2/Medium-Low	Minor spatial constraints or temporary obstruction; moderate instability	D = 5–10 s, SD ≈ 0.3–0.5, PA = 50–70°, BGCR = 0.3–0.5, AI = 0.6–0.8	Low–medium risk: slightly longer duration (6–15 s)/crowd size: 4–6 people/mild tremor/beginning of abnormal posture/approaching ground
0.73–0.85	L3/High	Environment shows clear design deficiencies or material hazards	D = 10–15 s, SD > 0.6, PA< 45°, BGCR > 0.6, AI < 0.6	Medium–high risk: sustained duration (16–25 s)/crowd size: 7–10 people/high frequency convulsions/falling or abnormal body angle/touching ground
0.86–1.00	L4/Very High	Structurally or spatially critical zone with severe accessibility failure	D > 15 s, SD > 0.8, PA < 25°, BGCR ≈ 1.0, AI < 0.3	Extremely high risk: uninterrupted duration (26–45 s)/crowd size > 11/high-frequency intense convulsions/fully lying down/heavy ground contact

The proposed risk-scoring framework was reviewed by two senior experts in building-safety engineering and facility management (each with over ten years of professional experience) to ensure semantic and practical consistency. Further validation involved thirty representative video segments independently annotated by five interdisciplinary experts specializing in building safety, ergonomics, and human behavior analysis. Inter-rater agreement was evaluated using Fleiss' Kappa, yielding $\kappa = 0.81$ ($p < 0.001$), which indicates substantial agreement. Spearman's rank correlation between expert ratings and model-generated scores was $\rho = 0.74$ ($p < 0.001$), confirming strong positive alignment between expert assessment and automated quantification.

In comparative experiments, alternative stratification strategies—including unequal-interval grouping and expert-only ordinal labeling—produced an average 7.9% reduction in classification accuracy, with Top-2 errors concentrated in adjacent risk levels. These findings confirm that the adopted four-level equal-width scheme provides an optimal trade-off between statistical discriminability, architectural interpretability, and practical management relevance.

(vii) Feature Selection, Dimensionality Reduction, and Classification. The complete feature matrix included 15 variables—five behavioral indicators plus ten categorical spatial encodings derived from architectural context. An RF-based Gini-importance filter first removed low-variance or highly collinear features ($r > 0.9$). Subsequently, Principal Component Analysis (PCA) was applied to the remaining features, reducing the dimensionality from 15 to 5 principal components explaining over 90% of total variance. This procedure performs genuine dimensionality reduction and enhances model generalization rather than identity transformation.

The resulting compact representation served as input to the Random Forest classifier (100 trees, maximum depth = 10). SHAP analysis was conducted on the trained RF model, and attributions were projected back to the original interpretable indicators (D, SD, PA, BGCR, AI) for visualization, ensuring consistency between model interpretability and semantic meaning.

(viii) Summary of Interpretability Framework. By integrating the quantitative indicators (*D*, *SD*, *PA*, BGCR, AI) with interpretable machine-learning techniques (RF + SHAP), the framework establishes a transparent linkage between observable human behaviors and architectural safety conditions. BGCR and *PA* capture individual posture-related risk; *SD* and AI represent spatial and accessibility factors; D characterizes temporal persistence. The combination yields predictive accuracy and actionable interpretability, providing a data-driven basis for design and management decisions in risk-informed feedback.

The Random Forest (RF) classifier was trained and evaluated using a standard hold-out method with an 80/20 train-test split. Eighty percent of the dataset was used for model construction, while the remaining twenty percent was reserved for generalization testing. Categorical variables (e.g., discretized space type) were encoded numerically using OHE, and all continuous features were normalized within [0, 1] to maintain consistency across heterogeneous spatial–behavioral inputs, as shown in Figure 5.

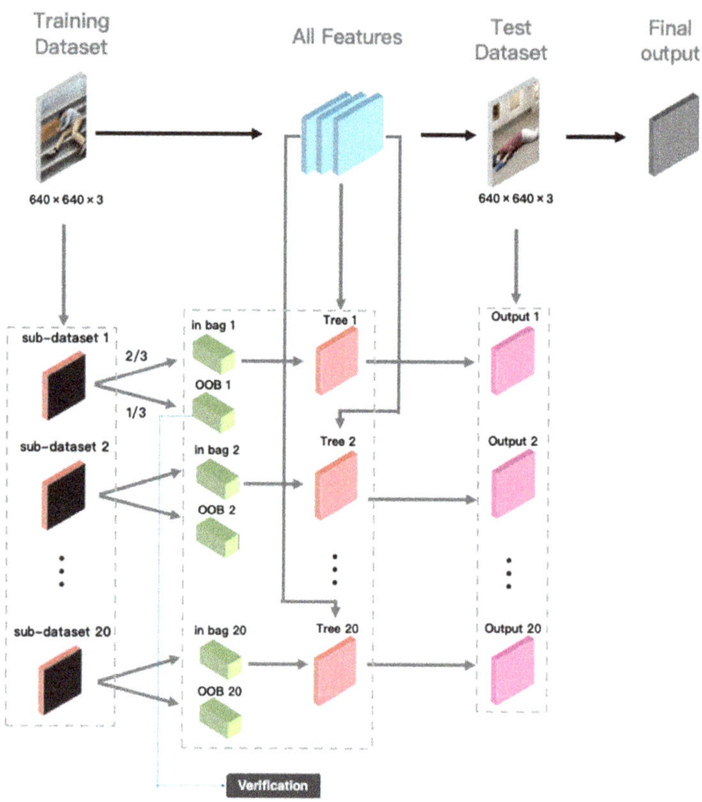

Figure 5. Random Forest Architecture. Workflow of the Random Forest-based risk-level prediction module integrated in the YOLOv11-Safe framework. The process begins with feature extraction from the training dataset (image size = 640 × 640 × 3) to produce a multi-dimensional feature set that includes geometric, temporal, and spatial descriptors. The dataset is randomly divided into multiple sub-datasets (bags) through bootstrapping, each used to train an independent decision tree (Tree 1–Tree 20). During inference, the test dataset follows the same preprocessing pipeline, and its extracted features are input to all trained trees. Each tree outputs an individual probability or class score (Output 1–Output 20), which are then aggregated by majority voting or averaging to generate the final risk-level prediction. The verification block ensures that the ensemble output aligns with previously validated safety labels. Blue blocks–feature extraction and aggregation stages. Green blocks–individual decision trees (Tree 1–Tree 20). Pink/purple blocks–prediction outputs from each tree. Gray block–final ensemble output. Dashed boxes–indicate grouped processes (bagging, dataset partitioning, or verification). Scale: Each sub-dataset contains approximately one-twentieth of the total samples, generated by random sampling with replacement to form 20 bags for ensemble learning.

To prevent overfitting and balance model complexity with accuracy, a grid search was conducted over multiple hyperparameter settings: (i) Number of trees (numTrees): 10, 50, 100, 150. (ii) Maximum depth (maxDepth): 2, 5, 10, 20. (iii) Maximum features (maxFeatures): one-third of total features, and $\sqrt{d}$ (where d is the feature dimension). (iv) Minimum samples per leaf (minSamplesLeaf): 1, 5, 10, 20, 50, 100.

After completing model training, SHAP analysis was employed to interpret feature contributions without requiring retraining, thereby avoiding additional bias. This interpretability workflow aimed to connect quantitative model outputs with spatial safety insights, revealing how each variable influences risk levels and, consequently, informs architectural design or management strategies. The workflow consisted of five complementary components: (i) Class-wise feature contribution: A SHAP summary plot (beeswarm plot) was generated to visualize the distribution of SHAP values across all samples, showing both the magnitude and direction (positive or negative) of each feature's effect on predicted safety levels. In this study, high SHAP values for body–ground contact ratio and low accessibility index strongly corresponded to L3–L4 classifications (high or very high risk), indicating areas lacking proper handrails, ramps, or non-slip surfaces. (ii) Global feature importance: The mean absolute SHAP values were aggregated and visualized in a bar chart to compare the relative importance of the five core features (event duration, spatial density, body posture angle, body–ground contact ratio, and accessibility index). This ranking revealed that body–ground contact ratio and accessibility index were dominant predictors of high-risk zones, while spatial density played a moderating role, especially in corridors and staircases where crowding elevates fall risk. (iii) Feature-level dependency analysis: Scatter plots of SHAP values against raw feature values were plotted to reveal marginal effects. Longer event durations and steeper body posture angles exhibited strong positive SHAP contributions, pushing predictions toward L3–L4 levels and identifying insufficient friction or poor illumination as spatial triggers. Conversely, low spatial density and high accessibility index (presence of ramps and unobstructed exits) produced negative SHAP values, aligning with safer spatial configurations (L1–L2). This analysis provided quantitative, design-relevant evidence explaining why particular areas were classified as high risk. (iv) Precision–Recall (PR) evaluation: To address the imbalance between high- and low-risk samples, five-fold cross-validation was conducted, and PR curves were plotted to evaluate detection performance under minority class conditions (L3–L4). This step ensured that areas classified as structurally unsafe were reliably identified without excessive false alarms. (v) ROC–AUC comparison: Multi-class ROC curves were generated, and area-under-curve (AUC) scores were computed for both training and test sets. Consistent AUC performance confirmed the stability and generalization of the RF model in predicting risk patterns across different campus building types.

All SHAP visualizations were implemented using the Python SHAP library (v0.44), combined with Matplotlib (v3.8) and Scikit-learn (v1.3). The integration of quantitative interpretability metrics with standard evaluation curves (PR and ROC–AUC) not only quantified feature influence but also strengthened the transparency, architectural relevance, and reproducibility of the proposed risk prediction and building-safety evaluation framework.

2.5. System Architecture and Application in Campus Spatial Scenarios

A conceptual system architecture is proposed to illustrate potential applications. The overarching objective of the proposed system is to enhance interior building safety, spatial risk awareness, and evidence-based design improvement in intelligent campus environments. To address spatially induced risks such as falls and wheelchair accidents occurring in corridors, staircases, ramps, and dormitories, we designed an integrated framework—B-SAFE (Building-Integrated Safety Feedback Framework)—that combines deep learning-based event detection with hierarchical IoT–Fog–Cloud computing and spatial analytics. The system is embedded within a multi-layer perception and decision-making architecture, with the following key capabilities: (i) A secure and scalable infrastructure for multi-source video and environmental data acquisition, supported by edge cameras and local processing nodes. (ii) A spatially aware system design that benefits both facility managers and campus administrators by transforming detected incidents into actionable architectural insights. (iii) Facilities for data storage, semantic annotation, and visualization of safety-critical events, enabling longitudinal safety tracking and identification of spatial design deficiencies. (iv) YOLOv11-Safe framework deployed at the edge and fog levels for behavioral detection, risk scoring, and spatial mapping of incident locations. (v) A comprehensive multi-layer architecture integrating IoT (perception), Fog (computation), Cloud (analysis), and Application (feedback), ensuring real-time flow of data and enabling immediate as well as long-term spatial interventions. Future work will focus on empirical validation in real-world campus settings.

As shown in Figure 6, the proposed B-SAFE architecture consists of four interconnected layers:

(i) Cloud Layer. The Cloud layer serves as the central analytical platform, aggregating data from multiple fog nodes for large-scale model training, updating, and cross-building risk comparison. It generates structural safety heatmaps, accessibility metrics, and long-term trend analyses, identifying high-risk architectural elements such as slippery surfaces or poorly illuminated corridors. The cloud platform also hosts the visualization dashboard, enabling planners to review spatial risk distributions and prioritize retrofit projects.

(ii) Fog Layer. Optimized YOLOv11-Safe framework are deployed on fog nodes (FN), incorporating enhanced attention mechanisms and multi-feature fusion. These nodes perform low-latency local detection of falls and wheelchair instability, extracting quantitative features such as duration, spatial density, posture angle, and body–ground contact ratio. Each fog node transmits summarized incident data to the fog master node (FMN), which coordinates computation across multiple buildings and synchronizes risk-level updates with the cloud. This distributed setup ensures high responsiveness while reducing bandwidth requirements.

(iii) IoT Layer (Perception). Comprising fixed cameras, environmental sensors, and smart access systems strategically placed in corridors, staircases, and dormitories, this layer captures real-time visual and contextual data. It provides the foundational stream for YOLO-based inference and spatial risk modeling, including environmental attributes (e.g., lighting, floor friction, and occupancy density) that affect building safety performance.

(iv) Application Layer. The Application layer delivers actionable insights to architects, facility managers, and safety personnel. Outputs include risk-level predictions, spatial heatmaps, and accessibility indices visualized through intuitive dashboards.

The B-SAFE framework aims to support evidence-based design improvement through spatial risk visualization and feedback rather than algorithmic optimization. Its function is diagnostic and decision-supportive, guiding architects and facility managers toward safer spatial configurations.

To validate the applicability of the YOLOv11-Safe and B-SAFE framework within real campus environments, seven representative architectural scenarios were selected across Guangdong University of Science and Technology, Dongguan City, Guangdong Province (Figure 7). These spaces encompass both living and teaching zones, covering diverse functional layouts and material conditions typical of higher-education buildings. Each scene was monitored using fixed smart cameras connected to the IoT-Fog-Cloud network, forming a real-time spatial safety dataset for model training and testing.

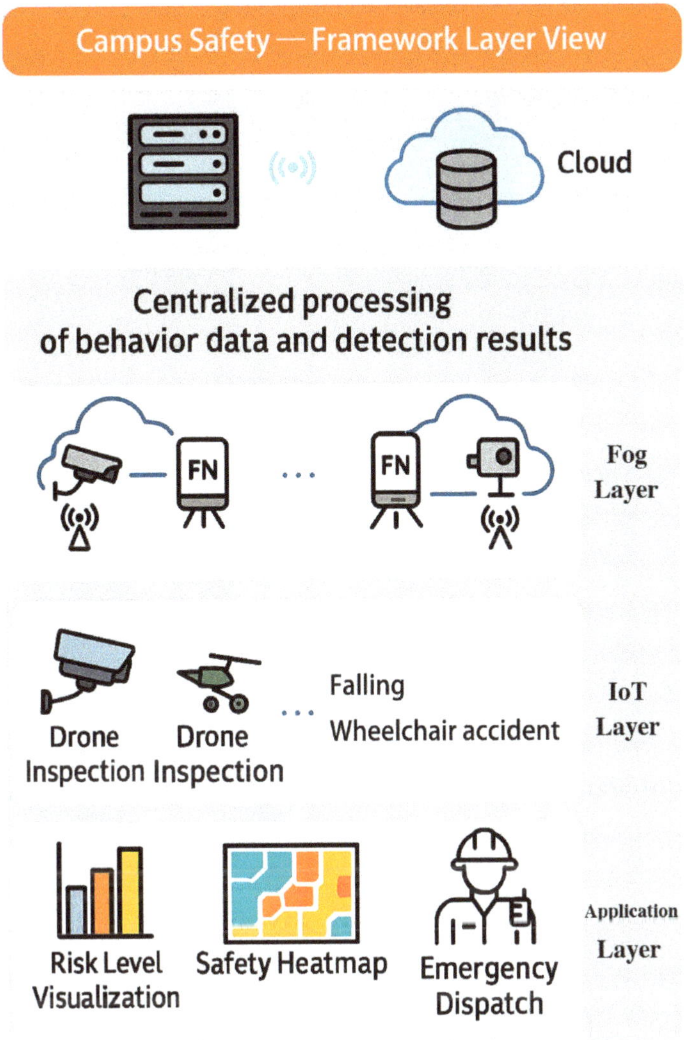

Figure 6. Layered architecture of the B-SAFE (Building-Integrated Safety Feedback) system for campus environments. The framework comprises four hierarchical layers that cooperate to detect, process, and visualize safety events in real time. IoT Layer: Front-end sensing modules such as surveillance cameras, drones, and environmental sensors collect multimodal data related to falling and wheelchair incidents. Fog Layer: Edge/fog nodes perform lightweight inference using embedded YOLOv11-Safe framework, enabling low-latency event recognition and preliminary risk scoring before cloud transmission. Cloud Layer: The centralized server aggregates detection results from multiple fog nodes, executes higher-level analytics, stores long-term records, and manages large-scale data synchronization. Application Layer: Safety-oriented interfaces convert processed data into actionable insights through risk-level visualization, heatmap generation, and emergency-dispatch modules, supporting rapid decision-making and continuous safety improvement. Scale: Each layer operates at a different temporal and spatial resolution: the IoT layer captures data at the millisecond frame rate, the fog layer processes events in seconds, and the cloud layer maintains long-term (hour-to-day) safety archives. FN = Fog Node; IoT = Internet of Things; B-SAFE = Building-Integrated Safety Feedback System.

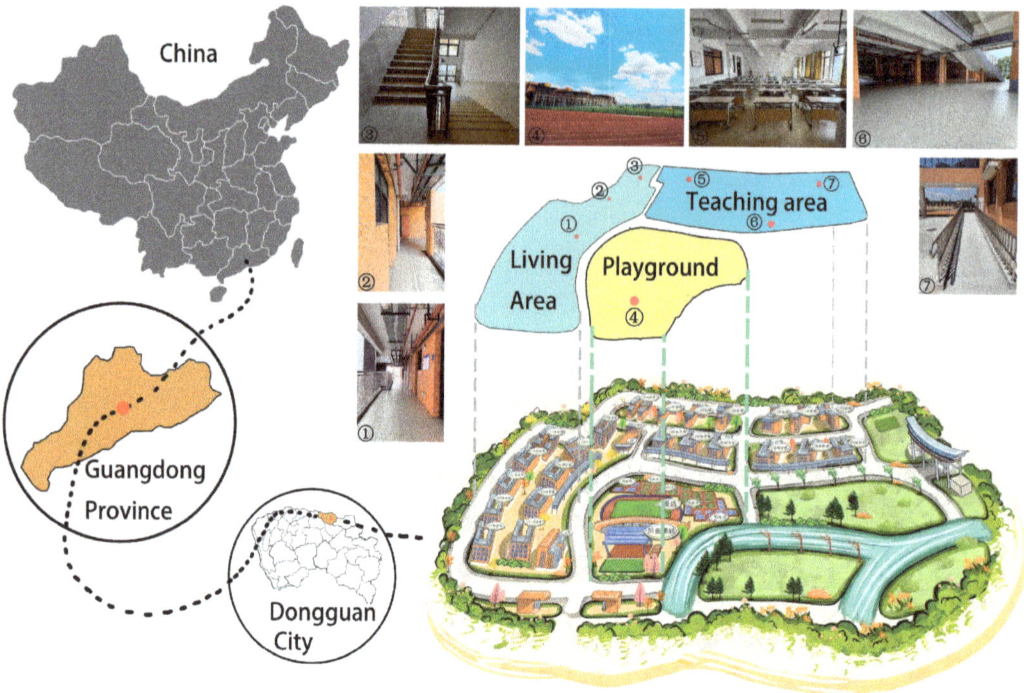

Figure 7. Spatial distribution and on-site deployment of the YOLOv11-Safe and B-SAFE framework in a university campus located in Dongguan City, Guangdong Province, China. The left side shows the geographic hierarchy (China → Guangdong Province → Dongguan City), while the right side illustrates seven representative application scenes across the campus: (1) dormitory corridor, (2) living-area walkway, (3) staircase node, (4) playground edge, (5) classroom interior, (6) teaching-block entrance hall, and (7) accessible ramp.

Scene 1: Dormitory Corridor. A narrow corridor with exposed pipelines and limited lighting was chosen to examine the effects of spatial density and accessibility index on near-fall detection. The system identified congestion points where furniture or belongings obstructed safe circulation.

Scene 2: Living-Area Walkway. This open corridor links dormitory units to the main plaza. Its smooth ceramic flooring and partial handrail coverage provided data for evaluating floor friction and handrail adequacy, variables strongly influencing the body-ground contact ratio.

Scene 3: Staircase Node. A multi-flight stairwell with irregular lighting conditions was used to test the sensitivity of the body posture angle indicator. YOLOv11-Safe captured descent-related imbalance behaviors, while SHAP analysis revealed that inadequate illumination and uneven riser height contributed to elevated risk scores (L3).

Scene 4: Playground Edge. The transition zone between the playground and teaching block often includes ramps and curbs. This site was utilized to evaluate wheelchair navigation and the accessibility index, emphasizing how gentle slope design and clear spatial markings mitigate instability risks.

Scene 5: Classroom Interior. A densely arranged classroom tested the influence of spatial density on movement safety. The model identified areas where furniture spacing below 0.9 m restricted wheelchair access, guiding design feedback for interior layout optimization.

Scene 6: Teaching-Block Entrance Hall. This wide but slippery entrance area was used to monitor fall events associated with surface materials. The analysis connected prolonged event duration and high body–ground contact ratio with low-friction tiles, supporting recommendations for surface replacement or anti-slip coating application.

Scene 7: Accessible Ramp. The outdoor ramp connecting the teaching and playground areas served as a benchmark for evaluating accessibility compliance. The B-SAFE feedback framework correlated low incident frequency and short event duration with effective gradient control and handrail design, confirming compliance with universal-design principles.

Collectively, these seven locations represent the functional and morphological diversity of campus building spaces. In the context of the seven campus spatial scenarios, five quantitative variables were designed as key measurement indicators to evaluate the spatial safety performance of building interiors. Each indicator was derived from YOLOv11-Safe detection outputs and spatial metadata, serving as an analytical bridge between behavioral incidents and architectural risk characteristics. Specifically, the event duration (D) is hypothesized to represent the persistence of unsafe states—longer durations may suggest insufficient friction, lighting imbalance, or delayed human response. The spatial density (SD) measures the proportion of occupied floor area relative to the total visible scene; higher SD values are expected in narrow or obstructed spaces (e.g., dormitory corridors or classrooms), indicating potential congestion or circulation inefficiency. The body posture angle (PA) quantifies the deviation of the human body axis from the vertical line, acting as an indirect measure of balance stability—smaller angles may correspond to greater fall likelihood on uneven surfaces or stair edges. The body–ground contact ratio (BGCR) estimates the proportion of body area in contact with the ground, used here as a proxy for impact severity and surface risk; higher BGCR values could indicate slippery materials or inappropriate floor gradients. Finally, the accessibility index (AI) reflects the spatial availability of safety-support features such as ramps, handrails, and unobstructed egress paths. Low AI scores are presumed to correspond to incomplete barrier-free design or ineffective circulation planning.

These variables jointly serve as quantitative criteria for spatial safety assessment, allowing subsequent stages of analysis (e.g., RF classification and SHAP interpretation) to determine how behavioral events correspond to architectural conditions. While this study does not aim to establish empirical conclusions at this stage, the proposed indicators define a consistent framework for linking human–environment interactions with building safety evaluation in university campuses.

3. Results

3.1. Ablation Experiments

To further evaluate the contribution of the proposed modules, ablation experiments were conducted on the improved SimAM attention mechanism and the NWD loss function, as summarized in Table 2. Each configuration was trained and evaluated under identical hyperparameter settings to ensure comparability. The reported metrics represent the mean $\pm$ standard deviation from three independent runs, minimizing random variation due to initialization and confirming statistical consistency.

Integrating the improved SimAM attention module led to a modest yet consistent improvement over the baseline, with an F1 score of 85.6 $\pm$ 0.7%, precision of 86.5 $\pm$ 0.8%, recall of 84.8 $\pm$ 1.0%, and mAP@50 of 91.2 $\pm$ 0.7%. These gains (+0.5% F1 and +1.1% mAP@50 relative to the baseline) indicate that enhanced spatial attention reduces background interference and strengthens feature discrimination, while the computational overhead remains negligible (+0.01 GFLOPs).

Table 2. Result of Ablation Experiments. All values represent the mean ± standard deviation across three independent runs (N = 100 test images). All experiments were conducted under identical input resolution (640 × 640) and batch size = 16 to ensure comparability.

Model	SimAM	NWD	F1 Score	p (%)	R (%)	mAP@50 (%)	GFLOPs	Params (MB)
Baseline			85.1 ± 0.7	85.7 ± 0.8	84.7 ± 1.0	90.1 ± 0.7	6.44	2.6
improved SimAM (attention mechanism)	√		85.6 ± 0.7	86.5 ± 0.8	84.8 ± 1.0	91.2 ± 0.7	6.45	2.6
NWD (loss function)		√	85.9 ± 0.8	89.3 ± 0.9	82.8 ± 1.2	91.35 ± 0.8	6.44	2.6
improved SimAM + NWD	√	√	86.9 ± 0.8	87.4 ± 0.9	86.3 ± 1.0	92.35 ± 0.8	6.45	2.6

When the NWD loss function was applied independently, the model achieved larger performance gains: F1 = 85.9 ± 0.8%, precision = 89.3 ± 0.9%, recall = 82.8 ± 1.2%, and mAP@50 = 91.35 ± 0.8%. This demonstrates that NWD provides a geometrically interpretable penalty for bounding-box regression, improving localization stability under scale variation and partial occlusion. The slight decrease in recall (−1.9%) suggests a trade-off between stricter box matching and sensitivity to positive samples.

Combining both SimAM and NWD yielded the best overall performance, achieving F1 = 86.9 ± 0.8%, precision = 87.4 ± 0.9%, recall = 86.3 ± 1.0%, and mAP@50 = 92.35 ± 0.8%. Compared with the baseline, these represent relative improvements of +1.8% in F1, +1.7% in precision, +1.6% in recall, and +2.25% in mAP@50, while maintaining the same parameter size (2.6 MB) and almost constant computation cost (6.45 GFLOPs).

Overall, the results show consistent and statistically stable improvements rather than random fluctuations. Both the improved SimAM and NWD modules contribute positively to detection accuracy, and their combination delivers the most balanced performance across all metrics. This confirms the complementary roles of fine-grained attention focusing and geometry-aware loss design in enhancing the accuracy–efficiency trade-off for safety-critical behavior detection.

3.2. Model Comparison

To ensure a fair comparison, all baseline detectors (Faster R-CNN, RT-DETR, YOLOv5n–YOLOv11n) were initialized from their publicly available COCO-pretrained weights, whereas YOLOv11-Safe using the stochastic gradient descent (SGD) optimizer with an initial learning rate of 1×10^{-2}, momentum = 0.9, and weight decay = 1×10^{-4}. The batch size was set to 16, and input images were resized to 640 × 640. Table 3 summarizes quantitative results in terms of accuracy, efficiency, and model complexity. Parameter counts and floating-point operations per second (FLOPs) were calculated using the PyTorch profiling tool with a batch size of 1 and input resolution of 640 × 640. All results represent full-precision (FP32) models without any quantization or compression applied.

Table 3. Result of Model Comparison. All baseline models (Faster R-CNN, RT-DETR, YOLOv5n–YOLOv11n) were fine-tuned from COCO-pretrained weights, whereas YOLOv11-Safe was trained from scratch using identical optimizer and learning-rate settings. Reported values correspond to the same test set (N = 100 images).

Model	F1 Score	p (%)	R (%)	mAP@50 (%)	GFLOPs	Params (MB)
Faster R-CNN	77.8	66.4	94.6	93.6	134.4	41.5
RT-DETR	79.6	82.3	77.1	79.6	54.1	19.0
YOLOv5n	81.7	85.6	78.1	86.9	7.2	2.5
YOLOv6n	73.6	85.1	64.8	85.1	11.9	4.2
YOLOv10n	73.0	78.6	68.0	78.5	8.4	2.7
YOLOv11n	85.1	85.7	84.7	90.1	6.44	2.6
YOLOv11-Safe	86.9	87.4	86.3	92.35	6.45	2.6

As a two-stage baseline, Faster R-CNN achieved strong recall (94.6%) and competitive mAP@50 (93.6%), demonstrating its capability to capture positive samples. However, its relatively low precision (66.4%) reduced the F1 score to 77.8%, indicating a higher false-positive rate. Moreover, Faster R-CNN incurred the heaviest computational cost (134.4 GFLOPs) and the largest parameter size (41.5 MB), limiting its suitability for real-time deployment in resource-constrained environments.

In comparison, RT-DETR provided a more balanced trade-off between accuracy and efficiency. It achieved 82.3% precision, 77.1% recall, and an F1 score of 79.6%, while significantly reducing computation to 54.1 GFLOPs and parameters to 19 MB. These results suggest that transformer-based detectors can offer an effective balance between accuracy and deployment efficiency.

Within the YOLO family, lightweight variants displayed clear advantages in performance–efficiency trade-offs. YOLOv5n achieved an F1 score of 81.7% (precision 85.6%, recall 78.1%) with only 7.2 GFLOPs and 2.5 MB parameters, providing a competitive lightweight baseline. YOLOv6n and YOLOv10n achieved moderate performance (F1 scores 73.6% and 73.0%), reflecting trade-offs between structural optimization and recognition stability.

YOLOv11n achieved the best overall performance among the original architectures, with the highest F1 (85.1%), precision (85.7%), recall (84.7%), and mAP@50 (90.1%), while maintaining extremely low computational complexity (6.44 GFLOPs) and parameter size (2.6 MB). The proposed YOLOv11-Safe, which integrates the improved SimAM attention and NWD loss modules, further improved the F1 score to 86.9%, precision to 87.4%, recall to 86.3%, and mAP@50 to 92.35%.

It should be noted that the SimAM and NWD components were implemented as lightweight plug-in modules. SimAM introduces a single activation function and several scalar operations without adding new convolutional layers, while NWD modifies only the loss function formulation. Consequently, both modules contribute less than 0.02 MB of additional parameters and no measurable change in FLOPs compared with YOLOv11n, confirming their computational efficiency.

Overall, the results indicate that while two-stage detectors (e.g., Faster R-CNN) excel in recall, lightweight one-stage detectors—particularly YOLOv11-Safe—achieve the best trade-off between accuracy, computational efficiency, and deployment feasibility, making them suitable for safety-critical monitoring in campus environments.

3.3. Performance Evaluation of CAM, Grad-CAM, XGrad-CAM, SSCAM for Deep Learning Model

To enhance the interpretability of YOLOv11-Safe in campus safety detection tasks, four widely used class activation mapping (CAM) methods—CAM, Grad-CAM, SSCAM, and XGrad-CAM—were employed to analyze the model's spatial attention and discriminative regions. As illustrated in Figure 8, attention heatmaps generated by the baseline YOLOv11n and the improved YOLOv11-Safe were compared under identical input conditions.

The visualization results show that YOLOv11-Safe consistently exhibits stronger semantic focus and boundary sensitivity across all four CAM methods, enabling more precise localization of safety-critical regions, such as areas around the human body and surrounding floor zone. Compared with YOLOv11n, the improved model produced more compact and contextually relevant attention maps, with reduced distraction from background clutter, illumination changes, or non-salient motion cues.

Notably, in Grad-CAM and SSCAM visualizations, the attention regions of YOLOv11-Safe were more accurately aligned with key behavioral indicators of unsafe conditions—such as body inclination, partial contact with the floor, or unstable wheelchair orientation—demonstrating its enhanced capacity to extract meaningful safety cues under complex campus environments. These results verify the contribution of the improved SimAM attention mechanism and NWD loss function in reinforcing spatial discrimination and geometric precision.

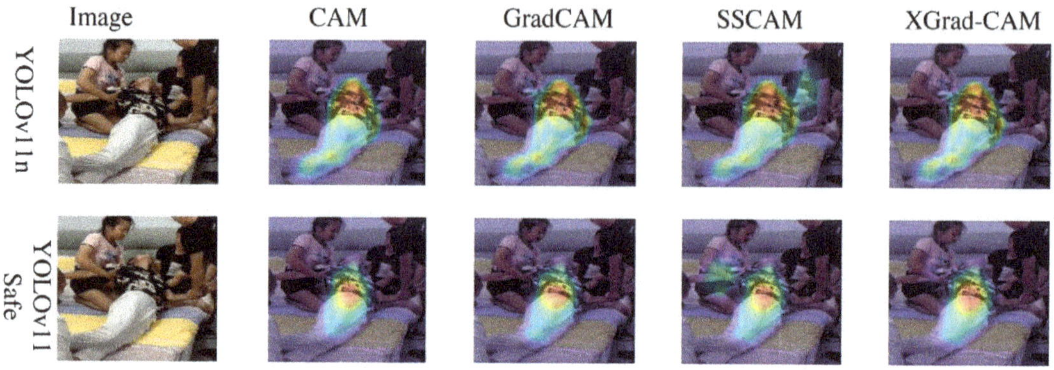

Figure 8. Visualization of attention maps for YOLOv11n and the proposed YOLOv11-Safe framework using four CAM variants (CAM, Grad-CAM, SSCAM, and XGrad-CAM) on representative campus-crisis detection samples. Each column corresponds to a different class-activation visualization technique, and each row represents model outputs under identical input conditions. Warmer colors (red/yellow) indicate regions with higher activation intensity and stronger model attention, whereas cooler colors (blue) denote areas of lower relevance. Compared with the baseline YOLOv11n, YOLOv11-Safe exhibits more concentrated and semantically aligned activation over the human–ground contact region, demonstrating improved focus and interpretability in identifying critical spatial cues related to fall events. Representative test sample was selected from seven spatial scenes (corridor, staircase, dormitory, classroom, etc.) to ensure contextual diversity.

Overall, the visualization findings confirm that YOLOv11-Safe not only improves the separation between safe and unsafe behaviors but also provides structured and interpretable decision cues for subsequent risk-level prediction. This transforms the system from an opaque detector into a transparent and explainable spatial–behavioral analysis framework, thereby strengthening its reliability and applicability for intelligent campus safety monitoring.

Although the YOLOv11-Safe framework can generally capture key spatial regions in most campus safety detection tasks and provide reliable visual explanations for subsequent risk-level prediction, its interpretability exhibits certain limitations under complex or ambiguous campus scenes. As illustrated in Figure 9, the input depicts a student experiencing a safety-critical event in a crowded corridor with multiple overlapping individuals, partial occlusion, and uneven illumination.

Under these conditions, the attention heatmaps generated by CAM, Grad-CAM, and SSCAM for YOLOv11-Safe primarily concentrated on central body areas (e.g., torso or brightly illuminated clothing) while showing insufficient activation around peripheral regions such as the arms, legs, or wheelchair boundaries. In particular, Grad-CAM visualizations produced narrowly focused hotspots that did not adequately capture body tilt, partial ground contact, or unstable wheelchair orientation—features that are crucial for security staff to determine whether an incident represents a genuine fall or instability event.

This observation suggests a semantic bias in the model's attention mechanism: it tends to emphasize high-contrast or central features while under-attending to low-contrast, peripheral, or occluded cues. Consequently, when unsafe behaviors are subtle, partially hidden, or occur within multi-person interactions, the current spatial attention module exhibits reduced sensitivity and interpretability.

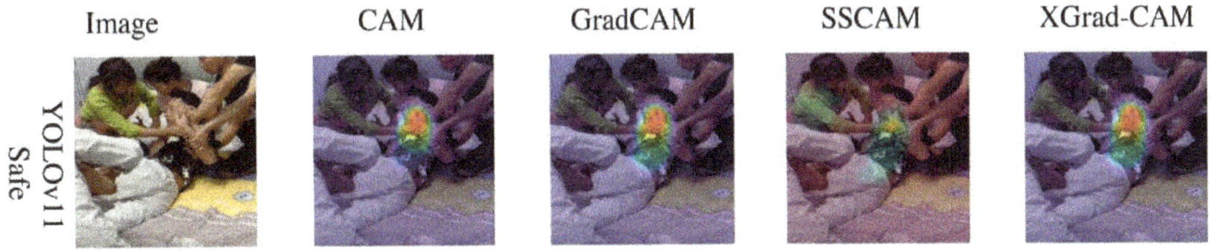

Figure 9. Visualization of YOLOv11-Safe attention behavior under occluded and ambiguous posture scenarios using four CAM variants (CAM, Grad-CAM, SSCAM, and XGrad-CAM). Each column represents a different attention visualization technique applied to the same image input. The heatmaps illustrate how occlusion and complex limb overlap reduce activation precision, causing the model to overemphasize high-contrast but semantically irrelevant regions. Warmer colors (red/yellow) denote stronger activation intensity, while cooler colors (blue/green) indicate lower attention weights. Despite these limitations, SSCAM and XGrad-CAM show relatively more stable spatial localization than CAM or Grad-CAM, highlighting the potential of advanced gradient-based visualization for diagnosing attention bias in complex campus environments. Representative test sample was selected from seven spatial scenes (corridor, staircase, dormitory, classroom, etc.) to ensure contextual diversity.

These findings reveal potential limitations for real-world deployment: while YOLOv11-Safe performs reliably for typical unsafe behaviors (e.g., clear falls, wheelchair instability), it may still suffer from incomplete attention coverage in crowded, occluded, or visually cluttered campus environments. This highlights the need for future improvements such as adaptive multi-scale attention, multi-modal integration (e.g., combining RGB with depth or audio data), and temporal modeling to achieve more comprehensive interpretability in complex safety-critical monitoring scenarios.

3.4. Random Forest Classification Results

Before conducting feature-level interpretability analysis, statistical aggregation of the experimental dataset was performed to identify spatial scenes with dominant variable characteristics. Among the seven monitored locations, Scene 1 (dormitory corridor) exhibited the most prominent distributional deviations across all five variables. Specifically, the average event duration (D) reached 8.7 s, exceeding the cross-scene mean by 41%, while spatial density (SD) averaged 0.46 due to narrow corridor width and the presence of fixed barriers. The body posture angle (PA) showed a higher standard deviation (±18°) than any other scene, suggesting frequent imbalance during movement. Meanwhile, the body–ground contact ratio (BGCR) attained a moderate-to-high level (0.58), indicating repeated near-fall or slip events. The accessibility index (AI) remained the lowest (0.42) because of limited handrail continuity and uneven floor joints near door thresholds (as shown in Figure 8). These combined factors resulted in Scene 1 contributing the largest number of L3–L4 samples (37.5% of all high-risk cases) in the dataset. This statistical dominance establishes Scene 1 as a representative high-risk spatial context for subsequent interpretability analysis. Accordingly, Random Forest classification and SHAP-based feature attribution were further applied to quantify how each variable influences the model's prediction of risk levels (L1–L4).

To ensure that the model not only achieves accurate classification but also provides interpretable decision logic, SHAP analysis was applied to the Random Forest risk-level predictor to quantify the marginal contribution of each spatial–behavioral feature across the four levels (L1–L4). As shown in Figure 10, the class specific SHAP bee swarm plots reveal distinct contribution patterns, indicating strong alignment between the model's prediction mechanism and established principles of architectural safety assessment.

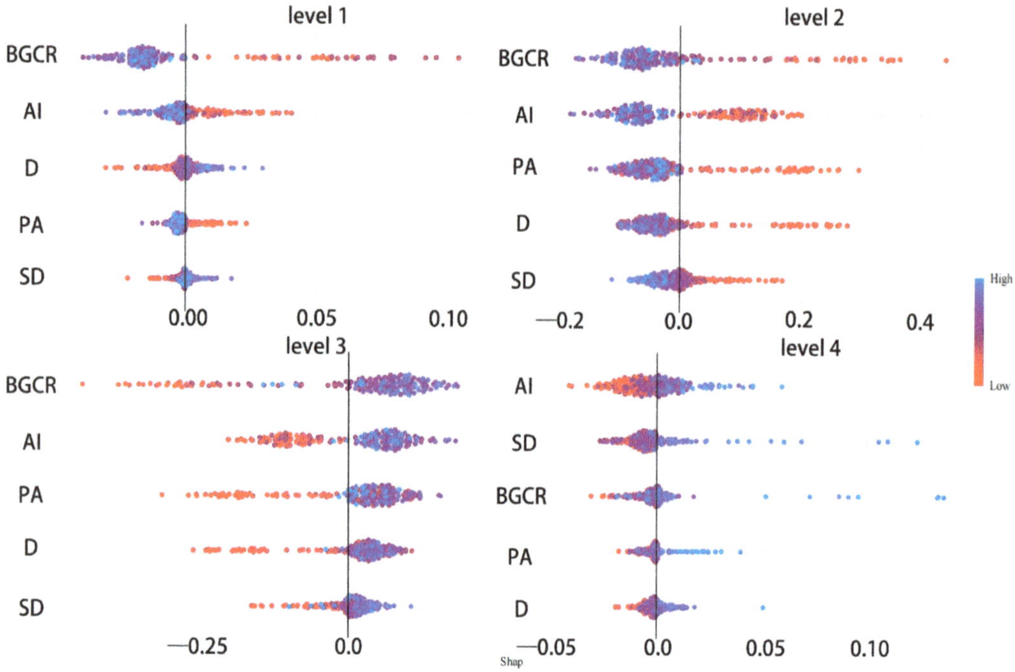

Figure 10. SHAP (SHapley Additive exPlanations) summary plots illustrating the contribution of each variable to the Random Forest risk-level prediction across four safety levels (Level 1–4). Each point represents one observation; its horizontal position corresponds to the SHAP value (magnitude of influence on the model output), while its color encodes the original feature value (red = high, blue = low). Positive SHAP values indicate features that increase the predicted risk level, whereas negative values contribute to lower predicted risk.

First, the body–ground contact ratio (BGCR) exhibited the highest absolute SHAP values across all risk levels, confirming its role as the most influential determinant of spatial safety. High BGCR values strongly shifted predictions toward L3–L4, reflecting conditions where complete body-floor contact occurs—an outcome typically associated with slippery surfaces or steep gradients. Conversely, low BGCR values stabilized predictions in L1–L2, corresponding to safe surface conditions and stable flooring materials.

Second, the accessibility index (AI) formed the next most important factor, exerting a strong negative correlation with predicted risk. Low AI values—characterized by absent handrails, blocked ramps, or narrow egress paths—amplified high-risk predictions, whereas high AI values mitigated risks by providing effective environmental support. This finding reinforces the preventive role of barrier-free and ergonomically optimized design in campus buildings.

Third, event duration (D) ranked in the mid-tier of influence, with longer events contributing positively to risk escalation. Prolonged durations suggest spatial conditions that delay recovery or impede mobility, such as low-friction flooring or inadequate lighting. In contrast, short-duration events were more frequently classified as low risk.

Fourth, body posture angle (PA) acted as a boundary-refinement feature, influencing predictions primarily at the transition between moderate and high-risk categories. Smaller PA values (greater deviation from vertical) reinforced high-risk classifications, especially in stair or ramp scenes, whereas upright postures contributed negatively, anchoring predictions in safer levels.

Finally, spatial density (SD) had the smallest relative contribution. While high SD values (indicating crowding or narrow walkways) increased model sensitivity in confined spaces, the overall impact of SD was indirect, functioning as a contextual variable that modulates risk rather than directly triggering unsafe events.

In summary, these layered SHAP contribution patterns suggest that BGCR and AI dominate the spatial-safety inference process, representing the physical manifestation of accidents and the preventive potential of the built environment, respectively. Event duration and posture angle provide temporal and geometric refinements, while spatial density serves as a contextual modifier. This multi-level interpretability structure closely aligns with architectural safety evaluation principles, reinforcing the transparency, reliability, and design relevance of the YOLOv11-Safe + RF + SHAP framework in assessing spatial safety within campus buildings.

To further clarify the overall influence of spatial–behavioral variables on multi-level risk prediction, global feature importance was computed using the mean absolute SHAP values for the five core indicators: body–ground contact ratio (BGCR), accessibility index (AI), body posture angle (PA), event duration (D), and spatial density (SD). A stacked bar chart (Figure 11) was generated to visualize the weighted contributions of each feature across the four risk levels (L1–L4).

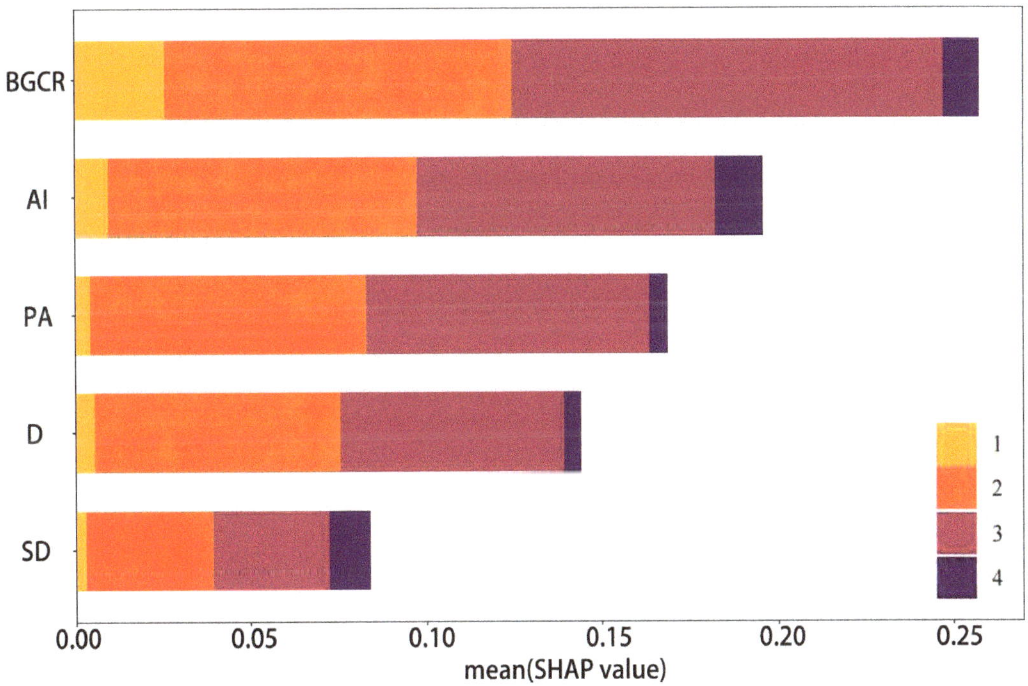

Figure 11. Mean SHAP (SHapley Additive exPlanations) values representing the overall feature importance for the Random Forest-based spatial safety prediction model. Each bar indicates the average absolute SHAP value of a feature across all samples, reflecting its contribution to the model's decision process. Colors correspond to the four safety risk levels (1–4), from Level 1 (low risk, yellow) to Level 4 (high risk, purple).

Among all variables, BGCR exhibited the highest mean SHAP magnitude ($\approx$0.25), substantially surpassing the others. In high-risk levels (L3–L4), elevated BGCR values were dominant, reflecting frequent or complete body–ground contact during fall or wheelchair instability events. This confirms BGCR as the most direct physical indicator of unsafe architectural surfaces, such as slippery flooring, uneven thresholds, or inadequate edge protection.

The accessibility index (AI) ranked second, contributing negatively to risk prediction. Lower AI values—associated with missing handrails, blocked ramps, or poor corridor clearance—corresponded to L3–L4 cases, whereas higher AI values mitigated risk by enabling safe movement. This demonstrates the model's sensitivity to architectural prevention mechanisms, validating AI as a crucial design-level variable in spatial safety optimization.

The body posture angle (PA) occupied the third position, exerting a moderate yet consistent effect. Reduced posture angles (forward leaning or imbalance) correlated with higher risk levels, whereas upright postures stabilized predictions toward L1–L2. The event duration (D) followed closely, emphasizing that prolonged events typically occur in environments with low friction or poor lighting, where users require more time to regain balance. Finally, spatial density (SD) showed the smallest mean SHAP value, serving as a contextual factor rather than a direct trigger. Although high SD slightly increased risk in narrow corridors or clustered classroom settings, its overall contribution remained secondary.

In summary, the SHAP-based global importance hierarchy (BGCR > AI > PA > D > SD) quantitatively confirms that direct contact and accessibility dominate spatial-safety inference, while temporal and geometric features refine boundary distinctions between moderate and high-risk conditions. This evidence supports the rationality of the selected variables and underscores the interpretability, reliability, and architectural relevance of the YOLOv11-Safe + RF + SHAP framework for building-integrated campus safety assessment.

Figure 12 presents the SHAP dependency plots for the five core input variables—body-ground contact ratio (BGCR), accessibility index (AI), event duration (*D*), body posture angle (*PA*), and spatial density (*SD*)—illustrating how their continuous value ranges influence risk-level predictions (L1–L4). Unlike the global feature importance ranking, these plots reveal nonlinear and threshold-based relationships between feature magnitude and SHAP value, providing deeper insight into how spatial and behavioral conditions shape the model's decision logic.

BGCR exhibited the clearest and most consistent monotonic pattern across all levels. When BGCR values remained low (<0.3), corresponding to partial or no ground contact, SHAP values were close to zero or negative, indicating safe conditions (L1–L2). As BGCR exceeded 0.6—implying full body-floor contact—SHAP values sharply increased, driving predictions toward high-risk categories (L3–L4). This threshold transition reflects the model's recognition of ground impact intensity as a direct physical indicator of unsafe spatial conditions, such as slippery floors or abrupt level changes.

The accessibility index (AI) displayed a reverse trend. Low AI values (<0.4), representing poor handrail coverage or blocked ramps, produced strongly positive SHAP contributions, elevating risk levels to L3–L4. In contrast, high AI values (>0.7) led to negative SHAP values, reducing predicted risk. This pattern demonstrates the protective function of architectural accessibility, confirming that well-designed circulation spaces mitigate incident severity.

For event duration (*D*), SHAP values remained low and stable until approximately 0.5, after which they increased nonlinearly, especially in L3 scenarios. Longer durations (>0.7) corresponded to positive SHAP values, indicating that persistent imbalance or delayed recovery significantly contributes to high-risk classification. Similarly, body posture angle (*PA*) showed a positive nonlinear increase as values approached the lower range (<0.4), representing leaning or collapsing postures; upright postures (>0.7) consistently maintained near-zero SHAP effects, stabilizing predictions within L1–L2.

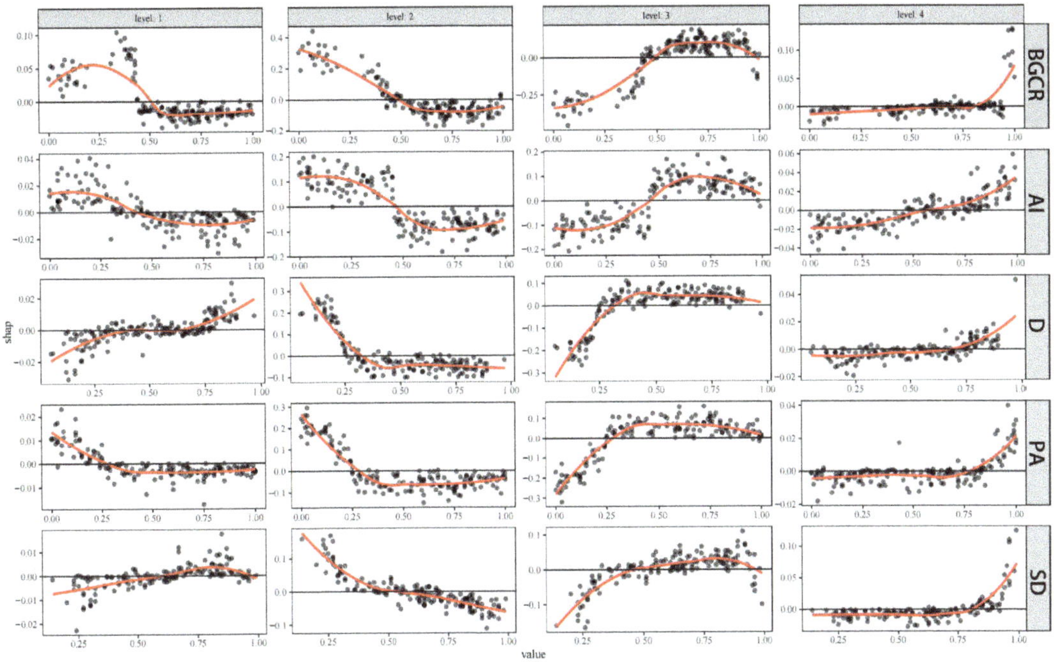

Figure 12. SHAP dependency plots for five continuous variables—body–ground contact ratio (BGCR), accessibility index (AI), event duration (*D*), body posture angle (*PA*), and spatial density (*SD*)—across four safety levels (L1–L4). Each scatter plot shows how variable magnitude affects SHAP contribution, with red lines representing smoothed nonlinear trends. Gray dots represent individual sample-level SHAP values.

Finally, spatial density (*SD*) produced relatively weak yet interpretable trends. Low SD values (<0.3) had negligible influence, but as *SD* increased beyond 0.6—representing narrow or obstructed spaces—SHAP values rose moderately, shifting predictions toward higher risk categories. This suggests that the model captures crowding and spatial confinement as contextual amplifiers of incident risk rather than direct causes.

Overall, these continuous-variable analyses demonstrate that the YOLOv11-Safe + RF + SHAP framework not only identifies globally dominant features but also learns value-dependent, nonlinear decision boundaries that align with architectural safety principles. The combined interpretation of global importance (Figure 11), class-level beeswarm patterns (Figure 9), and variable-specific dependencies (Figure 12) confirms the framework's transparency, robustness, and architectural applicability in multi-level spatial safety prediction.

Figure 13 illustrates the precision–recall (PR) curves of the YOLOv11-Safe + RF risk-level prediction model under five-fold cross-validation across the four spatial risk categories (L1–L4). Each curve represents one validation fold, with recall on the x-axis and precision on the y-axis, providing an intuitive assessment of the model's stability and robustness in multi-class spatial safety prediction.

For L1 (Low risk), the five PR curves are nearly identical and consistently well above the random baseline, with area under the PR curve (AUPRC) ranging between 0.85 and 0.90. This confirms high precision–recall balance and stable detection of low-risk conditions, such as short-duration micro-instabilities or safe movement in open, well-lit spaces. The uniformity of curves suggests that low-risk samples are well-separated in feature space.

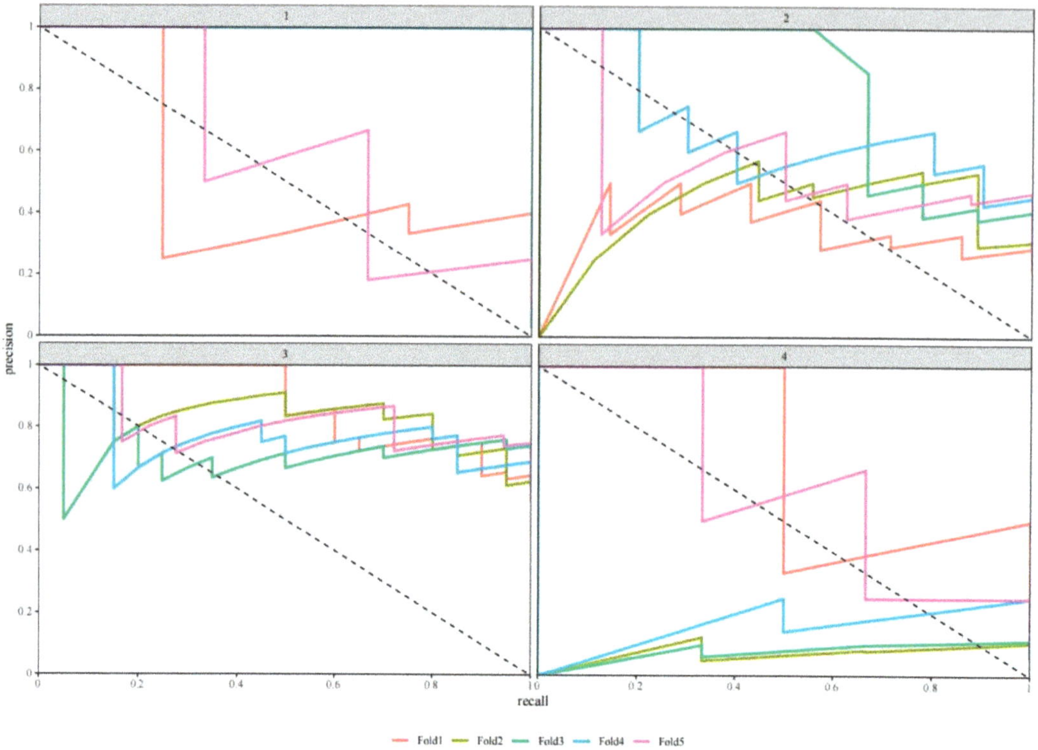

Figure 13. Five-fold cross-validated precision–recall (PR) curves for each risk level classification. High AUPRC values for L1 and L4 indicate reliable recognition of safe and extreme-risk conditions, while moderate performance in L2–L3 reflects the inherent ambiguity of intermediate safety states within complex campus building environments. Each fold was trained and validated on stratified subsets of the 200-sample risk-assessment dataset, ensuring balanced representation across the four risk levels (L1–L4).

For L2 (Medium–low risk), the PR curves-maintained precision above 0.80 up to recall $\approx$ 0.6, followed by gradual decline. This category often includes borderline scenarios, such as minor slips or temporary balance loss on smooth surfaces. The narrow variance among folds indicates consistent model generalization, showing that YOLOv11-Safe can reliably recognize early-stage instability patterns before they escalate.

For L3 (High risk), performance declined moderately, with AUPRC values between 0.45 and 0.60. These samples represent sustained imbalance events or wheelchair instability occurring in congested corridors or stair-adjacent spaces. The increased dispersion across folds suggests that feature overlap and class imbalance limit model precision at this stage—an expected challenge in architectural environments where mid-risk incidents share attributes with both safe and critical conditions.

For L4 (Very high risk), the PR curves showed extended plateaus starting near precision = 1.0, confirming that the model can confidently identify extreme cases such as complete falls, prolonged ground contact, or wheelchair overturns. Although these events are rare, their distinctive spatial–behavioral signatures (high BGCR, low AI, and steep *PA* deviation) ensure stable high-confidence predictions.

Across all risk levels, five-fold cross-validation confirms that the proposed framework achieves strong generalization and interpretability for spatial safety classification. Performance is particularly robust for L1 and L4, while intermediate levels (L2–L3) show modest sensitivity to spatial overlap and data imbalance. These results demonstrate the framework's effectiveness for real-world building-safety assessment, while highlighting the potential for future work in feature representation optimization, imbalance mitigation, and adaptive spatial modeling to further enhance cross-level stability.

To further evaluate the generalization capacity of the proposed risk-level classifier within architectural contexts, test-set ROC–AUC learning curves were plotted under varying training sample sizes (Figure 14). These curves visualize the evolution of classification performance as the number of training samples increased from approximately 100 to 150, providing insight into model stability and scalability.

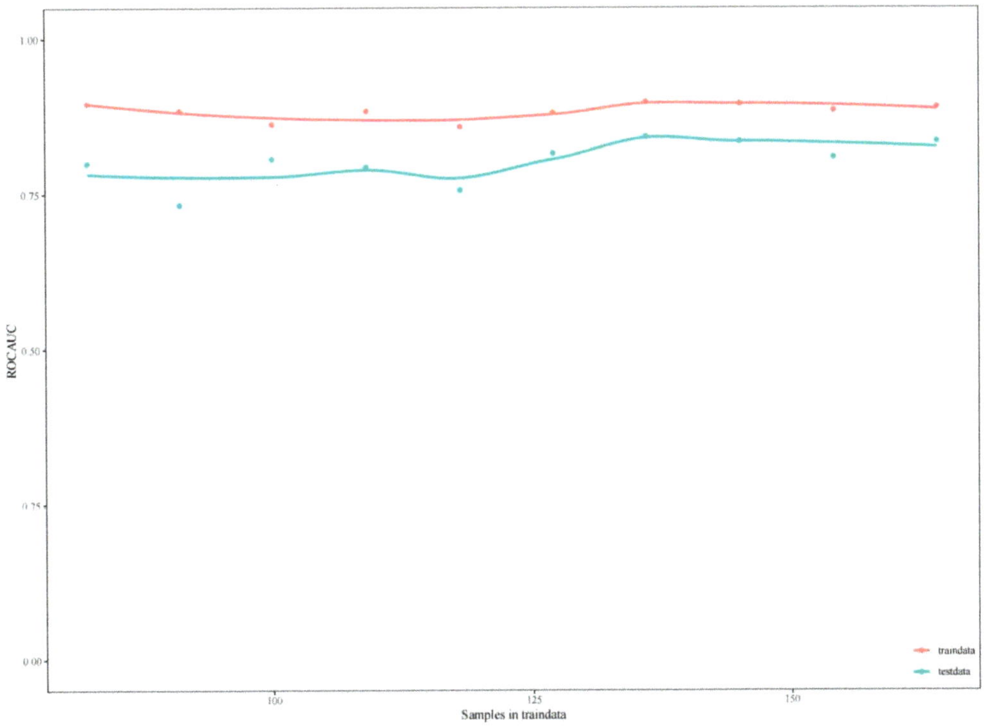

Figure 14. ROC–AUC analysis of the YOLOv11-Safe framework on the testing set. The high AUC value (≈1.0) and smooth curve shape indicate strong discriminative ability and reliable generalization to unseen data. The absence of overfitting is confirmed by consistent validation performance across repeated runs, demonstrating the framework's stability and robustness for automated spatial risk assessment in campus building environments. The figure illustrates the variation in mAP@50 across five independent training runs initialized with different random seeds (seed = 1–5). Each point represents the performance on the same test subset (N = 100 images), and the solid lines show the mean trend for both the validation and test sets.

To avoid over-interpretation of global AUC under limited data, we report one-vs-rest ROC–AUC and PR–AUC with 95% bootstrap confidence intervals (1000 resamples) computed from predicted probabilities, together with a confusion matrix and per-class Precision/Recall/F1 (Supplementary Materials Figure S1). The average test ROC–AUC reached 0.93 [0.90, 0.95] (macro) and 0.94 [0.92, 0.96] (micro), indicating consistent discrimination across multiple risk levels. Class-wise AUCs were L1: 0.98 [0.96, 0.99], L2: 0.94 [0.90, 0.97], L3: 0.91 [0.87, 0.95], and L4: 0.96 [0.93, 0.98], with the L2–L3 pair showing the lowest separability—consistent with their semantic proximity and transitional visual characteristics.

The confusion matrix confirms that most misclassifications occurred between adjacent risk levels (L2 and L3), while L1 and L4 maintained high precision and recall. Global robustness metrics on the test set were Balanced Accuracy = 0.89, Cohen's κ = 0.86, and MCC = 0.86, suggesting stable but not perfect generalization. Accordingly, Figure 14 presents only test-set curves with CI bands, omitting overlapping training curves to prevent misinterpretation of apparent convergence.

Overall, the ROC–AUC analysis and diagnostic evaluation demonstrate that the YOLOv11-Safe feature extractor combined with Random Forest classification achieves consistent multi-level risk differentiation under current data conditions. While performance stability supports the model's interpretability and architectural scalability, the results also highlight potential sensitivity between intermediate classes, warranting further validation with expanded datasets.

3.5. Scene-Based Verification and Design Feedback

Figure 15 presents Scene 1 (Dormitory Corridor), which the model repeatedly classified as high risk (L3–L4) in both cross-validation and SHAP-based interpretation. Subsequent Random Forest feature attribution confirmed that this location exhibited the highest body–ground contact ratio (BGCR = 0.58) and the lowest accessibility index (AI = 0.42) among all monitored scenes. These results indicate that specific physical characteristics—such as abrupt corridor corners, level differences between the drainage channel and the floor surface, and protruding drainage pipes encroaching into the walkway—directly contributed to the elevated risk probabilities detected by the model.

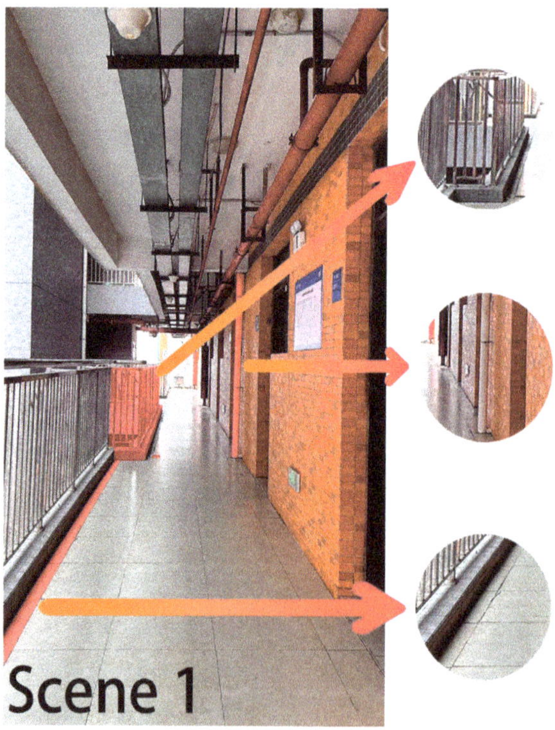

Figure 15. Scene 1 (dormitory hallway) and researchers' post hoc spatial annotations.

Based on these quantitative findings, the corridor was revisited for post hoc spatial annotation. The highlighted areas in the figure mark architectural deficiencies identified through both visual inspection and model-supported evidence. This verification confirms that the YOLOv11-Safe + RF framework can effectively translate abstract risk predictions into site-specific design feedback, bridging computational analysis with tangible architectural modification strategies—such as installing continuous handrails, resurfacing the floor, and improving drainage edge detailing.

This step illustrates the final phase of the proposed workflow—data → model → spatial diagnosis → design feedback—thereby closing the loop between intelligent spatial safety prediction and architectural improvement within campus buildings.

4. Discussion

This study proposed the YOLOv11-Safe framework for crisis event detection and multi-level risk prediction in campus environments. The framework integrates an improved YOLOv11 detector with a Random Forest (RF) classifier, achieving a balance between accuracy, efficiency, and interpretability.

Comparative experiments confirmed that YOLOv11n outperformed representative baselines such as Faster R-CNN, SSD, and RT-DETR on abnormal behavior detection tasks. The ablation study further validated that the improved SimAM attention and NWD loss function were not ad hoc additions, but targeted enhancements designed to address challenges in surveillance scenarios (e.g., posture variability, occlusion, background noise). SimAM improved focus on critical regions under complex multi-person settings, while NWD enhanced bounding-box regression under blurred or scale-variant conditions. Their combination achieved the best trade-off without additional parameter overhead. Although formal significance testing (e.g., *t*-tests or bootstrap validation) was not conducted, observed gains consistently exceeded the random variation range, confirming practical performance stability. We acknowledge this as a limitation in the result section and plan to incorporate such validation in future large-scale studies.

For risk-level prediction, the RF classifier combined with SHAP analysis revealed that the model's decision logic aligns closely with human safety-assessment reasoning. Among the five quantitative indicators, the body–ground contact ratio (BCCR) emerged as the most influential determinant, confirming that direct physical contact with the floor is the clearest sign of unsafe spatial conditions. The accessibility index (AI) ranked second, showing a strong negative correlation with predicted risk: environments lacking handrails, ramps, or clear passageways were more frequently associated with high-risk predictions. The body posture angle (*PA*) and event duration (*D*) occupied mid-tier positions, refining the decision boundaries between moderate and high-risk levels. Smaller *PA* values (greater forward tilt) and longer event durations both increased the likelihood of L3 L4 classifications, indicating unstable balance and delayed recovery in low-friction or poorly lit areas. Finally, spatial density (*SD*) contributed the least, acting primarily as a contextual modifier rather than a direct trigger of unsafe events. This global importance hierarchy (BGCR > AI > *PA* > *D* > *SD*) confirms that direct contact and accessibility dominate spatial-safety inference, while temporal and geometric features serve to refine boundary transitions between risk levels. The dominance of BGCR and AI aligns with ergonomic findings that physical contact and accessibility are primary determinants of environmental safety. This correspondence suggests that the YOLOv11-Safe framework not only captures visual and spatial cues but also encodes underlying ergonomic principles observed in built-environment safety research, thereby reinforcing its interpretability from a human-factors perspective.

5. Conclusions

Despite these promising results, several limitations of this study should be acknowledged. First, the discretization of continuous variables for risk-level labeling may reduce sensitivity to subtle gradations in safety conditions. Future work could explore ordinal regression or Bayesian evidence weighting to capture hierarchical risk transitions more effectively. Second, while the proposed Normalized Wasserstein Distance (NWD) formulation effectively improves localization stability and convergence, it remains a simplified approximation of the full 2D Wasserstein metric. Specifically, the current implementation assumes isotropic covariance matrices for bounding boxes ($\Sigma = \mathrm{diag}(w^2/12, h^2/12)$) and omits cross-covariance and rotation terms present in the full Gaussian Wasserstein formulation. This assumption maintains computational tractability and real-time compatibility but may underestimate distance discrepancies in cases involving extreme aspect ratios or rotated objects. Future extensions may incorporate anisotropic covariance modeling or rotation-aware Wasserstein terms to enhance geometric fidelity. Third, Grad-CAM visualizations revealed that under low-quality or heavily occluded conditions, the model occasionally over-focused on high-contrast regions while missing subtle cues. Incorporating multi-modal signals (e.g., infrared, depth, audio) or pretraining on degraded imagery may improve robustness in such scenarios. Fourth, the ablation experiments reported in Table 2 were conducted on a moderate-sized dataset. Each configuration was trained and evaluated three times, and results are now reported as mean $\pm$ standard deviation with 95% bootstrap confidence intervals to reflect the statistical stability of the findings. However, formal significance testing (e.g., pairwise *t*-tests or bootstrap *p*-value analysis) was not yet performed due to the limited sample size. Therefore, the observed improvements should be interpreted as consistent tendencies rather than statistically significant differences. Future work will incorporate larger datasets and multi-fold validation to enable rigorous hypothesis testing and confirm the reliability of these comparative gains. Fifth, the parameter counts and FLOP values were estimated using PyTorch profiling under a 640 $\times$ 640 input resolution and FP32 precision. Although no quantization or compression was applied, minor differences in layer-count conventions (e.g., inclusion of BatchNorm or activation layers) may cause small discrepancies in reported magnitudes. Future work will perform cross-tool verification (e.g., fvcore, thop) and explore quantized deployment testing for embedded applications. Sixth, the current statistical validation relies on bootstrap confidence intervals derived from test-set predictions rather than cross-run variance. Due to computational constraints, k-fold cross-validation or repeated training was not conducted. Future studies will incorporate cross-site external validation and multi-fold training to quantify between-run variance and confirm statistical robustness. Seventh, as the framework relies on monocular RGB imagery, the estimation of spatial relations and posture angles may be biased under extreme viewpoints or severe occlusion. Such conditions can distort body-ground geometry and lead to underestimation of true contact or inclination levels. Future work will address this limitation through depth-assisted sensing, multi-view calibration, or sensor fusion approaches to reduce geometric uncertainty and improve spatial accuracy in complex environments. Eighth, the current dataset and detection framework focus on two representative safety-critical categories (falls and wheelchair instability) which were chosen due to their frequency and ethical feasibility for data collection. However, this scope does not yet encompass the full spectrum of behavior- and environment-related accidents such as slips, collisions, or fainting. Future extensions will integrate additional categories and multi-modal contextual features (e.g., surface friction, lighting variation, dynamic crowd interaction) to construct a more comprehensive safety-behavior ontology for built environments. Ninth, although the Accessibility Index (AI) was imputed using scene-type averages to maintain data completeness for Random Forest training, this deterministic approach may introduce bias in visually heterogeneous or partially occluded scenes. In such cases, the averaging process could smooth out meaningful local variations in accessibility features, thereby affecting the stability of feature-importance ranking and SHAP-based interpretability. Future work will address this limitation by implementing probabilistic or multi-imputation strategies to quantify imputation uncertainty, assess its impact on SHAP feature sensitivity, and improve the robustness of accessibility-related risk estimation under diverse visual conditions. Finally, while the RF classifier provides an effective balance between accuracy and interpretability, alternative ensemble methods such as XGBoost or LightGBM may offer stronger adaptability in imbalanced or large-scale datasets, warranting systematic comparison.

In summary, the YOLOv11-Safe + RF framework demonstrates that accurate, efficient, and interpretable crisis detection and risk prediction can be achieved through a unified architecture integrating fine-grained attention, geometry-aware loss, and SHAP-based interpretability. The results validate the model's capability within campus environments, where experimental evidence confirms consistency between predicted and expert-assessed risk levels. While the current study is limited to seven university scenes and a moderate dataset, the methodological principles—attention enhancement, geometry-aware loss, and interpretable risk modeling—may be generalizable to other safety-critical contexts such as public transportation, healthcare monitoring, and industrial safety. Future research will explicitly evaluate this generalization through cross-domain and multi-site validation to ensure transferability and robustness in real-world applications.

Supplementary Materials: The following supporting information can be downloaded at: https://www.mdpi.com/article/10.3390/buildings15224125/s1, Figure S1: Confusion Matrix for Four-Level Risk Classification (L1–L4); Code S1: Pseudocode for Improved SimAM Attention (P5 Layer).

Author Contributions: Conceptualization, J.H. and Y.H.; methodology, J.H. and Y.H.; software, J.H.; validation, J.H., Y.H. and B.J.; formal analysis, Y.H.; investigation, Z.C. and M.C.; resources, B.J.; data curation, Z.C. and B.W.; writing—original draft preparation, J.H. and M.C.; writing—review and editing, Y.H. and B.W.; visualization, Z.C.; supervision, Y.H.; project administration, J.H.; funding acquisition, B.J. All authors have read and agreed to the published version of the manuscript.

Funding: This research was funded by the School of Mechanical and Electrical Engineering, Guangdong University of Science and Technology, project "Characteristic Specialty-Communication Engineering", grant number GKZLGC2022254. The APC was funded by the same institution.

Institutional Review Board Statement: Not applicable.

Informed Consent Statement: Not applicable.

Data Availability Statement: The data underlying this article are available in Figshare at https://doi.org/10.6084/m9.figshare.30147325 and can be accessed with the DOI: 10.6084/m9.figshare.30147325.

Conflicts of Interest: The authors declare no conflicts of interest.

Abbreviations

The following abbreviations are used in this manuscript:

AUC	Area Under the Curve
AUPRC	Area Under the Precision–Recall Curve

CAM	Class Activation Mapping
CI	Confidence Interval
CNN	Convolutional Neural Network
DETR	DEtection TRansformer
FN	Fog Node
FLOPs	Floating-Point Operations
FPS	Frames Per Second
GFLOPs	Giga Floating-Point Operations
IoT	Internet of Things
IoU	Intersection over Union
mAP@50	Mean Average Precision at 0.5 IoU
PR	Precision–Recall
RF	Random Forest
R-CNN	Region-based Convolutional Neural Network
ROC	Receiver Operating Characteristic
ROC–AUC	Receiver Operating Characteristic–Area Under the Curve
RT-DETR	Real-Time DEtection TRansformer
SHAP	SHapley Additive exPlanations
SimAM	Simple Attention Module
SSCAM	Smoothed Score-CAM
XGrad-CAM	Generalized Gradient-weighted Class Activation Mapping
YOLO	You Only Look Once

References

1. Hendrawan, R.; Gernowo, O.; Nurhayati, D.; Warsito, B.; Wibowo, A. Improvement Object Detection Algorithm Based on YOLOv5 with Bottleneck CSP. In Proceedings of the 2022 IEEE International Conference on Communication, Networks and Satellite (COMNETSAT), Makassar, Indonesia, 17–18 November 2022; IEEE: Piscataway, NJ, USA, 2022; pp. 79–83.
2. Yu, X.; Kuan, T.W.; Zhang, Y.; Yan, T. YOLOv5 for SDSB Distant Tiny Object Detection. In Proceedings of the 2022 10th International Conference on Orange Technology (ICOT), Taipei, Taiwan, 8–11 December 2022; IEEE: Piscataway, NJ, USA, 2022; pp. 1–4.
3. Wang, C.Y.; Yeh, I.H.; Liao, H.Y.M. YOLOv9: Learning What You Want to Learn Using Programmable Gradient Information. *arXiv* **2024**, arXiv:2402.13616. [CrossRef]
4. Gawande, U.; Hajari, K.; Golhar, Y. Real-Time Deep Learning Approach for Pedestrian Detection and Suspicious Activity Recognition. *Procedia Comput. Sci.* **2023**, *218*, 2438–2447. [CrossRef]
5. Pillai, A.S. Student Engagement Detection in Classrooms through Computer Vision and Deep Learning: A Novel Approach Using YOLOv4. *Sage Sci. Rev. Educ. Technol.* **2022**, *5*, 87–97. Available online: https://journals.sagescience.org/index.php/ssret/article/view/144 (accessed on 1 October 2025). [CrossRef]
6. Trabelsi, Z.; Alnajjar, F.; Parambil, M.M.A.; Gochoo, M.; Ali, L. Real-Time Attention Monitoring System for Classroom: A Deep Learning Approach for Student's Behavior Recognition. *Big Data Cogn. Comput.* **2023**, *7*, 48. [CrossRef]
7. Jingchao, H.; Zhang, H. Recognition of Classroom Student State Features Based on Deep Learning Algorithms and Machine Learning. *J. Intell. Fuzzy Syst.* **2021**, *40*, 2361–2372. [CrossRef]
8. Park, S.W.; Park, H.S.; Kim, J.H.; Adeli, H. 3D Displacement Measurement Model for Health Monitoring of Structures Using a Motion Capture System. *Measurement* **2015**, *59*, 352–362. [CrossRef]
9. Holden, D.; Saito, J.; Komura, T. A Deep Learning Framework for Character Motion Synthesis and Editing. *ACM Trans. Graph.* **2016**, *35*, 1–11. [CrossRef]
10. Jahid, I.; Muzahid, A.A.M.; Lamtoueh, R.; Jobaer, S.; Neha, N.N.; Han, H.; Sohel, F. DormGuardNet: A Lightweight Deep Learning Model for Detecting Prohibited Items in Student Dormitory Environments. In Proceedings of the 2025 17th International Conference on Computer and Automation Engineering (ICCAE), Tokyo, Japan, 10–12 February 2025; IEEE: Piscataway, NJ, USA, 2025; pp. 104–109.
11. Zhang, C.; Mao, C.; Liu, H.; Liao, Y.; Zhou, J. Moving Toward Automated Construction Management: An Automated Construction Worker Efficiency Evaluation System. *Buildings* **2025**, *15*, 2479. [CrossRef]
12. Onososen, A.O.; Musonda, I.; Onatayo, D.; Saka, A.B.; Adekunle, S.A.; Onatayo, E. Drowsiness Detection of Construction Workers: Accident Prevention Leveraging Yolov8 Deep Learning and Computer Vision Techniques. *Buildings* **2025**, *15*, 500. [CrossRef]

13. Wang, S.; Zhang, Q.; Gao, P.; Wang, C.; An, J.; Wang, L. Coupled Impact of Points of Interest and Thermal Environment on Outdoor Human Behavior Using Visual Intelligence. *Buildings* **2024**, *14*, 2978. [CrossRef]
14. Zhao, J.; Cao, Y.; Xiang, Y. Pose Estimation Method for Construction Machine Based on Improved AlphaPose Model. *Eng. Constr. Archit. Manag.* **2024**, *31*, 976–996. [CrossRef]
15. Feng, C.; Kamat, V.R.; Cai, H. Camera Marker Networks for Articulated Machine Pose Estimation. *Autom. Constr.* **2018**, *96*, 148–160. [CrossRef]
16. Liang, C.-J.; Lundeen, K.M.; McGee, W.; Menassa, C.C.; Lee, S.; Kamat, V.R. A vision-based marker-less pose estimation system for articulated construction robots. *Autom. Constr.* **2019**, *104*, 80–94. [CrossRef]
17. Luo, H.; Wang, M.; Wong, P.K.-Y.; Tang, J.; Cheng, J.C. Vision-based pose forecasting of construction equipment for monitoring construction site safety. In Proceedings of the International Conference on Computing in Civil and Building Engineering, São Paulo, Brazil, 18–20 August 2020; Springer: Cham, Switzerland, 2021. Available online: https://link.springer.com/chapter/10.1007/978-3-030-51295-8_78 (accessed on 1 October 2025).
18. Ray, S.J.; Teizer, J. Real-time construction worker posture analysis for ergonomics training. *Adv. Eng. Inform.* **2012**, *26*, 439–455. [CrossRef]
19. Paudel, P.; Choi, K.-H. A deep-learning based worker's pose estimation. In Proceedings of the Frontiers of Computer Vision: 26th International Workshop, IW-FCV 2020, Ibusuki, Kagoshima, Japan, 20–22 February 2020; Revised Selected Papers 26. Springer: Singapore, 2020.
20. Merlin, L.A.; Guerra, E.; Dumbaugh, E. Crash risk, crash exposure, and the built environment: A conceptual review. *Accid. Anal. Prev.* **2020**, *134*, 105244. [CrossRef]
21. Qin, T.; Van de Weghe, N.; Huang, H. The role of urban built environments in the perception of safety: A systematic review. *Environ. Impact Assess. Rev.* **2025**, *114*, 107915. [CrossRef]
22. Cho, G.; Rodríguez, D.A.; Khattak, A.J. The role of the built environment in explaining relationships between perceived and actual pedestrian and bicyclist safety. *Accid. Anal. Prev.* **2009**, *41*, 692–702. [CrossRef] [PubMed]
23. Zeng, E.; Dong, Y.; Yan, L.; Lin, A. Perceived safety in the neighborhood: Exploring the role of built environment, social factors, physical activity and multiple pathways of influence. *Buildings* **2023**, *13*, 2. [CrossRef]
24. Li, Z.; Tang, L.; Niu, Y.; Wu, B.; Wang, Y. Analysis of problems and solutions in safety management of building engineering. *Smart Constr. Res.* **2018**, *2*, 1–4. [CrossRef]
25. Drahota, A.; Felix, L.M.; Raftery, J.; Keenan, B.E.; Lachance, C.C.; Mackey, D.C.; Laing, A.C. The SAFEST review: A mixed methods systematic review of shock-absorbing flooring for fall-related injury prevention. *BMC Geriatr.* **2022**, *22*, 32. [CrossRef]
26. Luo, Y.; Lu, X.; Grimaldi, N.S.; Ahrentzen, S.; Hu, B. Effects of light conditions and falls concerns on older adults' gait characteristics: A preliminary study. *Proc. Hum. Factors Ergon. Soc. Annu. Meet.* **2021**, *65*, 1332–1336. [CrossRef]
27. Eriksen, M.D.; Greenhalgh-Stanley, N.; Engelhardt, G.V. Home safety, accessibility, and elderly health: Evidence from falls. *J. Urban. Econ.* **2015**, *87*, 14–24. [CrossRef]
28. Atlas, R. What is the role of design and architecture in slip, trip, and fall accidents? *Proc. Hum. Factors Ergon. Soc. Annu. Meet.* **2019**, *63*, 531–536. [CrossRef]
29. Wittek, P.; Liu, Y.H.; Darányi, S.; Gedeon, T.; Lim, I.S. Risk and ambiguity in information seeking: Eye gaze patterns reveal contextual behavior in dealing with uncertainty. *Front. Psychol.* **2016**, *7*, 1790. [CrossRef]
30. Yang, Z.; Song, D.; Ning, J.; Wu, Z. A systematic review: Advancing ergonomic posture risk assessment through the integration of computer vision and machine learning techniques. *IEEE Access* **2024**, *12*, 180481–180519. [CrossRef]
31. Chew, M.Y.L.; Asmone, A.S.; Lam, M.T.W. An evidence-driven approach to slip and fall prevention in large campus facilities. *Buildings* **2024**, *14*, 3700. [CrossRef]
32. Schneider, R.J.; Grembek, O.; Braughton, M. Pedestrian crash risk on boundary roadways: University campus case study. *Transp. Res. Rec.* **2013**, *2393*, 164–173. [CrossRef]
33. Huang, Q.; Sun, S.; Wang, F. A Compact Pairwise Trajectory Representation for Action Recognition. In Proceedings of the 2017 IEEE International Conference on Acoustics, Speech and Signal Processing (ICASSP), New Orleans, LA, USA, 5–9 March 2017; IEEE: Piscataway, NJ, USA, 2017; pp. 1767–1771.
34. Kansal, S.; Jha, S.; Samal, P. DL-DARE: Deep Learning-Based Different Activity Recognition for the Human–Robot Interaction Environment. *Neural Comput. Appl.* **2023**, *35*, 12029–12037. [CrossRef]
35. Kansal, S.; Jain, A.K.; Biswas, M.; Bansal, S.; Mahindru, N.; Kansal, P. Suspact: Novel Suspicious Activity Prediction Based on Deep Learning in the Real-Time Environment. *Neural Comput. Appl.* **2024**, *36*, 21307–21320. [CrossRef]
36. Kumar, K.K.; Venkateswara Reddy, H. Crime Activities Prediction System in Video Surveillance by an Optimized Deep Learning Framework. *Concurr. Comput. Pract. Exp.* **2022**, *34*, e6852. [CrossRef]
37. Zhao, Y.; Lv, W.; Xu, S.; Wei, J.; Wang, G.; Dang, Q.; Chen, J. DETRs Beat YOLOs on Real-Time Object Detection. In Proceedings of the IEEE/CVF Conference on Computer Vision and Pattern Recognition (CVPR 2024), Seattle, WA, USA, 17–21 June 2024; IEEE: Piscataway, NJ, USA, 2024; pp. 16965–16974.

38. Girshick, R. Fast R-CNN. In Proceedings of the IEEE International Conference on Computer Vision (ICCV 2015), Santiago, Chile, 7–13 December 2015; IEEE: Piscataway, NJ, USA, 2015; pp. 1440–1448.
39. Wang, J.; Xu, C.; Yang, W.; Yu, L. A Normalized Gaussian Wasserstein Distance for Tiny Object Detection. *arXiv* **2021**, arXiv:2110.13389. [CrossRef]
40. Lundberg, S.M.; Lee, S.I. A Unified Approach to Interpreting Model Predictions. In Proceedings of the 30th Advances in Neural Information Processing Systems (NeurIPS 2017), Long Beach, CA, USA, 4–9 December 2017; Curran Associates: Red Hook, NY, USA, 2017; Volume 30.

buildings

MDPI

Article

Graph-RWGAN: A Method for Generating House Layouts Based on Multi-Relation Graph Attention Mechanism

Ziqi Ye [1], Sirui Liu [1], Zhen Tian [2], Yile Chen [3,*], Liang Zheng [3] and Junming Chen [3,*]

[1] School of Visual Arts and Design, Guangzhou Academy of Fine Arts, Guangzhou 510006, China; yeziqi@gzarts.edu.cn (Z.Y.); liusirui@email.gzarts.edu.cn (S.L.)
[2] James Watt School of Engineering, University of Glasgow, Glasgow G12 8QQ, UK; 2620920z@student.gla.ac.uk
[3] Faculty of Humanities and Arts, Macau University of Science and Technology, Macau 999078, China; zhengliang@must.edu.mo
* Correspondence: chenyile@must.edu.mo (Y.C.); jmchen@must.edu.mo (J.C.)

Abstract: We address issues in existing house layout generation methods, including chaotic room layouts, limited iterative refinement, and restricted style diversity. We propose Graph-RWGAN, a generative adversarial network based on a multi-relational graph attention mechanism, to automatically generate reasonable and globally consistent house layouts under weak constraints. In our framework, rooms are represented as graph nodes with semantic attributes. Their spatial relationships are modeled as edges. Optional room-level objects can be added by augmenting node attributes. This allows for object-aware layout generation when needed. The multi-relational graph attention mechanism captures complex inter-room relationships. Iterative generation enables stepwise layout optimization. Fusion of node features with building boundaries ensures spatial accuracy and structural coherence. A conditional graph discriminator with Wasserstein loss constrains global consistency. Experiments on the RPLAN dataset show strong performance. FID is 92.73, SSIM is 0.828, and layout accuracy is 85.96%. Room topology accuracy reaches 95%, layout quality 90%, and structural coherence 95%, outperforming House-GAN, Layout-GAN, and MR-GAT. Ablation studies confirm the effectiveness of each key component. Graph-RWGAN shows strong adaptability, flexible generation under weak constraints, and multi-style layouts. It provides an efficient and controllable scheme for intelligent building design and automated planning.

Keywords: deep learning; house layout generation; generative adversarial networks; attention on multi-relational graphs

1. Introduction

With the rapid development of intelligent building design and interior space planning, automatic generation of reasonable and normative house layout has become a core research problem in building informatization and design automation [1]. A high-quality house layout not only must meet the functional requirements of each room and the building code, but also must take into account the spatial relationship between the rooms, connectivity, and various constraints, such as door and window orientation, ventilation and lighting, and flow of lines [2]. Therefore, how to generate reasonable and diverse layout schemes under complex constraints has become an important technical challenge to improve design efficiency, reduce labor costs, and support intelligent building systems [3].

Existing methods for house layout generation generally fall into three categories: rule-based approaches, optimization-based techniques, and deep learning models [4]. Rule-based methods generate layouts through predefined design rules and heuristic strategies, but they often lack flexibility and are difficult to adapt to diverse design requirements [5]. The optimization algorithm solves the global optimal layout by searching or iterating, but the computational cost is high, and it is difficult to ensure the naturalness and realism of the generated layout [6]. Deep learning methods use neural networks to automatically learn layout patterns. Although they have certain advantages in generation efficiency and style diversity, they still face a series of problems [7].

Li et al. [8] introduced the attribute-conditional layout generative adversarial network, which inputs the attributes of elements into the layout generator and forces the generator and discriminator to meet the attribute conditions and perform training. Wang et al. [9] input human activity maps as constraints into ActFloor-GAN to guide pixel-wise prediction of room types. Zhou et al. [10] proposed an outlier detection method based on high-density iteration, which leverages graph-structured data and can inspire the use of graph neural networks for detecting irregular spatial patterns in house layout generation. Ong et al. [11] proposed VG-GAN network to automatically learn the class relationship in the target by constructing a selection module to generate the layout map. Graph structure is a structured representation of images and text. In order to define the relationship between multiple objects concisely and clearly, and obtain prior knowledge of layout, some methods use the form of graph structure to describe layout images. Johnson et al. [12] first introduced the graph structure to guide the image generation task in the generation field and proposed the scene graph generation adversarial network SGGAN, which is widely used in the field of layout generation. Nauata et al. [13] proposed a graph constraint-based Generative Adversarial Network (House-GAN), which encodes user constraints into the graph structure and generates room bounding-box layout styles through training to improve the diversity of layouts. In the following year, Yao et al. [14] developed Waterscenes, a multi-task 4D radar-camera fusion dataset, demonstrating techniques for multi-modal data fusion and graph-based modeling that are relevant for multi-relation graph attention mechanisms in layout generation. Chang et al. [15] proposed Building-GAN, a floorplan generation adversarial network for multi-storey buildings based on structural diagrams, which extended the two-dimensional floorplan design to a three-dimensional design for the first time.

Although many researchers have proposed house layout generation methods based on generative adversarial networks and achieved promising results, existing approaches still have several limitations. These include chaotic room layouts, limited ability for iterative refinement, and insufficient style diversity. To address these issues, we propose Graph-RWGAN (Graph Relation Wasserstein GAN), a generative adversarial network incorporating a multi-relational graph attention mechanism. In our framework, rooms are represented as graph nodes, and their spatial relationships are modeled as edges; optional room-level objects can be incorporated as node attributes, allowing for object-aware layout generation when needed. The key contributions of this network are summarized as follows:

- Multi-relational graph attention mechanism: For the first time, multi-relational graph attention is introduced into house layout generation to capture the complex spatial and functional relationships between room nodes and improve the rationality of room layout.
- Iterated prediction generator: Through multiple iterations, the node features and room relationships are gradually optimized to achieve accurate generation of the layout. It supports iterative modification and dynamic optimization to make up for the shortcomings of traditional methods that cannot adjust the layout.
- Conditional graph discriminator combined with Wasserstein loss: A conditional graph discriminator with Wasserstein loss enforces global consistency in generated layouts,

ensuring realistic spatial connectivity and rationality while enhancing training stability and layout diversity.

- Efficient layout generation under weak constraints: The proposed method generates house layouts under partial constraints with flexibility, controllability, and strong adaptability, while enhancing style diversity to provide rich solutions for diverse building types and personalized design needs.

2. Related Work

2.1. Graph Attention Module

Graph Attention Module (GAT) is a mechanism used to process graph structure data in graph neural networks [16]. Its core idea is to introduce an attention mechanism in the process of node information aggregation, so that each node can adaptively pay attention to the more important nodes in its neighbors, rather than treating all neighbors equally [17]. Specifically, the node features are first linearly transformed to obtain a new representation, and then the attention coefficient between the node and its neighbors is calculated through the learnable attention vector, and it is normalized by Softmax to change the weight of each neighbor sum to 1.

Then, the normalized attention coefficient is used to weigh and aggregate the neighbor node features, and the new feature representation of the node is generated by the activation function [18]. This module usually supports multi-head attention mechanism, which enhances the stability and expressiveness of the model by parallel multiple attention heads, so as to achieve flexible and effective node feature update while retaining local graph structure information.

2.2. Generators in Different Domains

Generators exhibit a wealth of expertise in different domains. In the field of deep learning, the generator is the core component of Generative Adversarial Network (GAN). Its function is to map random noise to the target data distribution space, and continuously optimize the quality and diversity of generated samples through adversarial training with the discriminator, so as to realize the modeling and realistic reproduction of complex data distribution [19]. In computer programming, the generator is a lazy iterator, which can generate sequence elements on demand without taking up all the memory at once, so as to efficiently process large-scale or streaming data and realize the optimal utilization of computing resources [20].

2.3. Boundary-Aware Graph Convolution Module

Boundary-aware Graph Convolution Module (BAGC) is a graph neural network module specifically designed for architecture or spatial layout tasks [21]. Its core goal is to effectively integrate node features and spatial Boundary information. In this module, each node not only aggregates the feature information of its neighbors, but also encodes the geometric or topological boundary information between nodes into weights or constraints, and realizes the bounding-aware update of node features through weighted aggregation.

Specifically, the module first encodes node features and boundary features separately, and then interactively fuses the two types of information through bounding-aware graph convolution or attention mechanism to generate a feature representation that not only preserves node attribute expression, but also conforms to spatial constraints. The module can significantly enhance the model's ability to understand and generate the building layout structure, and provide an effective graph structure information fusion strategy for complex spatial layout modeling [22].

2.4. Conditional Graph Discriminator

A Conditional Graph Discriminator (CGD) is a discriminative module for graph generation tasks, which is widely used in graph structure modeling based on Generative Adversarial Network (GAN) [23]. The core goal of the module is to effectively fuse conditional constraint information (such as node attributes, boundary features, or external control signals) while assessing the authenticity of the graph or subgraph to realize the conditional discrimination of the generated graph [24]. Specifically, the conditional graph discriminator generates graph-level or node-level authenticity scores by encoding the node features, edge features, and condition vectors of the input graph, and aggregating neighbor information by graph convolution or graph attention mechanism.

This design not only ensures the discrimination ability of the overall structure and local features of the generated graph, but also strictly restricts the generated graph to meet specific conditions, which makes the model have higher accuracy and controllability in building layout, molecular graph generation, and other structural graph data generation tasks, and it provides an effective conditional discrimination strategy for the generation of complex graph structure [25].

3. Preparatory Work

This section provides an overview of the fundamental concepts underlying our proposed framework, including (i) utilizing a multi-relation graph attention module to capture the complex relationships between room nodes; (ii) gradually inferring the complete room relation graph through iterative prediction generators and effectively integrating node features with architectural boundary information; (iii) introducing a conditional graph discriminator and combining it with the Wasserstein loss to impose global consistency constraints on the generated layout.

3.1. Captures the Relationship Between Room Nodes

In building-layout modeling, rooms often have multiple types of relationships, such as spatial adjacency, functional association, and shared boundaries [26]. To effectively capture these complex relationships, we use a Multi-Relational Graph Attention Module (MR-GAT), which considers different types of side information when updating node features.

Consider a house layout represented as a graph $G = (V, E)$, where each node $i \in V$ corresponds to a room. Each node is associated with a feature vector.

$$h_i = [t_i, a_i, f_i, o_i] \in \mathbb{R}^F, \tag{1}$$

where t_i encodes the room type (e.g., bedroom, kitchen), a_i represents approximate area, f_i captures functional properties (e.g., bathroom fixtures), and o_i optionally includes room-level object features such as furniture attributes. The edge set E encodes R different types of relations between rooms, forming the full graph representation G. These node features provide both semantic and spatial information, enabling the graph attention module to reason about room connectivity, functional constraints, and structural coherence.

For a given relation $r \in \{1, \ldots, R\}$, the attention score between node i and its neighbor $j \in \mathcal{N}_i^r$ is computed as follows:

$$s_{ij}^{(r)} = \sigma\left(\mathbf{a}_r^\top (W_r h_i - W_r h_j)\right), \tag{2}$$

where W_r is a relation-specific transformation matrix, $\mathbf{a}_r$ is a learnable attention vector, and $\sigma(\cdot)$ denotes a LeakyReLU activation. The difference of transformed features captures relational interactions more effectively than standard concatenation.

The updated node representation is obtained by aggregating information from all neighboring nodes and relation types:

$$h_i' = \text{Fuse}\Big(\{h_j | j \in \mathcal{N}_i^r, r = 1, \dots, R\}\Big), \tag{3}$$

where Fuse denotes a fusion operation, such as concatenation or weighted sum. This multi-relational attention mechanism allows for the model to capture interactions among room nodes across multiple dimensions, enhancing its capability to model complex topological relations in building layouts.

3.2. Iterative Prediction Generator for Complete Room Graph Inference

To infer a complete room relationship graph, we design an iterative prediction generator that progressively refines the topology through multiple prediction–update steps. At each iteration t, node embeddings $\{h_i^{(t)}\}$ are updated by aggregating multi-relational attention features, while the generator predicts missing edges between room nodes:

$$\hat{A}_{ij}^{(t)} = \sigma\Big((h_i^{(t)})^\top W h_j^{(t)}\Big), \tag{4}$$

where $\hat{A}_{ij}^{(t)}$ denotes the probability of a relationship between nodes i and j, and W is a learnable parameter matrix.

The predicted adjacency matrix $\hat{A}^{(t)}$ is used to update the current graph structure iteratively:

$$G^{(t+1)} = \text{Update}(G^{(t)}, \hat{A}^{(t)}), \tag{5}$$

where Update$(\cdot)$ integrates predicted edges into the graph. The process continues until convergence, yielding a globally consistent room relationship graph.

To ensure geometric validity and architectural plausibility, the final objective combines relation prediction loss with constraint-aware regularization:

$$\mathcal{L} = \mathcal{L}_{rel} + \lambda \cdot \mathcal{L}_{cons}, \tag{6}$$

where $\mathcal{L}_{rel}$ supervises edge prediction accuracy, and $\mathcal{L}_{cons}$ penalizes violations

3.3. Conditional Graph Discriminator with Wasserstein Loss

To enforce global consistency in the generated layouts, we introduce a conditional graph discriminator that evaluates the plausibility of predicted room relationship graphs under architectural constraints. The discriminator takes both the generated graph G_{gen} and conditioning information c as input, and outputs a scalar score indicating layout realism:

$$D : (G, c) \mapsto \mathbb{R}. \tag{7}$$

Following the Wasserstein GAN framework, the discriminator is optimized using the Wasserstein loss with gradient penalty:

$$\mathcal{L}_D = \mathbb{E}_{G_{gen}}[D(G_{gen}, c)] - \mathbb{E}_{G_{real}}[D(G_{real}, c)] + \lambda_{gp}\mathbb{E}_{\hat{G}}[(\|\nabla_{\hat{G}} D(\hat{G}, c)\|_2 - 1)^2], \tag{8}$$

where G_{real} denotes the ground-truth graph, and $\hat{G}$ is a linear interpolation between G_{gen} and G_{real}.

The generator is trained to minimize

$$\mathcal{L}_G = -\mathbb{E}_{G_{gen}}[D(G_{gen}, c)] + \alpha \cdot \mathcal{L}_{cons}, \tag{9}$$

where $\mathcal{L}_{cons}$ incorporates architectural constraints. By combining the conditional discriminator with Wasserstein loss, the model imposes a global consistency constraint, ensuring that the generated layouts achieve realistic spatial connectivity and structural rationality comparable to real architectural designs.

4. Methodology

4.1. Graph-RWGAN Network Framework

The framework of the proposed Graph-RWGAN (Graph Relation Wasserstein GAN), illustrated in Figure 1, is designed to generate reasonable and globally consistent house layouts under weak constraints. To address the limitations of existing methods—such as disordered room placement, lack of iterative refinement, and limited style diversity—this framework consists of three key components: a Multi-Relational Graph Attention module (MR-GAT) [27], an iterative prediction generator, and a conditional graph discriminator.

MR-GAT captures complex multi-relational information between room nodes and learns structural and functional associations through graph attention, providing fine-grained relational representations for layout generation. Building on the attention-encoded graph features, the iterative prediction generator progressively infers the complete room relationship graph, effectively fusing node features with architectural boundary information to produce accurate layouts. It also supports iterative refinement to resolve potential conflicts and improve layout rationality.

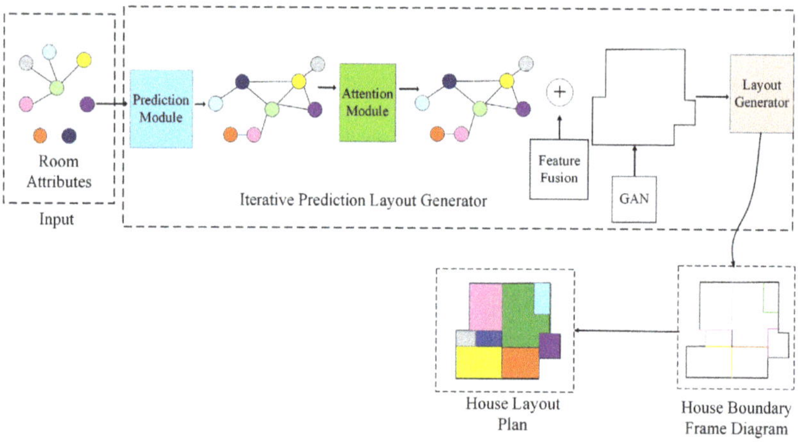

Figure 1. Graph-RWGAN network structure diagram.

The conditional graph discriminator, combined with Wasserstein loss, enforces global consistency constraints on the generated layouts, ensuring spatial connectivity and layout plausibility close to real designs, while enhancing the stability of the generation process and the diversity of layout styles. In practice, given room attributes and partial edge constraints, MR-GAT first encodes node features and relational information, then the iterative prediction generator infers the complete room relationship graph and generates a layout, and finally the conditional graph discriminator performs adversarial training to enforce global consistency. The resulting output is a coherent and realistic house layout under weak constraints. This framework not only improves layout rationality and structural coherence but also significantly enhances style diversity and controllability, providing an efficient, flexible, and adaptable solution for intelligent architectural design and automated planning.

4.2. Relationship Prediction Module

We define the constraint relation graph as $G = (O, E)$, where $O = \{o_1, \ldots, o_N\}$ is the set of room objects, each encoded as a 128-dimensional feature vector, and E represents the relations between room pairs. The adjacency tensor $A \in \{0,1\}^{N \times N \times M}$ encodes $M = 3$ edge types (including connections), with self-connections set as $A_{ii}^m = 1$.

The network input consists of initial room features $X = \{x_i\}$ and adjacency matrix A. The relation prediction module $\mathcal{F}_{rel}$ is a three-layer MLP (64-16-1) that predicts door and wall connection probabilities, denoted as p_{ij}^{door} and p_{ij}^{wall}. The overall connection probability between nodes i and j at iteration k is computed as follows:

$$P_{ij}^{(k)} = p_{ij}^{door} + p_{ij}^{wall} = \sigma(\mathcal{F}_{rel}(h_i^{(k)}, h_j^{(k)})), \tag{10}$$

where $h_i^{(k)}$ is the embedding of node i at iteration k, and σ denotes the sigmoid function.

The adjacency matrix is then updated iteratively:

$$A_{ij}^{(k+1)} = \begin{cases} 1, & P_{ij}^{(k)} \geq \tau, \\ 0, & \text{otherwise}, \end{cases} \tag{11}$$

where τ is a predefined threshold, and self-connections are preserved.

This iterative procedure allows the generator to progressively refine the room relation graph, effectively integrating node features and architectural boundary information to produce accurate house layouts under weak constraints.

4.3. Multi-Relation Graph Attention Module

As shown in Figure 2, the Multi-Relational Graph Attention (MR-GAT) module captures complex relationships between room nodes by adaptively learning attention weights.

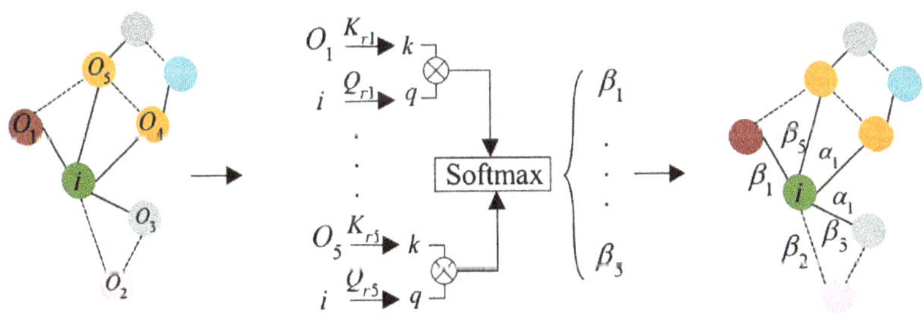

Figure 2. Multi-relational graph attention module.

The Multi-Relational Graph Attention (MR-GAT) module captures complex relationships between room nodes by adaptively learning attention weights. The constraint graph is decomposed into single-relation subgraphs. For each node i, multi-head attention evaluates the importance of neighbors $j \in \mathcal{N}_i^r$ under relation r and captures contextual attributes:

$$q_j^r = K_r o_j, \quad k_j^r = Q_r o_j, \tag{12}$$

$$\alpha_{ij}^r = \text{softmax}_j\left(\frac{(q_j^r)^\top k_i^r}{\sqrt{F}}\right), \quad j \in \mathcal{N}_i^r, \tag{13}$$

$$y_i^r = \sum_{j \in \mathcal{N}_i^r} \alpha_{ij}^r W_r o_j, \tag{14}$$

$$y_i^{r'} = W_m \, \text{concat}(y_i^{r_1}, \ldots, y_i^{r_m}), \tag{15}$$

where Q_r, K_r, and W_r are learnable matrices, and W_m projects concatenated vectors back to the original feature dimension D. Relation-specific weights β^r adjust the contribution of each relation:

$$\tilde{A}^r = \beta^r \cdot \text{Attention}(A^r), \tag{16}$$

$$h_i' = \sigma\left(\sum_r \sum_{j \in \mathcal{N}_i^r} \tilde{A}_{ij}^r W_o o_j\right), \tag{17}$$

where W_o is learnable, and σ is an activation function. This mechanism enables MR-GAT to effectively integrate information from multiple relation types while distinguishing key neighbors under different relational contexts.

4.4. Layout Generation Module

The layout generation module consists of two parts. The first part fuses architectural boundary features from a CNN with node features from the graph network to generate boundary box layouts [27]. The second part uses two decoders, similar to the relation prediction module, to predict door and wall connectivity, combining them to compute the connection probability between nodes:

$$P_{ij}^{r,(k)} = f_{\text{decoder}}(h_i^{(k)}, h_i^{(0)}, A), \tag{18}$$

where $h_i^{(0)}$ preserves initial node information, and f_{decoder} is the decoding network for relation r.

Binary cross-entropy losses L_{door}, L_{wall}, and L_{con} are applied for door, wall, and overall connectivity, respectively. The total prediction loss is

$$L_{\text{pred}} = \sum_{k=1}^{K} \sum_{r \in \{\text{door, wall, con}\}} \lambda_r L_r^{(k)}, \tag{19}$$

and the overall network loss combines this with the original loss L_{org}:

$$L_{\text{total}} = L_{\text{org}} + L_{\text{pred}}. \tag{20}$$

Node features are updated via multi-relation graph attention:

$$h_i^{(k+1)} = \sigma\Big(\sum_r \sum_{j \in \mathcal{N}_i^r} \alpha_{ij}^r W_r o_j\Big), \tag{21}$$

where α_{ij}^r refers to attention coefficients and W_r to learnable weights. The attention-weighted adjacency matrix is the following:

$$\tilde{A}^r = \sum_r \beta^r A^r W_r, \tag{22}$$

enabling effective aggregation of multi-relation information for accurate layout generation.

4.5. Conditional Graph Discriminator with Wasserstein Loss

The discriminator integrates a Multi-Relational Graph Attention (MR-GAT) module before the Conv-MPN layer to capture node features connected by different relations, allowing it to evaluate both layout realism and room connectivity. Let h_i be node features and A the adjacency matrix from real data. MR-GAT updates node features as follows:

$$h'_i = \sigma\Big(\sum_r \sum_{j \in \mathcal{N}_i^r} \alpha_{ij}^r W_r h_j\Big), \quad \tilde{A}^r = \sum_r \beta^r A^r W_r, \tag{23}$$

where α_{ij}^r denotes attention coefficients, and W_r denotes learnable weights. The updated features h'_i are fused with the boundary layout B:

$$x_{\text{in}} = \text{Flatten}([h'_1, \ldots, h'_N, B]), \quad x_{\text{feat}} = W_{\text{lin}}\, x_{\text{in}} \in \mathbb{R}^{8192}, \tag{24}$$

with W_{lin} as a linear layer. The Conv-MPN layers then process x_{feat}, and the discriminator loss is as follows:

$$\mathcal{L}_D = \mathbb{E}[D(x_{\text{real}})] - \mathbb{E}[D(x_{\text{fake}})], \tag{25}$$

ensuring both global layout consistency and correct inter-room connections, while the generator is trained to minimize the adversarial loss:

$$\mathcal{L}_G = -\mathbb{E}[D(x_{\text{fake}})] + \lambda_{\text{pred}}\, \mathcal{L}_{\text{pred}}, \tag{26}$$

where $\mathcal{L}_{\text{pred}}$ is the relation prediction loss, and λ_{pred} is a weighting factor. This formulation ensures that the generated layouts not only maintain global structural consistency but also preserve accurate room adjacency and connectivity, closely aligning with real design standards.

4.6. User Input and Weak Constraint Specification

To clarify how users interact with the proposed system, we define the concept of weak constraints and describe the minimum input requirements. In our framework, weak constraints refer to high-level and flexible specifications provided by users rather than precise geometric parameters. These typically include the number and types of rooms, approximate area ratios, and optional adjacency preferences. Such constraints allow users to express design intentions without detailed architectural drawings. Formally, let the input constraint set be denoted as follows:

$$C = \{c_i \mid c_i \in \mathcal{C}, i = 1, 2, \ldots, N\}, \tag{27}$$

where each c_i represents a soft constraint describing a semantic or spatial relationship (e.g., room type, size, or adjacency). Unlike hard constraints that must be strictly satisfied, weak constraints are incorporated into the optimization objective as penalty terms or probabilistic priors.

$$L^* = \arg\max_L P(L \mid C) = \arg\max_L \exp(-\mathcal{L}_{\text{gen}}(L, C)), \tag{28}$$

where $\mathcal{L}_{\text{gen}}$ denotes the generation loss, balancing structural consistency and constraint satisfaction. The minimum required input consists of a list of functional rooms and their rough size proportions. Based on these inputs, the system automatically infers reasonable spatial relationships and generates coherent layouts that satisfy the given constraints. Non-expert users can interact with the model using intuitive parameters, while professional architects can optionally provide more detailed spatial or relational hints to refine the design.

5. Experimental Setup

5.1. Dataset Introduction

The dataset used in this study is the real-world large-scale architectural floor plan dataset RPLAN, which serves as a benchmark for house layout generation. Figure 3 illustrates a few sample cases from the RPLAN dataset [28], showing the annotated room types and the locations of doors, windows, and walls. There are nine room types in total: master bedroom, secondary bedroom, living room, balcony, bathroom, kitchen, storage room, study, and dining room. These visualizations are only intended to provide readers with an understanding of the dataset and its typical layouts, rather than representing outputs of our model.

Figure 3. Cases in the RPLAN dataset.

5.2. Experimental Environment

To evaluate whether the proposed algorithm can generate reasonable and realistic layout designs under weak constraints, this section compares both quantitative and qualitative results on the RPLAN dataset against several popular methods, including House-GAN [29], LayoutGAN [30], MR-GAT [31], and LayoutDiffusion [32].

The model is implemented using the deep learning framework PyTorch 2.1.0, with experiments conducted on an Ubuntu 16.04 operating system. The model is optimized using the Adam optimizer with a learning rate of 0.0001 and a batch size of 32. Training and evaluation are conducted on the RPLAN dataset, producing layout images at a resolution of 256×256 pixels. The hyperparameters α and β are empirically set to 0.9 and 0.7, respectively, based on experimental calibration.

During fine-tuning, key hyperparameters such as learning rate, batch size, and loss weight coefficients are slightly adjusted within small ranges to maintain stable convergence and balance between reconstruction accuracy and visual realism. When adapting the model to different building types or scales, only lightweight fine-tuning of the decoder and relational reasoning modules is required, while the graph encoder remains largely fixed. All other settings and configurations remain consistent with the original model design described in the paper.

5.3. Evaluation Indicators

When generating house layouts with deep learning, researchers typically evaluate their methods using both qualitative and quantitative assessments. Qualitative evaluation relies on domain experts subjectively judging layout realism, which can vary due to individual perception. To objectively evaluate the results, we employ three quantitative metrics: Frechet Inception Distance (FID) to assess visual fidelity, Structural Similarity Index (SSIM) to measure structural consistency, and Inception Score (IS) to capture layout diversity and clarity.

(1) Frechet Inception Distance.

The Frechet Inception Distance (FID) evaluates the difference between feature distributions of real and generated images, reflecting the visual quality. Lower values indicate higher fidelity. It can be expressed as follows:

$$\mathrm{FID} = \|\mu_{\mathrm{real}} - \mu_{\mathrm{gen}}\|_2^2 + \mathrm{Tr}\left(\Sigma_{\mathrm{real}} + \Sigma_{\mathrm{gen}} - 2(\Sigma_{\mathrm{real}}\Sigma_{\mathrm{gen}})^{1/2}\right) \tag{29}$$

where μ_{real} and μ_{gen} are the mean feature vectors of real and generated images, Σ_{real} and Σ_{gen} are the corresponding covariance matrices, and $\mathrm{Tr}(\cdot)$ denotes the matrix trace.

(2) Structural Similarity Index.

The Structural Similarity Index (SSIM) is widely used to evaluate the perceptual similarity between two images. In this study, it quantifies how well the generated layouts preserve spatial consistency between rooms or between rooms and building boundaries. An alternative formulation of SSIM can be expressed as follows:

$$\mathrm{SSIM}(x,y) = 1 - \frac{\sqrt{(l(x)-l(y))^2+\epsilon_l}}{\sigma_l+\epsilon_l} \cdot \frac{\sqrt{(c(x)-c(y))^2+\epsilon_c}}{\sigma_c+\epsilon_c} \cdot \frac{\sqrt{(s(x)-s(y))^2+\epsilon_s}}{\sigma_s+\epsilon_s} \tag{30}$$

where $l(x)$ and $l(y)$ denote the average luminance of images x and y; $c(x)$ and $c(y)$ represent the local contrast; $s(x)$ and $s(y)$ represent the structural information; σ_l, σ_c, and σ_s are normalization constants; and ϵ_l, ϵ_c, and ϵ_s are small constants added for numerical stability. This formulation preserves the essence of SSIM by combining luminance, contrast, and structure similarity into a single score.

(3) Inception Score.

The Inception Score measures the quality and diversity of generated images. It uses a pretrained classifier to assess how confidently each image can be classified and how diverse the overall set of generated samples is. The IS is defined as

$$\mathrm{IS} = \exp\left(\mathbb{E}_{x\sim G}\left[D_{\mathrm{KL}}(p_c(y|x) \,\|\, p_m(y))\right]\right) \tag{31}$$

where $x \sim G$ denotes a sample drawn from the generator, $p_c(y|x)$ is the conditional class probability predicted by the classifier, and $p_m(y) = \mathbb{E}_{x\sim G}[p_c(y|x)]$ represents the marginal distribution across all generated samples

(4) Layout Element Classification Accuracy.

$$\mathrm{Acc}\ (\%) = \frac{N_{\mathrm{correct}}}{N_{\mathrm{total}}} \times 100 \tag{32}$$

where N_{correct} is the number of correctly classified elements in the generated layout, and N_{total} is the total number of elements in the layout.

(5) Room Adjacency Score.

$$\text{Room Adjacency Score} = \frac{1}{N_{\mathrm{pairs}}}\sum_{i,j}\mathbf{1}\{(r_i,r_j) \in \mathcal{A}_{\mathrm{valid}}\}, \tag{33}$$

where N_{pairs} is the total number of room pairs in the layout, r_i and r_j are rooms in the layout, and $\mathcal{A}_{\mathrm{valid}}$ is the set of valid room adjacency pairs based on architectural logic. $\mathbf{1}\{\cdot\}$ is an indicator function that returns 1 when the room pair (r_i, r_j) is valid, and 0 otherwise.

6. Evaluation and Analysis

6.1. Preprocessing of Data

In this approach, the dataset is initially represented as a constraint graph, where nodes correspond to individual rooms, and edges denote their spatial relationships. Each node is annotated with attributes that describe the room's characteristics, including type, functional properties, and other relevant semantic information. As illustrated in Figure 4, the edges are defined based on adjacency relationships rather than exact spatial orientations, emphasizing the connectivity between rooms rather than their precise geometric positions.

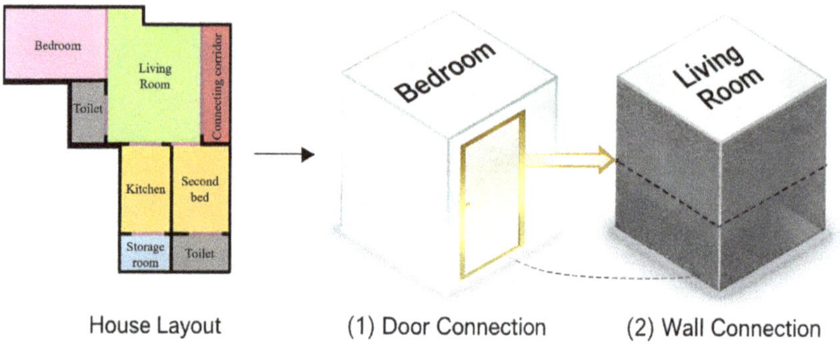

Figure 4. Diagram of edge relationships in dataset preprocessing.

Figure 4 shows the diagram of edge relationships in the dataset-preprocessing stage. To facilitate a clearer understanding, we distinguish three types of edges:

- Door connections: These edges represent rooms that are directly accessible via a door, allowing for movement or functional interaction between the spaces.
- Wall connections: These edges signify rooms that share a common wall, indicating a physical separation that prevents direct spatial flow between them but still establishes a structural connection.
- General connections: This category encompasses both door and wall connections, representing broader neighborhood relationships between rooms that might not necessarily require direct access but still form part of the layout's overall connectivity.

The graph-based representation thus encodes important structural and functional constraints of the environment. By focusing on connectivity—rather than precise geometrical positioning—the model effectively captures the relationships between rooms, ensuring that the spatial structure aligns with real-world architectural principles. This representation allows for efficient reasoning and analysis in the subsequent layout generation process.

A door connection indicates that two rooms are spatially accessible through a doorway, promoting interaction or movement between them. On the other hand, a wall connection reflects a barrier between two rooms, restricting direct spatial flow but still establishing a functional boundary. Both types of connections, along with the general connections, form the basis for modeling spatial constraints and relationships, enabling the model to produce layouts that respect the necessary functional and structural requirements.

6.2. Experimental Results

On the RPLAN test set, Graph-RWGAN outperforms House-GAN [29], Layout-GAN [30], and MR-GAT [31] across all metrics. Compared to House-GAN, Graph-RWGAN achieves a 15.14% reduction in FID, a 5.6% increase in SSIM, and a 5.68% improvement in Acc. In terms of Inception Score (IS), Graph-RWGAN also shows a superior performance with an increase of 9.47% compared to House-GAN. When compared to MR-GAT, Graph-RWGAN reduces FID by 5.24%, increases SSIM by 1.85%, and improves Acc by 4.36%. Additionally, Graph-RWGAN outperforms both House-GAN and MR-GAT in the Room Adjacency Score, reflecting a better capture of the spatial relationships and architectural consistency in the generated layouts. These results demonstrate that Graph-RWGAN better models room relationships under weak constraints, producing layouts with higher quality and greater structural coherence. The detailed values are provided in Table 1.

Table 1. Quantitative comparison of different methods on the RPLAN dataset.

Model	FID ↓	SSIM ↑	Acc (%) ↑	IS ↑	Room Adjacency Score ↑
House-GAN [29]	109.27	0.784	81.34	2.85	0.71
LayoutGAN [30]	175.36	0.525	68.22	2.12	0.60
MR-GAT [31]	94.69	0.813	82.37	3.05	0.74
Graph-RWGAN	92.73	0.828	85.96	3.12	0.78

Figure 5 compares house layout generation results from various models: (a) House-GAN, (b) LayoutGAN, (c) MR-GAT, (d) Ground Truth, and (e) Graph-RWGAN. The results are organized into three horizontal sets, each corresponding to the same incomplete graph input scenario, and each set is divided into "graph structure" and "layout" sections. The figure aims to show how different models transform structural graph information into spatial layouts, helping readers understand the generation process.

The graph structure section illustrates room topology using nodes and edges of different colors, representing spatial or functional relationships. Nodes correspond to rooms, and edges encode connectivity, adjacency, or other functional associations. The graph captures relational structure but not precise positions or sizes, so a single graph can correspond to multiple plausible layouts.

The layout section translates the graph into a specific geometric arrangement, with each room type represented by colored squares. The model infers numeric parameters such as room sizes and positions to generate a unique layout consistent with the graph. The legend maps colors to room types . This visualization enables side-by-side comparison, highlighting differences in room relationships, layout fidelity, and structural coherence across models.

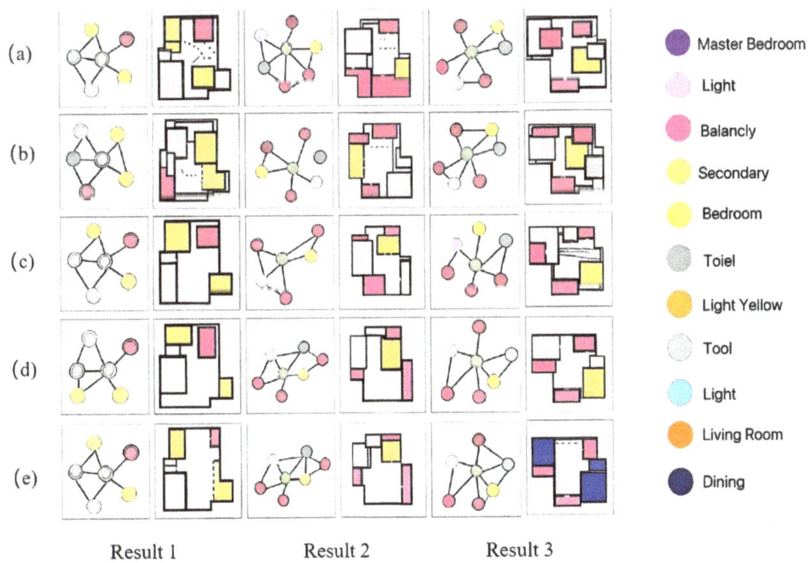

Figure 5. The results of different methods on the RPLAN dataset. ((**a**) House-GAN, (**b**) LayoutGAN, (**c**) MR-GAT, (**d**) Ground Truth, (**e**) Graph-RWGAN).

Table 2 presents a quantitative comparison of the performance of four models—House-GAN, LayoutGAN, MR-GAT, and Graph-RWGAN—across three evaluation criteria: Room Topology Accuracy, Layout Quality, and Structural Coherence. These criteria are rated on a percentage scale from 0% to 100%, with higher percentages indicating better performance. House-GAN scores 60% in Room Topology Accuracy, meaning it captures the general spatial relationships between rooms, but with some inaccuracies. Its Layout Quality is 70%, suggesting that the generated layouts are functional but could benefit from further optimization. The Structural Coherence score of 65% indicates that the relationships between rooms are somewhat coherent, though there are noticeable inconsistencies.

Table 2. Quantitative comparison of layout generation by different models.

Model	Room Topology Accuracy (%)	Layout Quality (%)	Structural Coherence (%)
House-GAN	60%	70%	65%
LayoutGAN	45%	50%	55%
MR-GAT	75%	85%	80%
Graph-RWGAN	95%	90%	95%

LayoutGAN performs relatively poorly across all metrics, scoring 45% in Room Topology Accuracy, indicating difficulty in modeling room relationships. Its Layout Quality score is 50%, showing that the generated layouts are of low quality, and the Structural Coherence score of 55% suggests weak structural relationships, resulting in less coherent overall layouts. MR-GAT performs better, achieving 75% in Room Topology Accuracy, reflecting a moderate ability to capture room relationships. It earns 85% for Layout Quality, signifying high-quality, space-efficient layouts, and 80% for Structural Coherence, indicating that the layouts are logically consistent, though not flawless. Finally, Graph-RWGAN outperforms all other models, with a 95% in Room Topology Accuracy, demonstrating its superior ability to model spatial relationships. Its Layout Quality score of 90% indicates that the layouts are highly functional and well-organized, while the Structural Coherence score of 95% reflects strong internal consistency and logical room relationships, making it the most reliable model for generating realistic house layouts.

As shown in Figure 6, where (A) represents the ground truth, (B) is generated by MA-GAN, (C) by MR-GAN, (D) by House-GAN, (E) by LayoutDiffusion, and (F) by our method. The layouts produced by both (E) and (F) align closely with the ground truth. However, while LayoutDiffusion (E) generates reasonable layouts, it does not produce a complete image, resulting in some missing details. In contrast, our method (F) not only preserves the overall spatial organization but also captures the detailed relationships among functional areas more accurately, producing fully realized and visually coherent layouts.

Technically, the advantage of our method lies in its precise modeling of spatial relationships, effectively capturing the topological constraints between regions. Through structural optimization and tailored loss functions, our approach enhances the generation of fine-grained details. Additionally, multi-scale feature fusion ensures coherence between local and global layouts. Collectively, these improvements enable our method to produce building layouts that are both more accurate and visually realistic compared to all baseline methods, including LayoutDiffusion.

Table 3 compares different methods on IoU, Position Error, Structural Consistency, and Visual Realism. Our method achieves the best performance across all metrics: IoU reaches 85.6%, which is 14.4% higher than the best baseline (House-GAN). Position Error is reduced to 5.2, nearly half of House-GAN's 10.3, showing superior accuracy in spatial alignment and object localization.

For Structural Consistency and Visual Realism, our method attains 0.92 and 0.95, outperforming most other methods. LayoutDiffusion achieves slightly higher Visual

Realism (0.96), but its overall performance is still below that of our approach. These results indicate that our method provides the best balance between geometric fidelity and perceptual quality, maintaining high structural consistency while generating realistic and visually appealing outputs.

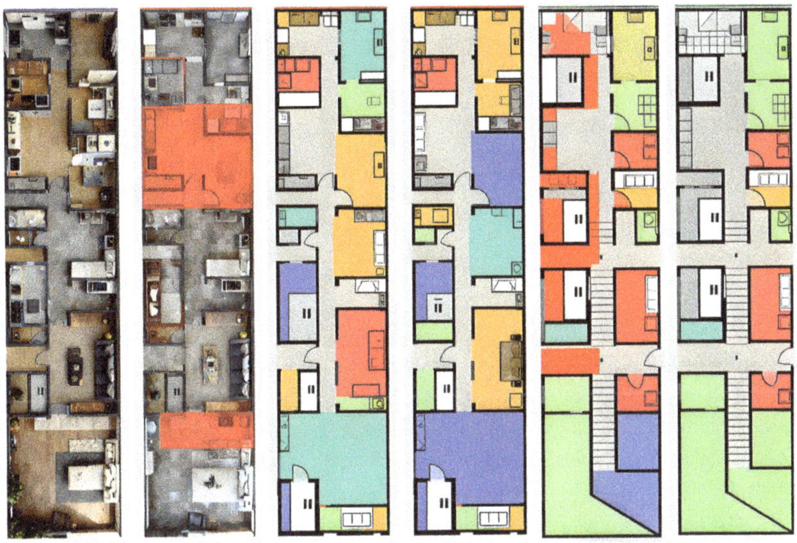

(A) (B) (C) (D) (E) (F)

Figure 6. Visual comparison of different methods for house layout generation. ((**A**) The ground truth, (**B**) MA-GAN, (**C**) MR-GAN, (**D**) House-GAN, (**E**) LayoutDiffusion, (**F**) Our Method).

Table 3. Quantitative comparison of different methods.

Method	IoU (%)	Position Error	Structural Consistency	Visual Realism
MA-GAN	62.3	15.8	0.68	0.72
MR-GAT	68.5	12.5	0.75	0.78
House-GAN	71.2	10.3	0.80	0.82
LayoutDiffusion	83.0	6.0	0.90	0.96
Our Method	85.6	5.2	0.92	0.95

6.3. *Ablation Study*

In this section, we perform an ablation study to assess the impact of each major component in the Graph RWGAN model. By selectively disabling or removing specific modules, we examine how these changes influence the model's overall performance [33]. Specifically, we investigate the effects of omitting:

- Iterative Learning Mechanism: This mechanism allows the model to progressively improve its predictions over time. By disabling this component, we observe a significant decrease in the model's ability to predict unknown relationships, as reflected in lower accuracy and poorer layout generation quality.
- Relation Prediction Module: This module plays a crucial role in predicting spatial relationships between rooms. Without it, the model's accuracy decreases, indicating the importance of precise relationship prediction for generating realistic and coherent layouts.
- Multi-Relation Graph Attention Module: This module consolidates information from multiple edge relationship types, improving the model's capacity to capture intricate

spatial dependencies. Its removal results in noticeable performance drops, especially in terms of structural consistency and the accuracy of room topology.

In Table 4, we compare the full model (Graph-RWGAN) with versions where one of the components has been removed. The analysis demonstrates that each module plays an essential role in the model's overall performance. Removing any of them results in a noticeable decrease in both layout quality and structural consistency.

Table 4. Ablation study on the RPLAN dataset.

Model Variant	FID ↓	SSIM ↑	Acc ↑
w/o Iterative Learning	112.87	0.729	74.17
w/o Relation Prediction Module	98.49	0.752	72.67
w/o Multi-Relation Graph Attention	109.15	0.693	69.29
Ours (Graph-RWGAN)	92.73	0.828	85.96

Table 4 presents the results of the ablation study on the RPLAN dataset, evaluating the contributions of the iterative learning mechanism, the relation prediction module, and the multi-relation graph attention module. When the iterative learning mechanism is removed, the accuracy (Acc) drops significantly, and both FID and SSIM decrease. Specifically, the accuracy decreases to 74.17%, while FID increases to 112.87, and SSIM drops to 0.729. This indicates that iterative learning allows the network to progessively capture more structural and semantic information from the input graph, which is essential for generating coherent and high-quality layouts.

Removing the relation prediction module, which effectively eliminates the prediction loss, leads to a further reduction in prediction accuracy. The accuracy decreases to 72.67%, FID rises to 98.49, and SSIM drops to 0.752, highlighting the critical role of the relation prediction module in modeling accurate edge relationships between rooms. Excluding the multi-relation graph attention module results in additional performance degradation, with an accuracy of 69.29%, FID increasing to 109.15, and SSIM decreasing to 0.693. This demonstrates its effectiveness in aggregating information from different types of edge relationships. Overall, the complete Graph-RWGAN model, which integrates all three components, achieves the best performance across all metrics, with an accuracy of 85.96%, FID of 92.73, and SSIM of 0.828, confirming the complementary nature and importance of these modules in improving layout prediction and visual quality.

6.4. Runtime and Efficiency Evaluation

In layout generation tasks, runtime efficiency is as important as the quality of generated layouts, especially for real-time applications or large-scale data processing, where efficiency and resource consumption are critical. In this evaluation, we compare the efficiency of different models in terms of runtime, memory consumption, and inference speed.

We first compare the computational time required for different models to generate the same layout. Graph-RWGAN performs the best overall, with an average generation time of 1.2 s. LayoutDiffusion requires 1.5 s per layout, slightly slower than Graph-RWGAN but faster than MA-GAN (1.8 s), MR-GAT (2.0 s), House-GAN (2.4 s), and LayoutGAN (3.1 s). This efficiency advantage of Graph-RWGAN is due to its more optimized structure and graph processing capabilities.

In terms of memory usage, LayoutDiffusion consumes 2.0 GB, slightly lower than Graph-RWGAN's 2.1 GB, indicating better memory efficiency for certain tasks. MA-GAN and MR-GAT consume 2.3 GB and 2.5 GB, respectively, while House-GAN and LayoutGAN require 3.5 GB and 4.2 GB. For inference speed, Graph-RWGAN remains superior, generating layouts in 0.8 s compared to LayoutDiffusion's 0.9 s, MA-GAN's 1.0 s, and MR-GAT's 1.2 s.

When testing large-scale data processing, Graph-RWGAN successfully generates 1000 layouts in 20 min. LayoutDiffusion requires 22 min, MA-GAN 25 min, and MR-GAT 28 min. Although LayoutDiffusion shows some advantages in memory consumption, its overall runtime and efficiency are still slightly behind Graph-RWGAN. The detailed results are presented in Table 5.

Table 5. Runtime and efficiency comparison of different models.

Model	Generation Time (s)	Memory Consumption (GB)	Inference Time (s)	Time for 1000 Layouts (min)
Graph-RWGAN	1.2	2.1	0.8	20
MA-GAN	1.8	2.3	1.0	25
MR-GAT	2.0	2.5	1.2	28
House-GAN	2.4	3.5	1.6	30
LayoutGAN	3.1	4.2	2.0	40
LayoutDiffusion	1.5	2.0	0.9	22

Overall, Graph-RWGAN demonstrates superior efficiency in all key areas, including computation time, memory consumption, and inference speed, making it ideal for real-time applications and large-scale data processing tasks. This efficiency allows it to deliver high-quality layouts while consuming fewer computational resources, providing a significant advantage over other models.

7. Conclusions

Graph-RWGAN (Graph Relation Wasserstein GAN) addresses key limitations of existing house layout generation methods, including disordered room arrangements, limited iterative refinement, and low style diversity. It leverages a multi-relational graph attention module, an iterative prediction generator, node-boundary feature fusion, and a conditional graph discriminator with Wasserstein loss to produce layouts with high structural coherence, accurate room connectivity, and diverse styles under weak constraints. While reliance on predefined graph representations may reduce flexibility for highly irregular layouts, and iterative graph inference can increase computational cost, Graph-RWGAN remains a robust and controllable framework, achieving significant improvements over previous methods.

Future research may focus on enhancing the local editability of generated layouts and enabling user interaction, allowing users to make fine-grained adjustments and personalized modifications during the generation process. Additionally, integrating more practical design constraints, such as lighting, ventilation, and furniture placement, could achieve multi-dimensional optimization closer to real-world architectural design. Furthermore, incorporating multi-objective generation and style transfer techniques may enrich layout styles and improve creativity, providing a more flexible, efficient, and controllable generative framework for intelligent building design and automated planning.

Author Contributions: Conceptualization, Z.Y. and J.C.; methodology, J.C.; software, J.C.; validation, Z.Y. and J.C.; formal analysis, Z.Y.; investigation, J.C.; resources, Z.Y.; data curation, Z.Y. and J.C.; writing—original draft, Z.Y., S.L., Z.T., Y.C., L.Z. and J.C.; writing—review and editing, Z.Y., S.L., Z.T., Y.C., L.Z. and J.C.; visualization, J.C.; supervision, Z.Y. and J.C.; project administration, J.C.; funding acquisition, J.C. All authors have read and agreed to the published version of the manuscript.

Funding: This research work are funded by Guangdong Provincial Education Science Planning Project (Higher Education Scientific Research Project): Research on Multisensory Communication of Animation in the Construction of Aesthetic Education in the Greater Bay Area (2021GXJK470); And the project FRG-25-064-FA and granted by the Research Fund of Macao University of Science and Technology (FRG-MUST).

Data Availability Statement: The original data presented in the study are openly available in ACM Digital Library at https://doi.org/10.1145/3528223.3530135.

Conflicts of Interest: The authors declare no conflicts of interest.

References

1. Zhao, Y.; Dong, J.; Wang, W.; Duan, H. A multi-typed multi-relational heterogeneous graph neural network model for complex networks. *Knowl.-Based Syst.* **2025**, *329*, 114291. [CrossRef]
2. Ai, W.; Liu, Y.; Wei, C.; Meng, T.; Shao, H.; He, Z.; Li, K. MFLM-GCN: Multi-Relation Fusion and Latent-Relation Mining Graph Convolutional Network for Entity Alignment. *Knowl.-Based Syst.* **2025**, *325*, 113974. [CrossRef]
3. Shakeel, H.M.; Iram, S.; Farid, H.M.A.; Hill, R.; Rehman, H. EPCDescriptor: A Multi-Attribute Visual Network Modeling of Housing Energy Performance. *Buildings* **2025**, *15*, 2929. [CrossRef]
4. Wang, F.; Liu, X.; Li, W.; Tian, W.; Jia, Z.; Ran, C. MRGCE: A Multi-Relational Graph Contrastive Enhancement Learning Framework for High-Value Patent Identification. *Scientometrics* **2025**, *130*, 10. [CrossRef]
5. Li, W.; Gao, Y.; Li, A.; Zhang, X.; Gu, J.; Liu, J. Sparse Subgraph Prediction Based on Adaptive Attention. *Appl. Sci.* **2023**, *13*, 8166. [CrossRef]
6. Zhou, X.; Zeng, Y.; Hao, Z.; Wang, H. A Robust Graph Attention Network with Dynamic Adjusted Graph. *Eng. Appl. Artif. Intell.* **2024**, *129*, 107619. [CrossRef]
7. Yin, H.; Zhong, J.; Wang, C.; Li, R.; Li, X. GS-InGAT: An Interaction Graph Attention Network with Global Semantic for Knowledge Graph Completion. *Expert Syst. Appl.* **2023**, *228*, 120380. [CrossRef]
8. Li, J.; Yang, J.; Zhang, J.; Liu, C.; Wang, C.; Xu, T. Attribute-Conditioned Layout GAN for Automatic Graphic Design. *IEEE Trans. Vis. Comput. Graph.* **2020**, *27*, 4039–4048. [CrossRef]
9. Wang, S.; Zeng, W.; Chen, X.; Ye, Y.; Qiao, Y.; Fu, C.-W. ActFloor-GAN: Activity-Guided Adversarial Networks for Human-Centric Floorplan Design. *IEEE Trans. Vis. Comput. Graph.* **2021**, *29*, 1610–1624. [CrossRef]
10. Zhou, Y.; Xia, H.; Yu, D.; Cheng, J.; Li, J. Outlier detection method based on high-density iteration. *Inf. Sci.* **2024**, *662*, 120286. [CrossRef]
11. Ong, Y.Z.; Zheng, L.; Feng, C.; Song, K. VG-GAN: Conditional GAN Framework for Graphical Design Generation. In Proceedings of the 2022 IEEE International Conference on Image Processing (ICIP), Bordeaux, France, 16–19 October 2022; IEEE: Piscataway, NJ, USA, 2022; pp. 781–785.
12. Johnson, J.; Gupta, A.; Li, F.-F. Image Generation from Scene Graphs. In Proceedings of the IEEE Conference on Computer Vision and Pattern Recognition, Salt Lake City, UT, USA, 18–23 June 2018; IEEE: Piscataway, NJ, USA, 2018; pp. 1219–1228.
13. Nauata, N.; Chang, K.H.; Cheng, C.Y.; Mori, G.; Furukawa, Y. House-GAN: Relational Generative Adversarial Networks for Graph-Constrained House Layout Generation. In Proceedings of the Computer Vision–ECCV 2020: 16th European Conference, Glasgow, UK, 23–28 August 2020; Part I; Springer International Publishing: Berlin/Heidelberg, Germany, 2020; pp. 162–177.
14. Yao, S.; Guan, R.; Wu, Z.; Ni, Y.; Huang, Z.; Liu, R.W.; Yue, Y.; Ding, W.; Lim, E.G.; Seo, H.; et al. Waterscenes: A multi-task 4D radar-camera fusion dataset and benchmarks for autonomous driving on water surfaces. *IEEE Trans. Intell. Transp. Syst.* **2024**, *25*, 16584–16598. [CrossRef]
15. Chang, K.H.; Cheng, C.Y.; Luo, J.; Murata, S.; Nourbakhsh, M.; Tsuji, Y. Building-GAN: Graph-Conditioned Architectural Volumetric Design Generation. In Proceedings of the IEEE/CVF International Conference on Computer Vision, Montreal, QC, Canada, 10–17 October 2021; pp. 11956–11965.
16. Pan, X.; Song, S.; Chen, Y.; Wang, L.; Huang, G. PLAM: A Plug-in Module for Flexible Graph Attention Learning. *Neurocomputing* **2022**, *480*, 76–88. [CrossRef]
17. Pan, T.; Zhang, L.; Song, Y.; Liu, Y. Hybrid Attention Compression Network With Light Graph Attention Module for Remote Sensing Images. *IEEE Geosci. Remote Sens. Lett.* **2023**, *20*, 6005605. [CrossRef]
18. Feng, Z.; Li, Y.; Chen, W.; Su, X.; Chen, J.; Li, J.; Liu, H.; Li, S. Infrared and Visible Image Fusion Based on Improved Latent Low-Rank and Unsharp Masks. *Spectrosc. Spectr. Anal.* **2025**, *45*, 2034–2044.
19. Wu, Y.; Wang, G.; Wang, Z.; Wang, H.; Li, Y. Triplet Attention Fusion Module: A Concise and Efficient Channel Attention Module for Medical Image Segmentation. *Biomed. Signal Process. Control* **2023**, *82*, 104515. [CrossRef]
20. Carneiro, P.M.R.; Vidal, J.V.; Rolo, P.; Peres, I.; Ferreira, J.A.F.; Kholkin, A.L.; Soares dos Santos, M.P. Instrumented Electromagnetic Generator: Optimized Performance by Automatic Self-Adaptation of the Generator Structure. *Mech. Syst. Signal Process.* **2022**, *171*, 108898. [CrossRef]
21. Du, J.; Li, B.; Yang, J. Boundary-aware Graph Convolutional Network for Building Roof Detection from High-Resolution Remote Sensed Imagery. *PFG J. Photogramm. Remote Sens. Geoinf. Sci.* **2025**, *93*, 443–455. [CrossRef]

22. Li, Q.; Sun, X.; Dong, J. Transductive zero-shot learning via knowledge graph and graph convolutional networks. *Sci. Rep.* **2025**, *15*, 28708. [CrossRef]
23. Liu, H.; Li, D.; Zeng, B.; Xu, Y. Learning discriminative features for multi-hop knowledge graph reasoning. *Appl. Intell.* **2025**, *55*, 468. [CrossRef]
24. Wen, X.; Yang, X.-H.; Ma, G.-F. ConDiff: Conditional graph diffusion model for recommendation. *Inf. Process. Manag.* **2026**, *63*, 104303. [CrossRef]
25. Hu, R.; Huang, Z.; Tang, Y.; van Kaick, O.; Zhang, H.; Huang, H. Graph2Plan: Learning Floorplan Generation from Layout Graphs. *arXiv* **2020**, arXiv:2004.13204. [CrossRef]
26. Wang, Z.; Fang, X.; Zhang, W.; Ding, X.; Wang, L.; Chen, C. Multi-relation spatiotemporal graph residual network model with multi-level feature attention: A novel approach for landslide displacement prediction. *J. Rock Mech. Geotech. Eng.* **2025**, *17*, 4211–4226. [CrossRef]
27. Nauata, N.; Hosseini, S.; Chang, K.-H.; Chu, H.; Cheng, C.-Y.; Furukawa, Y. House-GAN++: Generative Adversarial Layout Refinement Networks. *arXiv* **2021**, arXiv:2103.02574.
28. Wu, W.; Fu, X.-M.; Tang, R.; Wang, Y.; Qi, Y.-H.; Liu, L. Data-driven interior plan generation for residential buildings. *ACM Trans. Graph.* **2019**, *38*, 1–12. [CrossRef]
29. Nauata, N.; Chang, K.-H.; Cheng, C.-Y.; Mori, G.; Furukawa, Y. House-GAN: Relational generative adversarial networks for graph-constrained house layout generation. *arXiv* **2020**, arXiv:2003.06988.
30. Li, J.; Yang, J.; Hertzmann, A.; Zhang, J.; Xu, T. LayoutGAN: Generating graphic layouts with wireframe discriminators. *arXiv* **2019**, arXiv:1901.06767. [CrossRef]
31. Li, Z.; Zhao, Y.; Zhang, Y.; Wang, S.; Liu, H. Multi-relational graph attention networks for knowledge graph completion. *Knowl.-Based Syst.* **2022**, *251*, 109262. [CrossRef]
32. Zhang, J.; Guo, J.; Sun, S.; Lou, J.-G.; Zhang, D. LayoutDiffusion: Improving Graphic Layout Generation by Discrete Diffusion Probabilistic Models. *arXiv* **2023**, arXiv:2303.11589. [CrossRef]
33. Tang, H.; Zhang, Z.; Shi, H.; Li, B.; Shao, L.; Sebe, N.; Timofte, R.; Van Gool, L. Graph Transformer GANs for Graph-Constrained House Generation. *arXiv* **2023**, arXiv:2303.08225. [CrossRef]

buildings

MDPI

Article

A ConvLSTM-Based Hybrid Approach Integrating DyT and CBAM(T) for Residential Heating Load Forecast

Haibo Zhang, Xiaoxing Gao, Xuan Liu and Zhibin Liu *

College of Civil Engineering, Dalian Minzu University, Dalian 116650, China; haibozhang06@163.com (H.Z.); 15713299105@163.com (X.G.); liuxuan011105@163.com (X.L.)
* Correspondence: liuzhibin@dlnu.edu.cn

Abstract: Accurate forecasting of residential heating loads is crucial for guiding heating system control strategies and improving energy efficiency. In recent years, research on heating load forecasting has primarily focused on continuous district heating systems, and it often struggles to cope with the abrupt load fluctuations and irregular on/off schedules encountered in intermittent heating scenarios. To address these challenges, this study proposes a hybrid convolutional long short-term memory (ConvLSTM) model that replaces the conventional batch normalization layer with a Dynamic Tanh (DyT) activation function, enabling dynamic feature scaling and enhancing responsiveness to sudden load spikes. An improved channel–temporal attention mechanism, CBAM(T), is further incorporated to deeply capture the spatiotemporal relationships in multidimensional data and effectively handle the uncertainty of heating start–stop events. Using data from two heating seasons for households in a residential community in Dalian, China, we validate the performance of ConvLSTM-DyT-CBAM(T). The results show that the proposed model achieves the best predictive accuracy and strong generalization, confirming its effectiveness for intermittent heating load forecasting and highlighting its significance for guiding demand-responsive heating control strategies and for energy saving and emissions reduction.

Keywords: residential heating loads; intermittent heating; ConvLSTM; DyT; CBAM(T)

1. Introduction

Accurate forecasting of residential heating load is essential for efficient building energy management. During residential heating, the phenomenon of overheating is widespread and leads to significant energy waste [1]. Accurate heating load forecasting can guide the actual operation of heating systems, improve building energy efficiency, and effectively avoid the energy waste caused by overheating. In practice, residential heating load demand is influenced by multidimensional dynamic factors [2] (e.g., meteorological conditions, floor level, room temperature, and orientation), and heating demand varies substantially across dwellings. These combined influences render residential heating loads markedly nonlinear, nonstationary, and strongly disturbed, which increases the difficulty of load forecasting.

According to the current state of research, heating load forecasting methods can be categorized into physics-based, data-driven, and physics–data hybrid approaches [3,4]. Traditional physics-based models face challenges in quantifying dynamic variations associated with building thermal inertia and multidimensional dynamic factors, especially under intermittent heating scenarios where the load exhibits pronounced abrupt changes. In recent years, data driven methods have gradually become the mainstream direction for heating load forecasting [5,6]. By mining intrinsic patterns in historical operational data and capturing the effects of multidimensional features, these methods have demonstrated accurate predictive performance in heating load forecasting.

Heating load forecasting, characterized by a significant lag effect, is commonly treated as a time-series prediction task [7]. Data-driven approaches have led to the development of various machine learning and deep learning techniques tailored to such problems. Researchers have improved forecasting performance using both standalone models and hybrid architectures that incorporate attention mechanisms. Machine learning methods such as random forest (RF) [8], support vector regression (SVR) [9], and extreme gradient boosting trees (XGBoost) [10] are particularly effective in modeling nonlinear patterns. Wei et al. proposed an SVR-based supervised learning model optimized using six meta-heuristic algorithms, with the SVR-AEO variant achieving the highest accuracy in predicting residential heating and cooling loads [11]. Ritwik et al. developed a stacked ensemble combining XGBoost [12], decision trees (DT) [13], and RF, which outperformed SVR, k-nearest neighbor (KNN), and artificial neural networks (ANNs) [14]. Lu et al. introduced an AutoML-driven framework that automatically constructs optimal forecasting pipelines [15], demonstrating superior performance over ordinary least squares (OLS) [16], ICA-optimized ANN (ICA-ANN), and SSA-optimized multilayer perceptron (SSA-MLP) [17] on residential datasets. While machine learning models are effective for complex data, they often struggle to capture the spatiotemporal dependencies of heating loads, such as delays in thermal response and meteorological influences. Addressing these limitations is crucial for improving forecast accuracy.

Deep learning models outperform traditional machine learning methods in time-series forecasting by more effectively capturing long-term feature dependencies [18,19]. Architectures such as convolutional neural networks (CNNs) [20], gated recurrent units (GRUs) [21], long short-term memory networks (LSTM) [22], and transformers [23] demonstrate strong capabilities in handling time-series data and mitigating gradient-vanishing problems. Huang et al. introduced a regional heat load forecasting approach using a graph neural network (Ac-GRN), which integrates active deep learning with graph structures to model complex temporal patterns. Evaluated on a Danish district heating dataset for multistep prediction, the method outperformed 11 advanced models, including LSTM, in terms of accuracy, robustness, reliability, and computational efficiency [24]. Zhou et al. developed an air-conditioning heat-load forecasting model based on LSTM optimized via an improved sparrow search algorithm (ISSA-LSTM). By adaptively tuning hyperparameters, the model outperformed RNN, LSTM, and GRU on a custom dataset, significantly enhancing cooling-load prediction accuracy [25]. Zhu et al. proposed an adaptively regulated Kolmogorov–Arnold network (KAN), which adjusts its structure based on input data distribution to better capture local features and nonlinear patterns, maintaining high accuracy even with limited data [26].

Single models often struggle to capture the complex spatiotemporal dynamics of heating load data due to their limited and directional feature extraction capabilities. To address this issue, hybrid models have been developed, leveraging the complementary strengths of multiple architectures to enhance prediction accuracy. Wan et al. introduced the CNN-LSTM-A model to mitigate information loss from long input sequences. The model employs CNN for high-dimensional features and LSTM for capturing temporal dependencies. Experimental results showed that it outperformed conventional LSTM across all metrics [27]. Yan et al. proposed the feature-time transformer encoder-bi-LSTM (FTTrans-E-BL) model, an enhanced transformer-based architecture for short-term load forecasting, to address complex interdependencies in short-term multienergy load forecasting. Validated on real-world data, the model significantly outperformed conventional LSTM in predicting power, heating, and cooling loads [28]. Li et al. developed a convolutional gated recurrent unit (HR-CGRU) model for joint prediction of branch and aggregated district heating loads. It integrates temporal convolutional for feature extraction with a bidirectional GRU enhanced by multihead attention. By leveraging cross-hierarchical temporal dependencies, the model demonstrated superior performance on the combined heat and power load dataset [29].

In recent years, hybrid models have demonstrated good predictive accuracy in heating load forecasting; however, most studies have focused on continuous district heating systems, and research on residential heating load forecasting under intermittent heating scenarios remains relatively scarce, with notable limitations. The primary difference between intermittent and continuous heating is that, in the former, the heating duration, the length of the intermittent interval, and the start–stop times are not fixed, which makes load variations highly abrupt. Under intermittent heating, the switching of operating conditions at start and stop moments can induce short-term spikes in the load, causing severe fluctuations and greatly increasing the difficulty of prediction; even during low-load periods, basic network models may produce physically implausible negative predictions, exposing the shortcomings of existing methods in handling sudden changes in operating conditions. Moreover, although batch normalization, which is widely used in convolutional neural networks, can effectively alleviate gradient explosion and vanishing problems, its tendency to over-smooth time series can also affect the model's responsiveness to abrupt fluctuations, rendering it insufficiently sensitive to critical variations.

To address the above issues, this study is grounded in practical heating demand and, focusing on the multidimensional dynamic factors of the residential heating process, the operational characteristics of intermittent heating, and the limitations of conventional batch normalization, proposes a hybrid model based on ConvLSTM-DyT-CBAM(T) to effectively tackle the challenges of intermittent heating load forecasting and to enhance forecasting accuracy and stability. The main contributions of this study are as follows:

(1) Fully accounting for the diverse factors affecting heating load forecasting, this study constructs a multisource time-series dataset integrating indoor environmental variables, external weather conditions, and district heating network operating data.

(2) The Dynamic Tanh (DyT) activation function is adopted in place of the conventional batch normalization layer. DyT dynamically adjusts feature scaling according to the data distribution and, via a scaling factor α, precisely controls the activation range of the input, avoiding the over-smoothing and compression of extreme values by batch normalization and increasing the model's responsiveness to abrupt load fluctuations.

(3) The Channel–Spatial Attention Module (CBAM) is improved by introducing a Channel–Temporal Attention Module, CBAM(T), suited to time-series forecasting tasks. By focusing on heating start–stop conditions along the temporal dimension, the module enhances responsiveness to load spikes and convergence, effectively improving the predictive accuracy of intermittent heating loads.

2. Methodology

2.1. Data Preprocessing

The data preprocessing process involves three main steps: cleaning, transformation, and normalization. During cleaning, outliers are corrected, missing values are imputed, and noise is removed. Outlier detection is performed using the interquartile range (IQR) method [30], which identifies values beyond the calculated upper and lower bounds. The method is mathematically defined as follows:

$$l_{\text{upper}} = Q_3 + C \cdot IQR \quad (1)$$

$$l_{\text{lower}} = Q_1 - C \cdot IQR \quad (2)$$

$$IQR = Q_3 - Q_1 \quad (3)$$

where l_{upper} and l_{lower} denote the upper and lower bounds of the target variable, respectively; Q_1 and Q_3 represent the first and third quartiles of the target variable data, respectively; C signifies a scaling factor, commonly set to 1.5 for standard outliers or 3 for extreme cases [31]; and IQR denotes the interquartile range. This study employs $C = 3$ to detect extreme outliers.

Given the temporal nature of heating load data, identified outliers are not discarded. Instead, they are corrected using forward–backward mean imputation, consistent with the method applied for missing values. The corresponding mathematical expression is as follows:

$$\hat{x}_t = \frac{x_{t-1} + x_{t+1}}{2} \quad (4)$$

where $\hat{x}_t$ denotes the missing value at time t, and x_{t-1} and x_{t+1} are the known values at times $t-1$ and $t+1$, respectively. This interpolation method is applied only for short gaps; for large missing intervals, imputation is based on data from similar conditions in the surrounding week.

Following data cleaning, feature construction is performed by generating lagged variables, interaction, and date-based features, which substantially enhance model performance [32]. To reduce computational complexity and address dimensional inconsistencies across data types, Z-score normalization is applied. The corresponding formula is as follows:

$$\tilde{X} = \frac{X - \mu}{\sigma} \quad (5)$$

where $\tilde{X}$ represents the normalized data value; X denotes the original data value; μ refers to the mean of the data; and σ signifies the standard deviation.

2.2. eXtreme Gradient Boosting

XGBoost is an ensemble machine learning algorithm based on gradient-boosted decision trees (GBDT), which incrementally fits weak learners to residual errors and aggregates their output to approximate true values [33–35]. It incorporates second-order derivatives and regularization to prevent overfitting, enhance convergence, and improve generalization. The model optimizes the objective function in Equation (6) by iteratively refining residuals to improve predictive accuracy.

$$\mathcal{L}^{(t)} = \sum_{i=1}^{n} l\left(y_i, \hat{y}_i^{(t-1)} + f_t(x_i)\right) + \sum_{k=1}^{t} \Omega(f_k) \quad (6)$$

$$\Omega(f_k) = \gamma T + \frac{1}{2}\lambda \sum_{j=1}^{T} w_j^2 \quad (7)$$

where i indexes the ith sample, n denotes the total number of samples, and $l(y_i, \hat{y}_i)$ signifies the loss between the true value y_i and its prediction $\hat{y}_i$. $f_t(x_i)$ represents the weak learner added at the tth iteration, x_i refers to the feature vector of the ith sample, $\Omega(f_k)$ denotes the regularization term, T corresponds to the number of tree leaves, w_j represents the weight of the jth leaf, and γ and λ regulate model complexity and weight penalties to prevent overfitting.

During feature selection, this study employs the weight metric of XGBoost to assess feature importance. Features selected more frequently as split nodes are considered more influential, allowing efficient identification of relevant variables. The corresponding mathematical formulation is as follows:

$$\hat{I}_{\text{weight}}(j) = \frac{I_{\text{weight}}(j)}{\sum_k I_{\text{weight}}(k)} \times 100\% \tag{8}$$

where $\hat{I}_{\text{weight}}(i)$ represents the normalized importance of feature i based on its selection frequency across all split nodes, $I_{\text{weight}}(i)$ signifies the times feature i is used for splitting, and $\sum_k I_{\text{weight}}(k)$ denotes the total number of splits across all features.

2.3. *Convolutional Long Short-Term Memory (ConvLSTM)*

ConvLSTM combines the strengths of convolutional neural networks (CNNs) and long short-term memory (LSTM) networks [36,37]. While traditional LSTM effectively captures temporal dependencies, it lacks spatial feature extraction capability. ConvLSTM addresses this by replacing fully connected layers with convolutional kernels, enabling joint modeling of spatial and temporal patterns. To accurately represent spatiotemporal dependencies in residential heating load forecasting, this study adopts the ConvLSTM2D architecture for feature extraction.

The key distinction between ConvLSTM and standard LSTM is the replacement of fully connected operations with convolutional operations for updating the memory and hidden states. ConvLSTM consists of four convolution-based components: the input gate (i_t), forget gate (f_t), cell state (C_t), and output gate (O_t). The model is defined mathematically as follows:

$$i_t = \sigma(W_{xi} * X_t + W_{hi} * H_{t-1} + W_{ci} \circ C_{t-1} + b_i) \tag{9}$$

where the input gate i_t regulates the information retained in the cell state. It is derived by applying a sigmoid activation to the current input (X_t), the previous hidden state (H_{t-1}), and the prior cell state (C_{t-1}). The forget gate f_t determines the extent to which the previous cell state is retained. It is computed using a sigmoid function applied to the current input, the prior hidden state, and the previous cell state, as follows:

$$f_t = \sigma\left(W_{xf} * X_t + W_{hf} * H_{t-1} + W_{cf} \circ C_{t-1} + b_f\right) \tag{10}$$

The candidate cell state $\tilde{C}_t$ serves as temporary memory, generated by applying a Tanh function to the current input and previous hidden state. The updated cell state C_t is then computed by combining retained information from the previous state using the forget gate and incorporating the new candidate values, completing the memory update process as follows:

$$\tilde{C}_t = \tanh(W_{xc} * X_t + W_{hc} * H_{t-1} + b_c) \tag{11}$$

$$C_t = f_t \circ C_{t-1} + i_t \circ \tilde{C}_t \tag{12}$$

The output gate O_t regulates the information released from the cell by applying a sigmoid function to the current input, previous hidden state, and updated cell state, determining the final output.

$$O_t = \sigma(W_{xo} * X_t + W_{ho} * H_{t-1} + W_{co} \circ C_t + b_o) \tag{13}$$

The hidden state H_t is obtained by element-wise multiplying the output gate O_t with the cell state passed through a Tanh function:

$$h_t = O_t \circ \tanh(C_t) \tag{14}$$

In Equations (9)–(14), $*$ indicates convolution, $\circ$ denotes the Hadamard product, σ represents the sigmoid activation function, tanh signifies the hyperbolic tangent function, W and b denote the convolution weights and biases, respectively, and X_t, H_t, and C_t refer to the input, hidden state, and cell state at time t, respectively. In this study, the data sampling interval is 1 h, and the prediction target is the heating load at the next time step. The input features and window length are determined by a specific feature-selection method and hyperparameter optimization. Each sample window uses measured historical data as the initial condition. The model is trained iteratively according to the aforementioned ConvLSTM equations, and its architecture is shown in Figure 1.

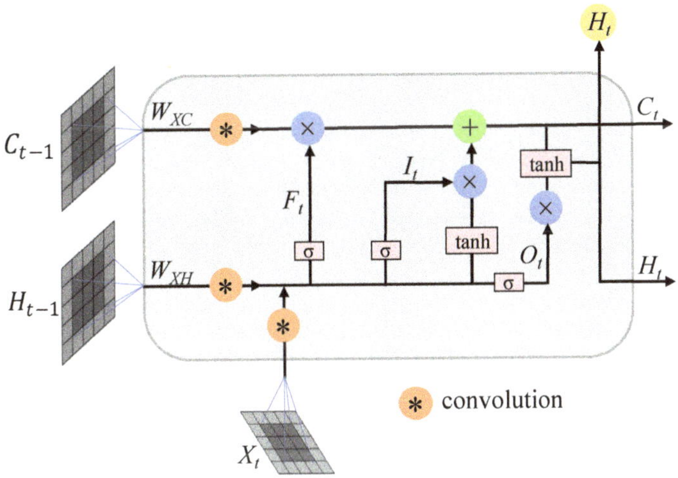

Figure 1. ConvLSTM architecture.

2.4. Dynamic Tanh (DyT)

The hyperbolic tangent (Tanh) function [38] is a widely used nonlinear activation function in neural networks. Its mathematical formulation is given as follows:

$$Tanh(x) = \frac{e^x - e^{-x}}{e^x + e^{-x}} \tag{15}$$

where x is the input value and e denotes the base of the natural logarithm. The Tanh activation function normalizes input values, reducing dimensional inconsistencies from multiscale data. However, under extreme input conditions, it may lead to gradient vanishing during training, thereby hindering convergence and slowing model optimization.

Dynamic Tanh (DyT), introduced by Zhu et al., serves as an alternative to traditional normalization layers by incorporating learnable parameters that modulate the output range of the Tanh function [39]. In this study, DyT is applied to residential heating load forecasting, where it dynamically adjusts to varying data distributions during training. This adaptation mitigates gradient vanishing and improves both the model's representational ability and training efficiency. The DyT activation function is defined as follows:

$$DyT(x) = \gamma * \tanh(\alpha x) + \beta \tag{16}$$

where α denotes a scalar parameter that can dynamically scale x at different magnitudes, and γ and β represent learnable channel-wise vector parameters.

2.5. Channel–Temporal Attention (CBAM(T))

The conventional convolution block attention module (CBAM) comprises channel and spatial attention submodules, which refine spatial feature representations to enhance focus on salient information [40]. However, the spatial component is suboptimal for data-driven regression. To better suit time-series forecasting, this study replaces spatial attention with temporal attention while retaining channel attention [41–43]. The architecture is illustrated in Figure 2, and the channel–attention mechanism is defined as follows:

$$\begin{aligned} M_c(F) &= \sigma\Big(MLP(AvgPool(F)) + MLP(MaxPool(F))\Big) \\ &= \sigma\Big(W_1(W_0F^c_{avg}) + W_1(W_0F^c_{max})\Big) \end{aligned} \tag{17}$$

where $M_c(F)$ represents the channel-weight vector; $F \in \mathbb{R}^{B\times T\times C}$ denotes the input feature tensor with batch size B, time steps T, and channels C; σ signifies the sigmoid activation function; $W_0 \in \mathbb{R}^{C\times C/r}$ and $W_1 \in \mathbb{R}^{C/r\times C}$ refer to the shared weights of a two-layer MLP; r corresponds to the channel reduction ratio; and F^c_{avg} and F^c_{max} are the temporal average max-pooled vectors derived from F, respectively.

The temporal attention module is a modified version of the spatial attention mechanism in the conventional CBAM, redesigned to capture temporal dependencies. Its formulation is as follows:

$$\begin{aligned} M_t(F) &= \sigma\Big(f^{1\times k}([AvgPool(F), MaxPool(F)])\Big) \\ &= \sigma\Big(f^{1\times k}([F^t_{avg}, F^t_{max}])\Big) \end{aligned} \tag{18}$$

where $M_t(F)$ represents the temporal weight vector; $f^{1\times k}$ denotes a one-dimensional convolution with kernel length k; F^t_{avg} and F^t_{max} signify the time series derived by applying average and max pooling, respectively, to the input feature F along the channel dimension. The revised temporal attention module applies a convolution across the temporal axis to aggregate critical information from neighboring time steps.

Feature recalibration integrates channel and temporal attention weights through element-wise multiplication with the input, enhancing salient spatiotemporal features while suppressing irrelevant ones. The corresponding formulation is as follows:

$$\tilde{F} = M_c(F) \otimes F \tag{19}$$

$$F^* = M_t(F) \otimes \tilde{F} \tag{20}$$

where $\tilde{F}$ signifies the channel-recalibrated feature, F^* represents the final output after applying both channel and temporal recalibration, and $\otimes$ denotes element-wise multiplication.

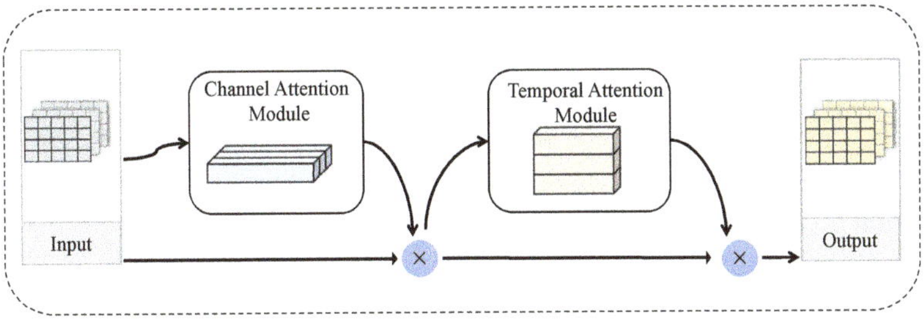

Figure 2. CBAM(T) architecture.

2.6. Optuna

Optuna is an advanced hyperparameter optimization framework known for its efficiency, flexibility, and ease of use [44]. It achieves rapid optimization through effective sampling and pruning strategies, dynamically defining the search space to identify optimal hyperparameter configurations. Compared with traditional methods such as grid and random search, Optuna significantly reduces both tuning time and the number of required trials.

During hyperparameter tuning, Optuna iteratively evaluates parameter combinations within a user-defined objective function and search space, aiming to optimize the objective value. It leverages the tree-structured Parzen estimator (TPE) [45] as its Bayesian optimization core, which models the probability distributions of high- and low-performing configurations to maximize expected improvement. This strategy enables efficient resource allocation toward the most promising hyperparameter sets and parameter combinations. The corresponding mathematical formulation is as follows:

$$EI_{y^*}(x) = \frac{l(x)}{g(x)} \tag{21}$$

where x represents a hyperparameter configuration; y denotes the evaluation metric (MSE) used in this study; y^* signifies the performance threshold distinguishing high- and low-quality results; and $l(x)$ and $g(x)$ denote the superior and inferior hyperparameter sets, respectively.

During optimization, Optuna utilizes the asynchronous successive halving algorithm (ASHA) to evaluate trial performance. Trials with metrics significantly below the historical median are terminated early to reduce computational cost. The detailed workflow is illustrated in Figure 3.

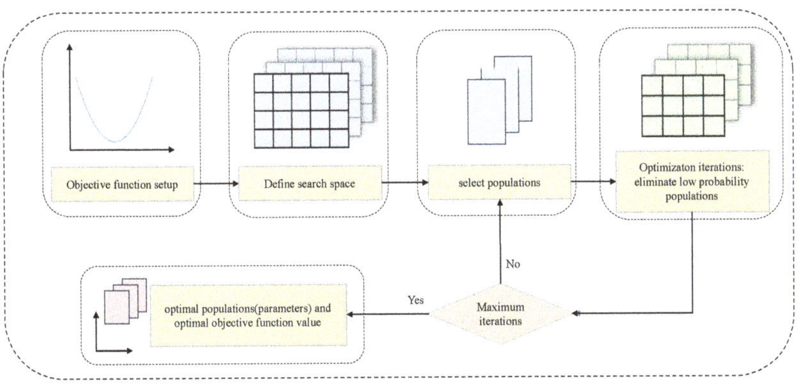

Figure 3. Optuna flowchart.

2.7. *ConvLSTM-DyT-CBAM(T)*

This study presents a ConvLSTM-DyT-CBAM(T)-based approach for residential heating load forecasting using multiscale time-series data. ConvLSTM serves as the core architecture, integrating the spatial feature extraction capabilities of convolutional neural networks (CNNs) with the temporal sequence modeling of long short-term memory (LSTM) networks to capture spatiotemporal dependencies. The DyT module replaces traditional normalization layers, enabling dynamic feature scaling across varying data distributions and enhancing the model's capacity to represent nonlinear patterns. In addition, the CBAM-based temporal attention mechanism, CBAM(T) module, introduces temporal attention to improve responsiveness to heating dynamics and load spikes. The model architecture is illustrated in Figure 4.

The multivariate heating-load time series is first preprocessed and transformed into multiscale tensors through down-sampling and feature engineering. A two-layer ConvLSTM network is then employed to jointly extract spatiotemporal features, with the DyT module replacing standard normalization to adaptively scale features across different time scales. A TimeDistributed layer flattens the feature maps while retaining temporal structure, after which the CBAM(T) module applies temporal attention to emphasize relevant information. Finally, globally pooled features are passed through another DyT layer for dynamic scaling and then by fully connected layers to produce a high-accuracy residential heating-load forecast.

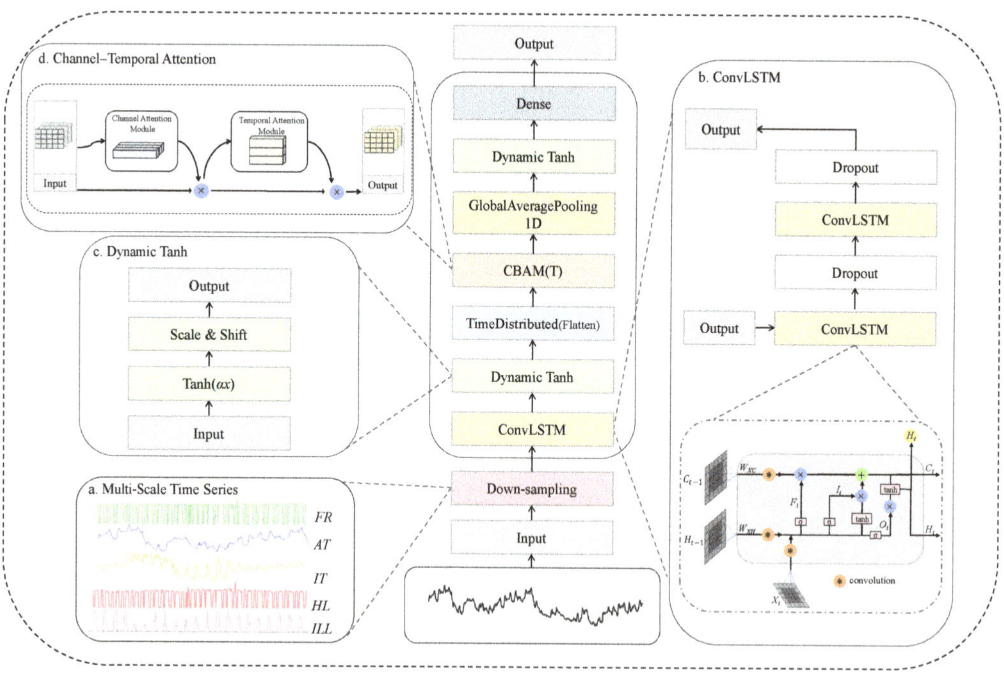

Figure 4. ConvLSTM-DyT-CBAM(T) architecture.

2.8. *Evaluation Metrics*

To assess forecasting accuracy, this study employs three evaluation metrics: mean absolute error (MAE), mean squared error (MSE), and the coefficient of determination (R^2), formulated as follows:

$$MAE = \frac{1}{n}\sum_{t=1}^{n}|y_t - \hat{y}_t| \tag{22}$$

$$MSE = \frac{1}{n}\sum_{t=1}^{n}(y_t - \hat{y}_t)^2 \tag{23}$$

$$R^2 = 1 - \frac{\sum_{t=1}^{n}(y_t - \hat{y}_t)^2}{\sum_{t=1}^{n}(y_t - \bar{y})^2} \tag{24}$$

where n denotes the number of predictions, y_t denotes the actual heating load, $\hat{y}_t$ represents the predicted heating load value, and $\bar{y}$ signifies the mean of the actual heating load.

MAE measures the average magnitude of absolute error; lower values indicate higher accuracy. MSE calculates the mean of squared deviations, penalizing larger errors more heavily; a smaller MSE reflects greater predictive precision. R^2 quantifies the proportion of variance in the target variable explained by the model; higher values signify a better fit.

3. Research Design

All experiments were implemented in Python 3.8 and executed on a 64-bit Windows 11 system with an Intel(R) Core (TM) i5-9300H CPU (base frequency 2.40 GHz), NVIDIA GeForce GTX 1650 GPU, and 32 GB of RAM, sourced from Changzhou, China.

3.1. Dataset and Preprocessing

The dataset used in this study was collected from a residential community in Dalian, Liaoning Province. Heat-meter collectors, rooftop weather stations, and indoor thermo-hygrometers recorded multiple feature parameters simultaneously. The data acquisition period covered the entire 2023 heating season, from 10:00 on 18 November 2023 to 23:00 on 24 March 2024, with hourly intervals, totaling 2822 samples. Of these, 70% were used for training, 10% for validation, and 20% for testing. Data acquisition details are summarized in Table 1.

Table 1. Multisource heterogeneous data collection information.

Data Category	Feature	Unit	Symbol
Heating Network Parameters	Heating load	kW	HL
	Instantaneous flow rate	m^3/h	IF
	Supply water temperature	°C	SWT
	Return water temperature	°C	RWT
Outdoor Meteorological Data	Air temperature	°C	AT
	Air relative humidity	%	AH
	Wind force (Beaufort scale)	Bft	WP
	Wind speed	m/s	WS
	Illuminance	Lux	ILL
Indoor Environmental Parameters	Indoor air temperature	°C	IT
	Indoor relative humidity	%	IH

Data collection was occasionally affected by transmission errors and losses due to power outages and signal disruptions. To address this, outliers were corrected and missing values were imputed by the time-series nature of the data (see Section 2.1). Subsequently, uniform normalization was applied to ensure consistency across feature scales.

To assess the model's performance over long time spans and under varying environmental conditions, this study employs cross–heating-season data as an independent validation set. Considering that seasonal changes and operational strategies may induce distribution shift, we diagnose distributional consistency for key variables using the overlap of the 5–95% coverage intervals and the Kolmogorov–Smirnov (KS) test. As shown in Table 2 and Figure 5, the coverage of all variables exceeds 98%, indicating that the validation samples do not fall outside the training empirical domain. However, the KS p-values are very small ($p < 0.001$), suggesting significant differences in the data distributions. This result indicates that even when variable ranges are broadly consistent, seasonal alternation and changes in operating strategies can still lead to sample-weight reallocation. On this basis, using cross-season independent validation enables a more objective assessment of the model's robustness and stability under distributional change.

Table 2. KS test results between training and independent validation sets.

Variable	Coverage	KS *p*-Value
HL	100%	7.82×10^{-11}
IF	100%	7.18×10^{-102}
SWT	100%	3.29×10^{-5}
IT	98.68%	4.17×10^{-13}
AT	100%	2.96×10^{-27}
ILL	100%	3.97×10^{-4}

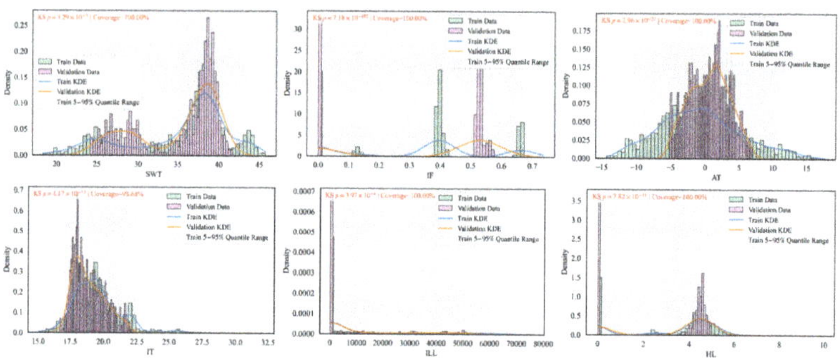

Figure 5. Distribution comparison between training and cross-season independent validation sets.

3.2. Feature Selection

During the feature-selection stage, a multistage screening procedure was employed to optimize the model inputs. First, feature importance was quantified using the XGBoost algorithm, with hyperparameters listed in Table 3 and results shown in Figure 6a. Heating-network operating parameters—supply water temperature (SWT), return water temperature (RWT), and instantaneous flow (IF)—exert a dominant influence on forecasting the heating load (HL), which is broadly consistent with thermodynamic principles. However, in practical heating operations, HL is computed from the energy-balance equation combining SWT, RWT, and IF; if all network parameters are included as inputs, the model tends to prioritize supply-side parameters while overlooking dynamic factors such as building thermal inertia, meteorological disturbances, and inter-apartment heat transfer. This leads to degeneration into a basic heat-balance calculator, forfeiting the ability to learn nonlinear dynamics and deviating from the intent of a data-driven model. Accordingly, a critical-parameter truncation strategy is adopted, retaining IF and SWT as representatives of network parameters to reserve more model capacity for learning other potentially relevant factors.

To validate the nonlinear relationships among variables, Spearman correlation analysis was conducted, as shown in Figure 6b. Results indicate a strong positive correlation between IF and HL, a significant negative correlation between air temperature (AT) and HL, and notable associations of indoor temperature (IT) and solar irradiance (ILL) with HL. To account for thermal inertia, the previous heating load HL_{t-1} was introduced as a lagged feature, ranking highly in both XGBoost importance and Spearman rank correlation coefficients. Based on multisource heterogeneous data integration, six key features were selected: IF, SWT, HL_{t-1}, AT, IT, and ILL, to capture the primary influencing dimensions.

Table 3. XGBoost parameter values.

#	Hyperparameter	Values
1	n_estimators	500
2	Learning Rate	0.005
3	Max_depth	5
4	subsample	0.8
5	Min_child_weight	1
6	Reg_λ	0

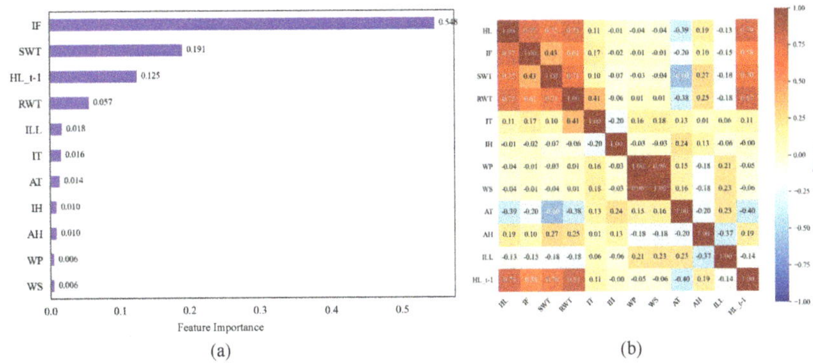

Figure 6. (**a**) XGBoost-based feature-importance ranking of the input variables; (**b**) Spearman rank-correlation heatmap among all variables.

3.3. Hyperparameter Optimization Based on Optuna

To ensure equitable hyperparameter tuning, all models in this study were optimized using the Optuna framework, with a unified hyperparameter search space and step sizes. Detailed hyperparameters are provided in Table 4.

Table 4. Hyperparameter-optimization search space.

#	Hyperparameter	Search Space	Step
1	Units1 or Filters1	[16, 512]	16
2	Units2 or Filters2	[16, 256]	16
3	Num_heads	[2, 8]	2
4	Key_dim	[16, 128]	16
3	Dropout1	[0, 0.5]	0.05
4	Dropout2	[0, 0.5]	0.05
5	Learning rate	$[1 \times 10^{-5}, 1 \times 10^{-2}]$	1×10^{-5}
6	Batch size	{16, 32, 64}	–
7	Time steps	[1, 6]	1
8	CBAM(T) ratio	[4, 16]	4
9	Temporal kernel	[3, 12]	3
10	DyT_alpha1	[0.1, 1.5]	0.1
11	DyT_alpha2	[0.1, 1.5]	0.1

Using the defined hyperparameter search space and time steps, the optimized hyperparameters obtained via Optuna are summarized in Table 5. Figure 7 presents a comparative analysis of the optimization results. Specifically, Figure 7a shows the MSE convergence trend across models, indicating that all achieved similar optimal values. Figure 7b visualizes the relationship between key model parameters (e.g., number of hidden units, filter size) and MSE using a contour plot, where contour density directly reflects the sensitivity of MSE to parameter changes. Darker colors indicate lower error; sparser contours and more contiguous low-error regions imply reduced sensitivity to hyperparameter perturbations. Figure 7c compares the optimal hyperparameter combinations across models using a histogram, highlighting differences in configuration strategies.

Table 5. Hyperparameter-optimization results across models.

Hyper Parameters	LSTM	CNN -Transformer	KAN	ConvLSTM	ConvLSTM -CBAM(T)	ConvLSTM -DyT	ConvLSTM -DyT -CBAM(T)
Units1/Filters1	448	256	208	288	224	224	192
Units2/Filters2	48	112	208	208	16	128	48
Num_heads	-	4	-	-	-	-	-
Key_dim	-	80	-	-	-	-	-
Dropout1	0.20	0.35	0.05	0.50	0.25	0.20	0.45
Dropout2	0.10	-	0.30	0.05	0.10	0.05	0.15
Learning_rate	0.001	0.002	0.001	0.003	0.007	0.002	0.002
Batch_size	16	16	16	64	32	32	16
Time_step	2	2	3	2	5	2	2
DyT_alpha1	-	-	-	-	-	1.9	0.5
DyT_alpha2	-	-	-	-	-	0.9	0.9
CBAM(T)_ratio	-	-	16	-	-	8	-
Temporal_kernel	-	-	6	-	6	-	6

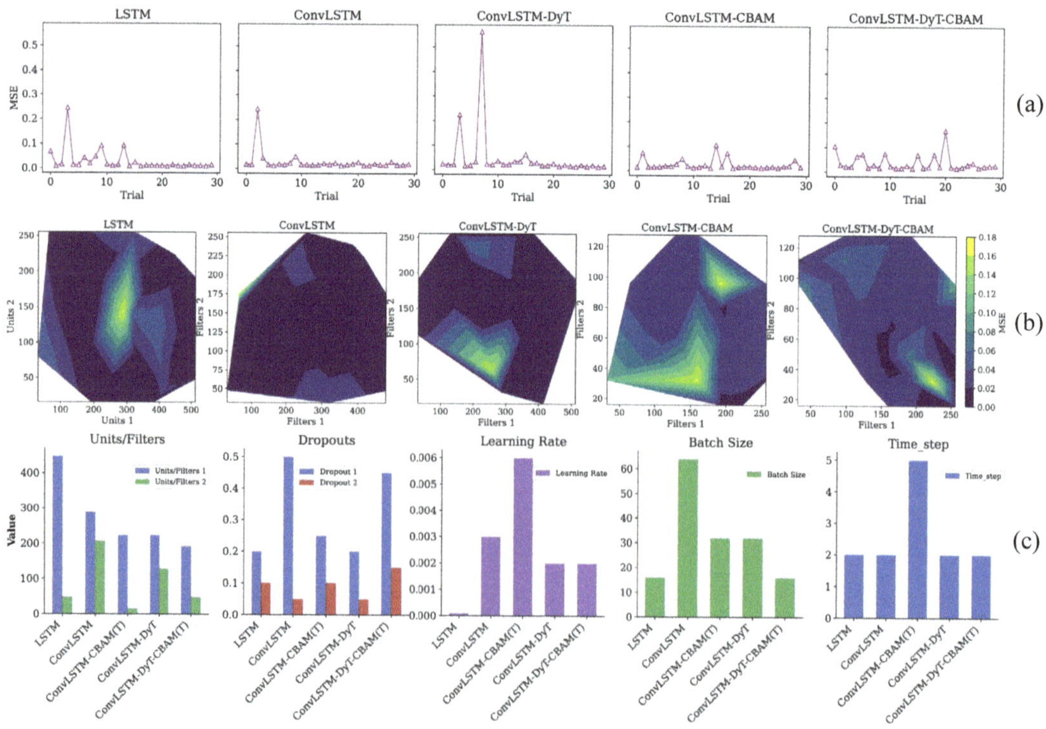

Figure 7. (**a**) MSE curves during hyperparameter optimization; (**b**) Hyperparameter comparison contour plots; (**c**) Optimal hyperparameters for each model.

4. Prediction Results and Analysis

4.1. Prediction Results of Residential Heating Load

To verify the effectiveness of the ConvLSTM-DyT-CBAM(T) model, ablation and comparative experiments were conducted, and two advanced models in this field—CNN-Transformer and the adaptive Kolmogorov–Arnold network (KAN)—were selected for comparison. To ensure experimental fairness, all models adopted the same hyperparameter optimization procedure, with the optimal configurations determined via Optuna. The specific hyperparameter settings for each model are provided in Table 5.

As shown in Table 6, the evaluation metrics for the different models include MSE, MAE, and R^2, while Figure 8 provides a visual comparison of these metrics across models. The results indicate that the LSTM model attains MSE, MAE, and R^2 values of 0.090, 0.216, and 0.972, respectively, revealing certain limitations in capturing the complex dependencies of multidimensional time series. Incorporating convolutional structures, ConvLSTM offers advantages in modeling spatiotemporal dependencies; its MSE, MAE, and R^2 are 0.065, 0.179, and 0.987, outperforming LSTM. Building on ConvLSTM, integrating the improved attention mechanism CBAM(T) enhances the model's responsiveness to load fluctuations; relative to ConvLSTM, its MSE and MAE are reduced by 7.69% and 6.28%, and R^2 increases by 0.1%. In the ConvLSTM–DyT model, replacing the conventional batch normalization layer with the dynamic Tanh activation avoids under-responsiveness to sudden and spiky load variations caused by over-smoothing. Compared to ConvLSTM, MSE and MAE decrease by 41.54% and 29.32%, and R^2 increases by 0.61%. As a comparison model, CNN–Transformer yields MSE, MAE, and R^2 of 0.251, 0.403, and 0.951, respectively, representing relatively weaker performance. Although this model often performs well on medium- to long-term forecasting tasks, it is highly data-dependent, and the limited data available for the present short-term heating-load task may have degraded its predictive performance. The KAN model performs strongly, with all metrics surpassing those of the ConvLSTM–CBAM(T) and ConvLSTM–DyT hybrids, achieving MSE, MAE, and R^2 of 0.052, 0.147, and 0.990, respectively, second only to the best model. The proposed ConvLSTM–DyT–CBAM(T) hybrid, which combines the advantages of temporal–channel attention and Dynamic Tanh, performs best: its MSE, MAE, and R^2 are 0.017, 0.082, and 0.997, representing reductions of 73.85% and 57.07% in MSE and MAE and an increase of 1.01% in R^2 relative to ConvLSTM. Even compared with KAN, MSE and MAE are further reduced by 67.31% and 44.22%, and R^2 increases by 0.71%.

Table 6. Comparison of prediction results across models.

#	Model	MSE	MAE	R^2
1	LSTM	0.090	0.216	0.972
2	CNN-Transformer	0.251	0.403	0.951
3	KAN	0.052	0.147	0.990
4	ConvLSTM	0.065	0.191	0.987
5	ConvLSTM-CBAM(T)	0.060	0.179	0.988
6	ConvLSTM-DyT	0.038	0.135	0.993
7	ConvLSTM-DyT-CBAM(T)	0.017	0.082	0.997

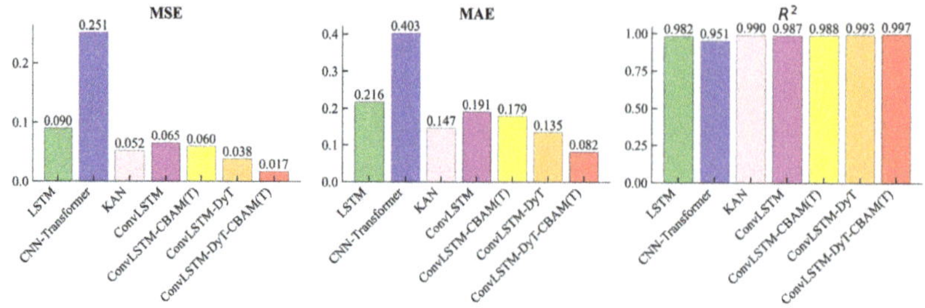

Figure 8. Comparison of evaluation metrics across models.

4.2. Analysis of Prediction Results

Figure 9 presents a comparison between each model's predicted heating loads and the ground-truth curve on the test set. The right-hand side provides locally magnified views of peak and valley segments to more clearly examine each model's grasp of heating start–stop instants and its responsiveness to spike-like load surges. Red and blue dashed boxes denote the regions magnified for peaks and valleys, respectively.

LSTM (green curve) markedly overestimates in peak regions and yields physically implausible negative values in valley regions, primarily because the model's relatively simple architecture struggles to capture abrupt fluctuations in residential heating loads under intermittent heating. With the introduction of convolutional structure, the ConvLSTM model (purple curve) achieves some improvement in predictive accuracy and is smoother overall; its peak fitting surpasses that of LSTM, yet negative predictions persist in valley regions, indicating limited ability to capture burst features in intermittent heating.

By replacing the conventional normalization layer with the Dynamic Tanh activation, the ConvLSTM-DyT model (orange curve) produces peak-region predictions that largely track the fluctuations of the true curve and essentially eliminate negative values in valleys; however, at the critical transition instants of heating on/off, prediction lag remains, and the curve does not fully and smoothly converge to zero.

Integrating the improved temporal–channel attention mechanism, the ConvLSTM-CBAM(T) model (yellow curve) is slightly inferior to ConvLSTM-DyT in peak-region fitting but superior in valley regions. Around the critical start–stop points, the valley-region predictions also basically converge smoothly to zero, indicating strong responsiveness.

As a comparison model, the CNN–Transformer (blue curve) performs well on medium- to long-term forecasting tasks; however, its global weighting characteristic tends to produce over-smoothing and phase delays for sparse start–stop transients. Consequently, it is difficult for the model to fit the fluctuation pattern in peak regions, and sizable negative-load predictions appear in valley periods. In addition, this model requires a larger amount of data and, thus, fails to fully exploit its advantages under the intermittent-heating scenario studied here. The KAN model (pink curve) can compose basis functions to fit smooth higher-order trends, but it lacks recurrent state and temporal gating to handle regime switching; although it shows good fitting performance at peaks, it still cannot overcome the problem of negative-load predictions in valley regions during heating on/off.

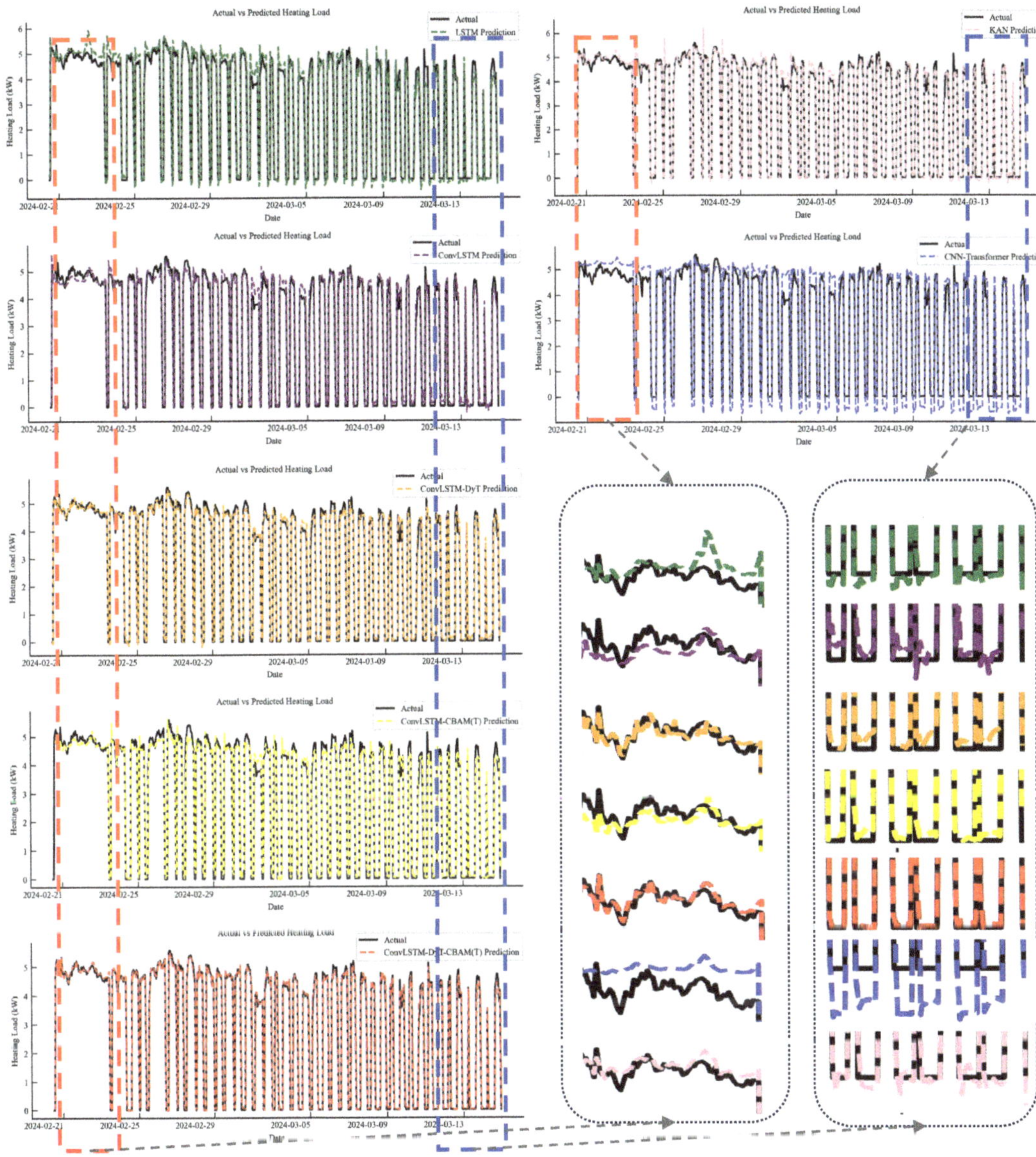

Figure 9. Comparison of heating-load prediction curves across models.

The proposed ConvLSTM-DyT-CBAM(T) model (red curve), combining the strengths of DyT and CBAM(T), performs best overall. It accurately fits the true heating-load curve in both peak and valley regions; in particular, at moments of rapid load change induced by intermittent heating, it responds quickly and precisely captures short-term fluctuation trends, clearly surpassing the other models. This is mainly attributable to the hybrid model's multimodule complementary advantages. The convolutional recurrence of ConvLSTM preserves phase and amplitude information between adjacent time steps within local spatiotemporal neighborhoods, which confers an advantage in handling the start–stop transients of intermittent heating; DyT dynamically adjusts feature scales, strengthening the model's ability to capture fine-grained features so that it can capture external information that more closely reflects reality and accurately predict the true behavior of the heating system, while avoiding the over-smoothing of features near zero load that would reduce model sensitivity and lead to negative load predictions. CBAM(T) highlights heating-load–related key variables along the channel dimension and focuses on start–stop windows and spike segments along the temporal dimension, thereby reducing the interference of slowly varying factors with the identification of abrupt changes. Working in concert, these three components effectively address the challenges of uncertain start–stop timing and inaccurate prediction of abrupt load changes in intermittent-heating scenarios, exhibiting excellent adaptability and precision under complex operating conditions.

Figure 10 presents the residual distributions of the forecasting models, where the vertical axis denotes residual density and the black line marks zero error. The ConvLSTM-DyT-CBAM(T) model shows the narrowest distribution and highest peak, reflecting minimal prediction error and superior stability compared with other models.

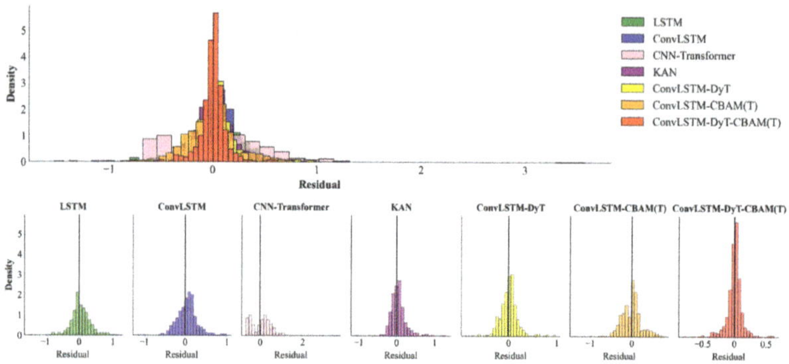

Figure 10. Comparison of residual distributions across models.

4.3. Analysis of Cross-Season Validation Results

Figures 11 and 12 present the accuracy metrics and the prediction–ground-truth comparison curves of the ConvLSTM-DyT-CBAM(T) model on the independent validation data. The MSE, MAE, and R^2 are 0.038, 0.149, and 0.991, respectively. Although the overall performance is slightly inferior to that on the test set, it remains at a high level of accuracy. In the comparative curves, the model largely preserves its test-set behavior: peak segments closely adhere to the true load curve and accurately track the amplitude, while valley segments rapidly approach and stably converge to zero. Combined with the earlier distribution-consistency diagnosis, the independent validation data exhibit covariate shift relative to the training set. Under this circumstance, the model still maintains good predictive accuracy, indicating strong robustness and generalization. These results provide solid support for subsequent engineering deployment.

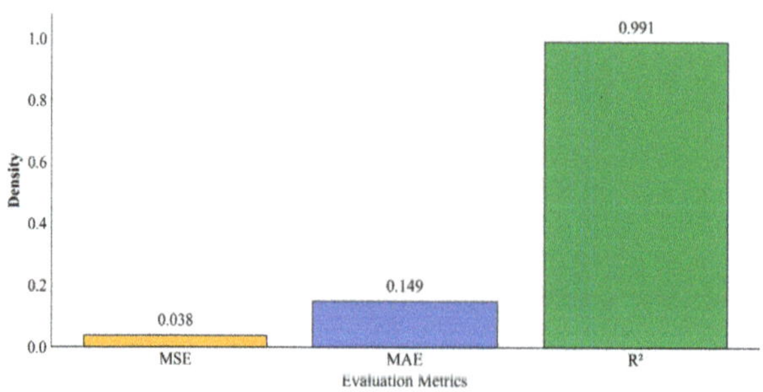

Figure 11. Evaluation metrics for cross-season validation.

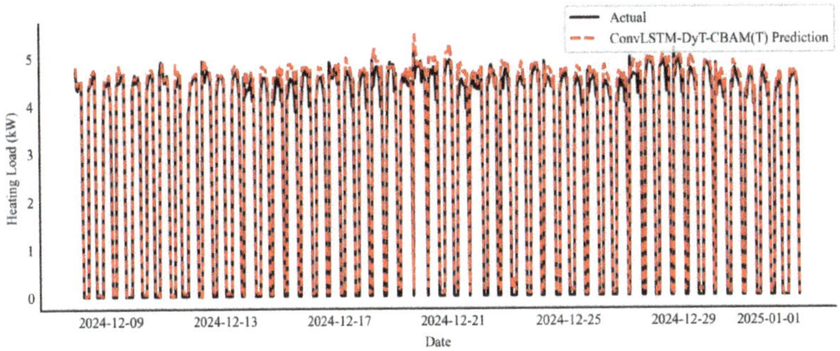

Figure 12. Comparison curves of cross-season heating-load prediction.

5. Conclusions

This study proposes a hybrid model, ConvLSTM-DyT-CBAM(T), that integrates a Dynamic Tanh activation function with a temporal–channel attention mechanism to effectively address the forecasting challenges posed by abrupt load fluctuations and irregular on/off timing in residential intermittent heating. Built on ConvLSTM, the model effectively captures the multidimensional spatiotemporal dependencies of heating-load sequences; on this basis, DyT replaces conventional batch normalization to flexibly adjust feature scaling and enhance sensitivity to load variability under intermittent heating; furthermore, the improved CBAM(T) focuses on temporal start–stop conditions, further strengthening the model's responsiveness to load spikes and convergence. The proposed approach both mines latent relationships in load data under intermittent heating and markedly improves the ability to cope with uncertain start–stop timing and abrupt load changes under such operating conditions.

Validated on datasets from two distinct heating seasons, the ConvLSTM-DyT-CBAM(T) model demonstrates high predictive accuracy and strong generalization, exhibiting stable and reliable performance in peak tracking, near-zero valley adherence, and start–stop transient response. The model provides robust technical support for heating-schedule optimization, energy-saving control, and residential thermal comfort in intermittent-heating scenarios.

The dataset used in this study originates from a single residential district. In future work, we will extend validation to buildings across diverse scenarios and explore the energy-saving benefits of regulating heating-network parameters under intermittent heating, so as to continuously enhance engineering practicality, stability, and reliability.

Author Contributions: Conceptualization, H.Z. and Z.L.; methodology, H.Z. and X.G.; software, H.Z.; validation, H.Z., X.G. and X.L.; formal analysis, X.G. and X.L.; investigation, X.G.; resources, Z.L.; data curation, H.Z. and X.L.; writing—original draft preparation, H.Z.; writing—review and editing, H.Z.; visualization, H.Z.; supervision, Z.L.; project administration, X.G.; funding acquisition, Z.L. All authors have read and agreed to the published version of the manuscript.

Funding: This research was supported by the Fundamental Research Funds for the Central Universities (No. 0444-20250037) and the Fundamental Research Funds for Liaoning Provincial Department of Education (No. LJ242412026001). We sincerely acknowledge these funding bodies for their financial support.

Data Availability Statement: The data that support the findings of this study are available on request from the corresponding author, upon reasonable request.

Conflicts of Interest: The authors declare that they have no known competing financial interests or personal relationships that could have appeared to influence the work reported in this paper.

References

1. Uerge-Vorsatz, D.; Cabeza, L.F.; Serrano, S.; Barreneche, C.; Petrichenko, K. Heating and cooling energy trends and drivers in buildings. *Renew. Sust. Energy Rev.* **2015**, *41*, 85–98. [CrossRef]
2. Dahl, M.; Brun, A.; Andresen, G.B. Using ensemble weather predictions in district heating operation and load forecasting. *Appl. Energy* **2017**, *193*, 455–465. [CrossRef]
3. Yu, H.; Zhong, F.; Du, Y.; Xie, X.; Wang, Y.; Zhang, X.; Huang, S. Short-term cooling and heating loads forecasting of building district energy system based on data-driven models. *Energy Build.* **2023**, *298*, 113513. [CrossRef]
4. Jiang, W.; Wang, P.; Ma, X.; Liu, Y. Development of a grey-box heat load prediction model by subspace identification method for heating building. *Build. Environ.* **2025**, *280*, 113119. [CrossRef]
5. Kazmi, H.; Fu, C.; Miller, C. Ten questions concerning data-driven modelling and forecasting of operational energy demand at building and urban scale. *Build. Environ.* **2023**, *239*, 110407. [CrossRef]
6. Quanwei, T.; Guijun, X.; Wenju, X. Cakformer: Transformer model for long-term heat load forecasting based on cauto-correlation and KAN. *Energy* **2025**, *324*, 135460. [CrossRef]
7. Zhao, J.; Li, J.; Shan, Y. Research on a forecasted load-and time delay-based model predictive control (MPC) district energy system model. *Energy Build.* **2021**, *231*, 110631. [CrossRef]
8. Xiao, Z.; Gang, W.; Yuan, J.; Zhang, Y.; Fan, C. Cooling load disaggregation using a NILM method based on random forest for smart buildings. *Sustain. Cities Soc.* **2021**, *74*, 103202. [CrossRef]
9. Liu, H.; Yu, J.; Dai, J.; Zhao, A.; Wang, M.; Zhou, M. Hybrid prediction model for cold load in large public buildings based on mean residual feedback and improved SVR. *Energy Build.* **2023**, *294*, 113229. [CrossRef]
10. Wei, Z.; Zhang, T.; Yue, B.; Ding, Y.; Xiao, R.; Wang, R.; Zhai, X. Prediction of residential district heating load based on machine learning: A case study. *Energy* **2021**, *231*, 120950. [CrossRef]
11. Cai, W.; Wen, X.; Li, C.; Shao, J.; Xu, J. Predicting the energy consumption in buildings using the optimized support vector regression model. *Energy* **2023**, *273*, 127188. [CrossRef]
12. Mohan, R.; Pachauri, N. An ensemble model for the energy consumption prediction of residential buildings. *Energy* **2025**, *314*, 134255. [CrossRef]
13. Yan, R.; Ma, Z.; Zhao, Y.; Kokogiannakis, G. A decision tree based data-driven diagnostic strategy for air handling units. *Energy Build.* **2016**, *133*, 37–45. [CrossRef]
14. Le, L.T.; Nguyen, H.; Dou, J.; Zhou, J. A comparative study of PSO-ANN, GA-ANN, ICA-ANN, and ABC-ANN in estimating the heating load of buildings' energy efficiency for smart city planning. *Appl. Sci.* **2019**, *9*, 2630. [CrossRef]
15. Lu, C.; Li, S.; Penaka, S.R.; Olofsson, T. Automated machine learning-based framework of heating and cooling load prediction for quick residential building design. *Energy* **2023**, *274*, 127334. [CrossRef]
16. Kavaklioglu, K. Robust modeling of heating and cooling loads using partial least squares towards efficient residential building design. *J. Build. Eng.* **2018**, *18*, 467–475. [CrossRef]
17. Guo, Z.; Moayedi, H.; Foong, L.K.; Bahiraei, M. Optimal modification of heating, ventilation, and air conditioning system performances in residential buildings using the integration of metaheuristic optimization and neural computing. *Energy Build.* **2020**, *214*, 109866. [CrossRef]
18. Al-Selwi, S.M.; Hassan, M.F.; Abdulkadir, S.J.; Muneer, A.; Sumiea, E.H.; Alqushaibi, A.; Ragab, M.G. RNN-LSTM: From applications to modeling techniques and beyond—systematic review. *J. King Saud Univ.-Comput. Inf. Sci.* **2024**, *36*, 102068. [CrossRef]
19. Wang, Z.; Liu, X.; Huang, Y.; Zhang, P.; Fu, Y. A multivariate time series graph neural network for district heat load forecasting. *Energy* **2023**, *278*, 127911. [CrossRef]
20. Chen, B.; Yang, W.; Yan, B.; Zhang, K. An advanced airport terminal cooling load forecasting model integrating SSA and CNN-transformer. *Energy Build.* **2024**, *309*, 114000. [CrossRef]
21. Zhao, Z.; Yun, S.; Jia, L.; Guo, J.; Meng, Y.; He, N.; Li, X.; Shi, J.; Yang, L. Hybrid VMD-CNN-GRU-based model for short-term forecasting of wind power considering spatio-temporal features. *Eng. Appl. Artif. Intell.* **2023**, *121*, 105982. [CrossRef]
22. Xue, G.; Qi, C.; Li, H.; Kong, X.; Song, J. Heating load prediction based on attention long short term memory: A case study of Xingtai. *Energy* **2020**, *203*, 117846. [CrossRef]
23. Ni, Z.; Zhang, C.; Karlsson, M.; Gong, S. A study of deep learning-based multi-horizon building energy forecasting. *Energy Build.* **2024**, *303*, 113810. [CrossRef]
24. Huang, Y.; Zhao, Y.; Wang, Z.; Liu, X.; Liu, H.; Fu, Y. Explainable district heat load forecasting with active deep learning. *Appl. Energy* **2023**, *350*, 121753. [CrossRef]

25. Zhou, M.; Wang, L.; Hu, F.; Zhu, Z.; Zhang, Q.; Kong, W.; Zhou, G.; Wu, C.; Cui, E. ISSA-LSTM: A new data-driven method of heat load forecasting for building air conditioning. *Energy Build.* **2024**, *321*, 114698. [CrossRef]
26. Zhu, J.; Yang, J.; Cui, X.; Peng, M.; Liang, X. A novel adaptive adjustment kolmogorov-arnold network for heat load prediction in district heating systems. *Appl. Therm. Eng.* **2025**, *274*, 126552. [CrossRef]
27. Wan, A.; Chang, Q.; AL-Bukhaiti, K.; He, J. Short-term power load forecasting for combined heat and power using CNN-LSTM enhanced by attention mechanism. *Energy* **2023**, *282*, 128274. [CrossRef]
28. Yan, Q.; Lu, Z.; Liu, H.; He, X.; Zhang, X.; Guo, J. An improved feature-time transformer encoder-bi-LSTM for short-term forecasting of user-level integrated energy loads. *Energy Build.* **2023**, *297*, 113396. [CrossRef]
29. Li, X.; Wang, S.; Chen, Z. Hierarchical reconciliation of convolutional gated recurrent units for unified forecasting of branched and aggregated district heating loads. *Energy* **2024**, *313*, 134097. [CrossRef]
30. Zhu, J.; Ge, Z.; Song, Z.; Gao, F. Review and big data perspectives on robust data mining approaches for industrial process modeling with outliers and missing data. *Annu. Rev. Control* **2018**, *46*, 107–133. [CrossRef]
31. Ding, Y.; Su, H.; Liu, K.; Wang, Q. Robust commissioning strategy for existing building cooling system based on quantification of load uncertainty. *Energy Build.* **2020**, *225*, 110295. [CrossRef]
32. Li, J.; Cheng, K.; Wang, S.; Morstatter, F.; Trevino, R.P.; Tang, J.; Liu, H. Feature selection: A data perspective. *ACM Comput. Surv.* **2018**, *50*, 94. [CrossRef]
33. Chen, T.; Guestrin, C. XGBoost: A scalable tree boosting system. In Proceedings of the 22nd ACM SIGKDD International Conference on Knowledge Discovery and Data Mining, San Francisco, CA, USA, 13–17 August 2016; pp. 785–794.
34. Wang, Z.; Hong, T.; Piette, M.A. Building thermal load prediction through shallow machine learning and deep learning. *Appl. Energy* **2020**, *263*, 114683. [CrossRef]
35. Li, Y.; Zhu, N.; Hou, Y. A novel hybrid model for building heat load forecasting based on multivariate empirical modal decomposition. *Build. Environ.* **2023**, *237*, 110317. [CrossRef]
36. Agga, A.; Abbou, A.; Labbadi, M.; El Houm, Y. Short-term self consumption PV plant power production forecasts based on hybrid CNN-LSTM, ConvLSTM models. *Renew. Energy* **2021**, *177*, 101–112. [CrossRef]
37. Shi, X.; Chen, Z.; Wang, H.; Yeung, D.-Y.; Wong, W.; Woo, W. Convolutional LSTM network: A machine learning approach for precipitation nowcasting. In Proceedings of the Advances in Neural Information Processing Systems (NIPS), Montreal, QC, Canada, 7–12 December 2015.
38. Parkes, E.J. Observations on the tanh–coth expansion method for finding solutions to nonlinear evolution equations. *Appl. Math. Comput.* **2010**, *217*, 1749–1754. [CrossRef]
39. Zhu, J.; Chen, X.; He, K.; LeCun, Y.; Liu, Z. Transformers without normalization. *arXiv* **2025**, arXiv:2503.10622. [CrossRef]
40. Woo, S.; Park, J.; Lee, J.-Y.; Kweon, I.S. CBAM: Convolutional block attention module. In Proceedings of the European Conference on Computer Vision (ECCV), Munich, Germany, 8–14 September 2018; pp. 3–19.
41. Nguyen, T.V.; Song, Z.; Yan, S. STAP: Spatial-temporal attention-aware pooling for action recognition. *IEEE Trans. Circuits Syst. Video Technol.* **2015**, *25*, 77–86. [CrossRef]
42. Wu, Y.; Tan, H.; Qin, L.; Ran, B.; Jiang, Z. A hybrid deep learning based traffic flow prediction method and its understanding. *Transp. Res. Pt. C-Emerg. Technol.* **2018**, *90*, 166–180. [CrossRef]
43. Cui, Z.; Ke, R.; Pu, Z.; Wang, Y. Stacked bidirectional and unidirectional LSTM recurrent neural network for forecasting network-wide traffic state with missing values. *Transp. Res. Pt. C-Emerg. Technol.* **2020**, *118*, 102674. [CrossRef]
44. Akiba, T.; Sano, S.; Yanase, T.; Ohta, T.; Koyama, M. Optuna: A next-generation hyperparameter optimization framework. In Proceedings of the 25th ACM SIGKDD International Conference on Knowledge Discovery & Data Mining, Anchorage, AK, USA, 4–8 August 2019; pp. 2623–2631.
45. Hoang-Phuong, N.; Liu, J.; Zio, E. A long-term prediction approach based on long short-term memory neural networks with automatic parameter optimization by tree-structured parzen estimator and applied to time-series data of NPP steam generators. *Appl. Soft Comput.* **2020**, *89*, 106116.

buildings

MDPI

Article

Renewal Design of Architectural Facade Features in the Shantou Xiaogongyuan Historic District Based on Deep Learning

Wanying Yan, Tukun Wang and Cuina Zhang *

Department of Civil and Intelligent Construction Engineering, School of Engineering, Shantou University, Shantou 515063, China; 21wyyan@alumni.stu.edu.cn (W.Y.); tkwang@stu.edu.cn (T.W.)
* Correspondence: cuinazhang@stu.edu.cn

Abstract: The Shantou Xiaogongyuan Historic District is a significant cultural symbol of the "Century-Old Commercial Port," embodying the historical memory of the Chaoshan diaspora culture and modern trade. However, amid rapid urbanization, the area faces challenges such as the degradation of architectural façade styles, the erosion of historical features, and inefficiencies in traditional restoration methods, often resulting in renovated façades that exhibit "form resemblance but spirit divergence." To address these issues, this study proposes a method integrating computer vision and generative design for historical building façade renewal. Focusing on the arcade buildings in the Xiaogongyuan District, an intelligent façade generation system was developed based on the pix2pix model, a type of Conditional Generative Adversarial Network (CGAN). A dataset of 200 annotated images was constructed from 200 field-collected façade samples, including Functional Semantic Labeling (FSL) diagrams and Building Elevation (BE) diagrams. After 800 training epochs, the model achieved stable convergence, with the generated schemes achieving compliance rates of 80% in style consistency, 60% in structural integrity, and 70% in authenticity. Additionally, a WeChat mini-program was developed, capable of generating façade drawings in an average of 3 s, significantly improving design efficiency. The generated elevations are highly compatible and can be directly imported into third-party modeling software for quick 3D visualization. In a practical application at the intersection of Shangping Road and Zhiping Road, the system generated design alternatives that balanced historical authenticity and modern functionality within hours, far surpassing the weeks required by traditional methods. This research establishes a reusable technical framework that quantifies traditional craftsmanship through artificial intelligence, offering a viable pathway for the cultural revitalization of the Xiaogongyuan District and a replicable systematic approach for AI-assisted renewal of historic urban areas.

Keywords: historical feature renewal; arcade buildings (Qi-lou); deep learning; building facade

1. Introduction

1.1. Research Background

The Xiaogongyuan Arcade Complex in Shantou exemplifies modern East–West architectural fusion. Emerging after Shantou's opening in 1860 and maturing by the 1930s around the Sun Yat-sen Memorial Pavilion as a commercial hub, it integrates Western structural and South Asian decorative elements within a Chaoshan spatial layout. This synthesis reflects the port city's commercial culture and the artistic achievements of its overseas Chinese community. They carry forward the exquisite decorative traditions of Chaoshan dwellings [1] and embody the historical memory of Shantou's "Century-Old Commercial Port."

As the Special Economic Zone was established, Shantou gradually expanded eastward, resulting in the loss of its central status and a dual decline in both physical and social structures [2]. Rapid urbanization has introduced uniform modern styles and commercialized simplification of decorative elements, resulting in homogenized arcade facades that are stripped of their original character. Concurrently, traditional building functions have become obsolete, while conventional restoration methods prove inefficient and risk damaging authentic features.

Artificial intelligence offers an efficient solution. For Xiaogongyuan's numerous and diverse buildings, deep learning can rapidly generate designs that maintain historical character while incorporating modern functions, significantly improving renovation efficiency. Compared to traditional methods, AI technology can learn and understand complex relationships from historical building data, providing more intelligent and scientific support for design decisions [3], with the potential to address core issues such as homogenization and inefficiency.

1.2. Literature Review

As a historical relic of Shantou, Xiaogongyuan Arcades not only carry the city's historical and cultural values but also hold significant importance for studying modern urban transformation through their spatial layout and road planning [4]. However, this district currently faces severe challenges, with widespread deterioration and numerous arcades requiring restoration [5]. Traditional design methods struggle to ensure design consistency and efficiency, failing to meet the demands of large-scale conservation and renewal.

At the level of historical conservation and contextual research, academia has extensively explored the formative causes and conservation strategies of arcade architecture. The diverse facades of the arcades in the Xiaogongyuan Historic District were shaped by multiple factors, including local climate and construction techniques [6]. In terms of conservation strategies, early government efforts focused on the restoration of individual structures and structural reinforcement [7], whereas current approaches emphasize adaptive reuse [8] and holistic conservation [9]. Scholars advocate that conservation work should integrate micro-regeneration [10], functional adaptation, and modern commercial-cultural tourism, aiming to balance cultural heritage with sustainable development while preserving historical context and collective memory. However, these conservation concepts still heavily rely on the experiential judgment of designers and lack efficient, quantifiable technical support, making it difficult to conduct rapid and scientific simulation and evaluation on a large scale, such as across the extensive Xiaogongyuan area.

In recent years, the emergence of artificial intelligence technologies has provided new perspectives and tools for this design dilemma. The academic community has begun exploring the application of AI technologies in the field of architectural heritage, driving the transformation of digital architectural research. For instance, technologies such as 3D scanning can efficiently complete the digital archiving of buildings, providing a high-precision data foundation for subsequent analysis [11]. Digitized architectural information can be leveraged by AI technologies to automatically detect structural pathologies [12] and systematically describe and analyze the cultural significance and conservation value of built heritage [13]. Furthermore, through the application of deep learning techniques, in-depth characteristic interpretation of traditional Chinese dwellings can be achieved, enabling more precise identification, description, and interpretation of these cultural attributes [14].

Deep learning is an important branch of machine learning, and since its breakthrough in 2012, it has achieved remarkable results in fields such as image recognition and natural language processing [15], with widespread applications in various aspects of digital architectural design. Among diverse deep learning models, Generative Adversarial Networks (GANs) initially gained attention for their exceptional image generation capabilities and efficiency [16], and are increasingly being utilized in architectural heritage conservation [17–21]. The framework employs artificial neural networks [22], primarily composed of a generator and a discriminator, which engage in mutual adversarial training and iterative optimization to ultimately produce highly realistic synthetic data [23]. Furthermore, various improved models derived from GANs, such as CycleGAN [24,25], CAIN-GAN [26], and pix2pix [27,28], have demonstrated powerful image generation and style transfer capabilities in the architectural domain, further expanding the application prospects of artificial intelligence in architectural design, conservation, and innovation.

The Conditional Generative Adversarial Network (CGAN) was proposed to achieve precise control over generation by integrating auxiliary information, thereby allowing for accurate regulation of the outputs. This characteristic makes it particularly suitable for the conservation and renewal of historic districts in the architectural field: CGAN accomplishes class-conditional image generation [29], capable of generating building facades based on input conditions such as historical district decorative styles [30], while effectively learning and summarizing architectural characteristics of specific cities to achieve precise contextual continuity and stylistic control [31]. By modifying the input dataset, CGAN can also accomplish the generation of floor plan designs [32]. CGAN demonstrate significant potential in empowering the digital restoration of architectural heritage. However, when confronted with the highly diverse stylistic decorations and complex components of historic buildings, the adaptability and refinement of current techniques require further development. This challenge precisely constitutes the main direction and untapped potential for future research.

1.3. Problem Statement and Objectives

As integral urban cultural heritage, historic districts preserve architectural features and historical narratives. However, urbanization threatens their distinctive character. In Shantou, located in the subtropical monsoon zone, climatic conditions like high temperatures, humidity, and frequent typhoons accelerate the weathering of arcade facades. The salt-rich coastal atmosphere further accelerates the deterioration of building materials [33]. Figure 1 shows the climatic data of the Shantou region. Against this backdrop, the facade features of Xiaogongyuan, as the study subject, are confronting a degradation crisis, manifesting as stylistic alienation and damage across six major facade elements. The replacement of original materials (such as glazed tiles and washed river sand) with modern alternatives has led to a loss of authentic texture; interrupted colonnade continuity and simplified window lintels reveal the dual challenges of fading craftsmanship and commercial compromise; and mechanical replication of decorative patterns and fading painted colors reflect the discontinuity of traditional techniques. These issues collectively constitute the "form resemblance but spirit divergence" crisis in Xiaogongyuan's facades. Figure 2 below illustrates these specifics.

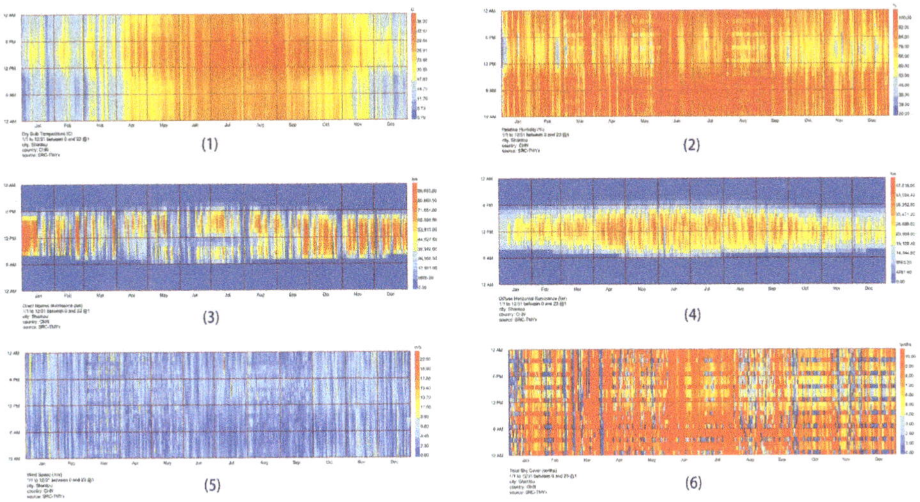

Figure 1. Climatic Data of Shantou Region. (**1**) Dry bulb temperature; (**2**) Relative humidity; (**3**) Direct normal illuminance; (**4**) Diffuse horizontal illuminance;(**5**) Wind speed; (**6**) Total sky cover.

Facade Elements	Existing Issues	Causation Analysis	Typical Cases
Colonnade	interrupted continuity	structural compromise for commerce	
	proportional imbalance of column shaft	improper modern reinforcement	
	chaotic hybridization of classical orders	unclear restoration standards	
	simplified modification of capital and shaft	lost craftsmanship of carving	
Window	window simplification	mass commercial production	
	fading stained glass	function-over-form modernism	
	arch repair with cement	lost red brick craft	
Wall	altered material colors	excessive facade uniformity	
	modern tiles replaced glazed tiles	Nanyang material shortage	
	cement spray replaced washed-sand finish	rapid repair demand	
Decorative Elements	faded polychrome painting	poor weather resistance of organic pigments	
	simplified decorative elements	reduced construction and maintenance costs	
	severely worn relief patterns	accelerated acid rain corrosion	
	rigid mechanically-carved motifs	mechanization replacing handicraft techniques	
Balcony	functional abandonment of recessed balconies	changed modern living habits	
	enclosure additions disrupting facade rhythm	inadequate regulation of resident self-building	
Cornice	lime sculpture peeling	weathering damage	
	damaged scroll patterns	few lime plaster artisans	
	signs covering original moldings	no signage rules	

Figure 2. Xiaogongyuan Arcade Facade Issue Analysis.

The historic district of Xiaogongyuan in Shantou exhibits widespread degradation of its arcade facades, compromising the authenticity and integrity of its historic appearance. Traditional restoration methods, which rely on manual surveying and subjective judgment, are not only inefficient but also prone to stylistic inconsistencies due to varying designer interpretations. In response, this study explores the application of deep learning and artificial intelligence technologies—specifically, a method based on Conditional Generative Adversarial Networks (CGANs)—to the facade renovation of Xiaogongyuan. By training on facade image datasets, the proposed system automatically generates architectural styles and performs style transfer, enabling rapid and visually consistent facade enhancement. This intelligent approach provides technical support for the preservation and innovative renewal of historic architecture in urban contexts. The specific research objectives are as follows:

(1) The ultimate goal of this study is to develop a deep learning-based auxiliary tool for architectural facade renovation. Integrating image recognition and style transfer, the tool will automatically generate diverse design options based on input conditions (e.g., architectural style, materials). It features a visual interface for real-time previews, allowing designers to quickly obtain tailored solutions, thereby reducing costs and enhancing efficiency through intelligent support.

(2) The model will also support the renovation of buildings in Xiaogongyuan by generating facade plans that align with its cultural, environmental, and renewal needs. Trained on collected facade images and data, it automatically learns features such as style, color, and materials to provide design solutions that balance historical continuity with contemporary adaptability.

(3) This research will optimize facade designs based on multiple needs such as residential living, tourism development, etc., improve the overall appearance of the district, promote the revitalization of historical and cultural districts, and achieve sustainable development of Shantou's urban image.

This study possesses significant theoretical and practical value. Theoretically, it merges AI with historic preservation, advancing deep learning applications in heritage digitization and providing new methodologies for style renewal through a CGAN-based facade model, thereby enriching specialized image generation scenarios and establishing a theory for digital stylistic inheritance. Practically, it tackles the style homogenization and character loss in Xiaogongyuan by employing an intelligent design method that utilizes a facade database and deep learning model to rapidly generate design proposals, which shortens design cycles, cuts costs, and ensures the retention of regional features in color, material, and style.

1.4. Research Framework

The research framework of this paper is shown in the figure below (Figure 3).

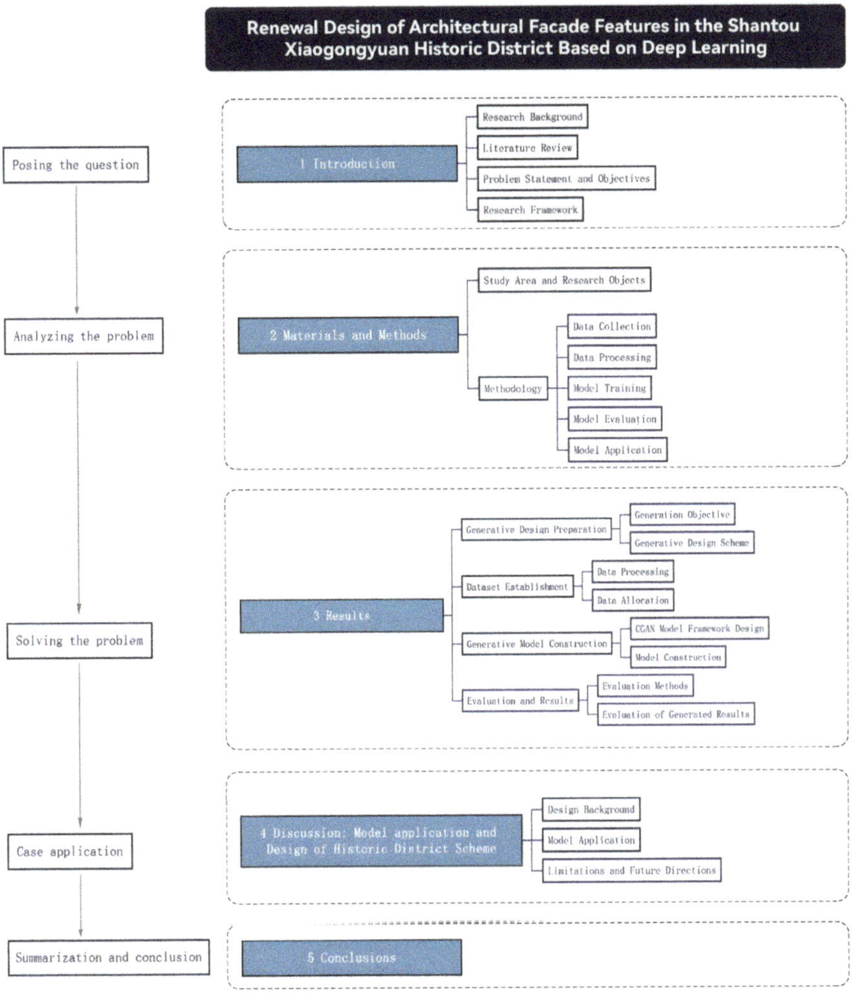

Figure 3. Research Framework Diagram.

2. Materials and Methods

2.1. Study Area and Research Objects

This study focuses on the Xiaogongyuan Historical District in Shantou as the core research area. The Xiaogongyuan Historical District can also generally refer to the old city of Shantou, located at the center of the old town. The area forms a street district that radiates from the Zhongshan Memorial Pavilion as the center, covering 12 major streets, including Guoping Road, Anping Road, etc. As one of China's largest and best-preserved modern arcade districts, the Xiaogongyuan Historical District showcases the architectural art of the early opening period of Shantou, with facades that blend multiple characteristics, including the Nanyang style and Chaozhou-Shantou traditional decorative craftsmanship. The selection of this area for the research is due not only to its well-preserved continuous arcade facades but also because its facades are undergoing systematic degradation, from material replacement to the homogenization of decorative features. This area, which has both cultural value and urgent restoration needs, provides a typical sample for exploring the use of artificial intelligence in assisting historical building renewal.

The Shantou arcade architectural complex is a typical representative of the fusion of modern Chinese and Western architectural cultures. Its architectural form and spatial organization exhibit unique composite characteristics. In terms of architectural style, the arcades incorporate Western architectural elements while adapting to local climatic conditions, resulting in a distinctive arcade typology. From a spatial perspective, the arcades demonstrate high commercial adaptability. Standardized bay and depth modules, combined with floor height design, create a vertical functional division of "shop below, residence above." The continuous covered corridors effectively address Shantou's hot summer climate, providing a comfortable pedestrian shopping environment while reflecting the functional ingenuity of modern commercial architecture. Along Xiaogongyuan Street, the arcades typically consist of 3–4 stories, forming a uniform skyline. The bay width predominantly ranges from 4 to 5 m, with depths of 2 to 2.5 m, creating uninterrupted, all-weather arcade streets [34]. The radial layout centered around Sun Yat-sen Memorial Pavilion enhances the visibility of arcade facades along the main streets. This uniform yet rhythmic skyline has become one of the most distinctive urban landscape features of Shantou's old town, shaping a highly characteristic spatial sequence in the historic district. The facade aesthetics of the Xiaogongyuan Historic District are illustrated in the figure below (Figure 4).

The eclectic characteristics of the Xiaogongyuan arcades serve as material witnesses to cultural collisions across different periods, concretely manifested through the coexistence of multiple architectural styles. This stylistic hybridity is not a mere patchwork but a sophisticated reorganization based on practical functionality and aesthetic expression. The ground level incorporates Nanyang-style open corridors, where continuous arcades combined with adjustable wooden louver systems form a passive climate adaptation strategy for subtropical conditions. The upper facades, meanwhile, skillfully employ simplified classical orders and decorative pediments, achieving a localized interpretation of Western architectural aesthetics through proportional adjustments. The stylistic features of the Xiaogongyuan arcade facades are detailed in the figure below (Figure 5).

The facades of Xiaogongyuan arcades predominantly adopt a tripartite design approach, reflecting rich cultural layering. The ground level features neatly arranged simplified European classical orders as structural supports. The middle section consists of walls, windows, and decorative elements, with some arcades incorporating recessed balconies at the top. The uppermost section is composed of parapets and pediments. Common column styles include simplified Tuscan and Ionic orders, typically with wider central spans. Bay divisions are determined by the number of columns, commonly appearing as single, double, triple, or quintuple bays. The wall surfaces in the middle section skillfully integrate diverse Western classical architectural styles with traditional Chaoshan decorative components. Wall materials exhibit rich variety, accompanied by windows of varying designs. Typical layouts feature one window per bay or two windows per bay. In most cases of single- or double-window configurations, the columns corresponding to the wall sections are embedded within the walls, employing quasi-Doric or square composite designs that enhance structural stability while maintaining aesthetic harmony. The uppermost pediments and parapets combine traditional Chaoshan patterns with Rococo influences, forming a unique and regionally distinctive decorative style that highlights the defining characteristics of Shantou's modern architecture. Specific arcade facade typologies are detailed in the figure below (Figure 6).

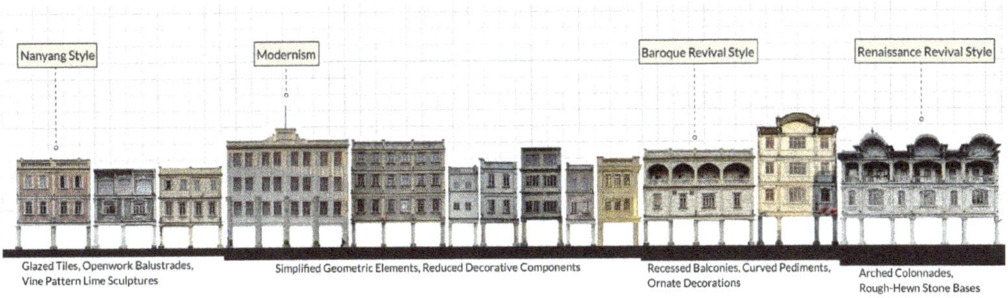

Figure 4. Architectural Facade Style Map of Shantou Xiaogongyuan Historic District.

Style Type	Core Elements	Typical Examples
Nanyang Style	Glazed Tiles, Openwork Balustrades, Vine Pattern Lime Sculptures	
Neoclassical Style	Triangular Pediments, Layered Ornamentation	
Renaissance Revival Style	Arched Colonnades, Rough-Hewn Stone Bases	
Baroque Revival Style	Recessed Balconies, Curved Pediments, Ornate Decorations	
Modernism	Simplified Geometric Elements, Reduced Decorative Components	

Figure 5. Summary Diagram of Arcade Facade Styles in Xiaogongyuan.

Facade Type	Typical Case	Characteristics
Single Bay		Composed of modular independent facades in symmetrical axial form; vertically designed with a tripartite composition; unified design between the second and third floors creates a visually elevated effect.
Double Bay (or Quadruple Bay)		Features an axially symmetrical balanced composition, typically with individual bays exhibiting axial symmetry, and double bays mirroring each other.
Multiple Bays (Triple Bay, Quintuple Bay, etc.)		Rhythmic continuous facade with overall axial symmetry. Typically, the central bay has a larger width while the side bays are narrower.
Special Type		The facade exhibits heterogeneous morphological expression, potentially featuring arcs or continuous curved surfaces.

Figure 6. Analysis Diagram of Arcade Facade Typology and Characteristics in Xiaogongyuan.

2.2. Methodology

This study adopts a deep learning-based research methodology comprising five steps: data collection, data processing, model training, model evaluation, and model application. The specific methods are as follows:

(1) Data Collection: Data collection serves as the foundation for constructing and operating deep learning models. This step aims to acquire image data of building facades in Shantou's Xiaogongyuan Historic District. Through field investigations of the district's architectural facade characteristics, it was observed that most arcades integrate traditional Chaoshan craftsmanship with Western architectural styles, forming a diverse composite typology. Their facades exhibit distinctive local features in colonnade forms, window and door styles, and decorative elements. For example, traditional decorative techniques such as stone carving and porcelain inlay from the Chaoshan region are reflected on some arcade facades. To ensure sample diversity and representativeness, the study selected facades along ten typical roads within the historic district (Yongping Road, Shengping Road, Anping Road, Zhiping Road, Guoping Road, Juping Road, Shangping Road, Zhenbang Road, Yongtai Road, and Xidi Road) as research objects, covering arcade buildings of different eras, styles, and functional types. The sample collection scope is illustrated in the figure below (Figure 7).

Figure 7. Research Sample Collection Scope Map.

Among these, Guoping Road and Anping Road—serving as the core commercial axes of the Xiaogongyuan Historic District—exhibit arcades with pronounced modern tendencies. Their facades feature abundant decorative elements, though commercial signage severely obstructs the architectural character. In contrast, Yongping Road and Juping Road demonstrate relatively uniform and continuous facade designs. Most arcades here are three-story structures of masonry and reinforced concrete, showcasing a hybrid style primarily characterized by the fusion of simplified Roman-style colonnades and modern architectural influences.

The arcades along Yongtai Road feature vibrant colors and diverse forms, with some three-story structures incorporating recessed balconies. Facade decorative elements predominantly draw from botanical and animal motifs, exhibiting distinct Nanyang stylistic influences. In contrast, Xidi Road's arcades are primarily characterized by eclecticism, with well-preserved facades showcasing clear decorative patterns. The columns feature simplified Ionic capitals and composite shafts. Beyond these dominant styles, the Xiaogongyuan arcade complex also incorporates diverse stylistic elements such as Baroque Revival and Nanyang styles, reflecting the architectural cultural fusion characteristic of modern Shantou as a treaty port, as well as the diversity and complexity of its arcade facades.

During field studies, cameras documented architectural styles, materials, and colors. The collected arcade facade images encompass typical styles while preserving historical authenticity and distinctive features. Data was digitized and categorized by building typology and style to form structured datasets for model training.

Additional data was gathered online via social media to understand public and expert perceptions of facades, informing model conditions. Literature reviews of academic papers and online resources provided theoretical foundations for applying deep learning to facade renewal design.

(2) Data Processing: After data collection, preprocessing is required for the acquired architectural facade photographs. This stage primarily involves preprocessing and annotation of the collected images. First, all image data must be cleaned and filtered to remove noise and low-quality images, ensuring that the image quality meets the requirements for model training. Second, technical software is used to perform correction and enhancement operations, processing the image data into two types. The first type is the Functional Semantic Labeling (FSL) diagram [30], which distinguishes architectural facade elements (such as walls, windows, columns, decorative components, etc.) and annotates each facade element using different color blocks. For example, wall sections are marked in yellow, windows in blue, columns in brown, and decorative components in orange. This color-block annotation method clearly identifies the different constituent elements of the facade, providing clear learning targets for the deep learning model and enabling it to more accurately learn the local facade style. The second type is the Building Exterior (BE) diagram [30], which displays the facade information of the arcade buildings, including decorative components, architectural structures, and materials. All processed image samples are uniformly standardized to a resolution of 512 × 512 pixels to ensure dimensional consistency of the input data for the model. The processed data will serve as training and validation sets for the training and optimization of the deep learning model, laying the data foundation for the subsequent generation of architectural facade renewal solutions.

(3) Model Training: After completing the data processing, this study uses the pix2pix algorithm model to generate historical building facade images with the specific style of a small park. The selection of pix2pix was based on several key considerations, primarily because this study possesses paired training data that aligns perfectly with the supervised learning framework of pix2pix. Furthermore, its deterministic output characteristic proves advantageous over stochastic generative models by better preserving the accuracy of key façade structures. Additionally, the model's technical maturity is well-demonstrated through its successful application in precedents of architectural image translation tasks. The model is an image translation model based on Conditional Generative Adversarial Networks (CGAN), which learns the mapping relationship between functional layout diagrams and real facade images through the adversarial training mechanism between the generator and the discriminator. During the model training process, the generator and discriminator continuously optimize parameters to minimize the differences, ensuring that the generated facade retains the accuracy of the original structure while incorporating authentic and natural architectural style characteristics, thus achieving the digital generation and preservation of the historical building's appearance.

(4) Model Evaluation: This study employs a comprehensive evaluation framework combining subjective and objective methods to systematically assess the performance of the trained pix2pix model. The subjective evaluation is conducted by a panel of 20 architecture students and supervising instructors through structured questionnaires and interviews. This process assesses both basic visual quality—including clarity, detail retention, and similarity to real images—and calculates a comprehensive Conservation Authenticity Score based on four weighted dimensions: Style Consistency (30%), evaluating adherence to characteristic decorative elements and compositional rules; Structural Integrity (25%), assessing logical correctness of load-bearing elements and structural relationships; Material Compatibility (25%), examining appropriate use and visual plausibility of traditional materials; and Weathering Pattern Retention (20%), quantifying preservation of natural aging characteristics that contribute to historical value. For objective evaluation, we employ dual approaches: monitoring the generator and discriminator loss curves to ensure training sta-

bility, and introducing the Structural Similarity Index (SSIM) to quantitatively measure the perceptual and structural fidelity between generated and reference images. The evaluation strategy is dynamically adjusted based on results, with model refinements made through methods such as augmenting training data or modifying loss functions when generated samples show significant deviations from real images in style or structure, ultimately ensuring the solutions meet both computational metrics and professional architectural standards.

(5) Model Application: This step applies the trained model to the facade renovation design of historical buildings in the small park area of Shantou. Specific applications include constructing an actual case test set, selecting typical damaged or stylistically mixed arcade facades in the district as input objects; generating multiple facade renovation schemes through the trained pix2pix model, including typical facade styles of the small park; placing the generated schemes into the NURBS surface modeling tool (Rhino7 (The detailed version of the software can be found in Appendix A)) for 3D visualization and assessing their coordination with the surrounding environment; comparing the facades generated by the pix2pix model with manually designed facades in terms of historical authenticity, coordination, time consumption, etc., and conducting a comprehensive evaluation; selecting the best scheme generated by the pix2pix model for further design refinement. If the facade scheme generated by the pix2pix model significantly outperforms the manually designed scheme in aspects such as timeliness, this method can be further extended to other areas of architectural heritage preservation and restoration, thereby greatly improving the efficiency and quality of professional design.

The research methodology flowchart is shown in the figure below (Figure 8).

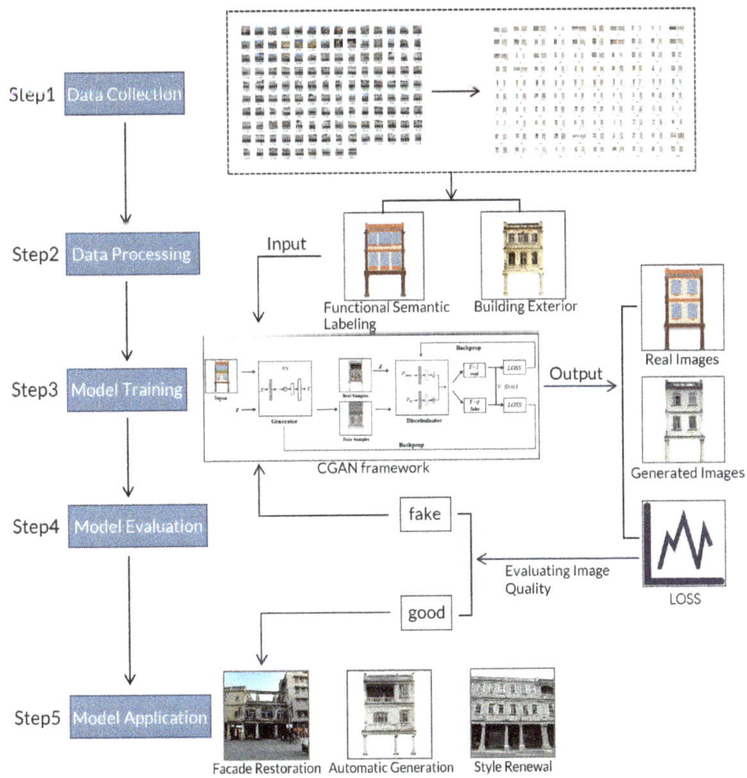

Figure 8. Research Process Flowchart.

3. Results

3.1. Generative Design Preparation

The generative design preparation phase involves systematically constructing the technical framework for facade generation design of Xiaogongyuan arcades. It primarily consists of three key steps: defining generation objectives, formulating generative technical solutions, and selecting hardware equipment and technologies that align with project requirements.

3.1.1. Generation Objective

The aim of this study is to develop a deep learning-based architectural facade generation design system, focusing on the facade renewal needs of the Xiaogongyuan Historic District in Shantou. By constructing a Conditional Generative Adversarial Network (CGAN) model, the system learns the style features, material expressions, and other aspects of the arcade facades, enabling it to automatically generate renewal plans that align with the historical characteristics upon inputting current architectural images. This generative model integrates image recognition and style transfer functions, allowing designers to preview diverse plans in real-time through a visual interface, which significantly improves design efficiency. While reducing renewal costs and enhancing design efficiency, the system achieves a multi-objective balance of historical authenticity preservation, functional adaptability improvement, and overall streetscape optimization, providing technical support for the sustainable development of the Xiaogongyuan Historic District.

3.1.2. Generative Design Scheme

The generation design process in this study consists of the following steps: The researchers first systematically collect and process data on the arcade facades of the Xiaogongyuan Historic District in Shantou, organizing the collected data into a dataset according to design requirements. This dataset will be used to train the CGAN model. During the model training process, users can observe the generation effects in real-time through a visual interface and make adjustments as needed. Once the generative model is capable of outputting facade designs that align with the characteristics of arcade facades, the model is considered trained and can be used as the Shantou Arcade Facade Generator. At this stage, designers only need to input plan images that the model can recognize, and the trained generative model will automatically output multiple arcade facade design options.

The generation process in this study follows a basic framework of "data preparation—model training—plan generation." Given the special requirements of historical building preservation, a manual review process is added during the output stage to ensure that the generated results not only meet modern functional needs but also maintain the authenticity of the historical style. This process provides a scalable technical path for the digital renewal of historic district facades. The design generation process is shown in the figure (Figure 9).

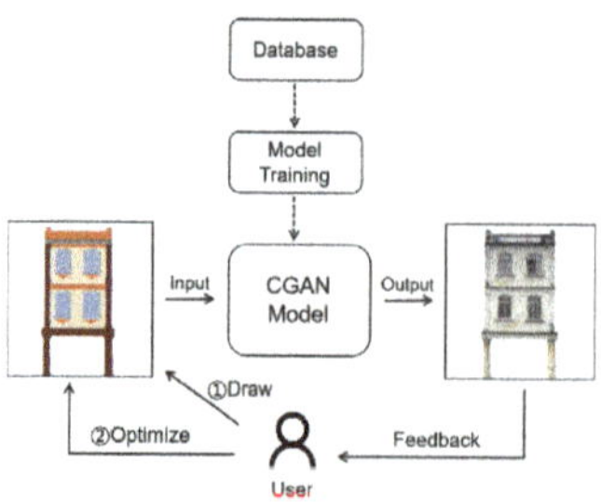

Figure 9. Generation Flowchart.

3.2. Dataset Establishment

3.2.1. Data Processing

The study selects the Xiaogongyuan Historic District in Shantou, Guangdong Province, as the research subject, and collects architectural facade image data through field research. During the research period, a total of 237 original facade images were obtained, of which 37 were discarded due to poor shooting angles or insufficient image resolution. Ultimately, 200 valid samples were retained. Building on this foundation, data augmentation techniques—including random horizontal flipping and scaling—were employed to construct a significantly larger and more diverse training set, thereby effectively enhancing the model's generalization capability and mitigating overfitting. All selected facade image samples must undergo the following processing steps:

(1) Valid samples will be imported into CAD and Adobe Photoshop 2021 for correction, brightness adjustment, and other processing to convert them into line drawings. During the preprocessing stage, temporary attachments (such as billboards) must first be removed to ensure the integrity of facade features. After processing, the samples undergo color separation treatment, using different color blocks to distinguish architectural facade elements, ultimately generating Functional Segmentation Layout (FSL) diagrams. Based on the compositional characteristics of Xiaogongyuan arcade facades, main facade elements are divided into four categories, marked with distinct colors in this study: colonnades in dark brown (R116, G69, B59), decorative components in bright orange (R213, G105, B66), windows in light blue (R141, G170, B200), and walls in light yellow (R234, G218, B192). Additionally, light gray (R146, G148, B147) for roofs and green (R125, G162, B95) for railings will serve as supplementary elements in the annotation. Specific color coding specifications are detailed in the figure below (Figure 10).

Facade Elements	Color	Legend
Colonnade	R116, G69, B59	
Decorative Elements	R213, G105, B66	
Window	R141, G170, B200	
Wall	R234, G218, B192	
Roof	R146, G148, 147	
Guardrail	R125, G162, B95	

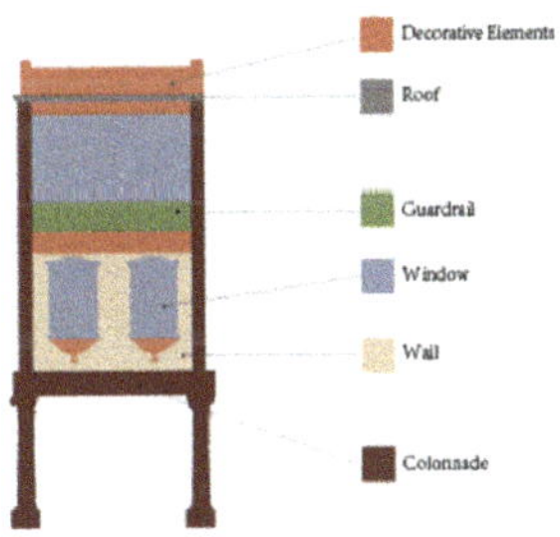

Figure 10. Color Coding Specification Table for Facade Elements of Xiaogongyuan Arcades.

According to the actual composition of the facade samples, each image may contain all or part of the annotated elements mentioned above. The color block annotations enhance the visual boundaries of different architectural elements through color contrast, reducing the model's sensitivity to interference from textures, shadows, and other factors, allowing the machine to focus more on extracting structural features.

(2) After completing color block annotation, architectural elevation drawings (BE) will also be retained as training objects. The processed elevation drawings clearly display important elements such as materials and textures of Xiaogongyuan arcade facades, facilitating post-machine learning generation. The processed image data are presented in the figure below (Figure 11).

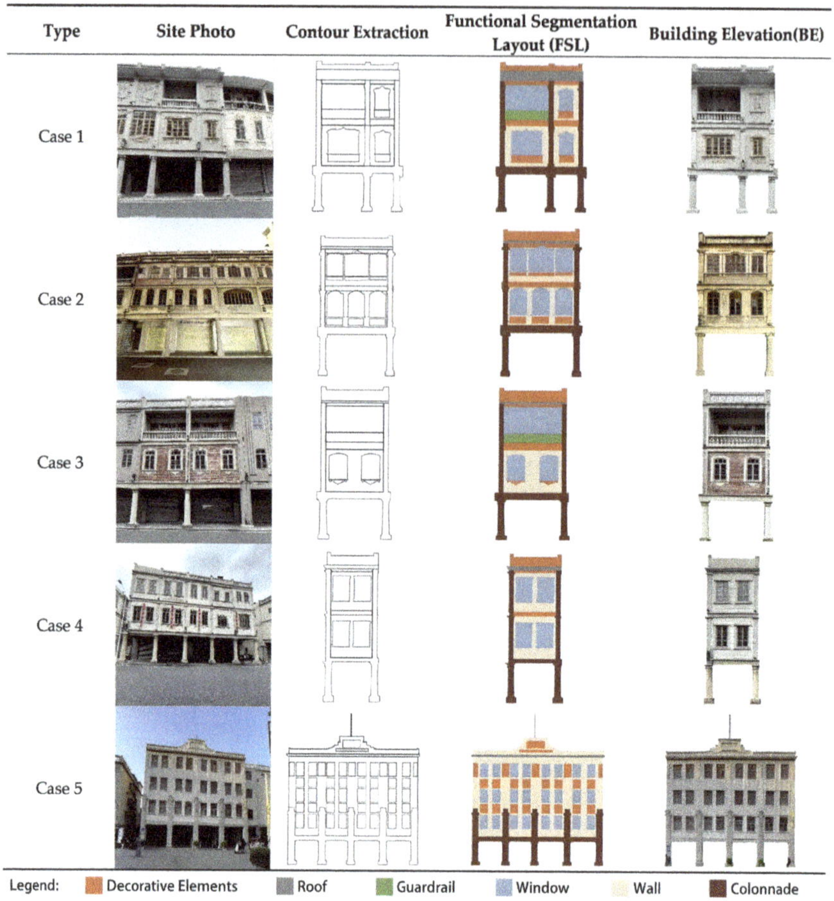

Figure 11. Classification Table of Facade Image Data for Xiaogongyuan Arcades.

Each valid sample will be processed into two types of images: Functional Segmentation Layout diagrams and Architectural Elevation drawings. The 200 samples will thus generate a total of 400 images after processing. The baseline data is shown in the figure below (Figure 12).

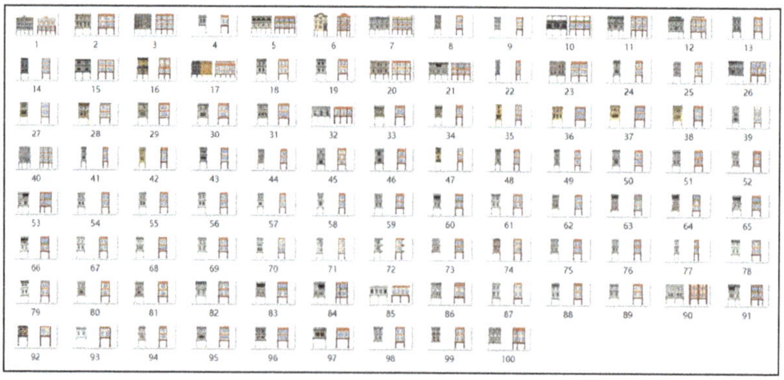

Figure 12. Base Data.

For the purpose of dataset generation, all original images are initially standardized to a resolution of 512 × 512 pixels. Subsequently, two such images are assembled side-by-side, yielding a composite image with a total resolution of 1024 × 512 pixels. Figure 13 illustrates a standardized format data sample.

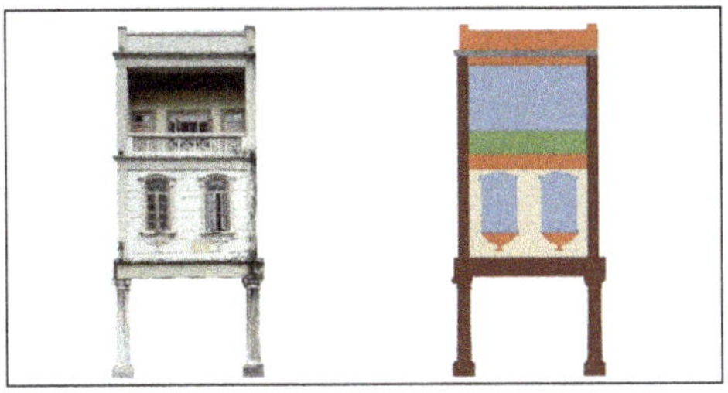

Figure 13. Standard-Format Data Diagram.

3.2.2. Data Allocation

Upon completion of dataset organization, the data will be divided into three subsets through sampling: training set, validation set, and test set. The training set comprises 60% of the total data volume, amounting to 120 samples, to ensure sufficient sample diversity. Both the validation set and test set each account for 20% of the total data, with 40 samples per set, aiming to support subsequent cross-validation.

3.3. Generative Model Construction

3.3.1. CGAN Model Framework Design

This study employs the pix2pix algorithm based on a Conditional Generative Adversarial Network (CGAN) model for training. This model demonstrates strong accuracy and stability in architectural image style transfer tasks, enabling accurate historical style transfer while preserving architectural structural features.

The pix2pix model generator adopts the U-Net [15] architecture to preserve the structural features of input images, ensuring that architectural details and overall contours remain intact during training. The discriminator utilizes the PatchGAN [15] architecture, focusing on authenticity judgment of local regions to enhance the quality of generated images. In the training mechanism, the generator receives input images and random vectors, learning to generate fake images; the discriminator compares generated images with real target images to assess authenticity. If the discriminator identifies an image as fake, it returns the difference between the fake and real images to the generator. During adversarial training, the generator continuously optimizes its parameters based on adversarial loss and L1 loss to reduce discrepancies with real target images, ultimately achieving precise style transfer from source to target images. The framework of this algorithmic model is shown in the figure below (Figure 14).

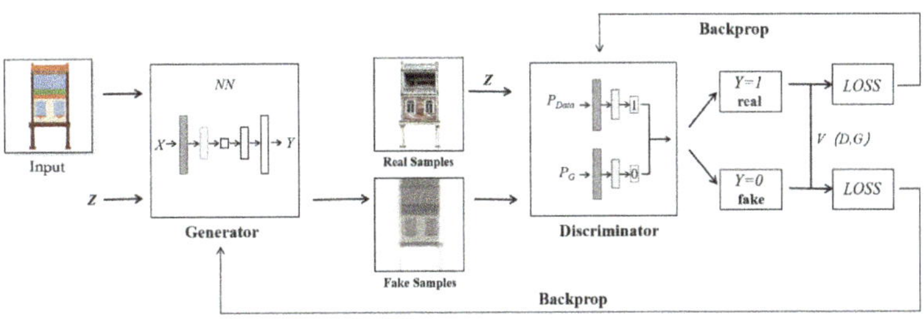

Figure 14. CGAN Algorithm Model Framework Diagram.

3.3.2. Model Construction

This study constructs an arcade facade generation model based on CGAN by leveraging open-source pix2pix algorithms from online platforms such as GitHub and CSDN, with model implementation conducted using tools including PyCharm 2024.3 (Community Edition) and PyTorch 1.8.1 (The detailed version of the software can be found in Appendix A). The operating system for this model setup is Windows 11.0.

The research conducted model training based on the previously established dataset of arcade facade images from Shantou Xiaogongyuan area. All samples underwent standardized preprocessing procedures, including key steps such as resolution unification and color space conversion. During the training process, parameters like learning rate and batch size were adjusted to achieve continuous optimization of model performance under GPU acceleration. The final generation results were evaluated to ensure compliance with historical preservation design requirements, providing reliable technical support for the digital renewal of historic districts.

3.4. Evaluation and Results

3.4.1. Evaluation Methods

This study primarily employs two methods to evaluate the results generated by the model, as detailed below:

(1). Subjective Evaluation: During the model iteration process, we formed an evaluation panel consisting of 20 architecture students and their supervising instructors. Through questionnaires and interview surveys, a subjective evaluation of the quality of the test-generated images was conducted. The evaluation criteria cover the following two aspects: first, the basic visual quality of the generated images, including clarity, detail retention, and similarity to real images; second, a comprehensive Conservation Authenticity Score, which is calculated as a weighted combination of four dimensions: Style Consistency (30%), Structural Integrity (25%), Material Compatibility (25%), and Retention of Weathering Patterns (20%).This study organizes a concise and universal set of evaluation criteria, as shown in Table 1.

(2). Objective Evaluation: We have introduced the Structural Similarity Index (SSIM). This metric assesses the visual fidelity of the generated results by comparing the luminance, contrast, and structural information between the generated image and the input color block reference image. By calculating and analyzing the SSIM values of both, we provide an objective quantitative basis for evaluating image generation quality.

Table 1. Quality Assessment Metrics for Generated Façades.

Evaluation Dimension	Evaluation Metric	Total Number
Basic Visual Quality	Clarity	The overall sharpness of the image, absence of blur, and the discernibility of key elements.
	Detail Retention	The ability of the generated image to reproduce fine features (e.g., textures, ornaments, edges) from the original image (or design intent).
	Similarity to Real Images	The realism of the overall visual effect, including lighting and color, and the degree of resemblance to real-world photographs.
Conservation Authenticity Score	Style Consistency (30%)	The degree to which the architectural style (e.g., period features, regional characteristics, formal language) of the generated result aligns with the target conservation style.
	Structural Integrity (25%)	The rationality of the structural logic, the correctness of component relationships, and the absence of structural errors or distortions.
	Material Compatibility (25%)	Whether the surface materials and textures presented in the generated building conform to historical authenticity or design intent, and the appropriateness of the transitions between different materials.
	Retention of Weathering Patterns (20%)	For historic buildings, the appropriate preservation and representation of characteristic traces of age (e.g., patina, fading, erosion), avoiding an "over-restored" appearance.

3.4.2. Evaluation of Generated Results

To evaluate the training effectiveness of the model, this study systematically monitored the changes in the loss functions of both the generator and discriminator. Figure 14 displays the evolution curves of four core metrics during the training process: generator adversarial loss (G_GAN), generator L1 reconstruction loss (G_L1), discriminator loss on real samples (D_real), and discriminator loss on generated samples (D_fake).

By analyzing the convergence trends and numerical relationships of these curves, the model's effectiveness in learning facade features can be revealed, and the validity of the training strategy can be verified. As shown in Figure 15, over 500 training epochs, both G_GAN and G_L1 gradually decreased with the training progress. Among them, G_L1 consistently dominated the training process, indicating that the model prioritized ensuring the accuracy of facade structures. The losses of D_real and D_fake eventually stabilized around 0.7 and 1.2, respectively. As initial training results, the model has established basic facade generation capabilities, though there remains room for optimization. The specific experimental model parameters are listed in Table 2 below.

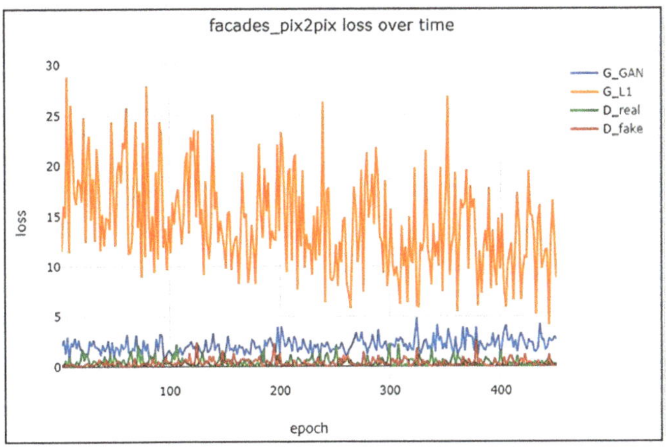

Figure 15. Training Log Loss Value Statistics Chart 1 for pix2pix Model.

Table 2. Parameter Value used in the First Study.

Parameter Name	Parameter Value
epochs	500
Batch size	4
Learning rate (D/G)	0.0002
channels	3
epochs decay	200
Weight decay	0.000
image size (W/H)	512
Beat 1	0.5
Beat 2	0.999

Based on the initial training results, further optimization of model parameters was conducted. In this training phase, the learning rate was reduced to 0.0001. A smaller learning rate ensures more stable optimization and prevents oscillations in the L1 loss. Simultaneously, the learning rate decay cycle was extended and the number of training epochs was increased, allowing the model more time for fine-tuning at lower learning rates, which contributed to a reduction in the L1 loss. The specific experimental model parameters are listed in Table 3 below. The training results are shown in the figure below (Figure 16).

Table 3. Parameter Value used in the Second Study.

Parameter Name	Parameter Value
epochs	800
Batch size	4
Learning rate (D/G)	0.0001
channels	3
epochs decay	300
Weight decay	0.000
image size (W/H)	512
Beat 1	0.5
Beat 2	0.999

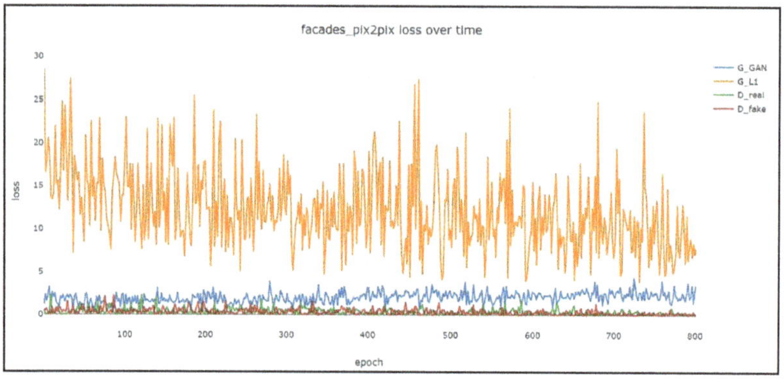

Figure 16. Training Log Loss Value Statistics Chart 2 for pix2pix Model.

Over 800 training epochs, all loss values exhibited typical dynamic equilibrium characteristics—the adversarial game between the generator and discriminator eventually stabilized. During the initial training phase (0–200 epochs), all loss functions fluctuated drastically, indicating the model was exploring the feature space. When training progressed to 201–450 epochs, the losses began to decrease, and the generator started to grasp the patterns of style transfer. Beyond 450 epochs, the curves gradually stabilized with minor oscillations and converged toward specific values. As training approached 800 epochs, G_L1 steadily converged to approximately 6.5, showing clear signs of convergence. The generator adversarial loss (G_GAN) and the discriminator loss on real samples (D_real) stabilized after 550 epochs, with G_GAN converging to around 2.3 and D_real eventually reaching 0.11. G_L1 remained significantly higher than other loss terms, ensuring that the generated images strictly preserved the structural features of the input facades. However, the secondary upward trend of G_L1 after 450 epochs may indicate that the model began to overfit localized damaged features in the training data. Based on the above analysis, the current model has achieved basic facade generation functionality, but it still exhibits insufficient learning of arcade-specific decorative elements. The evolution of generated images during the training process is shown in the figure below (Figure 17).

As shown in the table, the trained generative model has acquired reliable capability to reconstruct architectural facade styles, enabling relatively accurate restoration of the distinctive features of Xiaogongyuan's arcade buildings. By comparing images generated by the model with actual images, it is evident that the model not only maintains consistency in the overall facade style but also flexibly and effectively transfers and reshapes facade elements within the images, ensuring similarity and realism in details. The model demonstrates strong adaptability in style restoration, capable of handling complex architectural characteristics while achieving a sound balance between style consistency and element transformation. This characteristic endows the model with high practical value, allowing it to flexibly adjust and produce architectural facade effects that meet design requirements under different environments and needs.

Figure 17. Figure of Generated Image Variations Across Different Training Epochs. (Intermediate output images from the pix2pix model during the training iteration process. These images inherently demonstrate lower resolution as they represent non-final optimization stages.).

Although the generative model excels in restoring architectural facade styles and transferring elements, some limitations remain. The model still exhibits certain constraints when processing intricate details. Some decorative details (such as reliefs, carvings, etc.) may show slight distortions or blurring during the generation process, affecting the overall image refinement. Despite overall style consistency, the authenticity and detail richness of generated images may decline under specific lighting or viewing angles, particularly in complex background environments and subtle variations in building materials. Additionally, the model's generalization ability requires further enhancement. When confronted with different styles or entirely new architectural types, it may need more diverse data for additional training to improve its adaptability and accuracy.

Using the SSIM algorithm with the input scheme as the input image, we calculated its SSIM values against the real image, generated image. Some SSIM values are shown in the figure (Figure 18).

Figure 18. Exploration of SSIM Evaluation Method for Generated Facades of Arcade Buildings.

The SSIM comparison data show that the structural similarity between input images and real images remains in the range of 0.12–0.15, while the similarity between input images and generated images ranges from 0.10 to 0.14. The minimal differences in SSIM values (not exceeding 0.03) between generated and real images indicate the model's stable structural preservation capability.

Notably, the consistently low SSIM values may be related to the algorithm's inherent sensitivity to structural, luminance, and contrast features—when the generation task involves significant style transformation or material reconstruction, the pixel-level structural similarity might naturally remain low even if visual results meet expectations. However, whether this numerical characteristic truly reflects an inherent relationship between the algorithm's principles and the nature of the generation task requires further investigation.

4. Discussion: Model Application and Design of Historic District Scheme

4.1. Design Background

This study selects the intersection zone of Shangping Road and Zhiping Road as the experimental site. The arcade facades in this area show significant deterioration due to long-term neglect and require urgent renewal. Multiple architectural styles coexist here, including Baroque decorations, Nanyang-style colonnades, and modern renovation elements. Original decorative patterns are partially damaged or missing. This typical phenomenon of stylistic hybridity in historic districts provides ideal conditions to test the generative model's adaptability in complex real-world environments. The design scope is illustrated in the figure below (Figure 19).

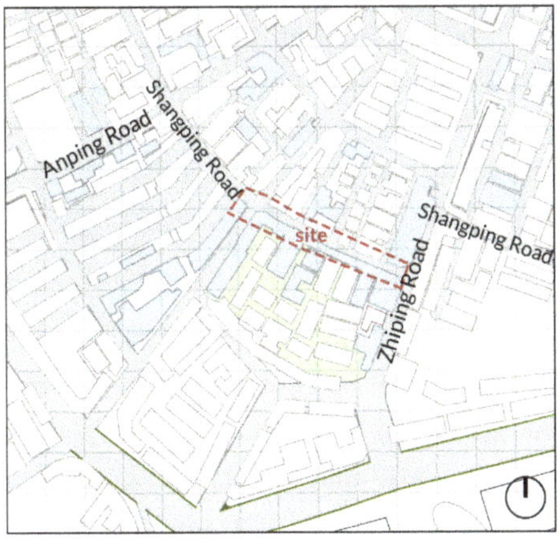

Figure 19. Design Area.

The complexity of the architectural style in this area is primarily reflected in three aspects: In terms of time, the reconstructions and extensions from different periods have formed distinct historical layering features; in terms of culture, the collision of Western architectural language with local construction techniques has given rise to a unique eclectic style; in terms of materiality, both natural aging and human interventions have jointly led to severe damage to the architectural details. This multi-layered complexity poses special challenges for digital generation technology—it needs to identify and deconstruct mixed stylistic elements while reasonably hypothesizing and reconstructing the architectural features based on the fragmented physical evidence. Existing rule-based computational design methods often struggle to address such non-standard historical architectural characteristics, while deep learning technology, by learning style patterns and construction logic from a large number of samples, offers a new solution to this problem. It is particularly noteworthy that the diverse cultural symbols placed on the facades of buildings in this area provide an excellent sample for testing the capabilities of generative models in cross-cultural architectural feature extraction and fusion.

4.2. Model Application

Upon completing model training, the study applied the model to assess its generative capabilities. Prior to application, 40 functional segmentation layout images absent from the original dataset were prepared and input into the trained model. The discrepancies between the generated images and real images were examined. By comparing structural integrity, detail accuracy, and overall fidelity between generated and actual images, the model's practical performance was evaluated, with its potential strengths and limitations analyzed. This process aimed to ensure the model could not only replicate features of known data but also demonstrate high generalization ability when confronted with entirely new samples.

The quality assessment of generated solutions is primarily based on the following three criteria:

(1) Structural Integrity: Load-bearing components such as colonnades and arches are completely generated without distortions like structural misalignment or fragmentation;

(2) Style Consistency: The output images maintain the stylistic features (e.g., decorative patterns, color schemes) of the input images used during model training;
(3) Generation Authenticity: The images generated by the model exhibit high fidelity in key characteristics such as facade morphology and proportional scales, ensuring realism.

A selection of generated images is presented in the figure below. (Figure 20).

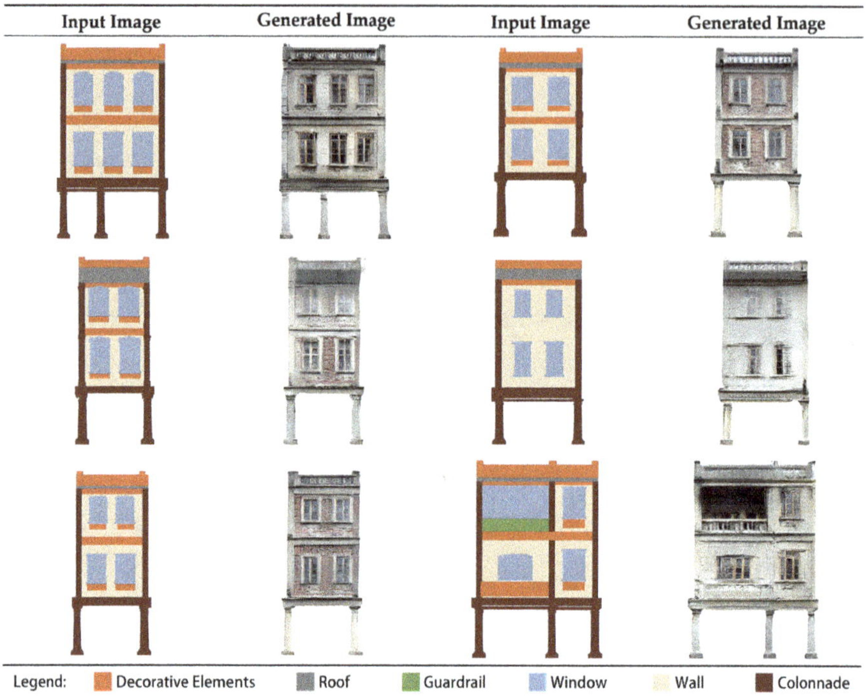

Figure 20. Figure of Generated Images.

As shown in the figure, the images generated by the model maintain high fidelity in core characteristics such as architectural facade morphology and proportional scales, with particularly accurate and refined handling of details like colonnades. The output images exhibit style consistency with the real images in the training set, demonstrating complete facade structures without issues such as fragmentation or structural misalignment. However, minor blurring occurred in some detailed decorations (e.g., relief textures) during the generation process, especially for complex elements such as pediment decorations, where resolution and detail representation require further improvement. Additionally, two remaining output images showed suboptimal quality failing to meet evaluation standards. Specific cases are detailed in the figure below (Figure 21).

Based on the aforementioned generated image results and evaluation criteria, the performance assessment of the generative model is presented in the table below (Table 4). Among the 40 test samples, style consistency achieved the highest compliance rate, while generation authenticity and structural integrity reached 70% and 60%, respectively. This indicates that the model performs optimally in restoring stylistic features but requires improvement in generating complex structures. When reproducing certain subtle decorative elements, the model may be constrained by the diversity and detail depth of the training data, resulting in reduced clarity and realism in these areas. To address these issues, subsequent optimization should enhance training focused on high-detail regions to improve overall image precision and detail richness.

Image ID	Input Image	Generated Image	Specific Issue
3			Incomplete image generation with blurred details.
5			Vague window opening contours with cluttered and indistinguishable content; structural misalignment, incomplete colonnade sections, and blurred facade details.

Legend: Decorative Elements Roof Guardrail Window Wall Colonnade

Figure 21. Figure of Non-Compliant Generated Image Quality. (Examples of unsuccessful iterations generated by the trained model, illustrating specific failure cases in style transfer.).

Table 4. Generated Model Performance Evaluation.

Evaluation Criteria	Number of Compliant Cases	Total Number	Compliance Rate
Structural Integrity	24	40	60%
Style Consistency	32	40	80%
Generation Authenticity	28	40	70%

Based on the generated imagery, this study has established a comprehensive technical pipeline from 2D generation to 3D modeling. First, develop a facade design mini-program (https://dorahsy.github.io/two/ (accessed on 1 December 2025)). The first step is to generate the facade image via the mini-program. As shown in the figure below (Figure 22), the left column displays the initially generated images based on the conditional generative adversarial network; the middle column presents the optimized facade images after style adjustments, where the completeness and style consistency of detailed decorative elements were significantly improved; the right column further transforms the optimized 2D images into editable 3D architectural models through AI techniques, attempting to explore dimensional expansion in the digital preservation of historic architecture. These three sets of visual outcomes form a technical chain of "facade generation—facade optimization—3D conversion." This exploration not only visually demonstrates the application potential of generative algorithms in architectural heritage conservation but also establishes a technical closed loop from 2D to 3D and from images to models, providing a verifiable methodological framework for the digital renewal of historic districts. Particularly noteworthy is that the parametric component design of the 3D models not only preserves the typical characteristics of historic architectural elements but also meets the technical requirements of modern building information management, demonstrating the innovative value of this study in the digital transmission of cultural heritage.

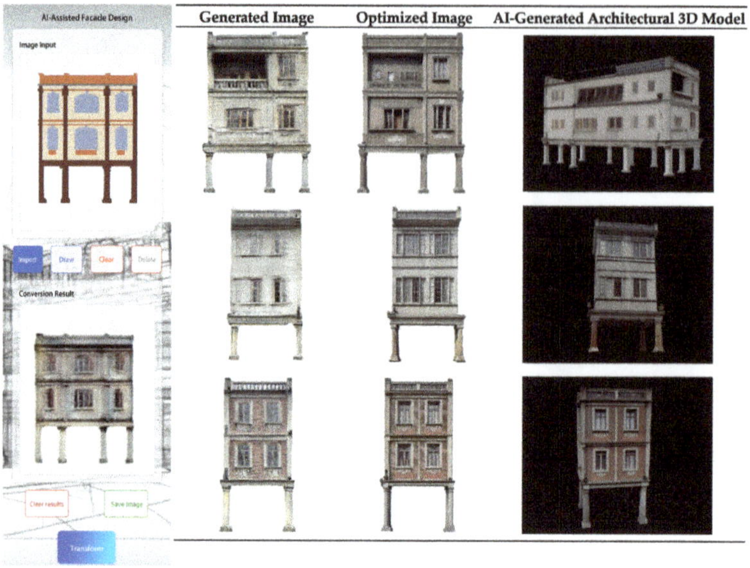

Figure 22. Optimized Generated Images Figure.

Relying on the aforementioned research results, this study conducted a renovation of the arcade facades at the intersection of Shangping Road and Zhiping Road. During the renovation process, it was necessary to update the original arcade features by integrating the cultural connotations of historic architecture with the functional requirements of modern buildings. The generative method based on conditional generative adversarial networks can meet the needs of accurate restoration and innovative optimization.

During the renovation, generative image technology was used to preliminarily restore the original facades, ensuring that the architectural appearance complies with historical features requirements. The renovated arcade features not only meet the functional demands of modern urban life but also preserve the unique charm of historic architecture, injecting new vitality into the intersection area of Shangping Road and Zhiping Road, and providing a new direction for the digital preservation and innovative renovation of architectural heritage. The figure below (Figure 23) shows the effect of the renovated arcade facades.

Figure 23. Design Rendering.

To further refine the facade renewal process, the study also produced detailed architectural elevation drawings. These drawings encompass various detailed designs of the renovated arcade facades, including structural elements, decorative features, window opening styles, etc., accurately representing the proportions, scales, and spatial layout after renovation. In practical projects, such drawings help construction teams better visualize the design intent, providing crucial reference for actual construction and ensuring the accurate translation of design concepts into physical architectural forms. The figures below (Figures 24 and 25) show the renovated arcade elevation drawings.

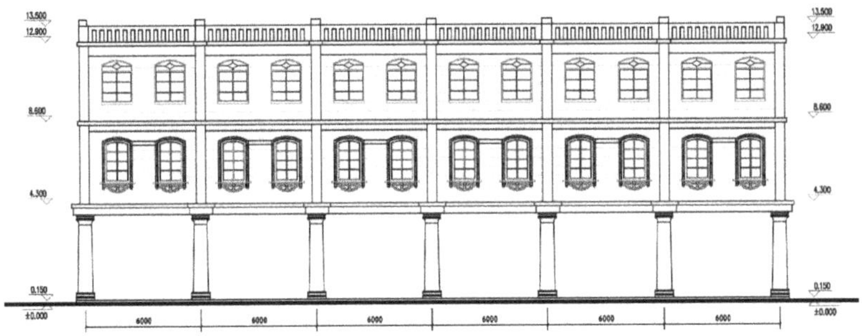

Figure 24. Renovated Arcade Facade Drawing No. 1 (1:200).

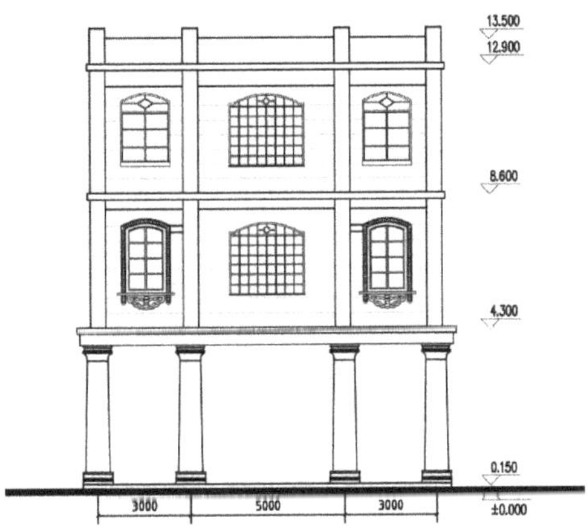

Figure 25. Renovated Arcade Facade Drawing No. 2 (1:200).

4.3. Limitations and Future Directions

This study on deep learning-based facade generation for Shantou's Xiaogongyuan arcades has several limitations. First, the limited dataset of 100 facade images restricts detailed decorative element capture, causing blurring in high-precision generation. However, some scholars like David Newton [35] argue such ambiguous outputs can inspire architectural creativity through intentional "incompleteness." Second, the pix2pix framework prioritizes structural restoration over material texture simulation and construction logic. These constraints, documented in the literature, prompted developments like pix2pix HD [36] for higher-resolution output. While 3D conversion was attempted, results lacked parametric details and practical applicability.

Future research could expand in the following directions:

(1) Enhancing datasets with high-precision scanning and multimodal deep learning;
(2) Integrating AR technologies for public participation and extending methodology to other historic Overseas Chinese districts. These advances could transition AI's role from tool to collaborative decision-maker in cultural heritage preservation and urban renewal.

5. Conclusions

This study focuses on the arcade building complex in the Xiaogongyuan historical district of Shantou and explores an intelligent design method based on deep learning technology to address the need for facade updates. By constructing a Conditional Generative Adversarial Network (CGAN) model, the study achieves automatic generation and style transfer for building facades, providing a new technological path for the digital preservation and renewal of historical districts. The research demonstrates the advantages and effectiveness of this method through data collection and processing, model construction and training, and evaluation of generation outcomes.

The innovative value of this study lies in both its methodology and practical application. This research is the first to systematically apply CGAN technology to the facade renewal of Shantou's arcades, offering a new approach and technical framework for the digital renewal of historical buildings. In addition, the generative model developed in this study enables the rapid output of multiple design alternatives, effectively alleviating issues such as style deviations and low efficiency in traditional restoration processes. In practical applications, the study establishes a digital "Xiaogongyuan Arcade Facade Database" through deep learning of decorative patterns and material characteristics, providing new ideas for the renewal and preservation of historical districts, while also offering a new technological platform for the gradually disappearing traditional crafts.

However, despite the achievements, there are still some limitations. In terms of data, the limited amount of data and insufficient coverage of specific decorative patterns may affect the comprehensiveness of the model, leading to issues such as low-resolution generated images. In terms of technology, while the generation results are promising, there is still room for improvement in simulating material textures with higher precision. In terms of application, the quality of 3D models is still suboptimal and requires further technical refinement.

Overall, this study demonstrates the feasibility and innovativeness of applying deep learning technology in the renewal of historical building facades. Its core value lies in using artificial intelligence to quantify the features and patterns of traditional craftsmanship, establishing a reusable technical framework that balances preservation efficiency with authenticity. This provides an empirical case for the application of AI in cultural heritage preservation. In the future, through cross-regional cooperation and continuous technological optimization, it will be possible to further promote the realization of a new model for historical district protection based on "digital preservation—intelligent renewal—sustainable revitalization," offering a new solution for the global preservation of similar cultural heritage.

Author Contributions: Conceptualization, W.Y., T.W. and C.Z.; methodology, W.Y., T.W. and C.Z.; software, W.Y. and T.W.; validation, W.Y. and T.W.; formal analysis, W.Y.; investigation, W.Y.; resources, W.Y. and T.W.; data curation, W.Y. and T.W.; writing—original draft preparation, W.Y.; writing—review and editing, W.Y., T.W. and C.Z.; visualization, W.Y. and T.W.; supervision, W.Y., T.W. and C.Z.; project administration, W.Y., T.W. and C.Z.; funding acquisition, W.Y., T.W. and C.Z. All authors have read and agreed to the published version of the manuscript.

Funding: This study was supported by 2025 Ministry of Education's Industry-University Collaborative Talent Cultivation Project, Research on practical teaching of virtual human living environment design driven by "AI + human factors" (2507053415); Shantou University Scientific Research Startup Funding Project (NTF22015); Shantou University's 2025 "Artificial Intelligence+" Higher Education Teaching Reform Key Project (STUAI2025006).

Data Availability Statement: The datasets used and analyzed during the current study are available from Wanying Yan (21wyyan@alumni.stu.edu.cn) on reasonable request. All historical facade photographs were taken by Wanying Yan specifically for this research project; all architectural drawings, diagrams, and generated images were produced by Wanying Yan; any previously published images have been properly cited and necessary permissions have been obtained.

Acknowledgments: We acknowledge the financial support from the 2025 Ministry of Education's Industry-University Collaborative Talent Cultivation Project, Shantou University Scientific Research Startup Funding Project, and Shantou University's 2025 "Artificial Intelligence+" Higher Education Teaching Reform Key Project. We also extend our gratitude to our classmates for their invaluable assistance and insightful discussions throughout this study.

Conflicts of Interest: The authors declare no conflicts of interest.

Appendix A

The hardware configuration of the experimental equipment used in this study is as follows: GPU—NVIDIA GeForce RTX 3060 Laptop; CPU—12th Gen Intel(R) Core(TM) i7-12700 HX with a base speed of 2.30 GHz; RAM—16 GB.

The primary software tools employed include Rhino7, Python 3.8.20, PyCharm 2024.3 (Community Edition), PyTorch 1.8.1, with CUDA version 12.6.65.

References

1. Xiao, Z. Analysis of the Cultural Characteristics of Modern Residential Architecture in Shantou. *Shanxi Archit.* **2016**, *42*, 19–20.
2. Yuan, Q.; Cai, T. The Dilemma of Historic District Reconstruction Based on the Perspective of Space Production: A Case Study of Small Park Historic District of Shantou. *Mod. Urban Res.* **2016**, *7*, 68–77.
3. Luo, D. Research on Evidence-Based Design of Building Facades in Historical and Cultural Neighborhoods Based on Machine Learning. Master's Thesis, Guangdong University of Technology, Guangzhou, China, 2025. [CrossRef]
4. Zhang, Y. Discussion on the implementation mode of protection and renovation of historical and cultural blocks. *World Arch.* **2022**, *12*, 73–77.
5. Wu, H. The Strategy of Construction Control Based on the Research About Urban Texture in Shantou Small Park District. Master's Thesis, South China University of Technology, Guangzhou, China, 2021. [CrossRef]
6. Ouyang, L.; Xie, S. Arcades and Road: Urban Planning in Shantou During the Republican Period. *Mod. Chin. Hist. Stud.* **2023**, *4*, 91–106+161.
7. Yi, G.; Wang, S.; Deng, X.; Yang, X.; Liu, J.; Zhang, Z. Structural reinforcement design of modern historical buildings in Trading Port Area of Shantou Small Park. *Build. Struct.* **2022**, *52*, 84–88.
8. Lyu, Q. Protection and Utilization Strategy of Arcade Buildings in Shantou Small Park Kailuan District from the Perspective of Contextual Continuance. Master's Thesis, South China University of Technology, Guangzhou, China, 2019. [CrossRef]
9. Wang, G.; Yi, L. Understanding of the value of arcade building and exploration of practice on restoration. *Urban Archit.* **2018**, *4*, 41–43.
10. Li, Q. Study on Protection and Renovation System of the Arcade Architecture in Lingnan Area—A Case of Protection and Renovation of Arcade Architecture in Zhaoan. Master's Thesis, South China University of Technology, Guangzhou, China, 2021. [CrossRef]
11. Cui, Y. Research on the Protection and Inheritance of Historical Architectural Cultural Heritage Empowered by Big Data 3D Scanning: A Case Study of Deqing Confucius Temple. In *Proceedings of the 2025 Higher Education Teaching Seminar (Volume 1)*; Henan Provincial Non-Government Education Association: Guangzhou, China, 2025; pp. 142–144.
12. Yu, Q.; Yuan, X.; Xu, L. Cross-Material Damage Detection and Analysis for Architectural Heritage Images. *Buildings* **2025**, *15*, 3100. [CrossRef]
13. Xu, J.; Ma, D.; Scaioni, M. Foreign Current Development of HBIM (Historic Building Information Modelling) for Built Heritage. *China Cult. Herit.* **2024**, *1*, 5–16.
14. Zu, X.; Gao, C.; Liu, Y.; Zhao, Z.; Hou, R.; Wang, Y. Machine intelligence for interpretation and preservation of built heritage. *Autom. Constr.* **2025**, *172*, 106055. [CrossRef]
15. Krizhevsky, A.; Sutskever, I.; Hinton, G.E. ImageNet classification with deep convolutional neural networks. *Commun. ACM* **2017**, *60*, 84–90. [CrossRef]
16. Chen, Y.; Wei, G. Application Review of Generative Adversarial Networks. *J. Tongji Univ. Nat. Sci. Ed.* **2020**, *48*, 591–601.
17. Karadag, I. Machine learning for conservation of architectural heritage. *Open House Int.* **2023**, *48*, 23–37. [CrossRef]
18. Basu, A.; Paul, S.; Ghosh, S.; Das, S.; Chanda, B.; Bhagvati, C.; Snasel, V. Digital restoration of cultural heritage with data-driven computing: A survey. *IEEE Access* **2023**, *11*, 53939–53977. [CrossRef]

19. Zhao, F.; Ren, H.; Sun, K.; Zhu, X. GAN-based heterogeneous network for ancient mural restoration. *Herit. Sci.* **2024**, *12*, 418. [CrossRef]
20. Maitin, A.M.; Nogales, A.; Delgado-Martos, E.; Intra Sidola, G.; Pesqueira-Calvo, C.; Furnieles, G.; García-Tejedor, ÁJ. Evaluating Activation Functions in GAN Models for Virtual Inpainting: A Path to Architectural Heritage Restoration. *Appl. Sci.* **2024**, *14*, 6854. [CrossRef]
21. Bachl, M.; Ferreira, D.C. City-GAN: Learning architectural styles using a custom Conditional GAN architecture. *arXiv* **2019**, arXiv:1907.05280.
22. Deng, L.; Yu, D. Deep learning: Methods and applications. *Found. Trends Signal Process.* **2014**, *7*, 197–387. [CrossRef]
23. Wang, Y. Research on Fuzzy Clustering Algorithm for Incomplete Data Based on Improved VAEGAN. Master's Thesis, Liaoning University, Shenyang, China, 2020. [CrossRef]
24. Ye, X.; Du, J.; Ye, Y. MasterplanGAN: Facilitating the smart rendering of urban master plans via generative adversarial networks. *Environ. Plan. B Urban Anal. City Sci.* **2022**, *49*, 794–814. [CrossRef]
25. Ji, W.; Guo, J.; Li, Y. Multi-head mutual-attention cyclegan for unpaired image-to-image translation. *IET Image Process.* **2020**, *14*, 2395–2402. [CrossRef]
26. Jiang, F.; Ma, J.; Webster, C.; Wang, W.; Cheng, J. Automated site planning using CAIN-GAN model. *Autom. Constr.* **2024**, *159*, 105286. [CrossRef]
27. Mostafavi, F.; Tahsildoost, M.; Zomorodian, Z.S.; Shahrestani, S.S. An interactive assessment framework for residential space layouts using pix2pix predictive model at the early-stage building design. *Smart Sustain. Built Environ.* **2024**, *13*, 809–827. [CrossRef]
28. Çiçek, S.; Turhan, G.D.; Taşer, A. Deterioration of pre-war and rehabilitation of post-war urbanscapes using generative adversarial networks. *Int. J. Archit. Comput.* **2023**, *21*, 695–711. [CrossRef]
29. Mirza, M.; Osindero, S. Conditional generative adversarial nets. *arXiv* **2014**, arXiv:1411.1784. [CrossRef]
30. Lin, H.; Huang, L.; Chen, Y.; Zheng, L.; Huang, M.; Chen, Y. Research on the Application of CGAN in the Design of Historic Building Facades in Urban Renewal—Taking Fujian Putian Historic Districts as an Example. *Buildings* **2023**, *13*, 1478. [CrossRef]
31. Zhang, L.; Zheng, L.; Chen, Y.; Huang, L.; Zhou, S. CGAN-assisted renovation of the styles and features of street facades—A case study of the Wuyi area in Fujian, China. *Sustainability* **2022**, *14*, 16575. [CrossRef]
32. Qu, G.; Xue, B. Generative design method of landscape functional layout in residential areas based on conditional generative adversarial nets. *Low Temp. Archit. Technol.* **2022**, *44*, 5–9+14.
33. Huang, Y.; Luo, K.; Qin, L. Impact of Chloride Ions on Durability of Concrete Structures in Marine Environment and Protective Measures. *China Sci. Technol. Inf.* **2022**, *19*, 114–116.
34. Shen, L. Traditional Landscape Features of the Xiaogongyuan Historic District in Shantou City. *Chaoshang* **2013**, *3*, 19–23.
35. Newton, D. Generative deep learning in architectural design. *Technol. | Archit.+Des.* **2019**, *3*, 176–189. [CrossRef]
36. Wang, T.; Liu, M.; Zhu, J.; Tao, A.; Kautz, J.; Catanzaro, B. High-resolution image synthesis and semantic manipulation with conditional Gans. In Proceedings of the IEEE Conference on Computer Vision and Pattern Recognition, Salt Lake City, UT, USA, 18–23 June 2018; pp. 8798–8807.

buildings

MDPI

Article

AI-Based Pre-Renewal Design for Historic Building Facades: An AIGC–LoRA Framework with Collaborative Assessment

Wen Duan [1], Jiacheng Rao [1], Jiarong Zhao [1], Nan Tao [1,*] and Jiangpeng Chen [2]

[1] Faculty of Architecture and City Planning, Kunming University of Science and Technology, Chenggong District, Kunming 650500, China; 20180055@kust.edu.cn (W.D.); raojiacheng@stu.kust.edu.cn (J.R.); yiyi@stu.kust.edu.cn (J.Z.)
[2] ZGC Industry Institute, Haidian District, Beijing 100089, China; chenjiangpeng@zgcresearch.cn
* Correspondence: taonan5599@kust.edu.cn

Abstract: Historic conservation areas face the challenge of balancing heritage preservation with modern adaptation, often resulting in irreversible risks. AIGC technology offers an effective solution to mitigate these renewal risks. Current methods struggle with three bottlenecks: a lack of high-quality datasets, difficulty integrating expert and public preferences, and generating diverse proposals under complex preservation rules. This study proposes an AI-driven pre-renewal framework for building facades, which involves (1) virtual pre-renewal design using a large language model (LLM) to generate facade proposals based on "non-change," "permissible," and "prohibited" rules; (2) a multi-stakeholder evaluation system integrating expert and public judgments via the Bradley–Terry model; and (3) LoRA fine-tuning of Stable Diffusion XL to optimize facade generation. In the case study of the Shangxijie and Xiaxijie Historic Conservation Area of Jincheng Ancient Town, the framework was implemented in three stages. First, LLM-generated facades addressed data scarcity by adhering to preservation constraints. Second, an online platform integrated expert and public evaluations to refine the training dataset. Finally, LoRA fine-tuning improved the model's contextual fidelity and stylistic coherence. Quantitative analysis showed that LoRA models outperformed the base model in authenticity and fidelity. Historic models achieved the highest fidelity (FID = 23.4, SSIM = 0.918, CLIPScore = 0.842), Style-Coordinated models performed stably (composite score = 0.82 ± 0.05, SSIM = 0.884), and Style-Incompatible models showed greater variability (mean = 0.78, SD = 0.09). The expert–public collaborative mechanism validated the iterative "generate–evaluate–refine" workflow as a sustainable approach for heritage facade renewal.

Keywords: historic conservation area; building facade renewal; pre-renewal design; collaborative assessment; LLM; AIGC; LoRA; Stable Diffusion XL

1. Introduction

Since the beginning of the 21st century, rapid urbanization and the global demand for modern functionality have placed historic conservation areas under dual pressures—preservation and renewal [1]. These areas serve as crucial carriers of urban memory and cultural continuity. They must not only maintain their traditional spatial patterns and architectural features but also respond to the modernization needs of residents' daily lives and public services [2]. However, due to aging buildings and insufficient infrastructure, large-scale spontaneous renovations and demolitions have occurred in many areas [3]. Although such activities have improved usability and living conditions, they have often weakened the continuity and integrity of historic characteristics, thereby posing a common challenge for urban heritage conservation worldwide.

In China, this issue has been particularly pronounced. Driven by rapid urbanization and population growth, many historic conservation areas have undergone drastic spatial reconstruction within a short period [4]. Traditional dwellings have been damaged or altered in the process of adapting to modern functional demands, while incompatible modern structures have intruded on a large scale, leading to fragmented streetscapes and a weakened sense of cultural identity [5]. To address this challenge, a series of policies, including the Guidelines for the Conservation and Enhancement of Historic Cities and Districts (Trial Edition), have emphasized preserving the overall historic environment while encouraging facade renewal to achieve harmony between the old and the new [6].

In parallel with these policy developments, digital technology has emerged as a key driver in the modernization of conservation practices. Recent advances in digital modeling and AI-based generative design have significantly transformed the research and practice of facade renewal in historic conservation areas. Among them, Historic Building Information Modeling (HBIM) and digital twin systems have become essential tools for recording the geometric, material, and historical features of facades with high precision. By integrating laser scanning, UAV photogrammetry, and archival images, HBIM enables accurate modeling for deterioration analysis and conservation planning [7]. These methods have greatly enhanced precision and scientific rigor, but their modeling and validation processes remain complex and time-consuming, limiting their applicability to fine-scale restoration projects. Moreover, existing studies suggest that HBIM applications still focus primarily on documentation rather than the adaptive or regenerative design of historical elements [7]. In practice, conservation decisions are largely expert-driven, with limited participation from local residents—despite historic conservation areas being living heritage spaces inhabited by active communities [8].

In recent years, AI-generated content (AIGC) technologies based on Generative Adversarial Networks (GANs) and Diffusion Models have opened new possibilities for facade renewal in heritage contexts. These methods can generate facade styles consistent with historical contexts [9], and they can also produce functional design variations aligned with users' aesthetic preferences and local spatial conditions [10]. However, most AIGC applications remain at the proof-of-concept or visual generation stage, and they still lack a systematic, project-oriented framework that incorporates collaborative assessment. Moreover, current AIGC models require large quantities of high-quality image data for effective training, yet most historic conservation areas suffer from material deterioration and limited photographic documentation, making it difficult to build sufficiently comprehensive datasets for model fine-tuning and validation. To bridge this gap, this study proposes an AIGC–LoRA-based framework for the virtual pre-renewal of historic building facades, aiming to establish an applicable, participatory, and iterative generation–evaluation system that balances authenticity, coordination, and practicality.

Despite significant progress in digital documentation and AI-assisted visualization, current approaches to facade renewal in historic conservation areas continue to face three major challenges.

First, the scarcity of high-quality and comprehensive datasets in historic conservation areas limits the effectiveness of AIGC models. Incomplete or degraded photographic records hinder the models' ability to capture local architectural semantics and material diversity, resulting in reduced generative fidelity and contextual adaptability.

Second, most generative processes lack controllable, rule-based mechanisms to ensure heritage authenticity and spatial consistency during virtual pre-renewal. Without semantic constraints or domain-specific guidance, outcomes generated may appear traditional but deviate from genuine historical logic.

Third, there is no integrated framework that effectively combines expert judgment, public preferences, and AI-based quantitative evaluation within a transparent and iterative assessment process. This absence limits the interpretability, inclusiveness, and reliability of AI-assisted renewal decisions.

To address these limitations, this study proposes an AI-based framework for the virtual pre-renewal of heritage facades, integrating semantic control through language-model-guided generation, Collaborative Assessment, and LoRA-based fine-tuning for domain adaptation. The framework establishes an iterative "generate–evaluate–regenerate" process that bridges technological innovation with cultural authenticity and participatory inclusiveness.

Accordingly, this research is guided by the following key questions:

RQ1: How can language-model-guided generative systems be designed to produce controllable and heritage-compliant virtual pre-renewal schemes, while addressing the challenge of limited and incomplete image datasets in historic conservation areas?

RQ2: How can a collaborative assessment mechanism be developed to effectively integrate expert judgment, public preferences, and AI-based quantitative indicators, thereby optimizing model performance and supporting evidence-based design selection?

RQ3: How can LoRA fine-tuning, informed by collaborative assessment feedback, be applied to enhance the contextual fidelity, stylistic coherence, and adaptive performance of generative outcomes, establishing a closed-loop workflow of "generation–assessment–regeneration"?

These research questions form the conceptual foundation of this paper. The subsequent sections review existing studies on heritage facade renewal, AIGC-based design methods, and participatory co-creation frameworks, before presenting the proposed methodological framework.

2. Literature Review

2.1. Heritage Facade Renewal Approaches

Since the twentieth century, international heritage conservation has gradually evolved from a material-based approach to a value oriented paradigm, establishing authenticity and compatibility as core principles for architectural renewal. The Venice Charter first articulated the concepts of authenticity and integrity [11], laying the foundation for the strict preservationist doctrine that sought to restore monuments to a recognizable earlier state while opposing falsification and excessive reconstruction. This philosophy was later expanded by the Nara Document on Authenticity, which emphasized that the interpretation of authenticity must be contextualized within diverse cultural settings [12]. As conservation philosophy shifted toward adaptive reuse and urban revitalization, the Burra Charter advanced the principle of "cultural significance first," allowing new design interventions that remain coordinated in material, form, and scale, yet are visually distinguishable from the original structure [13]. This more flexible and inclusive approach integrates contemporary design and use while maintaining visual harmony within historic contexts, reflecting the broader evolution of global conservation theory.

Building on international experience, China has gradually developed a localized and multi-layered heritage protection framework. The designation of the first National Historic and Cultural Cities marked the institutionalization of holistic conservation, while the Regulations on the Protection of Historic and Cultural Cities, Towns, and Villages codified the principles of authenticity and integrity [14]. Nationwide surveys and classification standards for historic buildings further reinforced the principle of "preserve all that should be preserved" [15]. Major cities such as Beijing, Shanghai, and Guangzhou have issued façade design guidelines emphasizing "visual harmony from afar, recognizability up close" [16], thereby establishing a comprehensive management framework across legal, planning, and technical dimensions.

Meanwhile, renewal strategies worldwide have continued to diversify. In the United States, the Secretary of the Interior's Standards require that new structures be "style-coordinated yet distinguishable," ensuring coherence in scale and detail while avoiding historical imitation [17]. In Europe—particularly Italy—the concept of symbiotic design encourages dialogue between old and new through proportion, rhythm, and material correspondence [18]. Together, these approaches reflect a global shift from static preservation to dynamic renewal. In line with this trend, China's recent practices have increasingly adopted micro-scale and incremental interventions to revitalize historic environments, as exemplified by Beijing's Yangmeizhu Xiejie, Shanghai's Xintiandi, and Guangzhou's Yongqing Fang [19].

Technological innovation has further transformed facade renewal from craftsmanship-based repair to scientific and intelligent management. BIM and HBIM enable precise digital documentation and structural diagnostics [20]; parametric design generates contextually responsive renewal schemes [21]; AI-based models such as GANs and diffusion frameworks assist in facade restoration and generative design [22]; and AR integrated with IoT monitoring facilitates dynamic, continuous maintenance [23]. Collectively, these advancements illustrate a global shift toward systematic, data-driven, and integrative approaches—anchored in the enduring principles of authenticity and compatibility, yet empowered by technological innovation and creative design to rejuvenate urban heritage.

Overall, the evolution of global heritage conservation reflects a fundamental shift from material restoration to value transmission, and from static preservation to dynamic renewal. Authenticity and compatibility remain the core principles underpinning facade conservation theory. With the advancement of digital technologies, heritage conservation is transitioning from experience-based practices to data-driven and intelligent management, fostering the integration of traditional craftsmanship with technological innovation and promoting the sustainable regeneration of historic urban spaces while preserving cultural continuity.

2.2. AIGC in Architectural Design

Artificial Intelligence-Generated Content (AIGC) is increasingly integrated into architectural design, offering new methodologies for heritage renewal in historic urban contexts. Through image generation and style transfer techniques, AIGC enables designers to efficiently produce diverse visual outcomes that expand creative exploration during the early conceptual stages of design [24]. Compared to traditional manual drawing and modeling, AIGC demonstrates superior efficiency, diversity, and creativity, effectively broadening the design space [25].

The development of AIGC in architecture has evolved from Generative Adversarial Networks (GANs) to Diffusion Models. Early GAN-based methods validated the feasibility of AI-assisted design but were limited by unstable training and inadequate structural control [26]. Diffusion Models, which generate images through progressive denoising, have since achieved higher fidelity and consistency than GANs [27]. Building on this, Stable Diffusion combined with text–image embedding (e.g., CLIP) enables the generation of high-quality architectural imagery guided by textual prompts. Zhang et al. [28] integrated this framework with expert systems to generate culturally distinctive arcade facades for heritage renewal using Low-Rank Adaptation (LoRA) and ControlNet, achieving precise control over composition and style with improved authenticity and diversity.

Prompt engineering has become a crucial factor influencing generative quality. Proper textual descriptions can enhance stylistic coherence and semantic alignment [29]. Researchers emphasize the use of architectural terminology and stylistic vocabulary to guide model outputs toward greater architectural logic [30]. The combination of LoRA and ControlNet allows designers to maintain structural constraints while flexibly adapting stylistic variations, balancing authenticity with creativity [31,32]. Furthermore, AIGC-driven human–AI co-design workflows have proven effective in improving design efficiency and outcome quality in heritage regeneration projects [33].

Nevertheless, the application of AIGC in heritage renewal remains constrained. Architectural design entails complex spatial and semantic logic, and general-purpose models without domain fine-tuning struggle to produce structurally valid or culturally appropriate results [34]. The scarcity of heritage-specific datasets further limits model generalization and contextual sensitivity [35]. Nevertheless, the application of AIGC in heritage renewal remains limited. Architectural design involves intricate spatial hierarchies and cultural semantics, yet current models—trained primarily on generic internet datasets—lack the contextual awareness required to reproduce historically coherent and culturally grounded results. The absence of curated, high-quality heritage datasets further constrains model generalization, as most available images suffer from inconsistent viewpoints, low resolution, and insufficient semantic labeling [35]. Recent studies have emphasized that overcoming these limitations requires integrating data-driven learning with domain expertise. Approaches such as semantic-guided generation, few-shot fine-tuning, and multimodal data fusion have been proposed to improve AIGC's contextual fidelity and cultural consistency in heritage-oriented design tasks [36].

In summary, AIGC has transformed architectural design by enabling efficient and diverse visual generation through diffusion-based frameworks, LoRA, and ControlNet. Yet, its application in heritage renewal remains limited due to the lack of high-quality, domain specific datasets and the challenge of aligning spatial logic with cultural semantics. Future progress depends on integrating data-driven intelligence with expert knowledge to achieve context-aware and culturally coherent design generation.

2.3. Participatory Evaluation and Co-Creation Frameworks

Public participation has become a central paradigm in contemporary heritage conservation and urban renewal. International policy and research indicate a paradigm shift from expert-dominated governance toward collaborative assessment and community co-management. Foundational documents such as the UNESCO Recommendation on the Historic Urban Landscape [37], the Washington Charter [38], and the Faro Convention [39] place the public at the core of heritage identification, protection, and use, emphasizing participation as essential for legitimacy, transparency, and local identity. Studies have shown that public involvement in interpreting heritage values, maintaining collective memory, and monitoring governance creates a positive cycle of "awareness–participation–preservation," thereby reinforcing social cohesion and sustainability [40].

In methodological terms, participation now follows a dual-track model combining offline and online approaches. Offline participation—such as public hearings, workshops, exhibitions, and community co-design—facilitates in-person dialogue and the integration of local knowledge, supporting issue identification and consensus building [41]. However, these processes often involve high organizational costs and limited coverage. In contrast, online participation—including web-based surveys, participatory mapping (PPGIS), digital voting, and integrated platforms [42]—offers broader outreach, lower costs, and faster feedback. Yet, digital participation is prone to biases stemming from questionnaire design, digital divides, and opinion polarization, thus requiring expert moderation to ensure validity [43]. Emerging digital co-creation tools, such as AI-based visualization and virtual reality simulation [44], enhance comprehension and engagement by making design proposals more intuitive. A hybrid model of "online visualization–offline deliberation–online feedback" has proven most effective for forming iterative and inclusive participation loops [45].

In China, public participation is transitioning from experimental to institutionalized practice. While planning announcements and opinion consultations remain dominant, small-scale "micro-renewal" projects are gaining traction [46,47]. Community workshops and dialogues in historic conservation areas have demonstrated positive impacts on design refinement, while online platforms expand information disclosure and feedback channels [48]. Nevertheless, several challenges persist: limited participation channels and coverage marginalize vulnerable groups [49]; engagement often remains consultative rather than collaborative [50]; and feedback mechanisms are weak, making it difficult to translate public input into policy or design outcomes [51]. Moreover, public understanding of heritage values remains shallow, with discussions focusing on environmental improvement and livelihoods rather than cultural continuity [52].

To address these challenges, researchers propose several strategies. First, participation should be advanced to the early stages of planning, forming a continuous process that spans investigation, design, and evaluation. Second, online–offline integration should be strengthened, using digital platforms to enhance transparency and efficiency while fostering consensus through face-to-face deliberation. Third, public education, community archiving, and visualization tools can improve understanding and expression. Finally, a systematic adoption and evaluation mechanism should be established, specifying opinion-processing procedures, responsible bodies, and supervision indicators within planning documents to ensure traceable and accountable outcomes. Through these measures, participation can evolve from passive consultation to active co-governance, achieving synergy between cultural value, social benefit, and governance efficiency.

Overall, public participation has become a key driver in heritage conservation and urban renewal. Internationally, people-centered governance frameworks have matured, while in China, participation is evolving toward institutionalized and digitalized practices. However, limited inclusiveness, weak feedback mechanisms, and insufficient cultural awareness continue to constrain its effectiveness. Strengthening public education, visualization, and online–offline integration can foster more inclusive and sustainable models of collaborative heritage governance.

3. Methodology

3.1. Technical Framework

The preceding review reveals that studies on heritage facade renewal have advanced along three paths: theoretical exploration of authenticity and compatibility, the technological evolution of AIGC-based generative design, and the emergence of participatory conservation frameworks. However, these strands remain largely disconnected. Existing research often isolates cultural interpretation from computational control and stakeholder evaluation, while data scarcity continues to limit the contextual fidelity of AI-generated results. To bridge these gaps, this study proposes an AI-based pre-renewal framework integrating semantic control, collaborative assessment, and LoRA fine-tuning to establish a controllable, inclusive, and context-aware workflow for adaptive facade renewal in historic conservation areas. To realize this objective in practice, the study constructs a technical framework that translates the proposed concept into an operable and verifiable methodology.

This study establishes an integrated methodological framework for AI-based facade pre-renewal, consisting of three interrelated stages: virtual pre-renewal, collaborative assessment, and LoRA fine-tuning (Figure 1). The framework aims to simulate and optimize facade renewal schemes before physical implementation, forming an iterative "generate–evaluate–regenerate" process that bridges cultural, aesthetic, and computational dimensions.

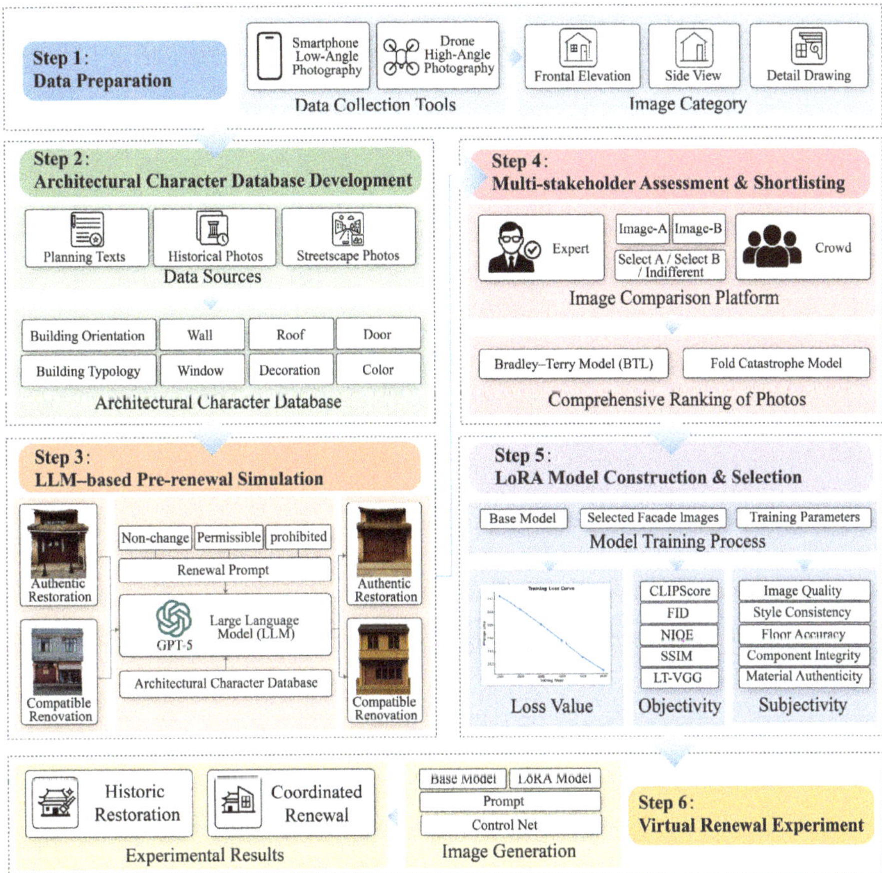

Figure 1. Technical Framework Diagram.

In the virtual pre-renewal stage, a language-model-guided multimodal generative framework [53] was employed to perform semantically controlled image generation under a rule-based constraint system defining "non-change," "permissible," and "prohibited" elements [54]. The language model interprets architectural descriptions and translates them into structured prompts that guide the visual generator, ensuring semantic alignment and contextual authenticity. Two generative modes: Authentic Restoration and Compatible Renovation [55], based on a review of heritage conservation approaches, are regarded as core principles of historic preservation, with authenticity and compatibility remaining central to facade conservation theory. Authentic Restoration focuses on preserving the original form and materials of the building, ensuring historical accuracy and integrity, and emphasizing fidelity to the historical context. On the other hand, Compatible Renovation allows for moderate modern design interventions that maintain harmony with the historical environment, while ensuring that new design elements complement the original structure without compromising its cultural significance.

In the evaluation stage, an online pairwise-comparison platform was developed to integrate subjective and objective assessments. Expert and public preferences were quantified using the Bradley–Terry paired comparison model [56], while objective indicators—including CLIPScore [57], FID [58], NIQE [59], SSIM [60], and VGG-sim [61]—were used to evaluate semantic consistency, visual quality, and structural fidelity. A Fold Catastrophe model [62] was then applied to synthesize expert and public evaluations, identifying the optimal subset of generated samples for model optimization.

In the LoRA fine-tuning stage, the Stable Diffusion XL 1.0 model served as the visual backbone and was domain-adapted through Low-Rank Adaptation (LoRA) [63]. The training process incorporated LLM-driven prompt encodings to enhance semantic control and cross-modal alignment. Hyperparameters were systematically configured, and loss convergence was monitored using the Adam optimizer [64] to ensure training stability. Through parameter exploration and multi-metric validation, controllable LoRA parameter ranges were established, enabling high-fidelity, context-sensitive, and text-guided facade generation that aligns with expert standards and public aesthetic expectations.

3.2. Data Preparation

Building facades served as the primary objects of data acquisition. Photographs focused mainly on frontal elevations, supplemented by side views when spatial conditions allowed. A unified shooting protocol was followed to ensure data consistency: images were taken under even lighting and minimal obstruction, with cameras aligned perpendicularly to minimize perspective distortion. Key architectural elements such as windows, eaves, and decorative details were additionally documented through close-up photography to support detailed morphological analysis.

To achieve comprehensive coverage and precision, a combination of drone and hand-held photography was adopted. Aerial and oblique images were captured using a drone, while ground-level details were recorded with a smartphone camera. This hybrid method ensured both high-resolution imaging and realistic spatial representation, balancing macro-level coverage with fine-grained accuracy.

Based on the collected imagery, a facade dataset was established to document the architectural features of buildings within the study area. Each record includes basic architectural information such as building type, material composition, roof form, opening ratio, decorative style, and color scheme, forming a structured system of visual and descriptive data.

To ensure the accuracy and contextual reliability of the dataset, field survey data were cross-validated with historical photographs, archival materials, and official planning documents. Historical images and local chronicles were used to verify facade authenticity and trace the evolution of architectural forms, while heritage protection plans and design guidelines provided references for typological and stylistic consistency.

Building upon this foundation, a Facade Material and Style Library was developed as a structured repository integrating both visual and semantic information. Characteristic architectural elements extracted from field data, historical imagery, and planning documents were encoded into metadata for subsequent use in prompt construction and model training.

3.3. Analytical and Modeling Methods

3.3.1. LLM-Based Facade Pre-Renewal Generation

In recent years, the protection and renewal of historic urban areas in China have placed growing emphasis on authenticity and stylistic compatibility. National guidelines call for "rectifying stylistically incompatible buildings and landscapes," while prohibiting practices such as "demolishing the genuine to build the fake" and "creating pseudo-historic sceneries," to avoid superficial imitation and excessive stylization in so-called "ancient-style streets" [65,66]. Within this policy context, new approaches are needed to balance community needs, modern functionality, and the continuity of historical character, ensuring the sustainable renewal of heritage environments.

To meet this challenge, the study proposes an LLM-guided diffusion framework for facade pre-renewal generation. In this approach, a large language model (LLM) acts as a semantic controller, guiding the diffusion process and replacing manually written prompts. The model interprets structured metadata from the facade library—covering materials, roof forms, openings, decorative elements, and color schemes—and converts them into context-aware textual descriptions that align with architectural and cultural semantics.

In this study, the large language model (LLM) used is GPT-5, accessed via the official OpenAI ChatGPT Plus platform (2025 release). GPT-5 was selected for its enhanced contextual reasoning and prompt-generation capabilities, which enable precise semantic interpretation of architectural metadata and coherent generation of context-aware design descriptions.

A rule-based constraint system ensures historical integrity and structural logic by classifying elements as non-change, permissible, or prohibited. Non-change components (e.g., proportions, roof form) preserve authenticity; permissible ones (e.g., color or texture) allow limited contextual adaptation; and prohibited actions prevent distortion of heritage features.

Based on this structure, two generative paths were defined: Authentic Restoration and Compatible Renovation. These reflect the dual principles of faithful preservation and contextual adaptation, guiding the model to balance historical fidelity with functional renewal and to align generative outcomes with the cultural values of heritage protection.

Finally, LLM-generated prompts were integrated into the diffusion model, whose semantic modules refined composition, material expression, and detailing. The generated pre-renewal images demonstrated strong semantic coherence, material realism, and contextual harmony, validating the feasibility of the LLM-guided approach as a low-risk, visual, and data-supported pathway for facade restoration in historic conservation areas.

3.3.2. Collaborative Assessment and Integration

An online scoring platform based on pairwise comparison was developed to implement this system. The platform was built using the React + Vite front-end framework and connected to Google Apps Script and Google Sheets for real-time data interaction and storage. It dynamically displayed two candidate images side by side, allowing evaluators to select one of three options—"Left image is better," "Right image is better," or "Both are similar." To reduce evaluation fatigue while maintaining balanced participation, the task allocation module automatically controlled the frequency of image appearances, ensured even distribution across evaluators, and prevented duplicate matchings. In addition, several standard and repeated samples were embedded for quality control to ensure the reliability and consistency of the collected evaluations (Figure 2).

During the data modeling stage, the Bradley–Terry model was applied separately to expert and public evaluations to estimate the latent preference strength of each image. For any image pair (i,j), if image i was preferred, $y_{ij} = 1$; otherwise, $y_{ij} = 0$. The corresponding log-likelihood function was expressed as follows:

$$\ell(\mu) = \sum_{(i,j)\in D} [y_{ij} \log \sigma(\mu_i - \mu_j) + (1 - y_{ij}) \log \sigma(\mu_j - \mu_i)], \sigma(x) = \frac{1}{1 + e^{-x}}$$

The resulting strength vectors of expert and public evaluations, μ^E and μ^P, were normalized using min–max scaling to ensure comparability:

$$e_i = \frac{\mu_i^{(E)} - \min(\mu^{(E)})}{\max(\mu^{(E)}) - \min(\mu^{(E)})}, p_i = \frac{\mu_i^{(P)} - \min(\mu^{(P)})}{\max(\mu^{(P)}) - \min(\mu^{(P)})}$$

To synthesize these two evaluation sources, a Fold Catastrophe model was employed to integrate expert and public judgments. Its potential function was defined as follows:

$$V(x) = -\frac{x^3}{3} + e_i x + p_i$$

where x represents the state variable, and (e_i, p_i) denote the normalized expert and public scores, respectively. The equilibrium condition

$$\frac{\partial V}{\partial x} = 0$$

yields the comprehensive state value S_i for each image, from which a ranking was derived. The resulting composite scores were then used to select the optimal subset of facade images for subsequent LoRA fine-tuning and training, effectively linking perceptual evaluation with computational optimization.

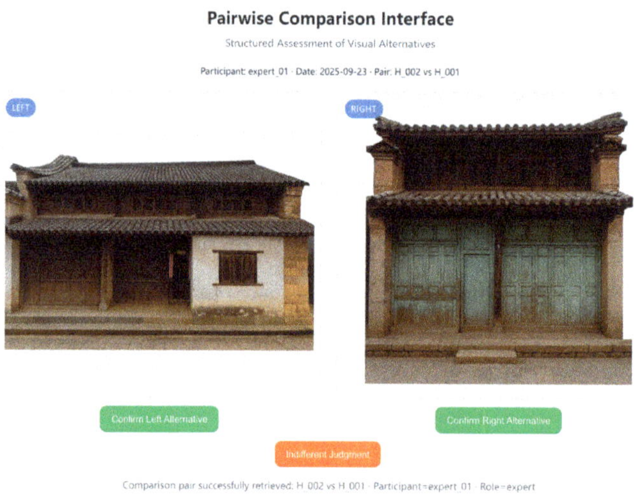

Figure 2. Interactive Scoring Platform.

3.3.3. LoRA Model Evaluation and Selection

To fully capitalize on the outcomes of the aforementioned collaborative assessment and to account for the dataset scale and model adaptability, this study adopted the LoRA (Low-Rank Adaptation) framework for model optimization and fine-tuning. LoRA is capable of generating facade images that meet the required standards while ensuring strong generalizability and preserving historical authenticity. To optimize the generative performance and perceptual authenticity of the LoRA models, a two-stage evaluation system was established, integrating both objective indicators and expert subjective assessments. In this context, "evaluation" refers to two complementary dimensions: data-based quantitative measurement and expert-driven qualitative judgment, which were analyzed and integrated sequentially to ensure methodological transparency.

Independent validation sets (excluded from training) were used to evaluate three categories of LoRA models. Each model was tested across weight intervals from 0.4 to 1.0 (with a step size of 0.1), and the optimal weight was determined based on the minimum Fréchet Inception Distance (FID) and maximum Structural Similarity Index (SSIM). Using these optimal weights, regenerated samples were analyzed through five normalized objective dimensions—semantic alignment, distribution similarity, image naturalness, structural consistency, and local detail fidelity—to ensure comprehensive comparison across models.

The composite objective score for each model was calculated using a weighted summation method:

$$S_i = \sum_{j=1}^{m} w_j \cdot X_{ij},\ w_j = \frac{1}{m}, m = 5$$

where X_{ij} denotes the normalized value of the j-th indicator for the i-th model. Models with higher comprehensive scores were ranked, and the top candidates from each category were selected for expert evaluation.

Subsequently, experts specializing in architectural heritage conservation conducted detailed evaluations using the Stable Diffusion + ControlNet framework. Each LoRA model was tested within its optimal weight range to generate comparative facade images. Experts scored the outputs across five dimensions—image quality, style consistency, floor count accuracy, component integrity, and material authenticity—based on standardized assessment criteria (Table 1).

Table 1. Evaluation Metrics for LoRA Model Scoring.

Evaluation Metric	Description	Scoring Criteria (1–5)
Image Quality	Clarity, detail rendition, and overall visual realism.	5 = Ultra-HD, flawless details; 4 = Clear, minor noise; 3 = Acceptable, noticeable defects; 2 = Blurry, detail loss; 1 = Severely distorted.
Style Consistency	Alignment of generated facades with the target Historic/Style-Coordinated/Style-Incompatible typology.	5 = Fully consistent; 4 = Key features preserved; 3 = Partially consistent; 2 = Stylistically confused; 1 = Completely inconsistent.
Floor Count Accuracy	Accurate number of storeys relative to historical reference.	5 = Exact match; 4 = Error $\leq$ 1 storey; 3 = Error $\leq$ 2 storeys; 2 = Error $\geq$ 3 storeys; 1 = Completely inconsistent.
Component Integrity	Completeness and refinement of architectural elements (doors, windows, eaves, carvings).	5 = All intact, no distortion; 4 = Minor simplifications; 3 = Key elements missing; 2 = Multiple errors; 1 = Completely absent.
Material Authenticity	Fidelity of material textures and colors to historical references.	5 = Fully authentic; 4 = Major materials accurate; 3 = Partial distortions; 2 = Mixed or confused materials; 1 = Entirely inaccurate.

Finally, models demonstrating an optimal balance between quantitative performance and expert-based evaluation were selected as the final candidates for subsequent generative experiments and validation.

4. Case Study

4.1. Study Area

The Shangxijie and Xiaxijie Historic Conservation Areas of Jincheng Ancient Town are located in the southwest of Jincheng Town, Jinning District, Kunming City, Yunnan Province, on the southeastern shore of Dianchi Lake. This area is a significant birthplace of ancient Dian culture and one of the three core Historic and Cultural Conservation Areas designated in the Conservation Plan for the Historic and Cultural City of Kunming. The protected zone covers 42.7 hectares, with 9.76 hectares designated as the core protection zone (Figure 3).

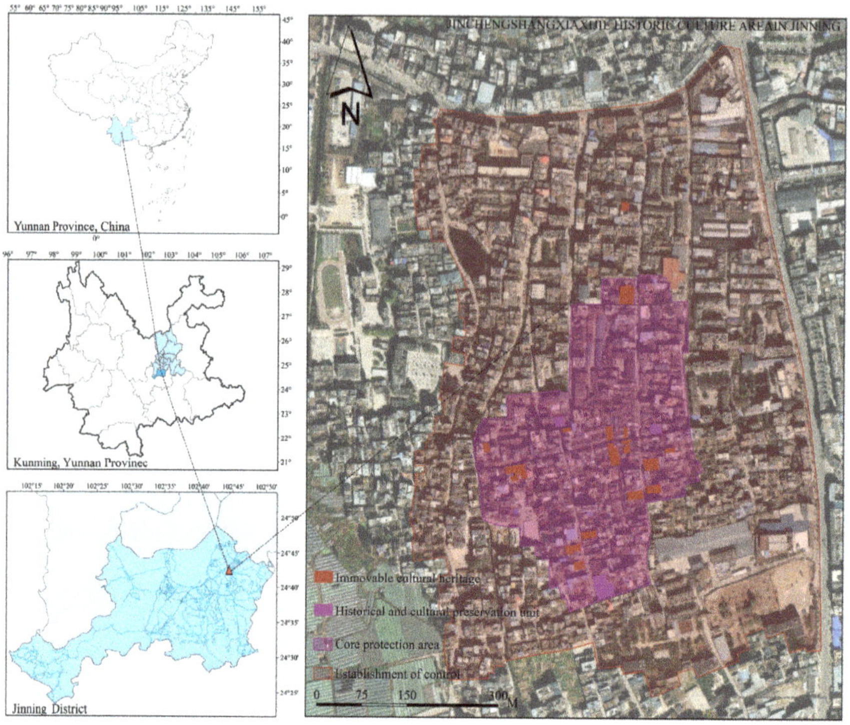

Figure 3. Location Map of the Study Area.

The area generally retains the traditional street pattern and courtyard-style dwellings, though the overall preservation condition is moderate. Many historic buildings show structural damage and roof collapse due to long-term neglect. The traditional "Yikeyin" courtyard layout can still be identified in some streets, forming the characteristic pattern of "shops in the front and residences in the back." The main building materials include rammed earth, grey bricks, and timber, with gable or hip roofs covered by small grey tiles. While some original details—such as eaves, window lattices, and door frames—are still visible, the overall architectural integrity has been partially compromised. The prevailing tones of grey-blue, ochre, and wood evoke a faint sense of tranquility and historical simplicity.

At the same time, modern development is increasingly encroaching on the area. A significant number of reinforced concrete residences have emerged around the periphery of the Historic and Cultural Conservation Areas. Some streets feature buildings with flat roofs, tiled facades, or bright-colored coatings, resulting in significant stylistic incompatibility with traditional architecture.

Due to its distinctive historical significance and the relatively deteriorated preservation condition, Jincheng Ancient Town—where historical heritage coexists with modern development—serves as an exemplary site that reflects a critical challenge in AIGC research: the scarcity of high-quality image data for model training. The coexistence of well-preserved historical elements and incompatible modern interventions provides an ideal experimental environment for investigating facade generation and conservation strategies, while also establishing a solid empirical foundation for this study's application and validation in the context of historic building renewal.

According to the Conservation Plan for the Shangxijie and Xiaxijie Historic Conservation Area of Jincheng Ancient Town in Jinning District, the buildings within the planning boundary are classified into three categories:

(1) Historic Buildings, referring to those with historical, scientific, or artistic value that reflect the city's traditional features. Many are in poor condition due to prolonged disrepair.
(2) Style-Coordinated Buildings, which are compatible with the historic landscape and include traditional timber–earth dwellings, early-period brick buildings representing specific historical stages of Jincheng, and buildings with moderate historical characteristics.
(3) Style-Incompatible Buildings, which conflict with the historic setting due to excessive height (over four stories) or inappropriate façade treatment lacking traditional elements. These buildings have been rapidly increasing in number around the protected zone.

Based on the latest conservation survey, the proportions of the three building categories are approximately as follows: Historic Buildings 22%, Style-Coordinated Buildings 19.5%, and Style-Incompatible Buildings 58.5%. Overall, the area demonstrates a spatial pattern in which a relatively intact historic core is encircled by increasingly incompatible peripheral developments (Figures 4 and 5). These results further highlight the coexistence of modern and traditional architecture within the Jinning historic district, underscoring the urgent need for systematic strategies of conservation and renewal. In this study, Authentic Restoration is applied to deteriorated historic buildings to recover their structural integrity and historical authenticity, while Compatible Renovation is adopted for the numerous style-incompatible buildings to achieve visual coherence and contextual harmony through façade coordination and adaptive redesign.

Figure 4. Aerial View of the Study Area in Its Current Condition.

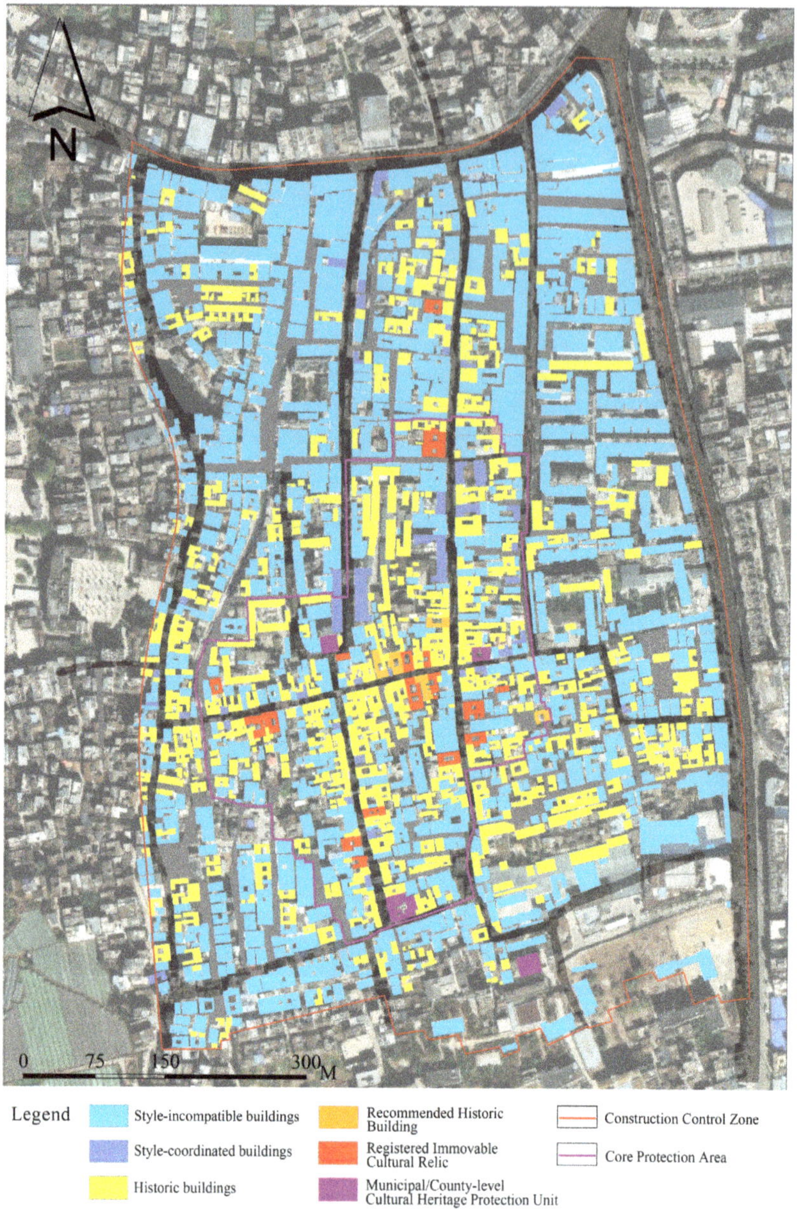

Figure 5. Architectural Characteristics of the Study Area.

4.2. Architectural Characterization and Facade Dataset Construction

The documentation of building facades in the Shangxijie and Xiaxijie Historic Conservation Area revealed a rich diversity of architectural typologies reflecting different periods of development and varying degrees of stylistic integrity. A total of 414 facade samples were identified, comprising 263 Historic, 64 Style-Coordinated, and 87 Style-Incompatible buildings. These images formed a comprehensive visual and structural record of the district's architectural landscape.

In terms of equipment, a combination of drone and smartphone photography was adopted to ensure multi-angle data acquisition. A DJI Mini 3 Pro (DJI, Shenzhen, China) was used to capture aerial and oblique perspectives, while an OPPO Find X8 Pro (OPPO, Dongguan, China) documented street-level details. This configuration ensured high-resolution and realistic imaging, balancing macro-level coverage with micro-scale precision.

Analysis of the collected data showed that historic and Style-Coordinated buildings generally preserved traditional features such as rammed-earth or grey-brick walls, timber doors and lattice windows, and grey-tiled gable roofs. Their balanced proportions, decorative eaves, and subdued color tones reflected the calm and coherent visual identity typical of the Dian cultural region. In contrast, Style-Incompatible buildings often exhibited bright coatings, metal roofing, and oversized billboards, disrupting the spatial rhythm and skyline continuity of the historic streetscape.

To ensure contextual accuracy, the interpretation of architectural features and classification criteria was cross-referenced with the Conservation and Development Plan for the Shangxijie and Xiaxijie Historic Conservation Area in Jinning District, Jincheng Ancient Town. This planning document provided regulatory definitions of architectural typology, material guidelines, and conservation zones, which were used to validate and standardize the dataset structure.

Building upon these findings, a Facade Material and Style Library was established to digitally archive and categorize the architectural characteristics of the area (Figure 6). The library integrates visual samples with semantic metadata describing material textures, roof forms, opening ratios, decorative patterns, and chromatic schemes, forming a dual-layered database of imagery and structured information. This dataset not only preserves the distinctive material and stylistic features of Jincheng Ancient Town, but also provides a systematic representation of local facade typologies and enables subsequent AI-based modeling to achieve context-aware and culturally grounded generation.

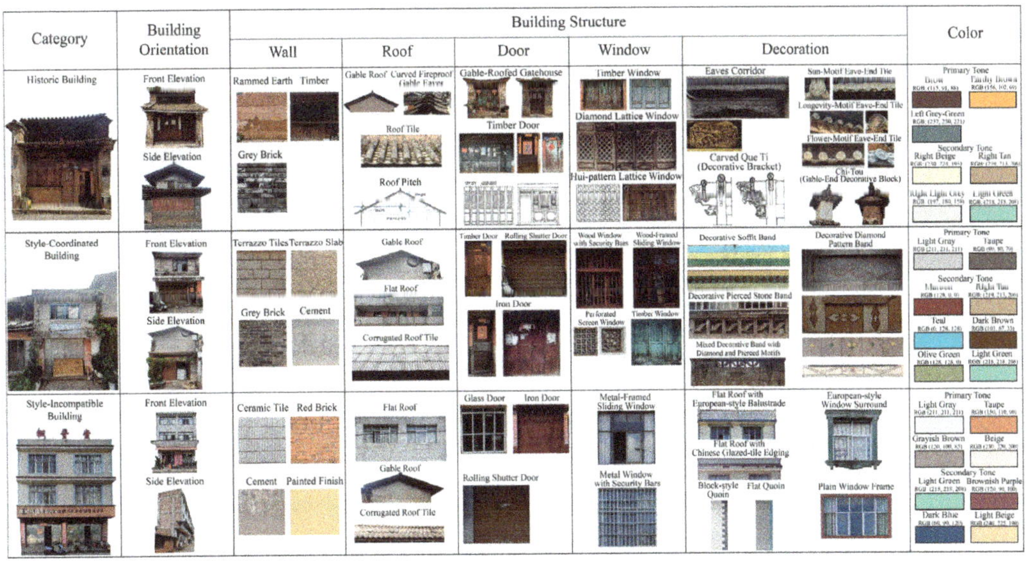

Figure 6. Facade Material and Style Library.

4.3. Semantic-Controlled Facade Pre-Renewal Results

4.3.1. Authentic Restoration

For the historic and Style-Coordinated buildings in the Shangxijie and Xiaxijie areas of Jincheng Ancient Town, the study adopted an authentic restoration approach. The original facade photographs were used as the primary inputs, accompanied by reference images containing the invariable elements as auxiliary guidance. Within the prompt, three categories of rules—non-change, permissible, and prohibited—were explicitly defined.

The non-change elements included the architectural style, wall materials, roof form, positions and proportions of door and window openings, and the locations and rhythms of decorative components such as eaves, moldings, tiles, and carvings, as well as the overall color palette. The permissible actions were limited to the removal of non-historic attachments such as wires, advertisements, vehicles, and security grilles, along with minor same-material repairs and light surface cleaning, unified under a soft diffuse lighting effect. The prohibited operations included the addition or removal of components, replacement of materials, or alteration of hue.

Based on these rule constraints, the model performed controlled editing and output the updated images (Figure 7). In total, 178 restored images of Historic buildings and 86 restored images of Style-Coordinated buildings were generated.

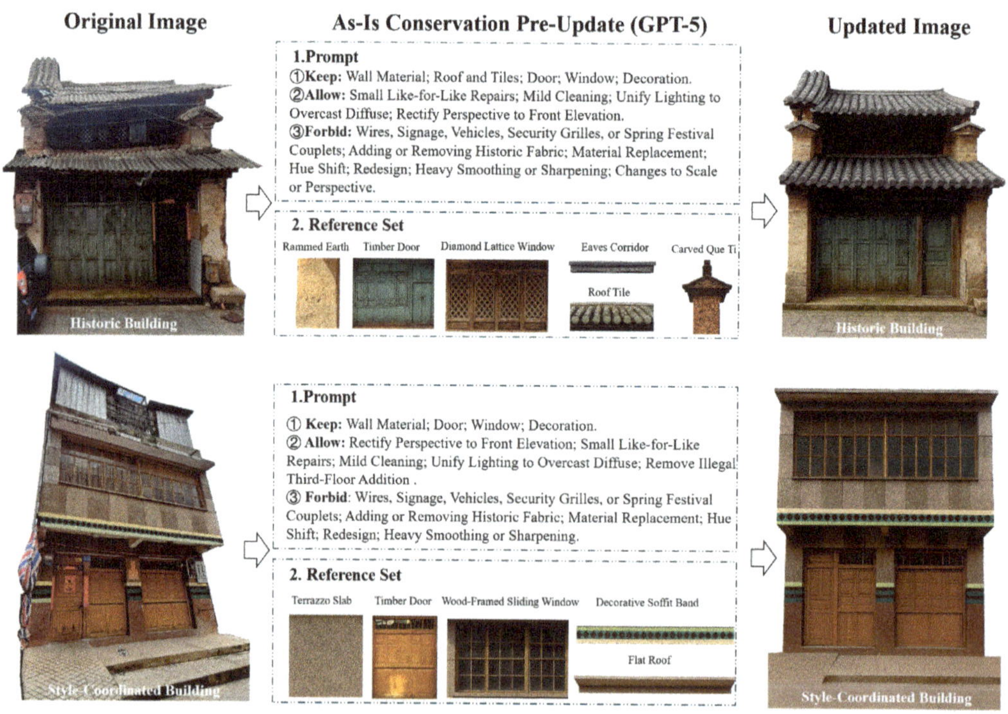

Figure 7. Demonstration of the Authentic Restoration Process.

4.3.2. Compatible Renovation

For stylistically incompatible buildings in the Shangxijie and Xiaxijie areas of Jincheng Ancient Town, this study adopted a compatible renovation strategy aimed at balancing modern functional needs with the continuity of historical character. The approach integrated findings from resident surveys, on-site observations, and expert assessments to guide the generation and optimization of facade renovation schemes.

A total of 21 residents representing different age groups were interviewed to identify major concerns regarding facade renewal. The results revealed three dominant priorities: economic efficiency, sense of historicity, and comfort. A structured questionnaire was subsequently distributed, yielding 63 valid responses. The mean scores for each dimension were 4.20 for economic efficiency, 3.93 for historicity, and 3.74 for comfort, indicating that residents prioritized cost-effectiveness while still valuing heritage continuity and moderate improvements in livability.

Field investigations further showed that stylistically incompatible buildings were typically characterized by colored steel roofs, oversized billboards, low-cost facade materials, and additional stories, which, although economical, disrupted the spatial continuity and skyline integrity of the historic district. In contrast, traditional architectural features—such as grey-tiled pitched roofs, rammed-earth or brick walls, and timber windows and doors—conveyed a coherent historic characteristic and served as references for compatible renovation.

Building on these findings, a prompt-generation logic framework was developed to translate public preferences and expert evaluations into multi-objective design parameters. The framework defines four renovation orientations—Heritage-first, Cost-first, Comfort-first, and Integrated—which collectively reflect the trade-offs among authenticity, economy, and livability. By regulating key facade components such as materials, windows, doors, and roof eaves, this framework produces differentiated renovation schemes that balance stylistic harmony with functional adaptation (Table 2).

Table 2. Prompt Generation Logic for Compatible Renovation.

Dimension	Heritage-First Prompt	Cost-First Prompt	Comfort-First Prompt	Integrated Prompt
Facade Materials	Traditional brick or rammed earth; restore texture with local repair.	Uniform paint (grey, beige, wood tone); low-cost coatings.	Insulated, eco-efficient finishes for comfort.	Wood-tone aluminum panels; moderate cost and insulation.
Windows	Timber lattice windows in traditional proportions.	Aluminum frames with wood finish; simple divisions.	Large double-glazed windows for light and soundproofing.	Wood-tone aluminum lattice windows; balance cost and daylight.
Doors	Timber or historically styled glass doors.	Simple, solid or faux-wood doors for economy.	Composite insulated or large glass doors	Wood-lattice or faux-wood doors combining economy and heritage.
Roof & Eaves	Grey/blue tiled roof with curved eaves restored.	Lightweight imitation tiles; short overhangs.	Insulated roof with drainage and shading.	Lightweight grey tiles; simplified eaves for tradition and efficiency.

Guided by this framework, the model successfully generated a total of 300 facade renovation images of Style-Incompatible buildings, distributed across four renovation orientations—Heritage-first (75 images), Cost-first (78 images), Comfort-first (72 images), and Integrated (75 images)—that were contextually coherent, economically viable, and visually harmonious (Figure 8).

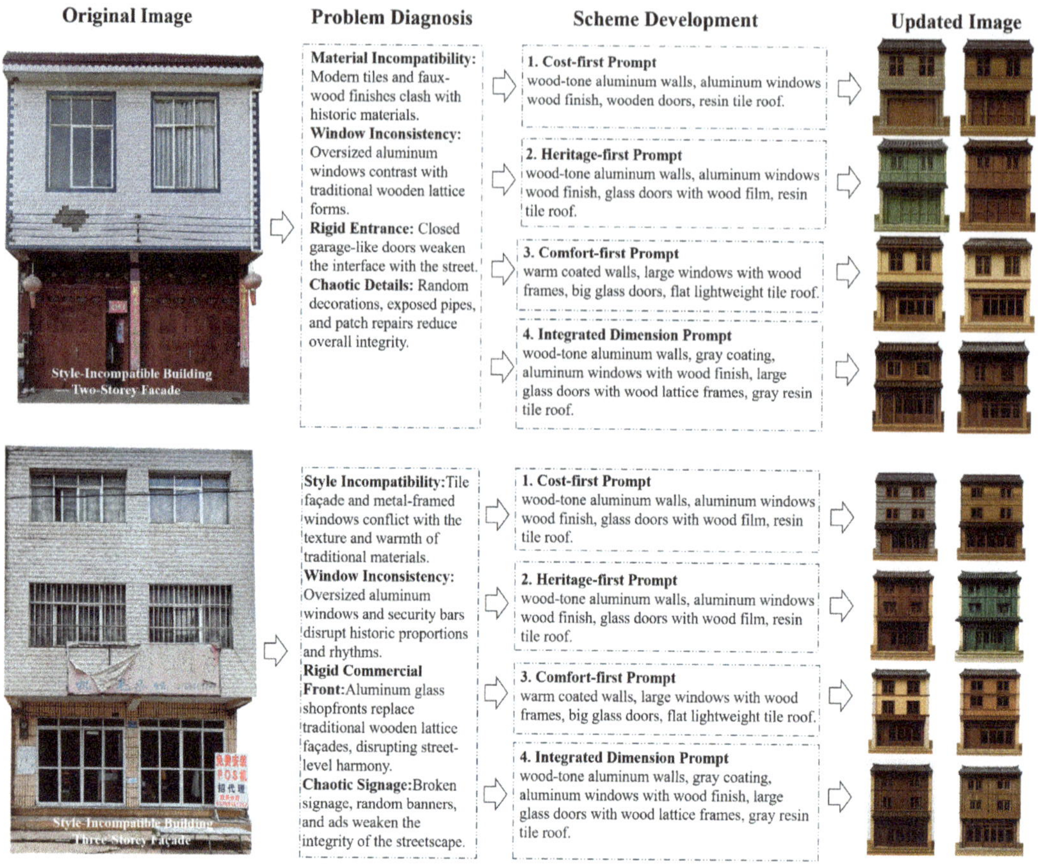

Figure 8. Demonstration of the Compatible Renovation Process.

4.4. Collaborative Assessment Results

In the collaborative assessment phase, a comparative scoring system was used to assess the visual quality and stylistic compatibility of the generated facade images. A total of 564 samples were evaluated, including 178 images of Historic buildings, 86 of Style-Coordinated buildings, and 300 of stylistically incompatible buildings.

A total of 44 participants took part in the evaluation, consisting of 6 experts in architecture and heritage conservation and 38 public representatives. Each participant completed multiple pairwise comparison tasks, selecting the preferred image or indicating similarity between two alternatives. To ensure both coverage and evaluation robustness, each image appeared in approximately 16 comparison pairs, resulting in a total of 4512 comparison tasks. On average, public participants completed around 72 comparisons each, while experts completed approximately 296, ensuring balanced participation and representative sampling across all building categories.

The evaluation results from experts and the public were analyzed independently and then normalized to ensure comparability. The findings revealed clear differences in judgment criteria: experts emphasized authenticity and stylistic integrity, while the public prioritized visual harmony and perceptual comfort. The integrated ranking combined both perspectives, using authenticity, stylistic coherence, and overall visual quality as core evaluation dimensions.

Based on the aggregated scores and manual verification, a total of 120 images of Historic buildings (204 components), 62 of Style-Coordinated buildings (126 components), and 192 images of Style-Incompatible buildings (328 components) were finally selected, further categorized into four renovation orientations—48 Heritage-first, 46 Cost-first, 49 Comfort-first, and 49 Integrated schemes. These curated datasets provide a high-quality foundation for LoRA fine-tuning and model training, ensuring balanced representation and stylistic diversity across categories.

4.5. LoRA Model Optimization and Performance Evaluation

4.5.1. LoRA Training and Convergence Results

The filtered image dataset was used for model training based on the Stable Diffusion XL (SDXL 1.0) model, with domain adaptation conducted using the LoRA technique [65]. The key parameters were configured as follows: rank = 128, α = 128, learning rate = 1×10^{-4}, batch size = 2, and image resolution = 1024 × 1024. These settings were systematically determined through multiple experimental trials and were consistent with adaptive fine-tuning strategies commonly applied in domain-specific LoRA models for architectural image generation.

A higher rank (r = 128) was employed to better capture the complex spatial configurations and stylistic details characteristic of historic facades [65], while the learning rate (1×10^{-4}) was selected to ensure stable and gradual convergence during fine-tuning [66]. The 1024 × 1024 resolution fully leveraged SDXL's native capacity for high-fidelity image generation [63], and the small batch size of 2 provided an optimal trade-off between GPU memory efficiency and overfitting prevention in small-scale datasets [67]. Collectively, these configurations achieved a balance between generative quality, computational efficiency, and model stability.

Training was performed on a workstation equipped with one 48 GB vGPU, 25 Intel Xeon Platinum 8481C vCPUs, and 90 GB RAM. The total number of iterations was adjusted to account for dataset scale and complexity: 9700 for Historic buildings, 6000 for Style-Coordinated buildings, and 15,600 for Style-Incompatible buildings. To ensure reproducibility and comparability, model checkpoints were saved at regular intervals, and five LoRA weight files were retained for each model upon completion.

During training, all three models demonstrated a smooth and continuous decline in loss values before reaching convergence, indicating stable optimization performance. Specifically, the loss value of the Historic model decreased from 0.063 at 1000 steps to 0.017 at 9700 steps; the Style-Coordinated model declined from 0.076 to 0.025 over 6000 steps, and the Style-Incompatible model dropped from 0.062 to 0.018 over 15,600 steps, stabilizing after approximately 13,000 steps. The absence of abnormal oscillations across all three loss curves further confirmed the robustness, convergence stability, and training effectiveness of the LoRA-based fine-tuning process (Figure 9).

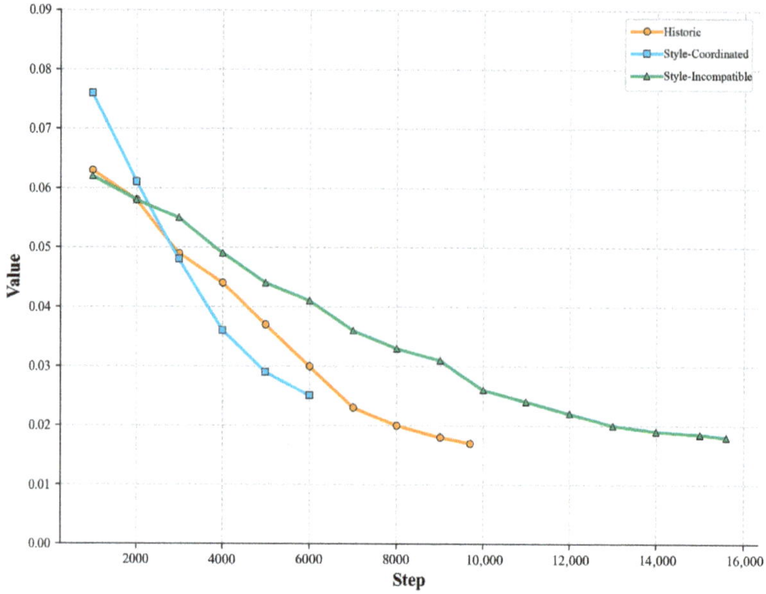

Figure 9. LoRA Model Loss Curves During Training.

4.5.2. Objective Indicator Evaluation

This section presents a systematic quantitative evaluation of the generative performance of three categories of LoRA models—Historic, Style-Coordinated, and Style-Incompatible. The model outputs were analyzed across five normalized objective dimensions: semantic consistency (CLIPScore), distribution similarity (FID), image naturalness (NIQE), structural fidelity (SSIM), and local texture similarity (VGG feature similarity). A weighted composite score was then used to assess overall model performance (Figure 10).

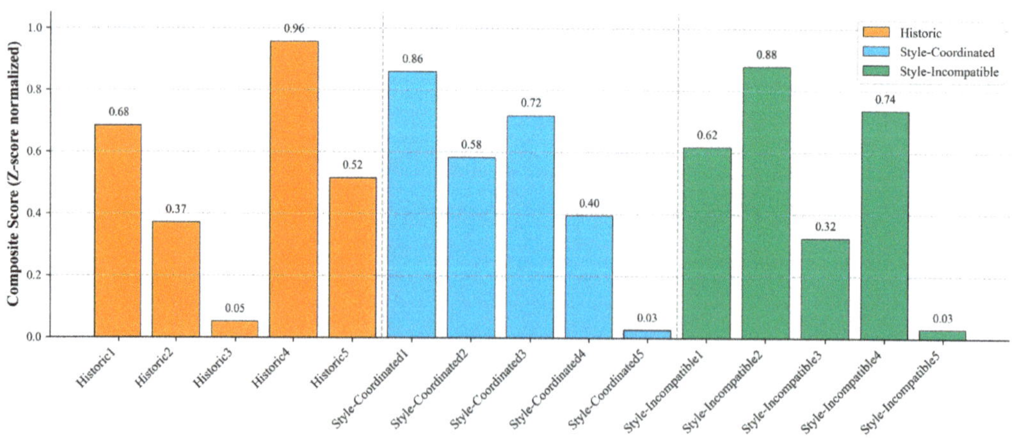

Figure 10. Bar Chart of Objective Indicator Evaluation for LoRA Model Selection.

In the Historic category, the models exhibited noticeable performance differentiation. Historic4 (0.96) and Historic1 (0.68) achieved the highest composite scores, indicating strong performance in structural fidelity and textural reconstruction. Specifically, Historic4 obtained the lowest FID (23.4), the highest SSIM (0.918), and a high CLIPScore (0.842), demonstrating superior semantic alignment and realistic material restoration, particularly in eaves, window frames, and decorative details. In contrast, Historic2 (0.37) and Historic3 (0.05) showed relatively weak generalization, with minor overfitting and inconsistent fine-detail generation.

The Style-Coordinated models achieved relatively balanced performance with minimal fluctuation (mean = 0.82, SD = 0.05), reflecting strong stability and coherent structural logic. Style-Coordinated1 (0.86) and Style-Coordinated3 (0.72) recorded higher scores, with an average CLIPScore of 0.815 and SSIM of 0.884, suggesting these models effectively maintained semantic accuracy while reproducing the stylistic harmony of traditional and modern architectural features.

The Style-Incompatible models showed the largest internal variation (mean = 0.78, SD = 0.09), indicating inconsistent performance across sub-models. Style-Incompatible2 (0.88) and Style-Incompatible4 (0.74) achieved favorable FID (27.1–29.4) and CLIPScore (0.826–0.835) values, demonstrating robust semantic alignment and adaptability under complex morphological and material conditions. However, Style-Incompatible3 (0.32) and Style-Incompatible5 (0.03) displayed high NIQE (4.1–4.5) and low SSIM (≈0.83), revealing limitations in texture realism and geometric precision.

Overall, the comparative analysis highlights clear differences among model categories: (1) Historic models achieved the highest overall fidelity but showed internal variability. (2) Style-Coordinated models maintained stable and balanced generative behavior. (3) Style-Incompatible models displayed weaker generalization and greater volatility due to data heterogeneity.

Based on composite performance and consistency, six representative models—Historic1 (0.68), Historic4 (0.96), Style-Coordinated1 (0.86), Style-Coordinated3 (0.72), Style-Incompatible2 (0.88), and Style-Incompatible4 (0.74)—were selected for subsequent expert-based subjective evaluation.

4.5.3. Expert Subjective Evaluation

Based on the comprehensive results of the objective evaluation, candidate LoRA models from each category were selected for the next stage of expert subjective assessment: Historic1 (0.68), Historic4 (0.96), Style-Coordinated1 (0.86), Style-Coordinated3 (0.72), Style-Incompatible2 (0.88), and Style-Incompatible4 (0.74).

Subsequently, using the text-to-image generation function of Stable Diffusion Web-UI combined with ControlNet, building images resembling the streetscape of the historic conservation area were generated. On this basis, the XYZ Plot tool was employed, where the LoRA weight (0.4–1.6) and the LoRA model were assigned as the X-axis and Y-axis, respectively. Descriptive keywords regarding architectural style, number of floors, components, facilities, and materials were included in the prompts. In total, 156 comparative image groups were produced (Figure 11), serving to determine the optimal weight range for each model.

Images generated within the determined optimal weight intervals were then used for expert evaluation. During the inference phase, different LoRA models exhibited distinct sensitivity ranges to weight adjustments. Specifically, the Historic1 model achieved the best generation quality at a weight of 0.9, while the Historic4 model performed optimally within the 1.2–1.4 range. Both Style-Coordinated1 and Style-Coordinated4 models showed their best results around 0.9.

For the Style-Incompatible2 series, the optimal weight ranges for the four generation orientations were as follows: Cost-first at 0.8, Heritage-first also at 0.8, Comfort-first within 1.0–1.1, and Integrated-first within 0.7–1.0. Similarly, the Style-Incompatible4 series demonstrated the following optimal ranges: Cost-first 0.7–0.9, Heritage-first 1.2, Comfort-first 0.6–1.0, and Integrated-first 0.8–0.9.

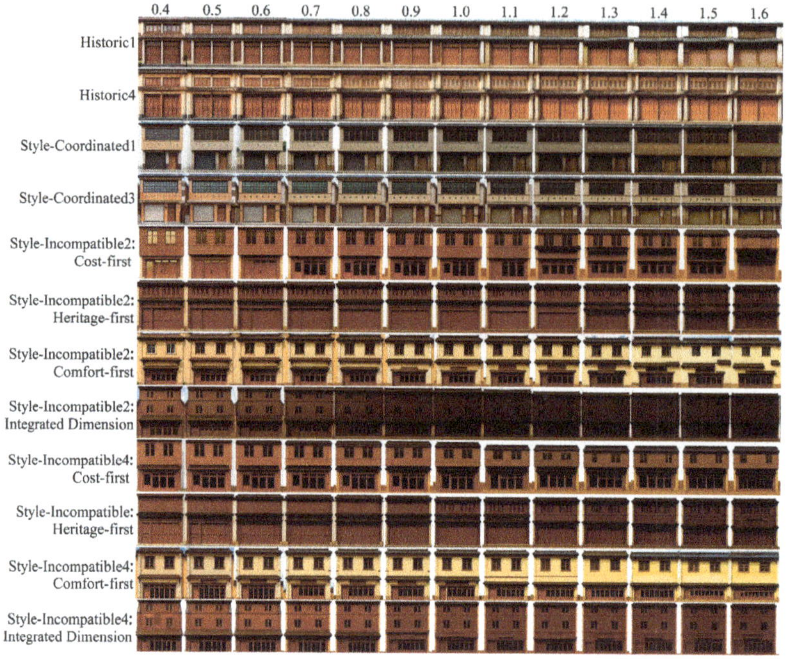

Figure 11. LoRA Model XY Plot.

Overall, the optimal weight ranges differed significantly across models, indicating that various facade orientation types exhibited distinct sensitivities in generation stability and detail fidelity.

According to the expert scoring results, which were provided by six experts specializing in architectural heritage conservation, the optimal models for each building category were identified (Figure 12) and were selected as the primary subjects for subsequent generation and validation experiments.

In the Historic building category, the Historic4 model achieved the highest average score (4.8) across "Image Quality," "Style Consistency," and "Component Integrity," demonstrating strong realism and detail fidelity; thus, it was selected as the representative model for this category. The Historic1 model, though performing steadily in "Floor Count Accuracy," received a slightly lower overall score (4.07) and was retained as a comparative model for analysis.

In the Style-Coordinated building category, the Style-Coordinated1 model achieved the best overall performance with an average score of 4.03. Its generated results showed high consistency with real samples in roof proportion, facade structure, and color coordination, and it was therefore selected as the main model for subsequent experiments. The Style-Coordinated3 model received a lower score (3.50) and was used for comparative evaluation of style adaptability under different LoRA weight settings.

For the Style-Incompatible building category, considering the differences among the four renovation orientations (Cost-first, Heritage-first, Comfort-first, and Integrated-first), comparative screening was conducted within two model sets.

In the Cost-first path, the Style-Incompatible2–Cost model (3.73) scored higher than Style-Incompatible4–Cost (3.30), demonstrating a better balance between facade coordination and cost efficiency, and was thus identified as the representative model for the Cost-first scenario.

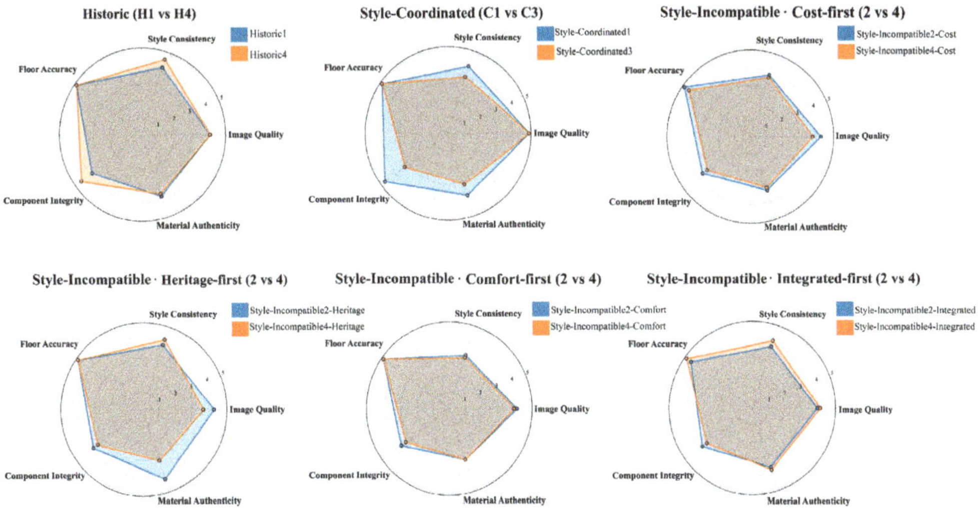

Figure 12. Radar Chart of Expert Subjective Evaluation for LoRA Models.

In the Heritage-first path, the Style-Incompatible2–Heritage model outperformed Style-Incompatible4–Heritage in both Historical Consistency and Material Authenticity, achieving the highest composite score (4.17) and was selected as the representative model for this orientation. Its generated facades exhibited improved alignment with traditional materials, roof forms, and decorative details, effectively restoring historical characteristics while preserving structural integrity.

In the Comfort-first setting, the Style-Incompatible2–Comfort model demonstrated greater stability in Floor Count Accuracy and Component Detail Performance, with an average score of 3.77, and was selected as the primary model for this path.

For the Integrated-first approach, the Style-Incompatible4–Integrated model achieved better performance in stylistic balance and structural coherence (3.53 vs. 4.00) compared with Style-Incompatible2–Integrated, and was therefore identified as the optimal model for the Integrated-first renovation scenario.

In summary, by integrating objective indicators with expert evaluations, the study selected Historic4 as the optimal model for Historic buildings and Style-Coordinated1 for Style-Coordinated buildings. For Style-Incompatible buildings, four representative models were identified according to different renovation orientations: Style-Incompatible2–Cost, Style-Incompatible2–Heritage, Style-Incompatible2–Comfort, and Style-Incompatible4–Integrated. These models achieved the best overall balance of visual realism, structural consistency, and stylistic adaptability, serving as the final set for subsequent generative validation.

4.6. Experimental Results and Visual Demonstration

4.6.1. Visual Results of Authentic Restoration

Based on the streets and alleys in the Shangxijie and Xiaxijie areas of Jincheng Ancient Town—where historic and Style-Coordinated buildings were densely distributed—this study applied the trained LoRA models to explore refined restoration and original appearance reproduction strategies for Historic buildings. By identifying and categorizing facade problems, targeted restoration strategies were developed. The Lineart (line-drawing control) and Depth (depth-aware) modules in ControlNet were integrated to precisely constrain the basic contours and structural hierarchy of facades, ensuring accurate spatial scale and structural relationships during the generation process.

Subsequently, utilizing the Text-to-Image and Image-to-Image inpainting functions of the Stable Diffusion WebUI platform, context-specific and historically consistent restoration schemes were generated (Figure 13).

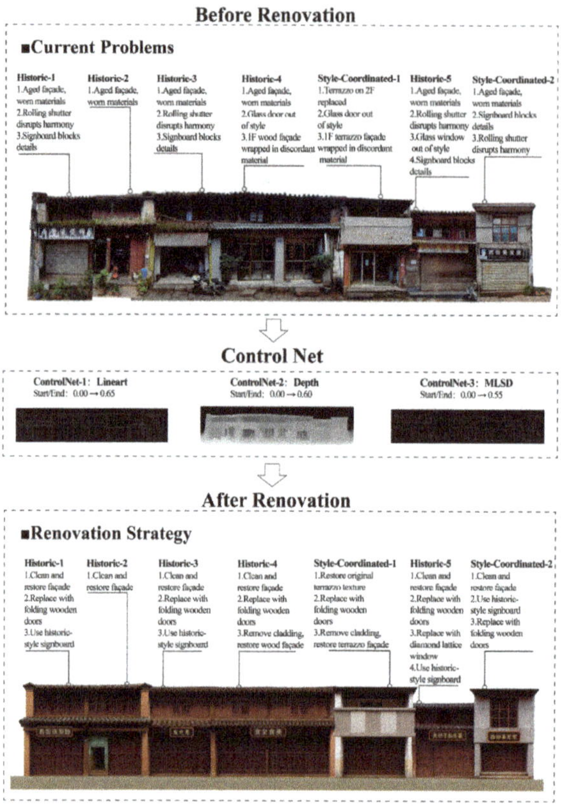

Figure 13. Process of Facade Restoration in Historic Conservation Area.

The experimental results demonstrate that the combined application of LoRA and ControlNet effectively restored facade materials, colors, and architectural components while maintaining the original scale, proportions, and structural logic of the buildings. This approach significantly enhanced both the authenticity and overall stylistic coherence of historic architecture.

4.6.2. Visual Results of Compatible Renovation

Focusing on the streets and alleys within the Shangxijie and Xiaxijie areas of Jincheng Ancient Town—where Style-Incompatible buildings were concentrated—this study applied the trained LoRA models to explore the feasibility and adaptability of generative technologies in the protective renewal of historic conservation areas. Through on-site investigations and resident interviews, the study systematically analyzed homeowners' preferences and intentions regarding facade renovation, including architectural modification directions, facade form, color and material selection, and component supplementation. These findings provided resident-oriented foundations for subsequent renovation design.

Building upon this, the Lineart (line-drawing control) and Depth (depth-aware) modules in ControlNet were employed to precisely constrain the base contours and structural hierarchy of facades, ensuring the accurate preservation of spatial scale and structural logic during image generation. Using the Text-to-Image and Image-to-Image functionalities of the Stable Diffusion WebUI platform, multiple facade renovation schemes were generated that integrated both LoRA model knowledge and resident feedback (Figure 14).

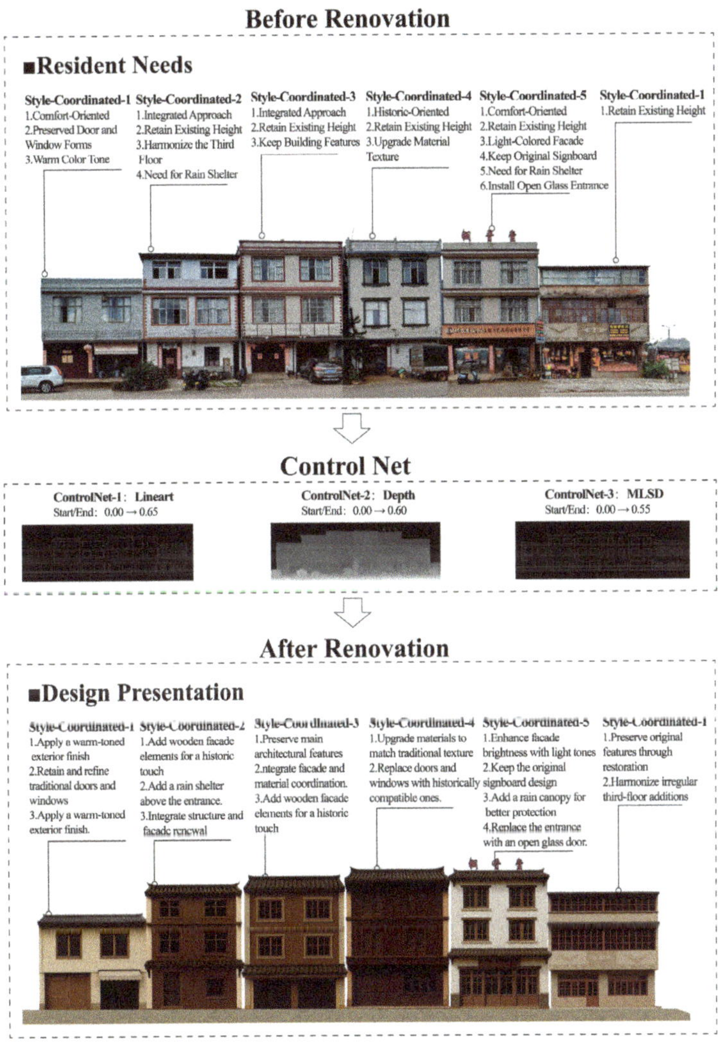

Figure 14. Process of Facade Harmonization in Historic Conservation Area.

The experimental results showed that the combined application of LoRA and ControlNet not only improved stylistic incoherence and visual degradation issues within the streetscape but also incorporated community feedback into the generative mechanism—achieving a shift from technological generation to collaborative generation. Due to the structural constraints of self-built residential houses, large-scale alterations to the buildings were not feasible. Accordingly, the model preserved door and window proportions at a detailed level while automatically optimizing eave moldings and decorative components.

The generated results exhibited higher rationality and consistency in terms of color and material selection. Overall, the renovation schemes reflected residents' subjective intentions and functional needs, while maintaining strong visual and cultural coherence with the historic conservation area's spatial texture—realizing a human-centered and context-sensitive facade renewal driven by generative AI technologies.

5. Findings and Discussion

This study demonstrates that the integration of AIGC and LoRA technologies can substantially transform the workflow of facade renewal in historic conservation areas, establishing a closed-loop framework of generation–evaluation–regeneration. The discussion below interprets the theoretical implications, methodological contributions, and cultural adaptability of the framework, followed by a reflection on its limitations and broader significance.

5.1. Theoretical Implications of AIGC in Urban Design

Artificial Intelligence-Generated Content (AIGC) is redefining how design knowledge is generated and communicated within the field of urban design. Rather than functioning solely as a visualization tool, AIGC serves as a computational reasoning medium that links semantic intentions with spatial representations through algorithmic processes. This transformation is not merely technical—it fundamentally alters how designers conceive, test, and refine ideas, turning design from a linear depiction into a dynamic interaction among human cognition, contextual constraints, and generative algorithms.

The workflow developed in this study illustrates this transformation in practice. Through rule-based generation and iterative feedback, AIGC enables designers to rapidly explore multiple facade design schemes that correspond to diverse architectural contexts and stylistic requirements. The "non-change–permissible–prohibited" constraint system and semantic prompting framework translate conservation principles into operational parameters, allowing the model to generate outcomes that are both contextually coherent and methodologically transparent. Each generated result thus becomes a dual entity—simultaneously a design proposal and a reflection of design logic—making the creative process more interpretable, reproducible, and adaptable.

Theoretically, these characteristics reposition AIGC from a supportive visualization instrument to a knowledge-generating mechanism. By externalizing tacit design understanding—such as proportion, form, and stylistic harmony—into measurable outputs, AIGC contributes to the development of a computational epistemology of design. Within heritage-oriented projects, this capacity for rapid and context-sensitive generation bridges technological innovation with cultural interpretation, demonstrating that AIGC can function as an active mediator between design intentionality, authenticity of context, and algorithmic intelligence. In doing so, it expands the theoretical framework of digital heritage design, offering new possibilities for both reflective practice and inclusive decision-making.

Compared with previous studies that primarily regarded AIGC as a visual simulation or auxiliary design tool [68,69], this study emphasizes its epistemological potential as a reasoning and mediation mechanism that transforms implicit design logic into explicit computational structures, thereby extending earlier digital design research focused on representational efficiency by introducing a new dimension of cognitive transparency and design reproducibility.

In doing so, this study expands the theoretical framework of digital heritage design while acknowledging that the epistemic depth of AIGC models remains limited by dataset representativeness and designers' interpretive capacity. Future research should therefore focus on adaptive semantic control, multimodal knowledge embedding, and cross-cultural datasets to further enhance AIGC's contextual understanding and theoretical robustness.

5.2. Collaborative Assessment and Collective Decision-Making

The introduction of a collaborative assessment mechanism fundamentally reshapes the decision-making logic in heritage restoration and urban design. Unlike conventional expert-centered reviews, which often privilege professional judgment and aesthetic authority, the collaborative assessment framework developed in this study integrates experts, community representatives, and quantitative evaluation into a unified, iterative process. This structure transforms evaluation from a one-time verification procedure into a participatory feedback loop, aligning professional and social perspectives within a shared design environment.

By synthesizing expert and public perspectives through the Bradley–Terry model and Fold Catastrophe integration, the system effectively bridges professional judgment and public perception, establishing a new form of hybrid rationality in design decision-making. This mechanism moves beyond traditional binary evaluation ("expert vs. public") toward a multi-dimensional negotiation process, where collective value judgments are continuously recalibrated through participatory interaction and iterative feedback.

From a theoretical standpoint, this approach reflects the growing trend toward distributed and collaborative intelligence in design practice. Decision-making authority is no longer concentrated in a single disciplinary domain but is dynamically shared among different stakeholders. The collaborative assessment framework thus embodies the principle of knowledge co-construction, wherein expertise, public perception, and contextual understanding interact to generate new design possibilities. Compared with previous participatory evaluation studies that mainly relied on qualitative consensus or expert-led weighting methods [70,71], the present framework advances this direction by establishing a multi-source participatory data platform that empowers the public to co-create and share design information in the protection and renewal of historic cultural districts, thereby reinforcing social consensus and highlighting the constructive role of community forces in spatial governance.

Moreover, this framework enhances transparency and inclusiveness in the design process. The explicit incorporation of stakeholder feedback reduces cognitive bias, makes evaluation criteria traceable, and fosters social consensus in culturally sensitive environments. However, its implementation also faces limitations. The balance of stakeholder representation remains context-dependent, and the weighting of expert, public, and data-driven inputs may vary across projects. Future research could focus on constructing AI-assisted 3D modeling environments combined with VR/AR-based immersive interfaces, allowing experts and the public to collaboratively explore and modify design alternatives in real time, thereby deepening interactive engagement and spatial understanding in the renewal process.

5.3. Cultural Adaptability and Contextualization of AIGC

The relationship between technology and culture presents a core challenge in applying AIGC to the restoration of historic environments. Cultural adaptability refers not only to stylistic alignment but also to a model's capacity to interpret and reproduce the aesthetic logic and values embedded in specific cultural contexts. In heritage design, restoration is both a technical renewal and a cognitive process of sustaining meaning and identity. Thus, technological interventions must respect the semantics of place, avoiding visual homogenization and symbolic dilution.

This study demonstrates that localized data training and LoRA fine-tuning markedly enhance cultural adaptability. While the base diffusion model trained on global datasets generated visually convincing but culturally neutral results, domain adaptation with local facade data improved structural coherence and stylistic fidelity, better reflecting regional architectural traditions. Cultural authenticity in AIGC is therefore not a passive product of scale but an active process of contextual calibration and semantic constraint.

Previous AI- and AIGC-based studies on architectural preservation [72–74] have primarily focused on well-preserved historic conservation areas, making their methods less applicable in environments where buildings are deteriorated or poorly documented. The current state of degradation in historic buildings presents significant challenges for the construction of training datasets. In contrast, this study constructs a high-quality, locally generated facade dataset based on large language models, which strengthens the contextual foundation and cultural specificity. This dataset enhances AIGC's adaptability in fragmented heritage environments and supports culturally sensitive model training and interpretation.

Technologically and theoretically, AIGC serves as a medium of cultural translation, transforming intangible heritage values into computable design logic. By embedding local knowledge through data selection, rule-based constraints, and prompt engineering, digital restoration becomes a site where algorithmic precision and cultural interpretation converge. Unlike prior studies that viewed AI as a visualization tool, this research frames AIGC as a culturally adaptive generative framework capable of integrating local semantics within a global model. Future research could further develop cross-cultural semantic alignment and context-aware generative models, enabling AIGC to interpret and reproduce diverse architectural traditions with greater cultural fidelity and contextual sensitivity.

6. Conclusions

This study proposes and validates an AI-based pre-renewal framework for historic building facades, integrating LLM-guided diffusion generation, collaborative assessment, and LoRA fine-tuning within Stable Diffusion XL. Using the Shangxijie and Xiaxijie Historic Conservation Area in Jincheng Ancient Town, Kunming, as a case study, the framework establishes a closed-loop workflow of "generation–evaluation–regeneration," enabling low-risk, multi-path virtual experimentation before on-site implementation.

This study addresses three key research questions: First, the language-model-guided diffusion generation process enables controllable and heritage-compliant virtual pre-renewal, even with limited and incomplete image datasets from historic conservation areas. By interpreting structured metadata from the facade database, the large language model (LLM) automatically generates semantically coherent prompts, while the diffusion model synthesizes visuals under the "Non-change/Permissible/Prohibited" constraint rules, ensuring historical authenticity and controllability in data-scarce contexts. Second, the collaborative assessment framework effectively integrates expert judgment, public preferences, and AI-based quantitative metrics, forming a scientifically rigorous and socially inclusive evaluation system that enhances both the interpretability and quality of generated outcomes. Lastly, LoRA fine-tuning, driven by feedback from collaborative assessments, significantly improves the contextual fidelity, stylistic coherence, and adaptive performance of the generated results. Through an iterative "generation–evaluation–regeneration" process, the framework demonstrates strong generalizability and goal-oriented optimization across diverse facade typologies, providing a sustainable foundation for virtual pre-renewal applications.

The experimental results show that LoRA-tuned generative models can consistently produce high-fidelity, stylistically coherent restoration outcomes. The hybrid evaluation system, linking human judgment with computational metrics (CLIPScore, FID, NIQE, SSIM, and VGG-sim), proved essential for model optimization and design selection. The iterative feedback loop between expert knowledge and AI generation enhances both methodological transparency and cultural sensitivity in the design decision-making process.

Looking ahead, the framework offers a replicable digital platform for testing facade renewal strategies prior to physical intervention, allowing planners and communities to assess trade-offs between authenticity, cost, and comfort. Extending the framework to other urban and heritage contexts will help verify its generalizability and adaptability. Future research should focus on cross-regional validation, adaptive weighting in participatory evaluation, and bridging the current 2D–3D fidelity gap through HBIM-based integration and immersive VR/AR-enabled co-review environments. As LLMs, diffusion models, and LoRA techniques continue to evolve, and their integration with 3D modeling and collaborative design platforms is expected to further strengthen the sustainability, inclusiveness, and scientific rigor of heritage facade conservation.

Author Contributions: Conceptualization, W.D. and J.R.; Methodology, J.R.; Software, J.R.; Formal analysis, J.R.; Investigation, J.R., J.Z. and J.C.; Data curation, J.R.; Writing—original draft, J.R.; Writing—review and editing, W.D. and J.Z.; Visualization, J.R.; Supervision, W.D. and N.T.; Project administration, W.D. and N.T.; Funding acquisition, W.D. and N.T. All authors have read and agreed to the published version of the manuscript.

Funding: This research was funded by the National Natural Science Foundation of China (52208061), the National Natural Science Foundation of China (52408065), and the Yunnan Province Science and Technology Department (202401CF070132). The APC was funded by the National Natural Science Foundation of China (52208061), the National Natural Science Foundation of China (52408065), and the Yunnan Province Science and Technology Department (202401CF070132).

Data Availability Statement: The original contributions presented in this study are included in the article; further inquiries can be directed to the corresponding author.

Conflicts of Interest: The authors declare no conflicts of interest.

References

1. Wang, S.; Li, J.; Zhang, Y.; Chen, H. How to Achieve a Balance between Functional Improvement and Heritage Conservation? A Case Study on the Renewal of Old Beijing City. *J. Hous. Built Environ.* **2023**, *38*, 1837–1859. [CrossRef]
2. Yung, E.H.; Chan, E.H. The Social Needs of Residents in Conservation Areas: The Case of Hong Kong. *Habitat Int.* **2012**, *36*, 317–325.
3. Zhang, L.; Lu, P.; Lau, R.; Yan, L.; Li, X.; Yang, R.; Leung, H.H.; Chen, P.; Wang, X. Unique Traditional Villages on the Loess Plateau of China: Research on Their Landscape Genes and Sustainable Development. *Herit. Sci.* **2021**, *9*, 118. [CrossRef]
4. Cheng, J.; Yu, T. Historic Conservation in Rapid Urbanization: A Case Study of the Hankow Historic Concession Area. *Habitat Int.* **2017**, *61*, 38–48. [CrossRef]
5. Wu, F. *Planning for Growth: Urban and Regional Planning in China*; Routledge: London, UK, 2015.
6. ICOMOS China. *Principles for the Conservation of Heritage Sites in China*; China ICOMOS: Beijing, China, 2011.
7. López, F.J.; Lerones, P.M.; Llamas, J.; Gómez-García-Bermejo, J.; Zalama, E. A Review of Heritage Building Information Modeling (H-BIM). *Multimodal. Technol. Interact.* **2018**, *2*, 21. [CrossRef]
8. Yung, E.H.; Chan, E.H. Problem Issues of Public Participation in Built-Heritage Conservation. *Habitat Int.* **2011**, *35*, 589–596. [CrossRef]
9. Kuang, Z.; Li, C.; Zhang, J. Preserving Architectural Heritage in Urban Renewal: A Stable Diffusion Model Framework for Automated Historical Facade Generation. *Npj Herit. Sci.* **2024**, *13*, 256. [CrossRef]
10. Sun, C.; Zhou, Y.; Han, Y. Automatic generation of architecture facade for historical urban renovation using generative adversarial network. *Build. Environ.* **2022**, *212*, 108781. [CrossRef]
11. ICOMOS. *International Charter for the Conservation and Restoration of Monuments and Sites (The Venice Charter)*; ICOMOS: Paris, France, 1964.
12. ICOMOS. *The Nara Document on Authenticity*; ICOMOS: Nara, Japan, 1994.
13. Australia ICOMOS. *The Burra Charter: The Australia ICOMOS Charter for Places of Cultural Significance*; Australia ICOMOS: Burwood, Australia, 1999.

14. State Council of the People's Republic of China. *Regulations on the Protection of Famous Historical and Cultural Cities, Towns and Villages*; State Council: Beijing, China, 2008.
15. General Office of the CPC Central Committee; General Office of the State Council. *Opinions on Strengthening the Protection and Inheritance of Historical and Cultural Heritage in Urban and Rural Construction*; CPC: Beijing, China, 2021.
16. Huang, B. The Strategy of Protection and Repair of Building Facades in Historic Urban Cultural Districts. *China Archit. Met. Struct.* **2024**, *23*, 85–87. [CrossRef]
17. National Park Service. *The Secretary of the Interior's Standards for the Treatment of Historic Properties*; U.S. Department of the Interior: Washington, DC, USA, 2017.
18. Olivieri, F. Symbiotic Design for a Resilient Urban Ecosystem. *AGATHÓN—Int. J. Archit. Art Des.* **2022**, *11*, 11–20.
19. Wang, Y.; Li, Y. Participatory Micro-Regeneration: Governing Urban Renewal in China. *Plan. Perspect.* **2025**, *40*, 205–226.
20. Murphy, M.; McGovern, E.; Pavia, S. Historic Building Information Modelling (HBIM). *Struct. Surv.* **2013**, *31*, 311–327. [CrossRef]
21. Eilouti, B.H. Parametric-Based Framework for Generating Context-Sensitive Architectural Form. *Alex. Eng. J.* **2018**, *57*, 3413–3426.
22. Wan, D.; Zhao, R.; Zhang, S.; Liu, H.; Guo, L.; Li, P.; Ding, L. A Deep Learning-Based Approach to Generating Residential Building Façades via Pix2Pix. *Sustainability* **2023**, *15*, 1816. [CrossRef]
23. Bacco, M.; Barsocchi, P.; Cassará, P.; Germanese, D.; Gotta, A.; Leone, G.R.; Moroni, D.; Pascali, M.A.; Tampucci, M. Monitoring Ancient Buildings: Real Deployment of an IoT System Enhanced by UAVs and Virtual Reality. *IEEE Internet Things J.* **2020**, *7*, 5178–5188. [CrossRef]
24. Colace, F.; Gaeta, R.; Lorusso, A. New AI Challenges for Cultural Heritage Protection. *J. Cult. Herit.* **2025**, *75*, 168–193. [CrossRef]
25. Kampelopoulos, D.; Koutsogiannis, A. A Review of LLMs and Their Applications in the Architecture, Engineering, and Construction Industry. *Artif. Intell. Rev.* **2025**, *58*, 250. [CrossRef]
26. Goodfellow, I.J.; Pouget-Abadie, J.; Mirza, M.; Xu, B.; Warde-Farley, D.; Ozair, S.; Courville, A.; Bengio, Y. Generative Adversarial Nets. In *Advances in Neural Information Processing Systems 27 (NIPS 2014)*; MIT Press: Cambridge, MA, USA, 2014; pp. 672–2680.
27. Rombach, R.; Blattmann, A.; Lorenz, D.; Esser, P.; Ommer, B. High-Resolution Image Synthesis with Latent Diffusion Models. In Proceedings of the IEEE/CVF Conference on Computer Vision and Pattern Recognition 2022, New Orleans, LA, USA, 18–24 June 2022; pp. 10684–10695.
28. Zhang, J.; Huang, Y.; Li, Z.; Li, Y.; Yu, Z.; Li, M. Development of a Method for Commercial Style Transfer of Historical Architectural Façades Based on Stable Diffusion Models. *J. Imaging* **2024**, *10*, 165. [CrossRef]
29. Radford, A.; Kim, J.W.; Hallacy, C.; Ramesh, A.; Goh, G.; Agarwal, S.; Sastry, G.; Askell, A.; Mishkin, P.; Clark, J.; et al. Learning Transferable Visual Models from Natural Language Supervision. In Proceedings of the International Conference on Machine Learning 2021, Virtual (online), 18–24 July 2021; pp. 8748–8763.
30. Hao, Y.; Yao, Z.; Wu, R.; Bao, Y. Damage and restoration technology of historic buildings: A review. *npj Herit. Sci.* **2024**, *12*, 301. [CrossRef]
31. Hu, E.J.; Shen, Y.; Wallis, P.; Allen-Zhu, Z.; Li, Y.; Wang, S.; Wang, L.; Chen, W. LoRA: Low-Rank Adaptation of Large Language Models. *arXiv* **2021**, arXiv:2106.09685.
32. Isola, P.; Zhu, J.Y.; Zhou, T.; Efros, A.A. Image-to-Image Translation with Conditional Adversarial Networks. In Proceedings of the IEEE Conference on Computer Vision and Pattern Recognition 2017, Honolulu, HI, USA, 21–26 July 2017; pp. 1125–1134.
33. Zhu, J.Y.; Park, T.; Isola, P.; Efros, A.A. Unpaired Image-to-Image Translation Using Cycle-Consistent Adversarial Networks. In Proceedings of the IEEE International Conference on Computer Vision 2017, Honolulu, HI, USA, 21–26 July 2017; pp. 2223–2232.
34. Simonyan, K.; Zisserman, A. Very Deep Convolutional Networks for Large-Scale Image Recognition. *arXiv* **2014**, arXiv:1409.1556.
35. Zhang, R.; Isola, P.; Efros, A.A.; Shechtman, E.; Wang, O. The Unreasonable Effectiveness of Deep Features as a Perceptual Metric. In Proceedings of the IEEE Conference on Computer Vision and Pattern Recognition 2018, Salt Lake City, UT, USA, 18–22 June 2018; pp. 586–595.
36. Brown, T.; Mann, B.; Ryder, N.; Subbiah, M.; Kaplan, J.D.; Dhariwal, P.; Neelakantan, A.; Shyam, P.; Sastry, G.; Askell, A.; et al. Language Models Are Few-Shot Learners. *Adv. Neural Inf. Process. Syst.* **2020**, *33*, 1877–1901.
37. UNESCO. *Recommendation on the Historic Urban Landscape*; UNESCO: Paris, France, 2011.
38. ICOMOS. *Charter for the Conservation of Historic Towns and Urban Areas (Washington Charter)*; ICOMOS: Washington, DC, USA, 1987.
39. Council of Europe. *Framework Convention on the Value of Cultural Heritage for Society (Faro Convention)*; Council of Europe: Faro, Portugal, 2005.
40. Sanoff, H. *Community Participation Methods in Design and Planning*; John Wiley & Sons: New York, NY, USA, 2000.
41. Al-Kodmany, K. Visualization Tools and Methods in Community Planning: From Freehand Sketches to Virtual Reality. *J. Plan. Lit.* **2002**, *17*, 189–211. [CrossRef]
42. Arnheim, R. *The Dynamics of Architectural Form*; University of California Press: Berkeley, CA, USA, 1977.
43. Kelly, J. A Fourth Age of Conservation? *Archit. Rev.* **2019**, *1458*, 80–85.
44. Alexander, C. *The Timeless Way of Building*; Oxford University Press: New York, NY, USA, 1979.

45. Lynch, K. *The Image of the City*; MIT Press: Cambridge, MA, USA, 1960.
46. Zhang, J. Spatial and Temporal Patterns and Influential Factors of Ancient Architectural Heritage in Shanxi Province. *J. Asian Archit. Build. Eng.* **2025**, *24*, 1–18. [CrossRef]
47. Ho, K.; Song, Y. Deep-Learning-Based Image-to-Image Translation for Style Transfer of Building Facades. *Autom. Constr.* **2020**, *119*, 103338.
48. Manovich, L. *AI Aesthetics*; Strelka Press: Moscow, Russia, 2020.
49. Carpo, M. *The Second Digital Turn: Design Beyond Intelligence*; MIT Press: Cambridge, MA, USA, 2017.
50. OpenAI. GPT-4 Technical Report. *arXiv* **2023**, arXiv:2303.08774. [CrossRef]
51. Riegl, A. *The Modern Cult of Monuments: Its Character and Its Origin*; Forster, K.W.; Ghirardo, D., Translators; MIT Press: Cambridge, MA, USA, 1982.
52. Brandi, C. *Theory of Restoration*; Rockwell, C., Translator; Istituto Centrale per il Restauro: Rome, Italy, 2005.
53. Schuhmann, C.; Beaumont, R.; Vencu, R.; Gordon, C.; Wightman, R.; Cherti, M.; Coombes, T.; Katta, A.; Mullis, C.; Wortsman, M.; et al. LAION-5B: An Open Large-Scale Dataset for Training Next-Generation Image-Text Models. In Proceedings of the Thirty-sixth Conference on Neural Information Processing Systems Datasets and Benchmarks Track, New Orleans, LA, USA, 28 November–9 December 2022.
54. Abdennadher, S. Rule-Based Constraint Programming: Theory and Practice. In *Principles and Practice of Declarative Programming*; Springer: Berlin/Heidelberg, Germany, 2001; pp. 129–150.
55. Xu, Z.; Yang, C.; Wang, J. Organic renewal and dynamic preservation: A study on the protection and restoration concepts of modern historical buildings. *Archit. J.* **2015**, *10*, 100–103.
56. Bradley, R.A.; Terry, M.E. Rank Analysis of Incomplete Block Designs: I. The Method of Paired Comparisons. *Biometrika* **1952**, *39*, 324–345. [CrossRef]
57. Heusel, M.; Ramsauer, H.; Unterthiner, T.; Nessler, B.; Hochreiter, S. GANs Trained by a Two Time-Scale Update Rule Converge to a Local Nash Equilibrium. In *Advances in Neural Information Processing Systems 30 (NIPS 2017)*; MIT Press: Cambridge, MA, USA, 2017; pp. 6626–6637.
58. Mittal, A.; Moorthy, A.K.; Bovik, A.C. No-Reference Image Quality Assessment in the Spatial Domain. *IEEE Trans. Image Process.* **2012**, *21*, 4695–4708. [CrossRef]
59. Wang, Z.; Bovik, A.C.; Sheikh, H.R.; Simoncelli, E.P. Image Quality Assessment: From Error Visibility to Structural Similarity. *IEEE Trans. Image Process.* **2004**, *13*, 600–612. [CrossRef]
60. Aziz, M.; Rehman, U.; Danish, M.U.; Grolinger, K. Global-Local Image Perceptual Score (GLIPS): Evaluating Photorealistic Quality of AI-Generated Images. *arXiv* **2024**, arXiv:2405.09426. [CrossRef]
61. Dosovitskiy, A.; Beyer, L.; Kolesnikov, A.; Weissenborn, D.; Zhai, X.; Unterthiner, T.; Dehghani, M.; Minderer, M.; Heigold, G.; Gelly, S.; et al. An Image is Worth 16 × 16 Words: Transformers for Image Recognition at Scale. In Proceedings of the ICLR 2021, Online, 3–7 May 2021.
62. Zeeman, E.C. Catastrophe Theory. *Sci. Am.* **1976**, *234*, 65–83. [CrossRef]
63. Rombach, R.; Podell, D.; Blattmann, A.; Müller, J.; Dockhorn, T.; English, Z.; Penna, J.; Rombach, R. SDXL: Improving Latent Diffusion Models for High-Resolution Image Synthesis. In Proceedings of the IEEE/CVF Conference on Computer Vision and Pattern Recognition Workshops (CVPRW) 2024, Seattle, WA, USA, 17–21 June 2024; pp. 1201–1210. Available online: https://openreview.net/forum?id=di52zR8xgf (accessed on 17 November 2025).
64. Kingma, D.P.; Ba, J. Adam: A Method for Stochastic Optimization. *arXiv* **2014**, arXiv:1412.6980.
65. Chang, Q. A critique on the 'authenticity' of historical building restoration. *Times Archit.* **2009**, *5*, 118–121.
66. Smith, L.N. A Disciplined Approach to Neural Network Hyper-Parameters: Part 1—Learning Rate, Batch Size, Momentum, and Weight Decay. *arXiv* **2018**, arXiv:1803.09820.
67. Masters, D.; Luschi, C. Revisiting Small Batch Training for Deep Neural Networks. *arXiv* **2018**, arXiv:1804.07612, 2018. [CrossRef]
68. Xu, S.; Zhang, J.; Li, Y. Knowledge-Driven and Diffusion Model-Based Methods for Generating Historical Building Façades: A Case Study of Traditional Minnan Residences. *Information* **2024**, *15*, 344. [CrossRef]
69. Tan, L. Using Generative AI (Midjourney) to enhance divergent and convergent thinking in architectural design. *Int. J. Des. Creat. Innov.* 2024, *online first*.
70. Yung, H.K.; Chan, H.W. Critical social sustainability factors in urban conservation: The case of the Central Police Station compound in Hong Kong. *Facilities* **2012**, *30*, 396–416. [CrossRef]
71. Chan, E.; Lee, G. Critical factors for improving social sustainability of urban renewal projects. *Soc. Indic. Res.* **2008**, *85*, 243–256. [CrossRef]
72. Huang, S.; Wang, H. AI-Driven Architectural Heritage Preservation: Exploring Generative Models and Cultural Sensitivity. *J. Archit. Technol.* **2025**, *28*, 35–52.

73. Lu, Y.; Wu, J.; Wang, M.; Fu, J.; Xie, W.; Wang, P.; Zhao, P. Design transformation pathways for AI-generated Chinese traditional architectural images (SD LoRA & Canny). *Electronics* **2025**, *14*, 282.
74. Chen, L.; Yang, H.; Li, J.; Wang, Q.; Zhang, Y. Generative AI for Vernacular Architecture Restoration: A LoRA-Enhanced Diffusion Framework for Cultural Heritage Facades. *Front. Built Environ.* **2025**, *11*, 1564872.

buildings

MDPI

Article

A Modular AI Workflow for Architectural Facade Style Transfer: A Deep-Style Synergy Approach Based on ComfyUI and Flux Models

Chong Xu * and Chongbao Qu

College of Architecture and Urban Planning, Qingdao University of Technology, Qingdao 266033, China; chongbaoqu@outlook.com
* Correspondence: xuchong621@hotmail.com; Tel.: +86-133-5687-6280

Abstract

This study focuses on the transfer of architectural facade styles. Using the node-based visual deep learning platform ComfyUI, the system integrates the Flux Redux and Flux Depth models to establish a modular workflow. This workflow achieved style transfer of building facades guided by deep perception, encompassing key stages such as style feature extraction, depth information extraction, positive prompt input, and style image generation. The core innovation of this study lies in two aspects: Methodologically, a modular low-code visual workflow has been established. Through the coordinated operation of different modules, it ensures the visual stability of architectural forms during style conversion. In response to the novel challenges posed by generative AI in altering architectural forms, the evaluation framework innovatively introduces a "semantic inheritance degree" assessment system. This elevates the evaluation perspective beyond traditional "geometric similarity" to a new level of "semantic and imagery inheritance." It should be clarified that the framework proposed by this research primarily provides innovative tools for architectural education, early design exploration, and visualization analysis. This workflow introduces an efficient "style-space" cognitive and generative tool for teaching architectural design. Students can use this tool to rapidly conduct comparative experiments to generate multiple stylistic facades, intuitively grasping the intrinsic relationships among different styles and architectural volumes/spatial structures. This approach encourages bold formal exploration and deepens understanding of architectural formal language.

Keywords: style transfer; architectural design; ComfyUI; AI workflow

1. Introduction

Architectural facade design embodies cultural, temporal, and spatial aesthetic characteristics. In recent years, artificial intelligence has achieved significant breakthroughs through data-driven approaches [1]. With the emergence and mature application of concepts and technologies such as big data and cloud computing, AI has sparked a profound technological revolution worldwide [2]. In the field of architecture, artificial intelligence is transforming buildings and urban environments, making the design process faster, more efficient, and more sustainable [3].

With the advancement of AI and deep learning, style transfer has emerged as a valuable exploratory tool in architectural design. However, existing methods often rely heavily on programming techniques, pose significant operational barriers, and lack user-friendly design workflows. ComfyUI, an emerging visual AI workflow tool, uses a node-based workflow in which each step serves as a node. Upon completing a task at a node, its output feeds into the subsequent step, thereby executing the entire process [4]. As a node-based visual deep learning platform, it provides designers with a low-code prototyping approach.

In this study on architectural facade style transfer using the ComfyUI platform, the combined application of the Flux Redux and Flux Depth models forms the core technical foundation of the workflow. Flux Redux serves as a powerful foundational generative model, excelling in high-fidelity interpretation and fusion of complex style prompts. It generates images with exceptional consistency and rich detail, ensuring accurate and deep stylistic expression. Flux Depth distinguishes itself by introducing a depth-aware dimension, enabling precise extraction and reconstruction of three-dimensional spatial structural information from input architectural model images. During style transfer, the Flux Depth model constrains the generation process to ensure that transferred stylized textures strictly adhere to the original facade's geometric and compositional logic. This effectively preserves the architectural entity's structural authenticity and maintains the stability of spatial order. However, it must be clarified that the structural authenticity and spatial order stability discussed here are both defined at the level of visual representation. They are explicitly manifested in the architectural forms depicted in images taken before and after relocation, rather than in the building's structural integrity. Furthermore, in subsequent articles, all architectural terminology—such as "architecture," "structure," and "space"—refers exclusively to their visual representations in two-dimensional images. The scope of this research is strictly confined to visual perception and image generation, and does not address the physical properties of architectural entities, structural engineering, or actual construction.

2. Theoretical Background

2.1. The Development of AI Image Style Transfer Technology

Style transfer technology began developing after Gatys et al. proposed a convolutional neural network-based algorithm. This approach achieves high-quality artistic style transfer by separating and recombining the content and style representations of an image. However, its iterative optimization process is relatively slow [5]. Subsequently, Johnson et al. introduced perceptual loss functions, enabling feedforward networks to perform real-time style transfer and significantly improving efficiency [6]. Subsequent research by Cai et al. demonstrated that multidimensional neural networks can analyze complex data and uncover nonlinear relationships [7], which forms the very foundation of AI image style transfer technology.

Existing style transfer techniques primarily fall into two categories. The first involves optimization-based methods that iteratively refine input images to match the target style's feature representation, though these are computationally intensive. The second employs feedforward network-based approaches, utilizing pre-trained loss networks to define perceptual loss for rapid style transfer. For instance, Huang et al.'s Adaptive Instance Normalization (AdaIN) layer achieves real-time transfer of arbitrary styles by aligning feature statistics [8].

2.2. Application of AI Image Style Transfer Technology in the Architectural Field

Style transfer is currently applied in artistic creation, image enhancement, and architectural design. Within architectural design, its applications exhibit considerable diversity. For instance, the multidisciplinary design optimization algorithm (MDOA) and model-driven decision support system (MDSS) proposed by Cai et al. for optimizing architectural spatial performance demonstrate that algorithmic models have become practical tools for

https://doi.org/10.3390/buildings16030494

solving complex architectural problems such as spatial form optimization and aesthetic style transfer [9]. Lin et al. applied conditional generative adversarial networks (CGAN) in their research. By interpreting image data of historical district facades, they developed a method capable of autonomously designing specific facade ornamentation styles [10]. Meng et al. designed a replicable five-stage AIGC workflow to transfer the artistic style of Song Dynasty boundary painting into interior design, advancing the effective integration of traditional aesthetics with modern design concepts [11]. Chen et al. developed a technical pipeline deeply integrating Stable Diffusion (for texture style transfer) with NeRF (for 3D reconstruction), enabling the complete creation process from 2D stylized images to 3D high-fidelity models, demonstrating AI visual technology's deep application in architecture—transitioning from expression to construction [12]. Yan et al. employed a pix2pix algorithm based on conditional generative adversarial networks (CGAN) for training. This model demonstrated exceptional accuracy and stability in architectural image style transfer tasks, enabling precise historical style transfer while preserving structural features [13]. Chen et al. proposed a typology-guided fusion framework integrating LoRA and diffusion models. Its core innovation lies in systematically embedding architectural typology theory into generative AI training and inference workflows, transcending traditional visual mimicry to achieve "typological transcoding" of complex historical architectural styles [14]. Concurrently, Duan et al. adapted the Stable Diffusion XL model for domain-specific tasks using LoRA, enabling it to generate high-fidelity facade images consistent with historical architectural contexts. This demonstrated the optimization role of LoRA fine-tuning in architectural style image generation [15].

2.3. *Visual AI Platform*

Research on existing style transfer techniques reveals that current methods for architectural facade style transfer suffer from complex operations, high computational costs, and a lack of systematic attention to preserving the building's core form before and after style transfer. Based on this, this study explores solutions using the ComfyUI platform. ComfyUI significantly lowers the technical barrier for users through its intuitive visualization and modular interface. Its core strength lies in its node-based visual programming interface, which provides users with unprecedented fine-grained control and workflow reproducibility [16]. Simultaneously, addressing the issue of altered architectural form caused by generative AI during style transfer, this study proposes a semantic inheritance evaluation method to quantitatively assess the similarity of architectural form before and after style transfer.

The two AI models integrated with ComfyUI in this study significantly streamline the AI image generation workflow. The Flux Depth model employs algorithms to convert complex RGB images into single-channel depth maps. While reducing data dimensions and optimizing data structures for AI recognition, it effectively incorporates spatial information to process depth hierarchies within images. This feature is particularly crucial for architecture-related tasks, enabling architects to represent spatial structures and three-dimensional effects in architectural designs with greater accuracy.

Building on this foundation, the Flux Redux model performs feature extraction and mapping on depth maps and style reference images, based on textual requirements for the output image. Adversarial training optimizes parameters to precisely transfer visual elements, such as texture and color, from the style reference image to the depth map. This preserves the architectural model's content while imparting artistic style, producing high-quality style transfer renderings that meet diverse architectural design needs.

https://doi.org/10.3390/buildings16030494

3. Design Process

This section outlines the overall workflow structure and describes the role of each module.

3.1. System Architecture Design

The AI image generation system workflow developed in this study integrates four modules: style feature extraction, depth information extraction, forward prompt input, and image generation (Figure 1).

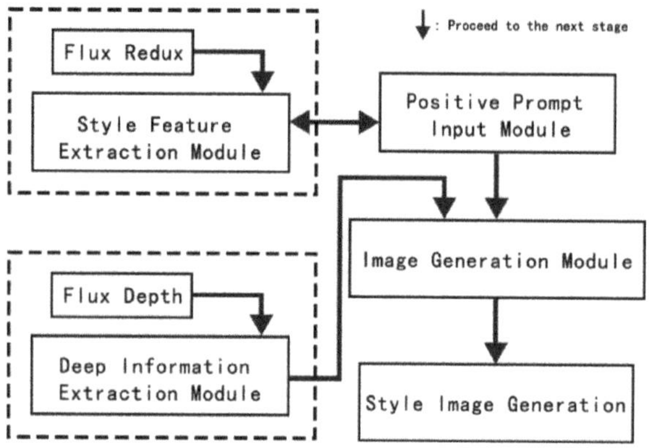

Figure 1. Schematic diagram of module composition. (Source: made by authors.)

Within this workflow, the system first uses the Flux Redux model to extract stylistic image feature vectors, enabling efficient capture of stylistic features. Concurrently, it integrates user-provided textual descriptions to guide subsequent semantic-level generation. Subsequently, the Flux Depth model analyzes the provided model images to interpret three-dimensional structure, generating depth maps to impose spatial geometric constraints. Based on this, multimodal features are encoded into a latent space for fusion, reconstructed into images via a decoder, and optimized through adversarial training to achieve high-quality image generation that balances stylistic consistency, semantic accuracy, and spatial plausibility.

3.2. Workflow Setup and Module Functions

3.2.1. Construction of the Style Feature Extraction Module Group

The construction of this module group first processes image files through the image loading and preprocessing module, converting them into a standardized tensor format. Additionally, this module generates image-related masks and ensures consistency in image format and dimensions.

Subsequently, the CLIP visual encoding module extracts deep semantic features from the images. Its model is dynamically loaded via a dedicated CLIP visual loader. The core innovation of this module group lies in introducing a style model application module. This module uses CLIP visual features as conditional input and, driven by pre-trained models (e.g., flux-redux-dev) loaded by the style model loader, performs stylistic adjustments and feature enhancement on the initial conditional data.

This module group systematically addresses the transformation and extraction of specific stylistic features from raw images (Figure 2).

https://doi.org/10.3390/buildings16030494

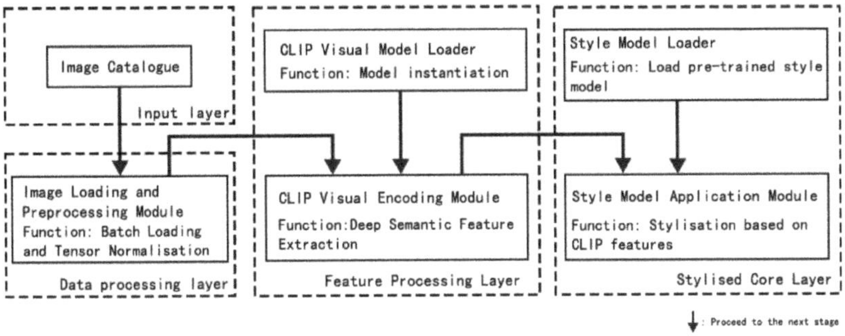

Figure 2. Hierarchical data flow and module dependencies within the style feature extraction module. (Source: made by authors).

3.2.2. Construction of the Depth Information Extraction Module Group

This module group adopts a serial architecture, sequentially processing images through loading, directional scaling, depth estimation, and visualization modules to achieve end-to-end processing from raw images to depth maps.

Specifically, the workflow sequentially traverses four specialized modules: The image loading module ensures uniformity of data input; The Image Oriented Scaling module performs geometric normalization of input images through configurable scaling standards and cropping strategies, providing downstream models with uniformly sized inputs. This significantly enhances the stability of subsequent processing and the comparability of results. The core DepthAnythingV2 depth preprocessor module generates high-precision, dense depth maps from preprocessed images. Its performance advantages directly benefit from the standardized data provided by preceding stages.

Additionally, to facilitate process monitoring and result validation, a depth map preview module is integrated at the end of the module suite. This optional component provides instant visual feedback and does not affect the generation of core depth data or subsequent style transfer tasks (Figure 3).

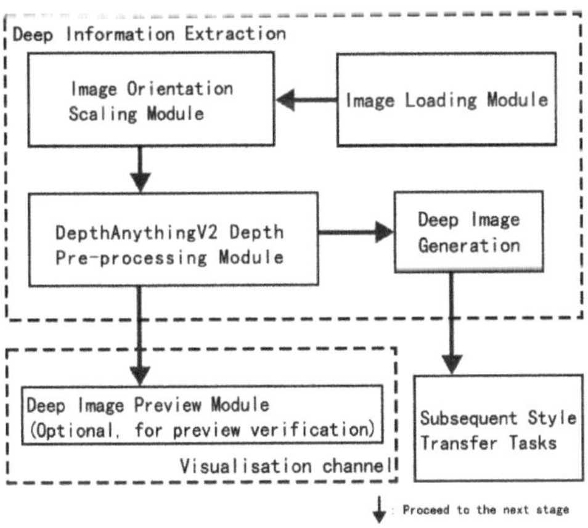

Figure 3. Serialized processing flow of the depth information extraction module. (Source: made by authors.)

3.2.3. Construction of the Prompt Input Module Group

This module group constitutes a dual-condition fusion prompt input system. In this design, the textual input "Generate a new architectural facade form based on the provided picture style" is converted into a semantic feature vector via the CLIP text encoder. Simultaneously, it receives visual style conditions from the style-feature-extraction module group. The core innovation lies in the Flux Guidance Module. Through its unique dual-input design, it fuses textual semantics and visual style features into unified guidance parameters, achieving dual-condition fusion of textual meaning and visual style.

Ultimately, this fused guidance signal is fed into the generative model to achieve controllable shaping of the generated content. This module group design systematizes the conditional encoding, fusion, and guidance process through clear modular interfaces, significantly enhancing the controllability and flexibility of the generation process (Figure 4).

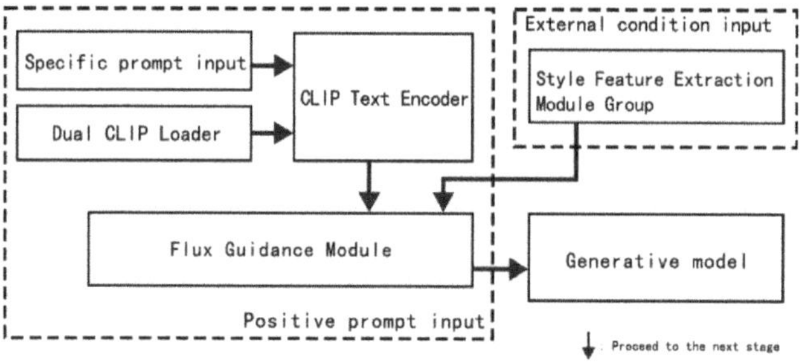

Figure 4. System flowchart of the forward prompt input module. (Source: made by authors).

3.2.4. Construction of the Image Generation Module Group

This module group constructs an end-to-end image generation system based on multi-condition guidance. At its core is the InstructPixToPix conditional module, which innovatively integrates conditional inputs from three independent module groups: favorable conditions from the style model, adverse conditions from the CLIP text encoder, and image inputs from the deep preprocessor module. These multimodal conditions are mapped to a unified latent space via the VAE encoder, forming a collaborative guidance signal.

The system adopts a modular loading architecture that dynamically manages VAE and UNet model resources via dedicated loaders. During generation, the K-sampler performs a controllable denoising process in the latent space using the UNet model and the integrated conditional signal, producing high-quality latent representations. Finally, the VAE decoder converts these latent representations back to the pixel space, with the preserved image module outputting the final result (Figure 5).

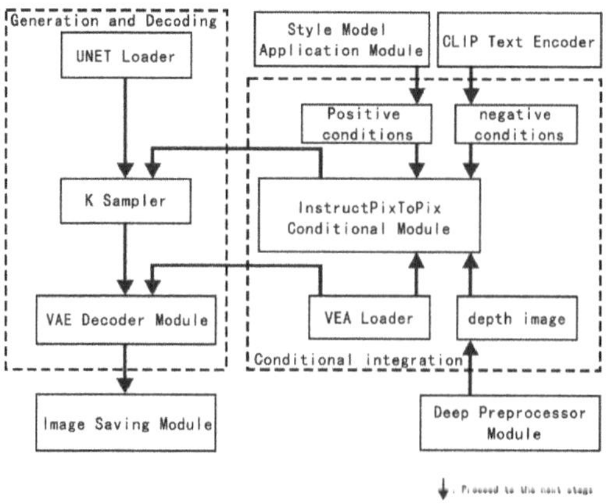

Figure 5. Image generation system flow. (Source: made by authors.)

https://doi.org/10.3390/buildings16030494

This completes the construction of the overall model (Figure 6). Additionally, to address inconsistent results caused by data-quality uncertainties in AI learning, we adopted the following solutions.

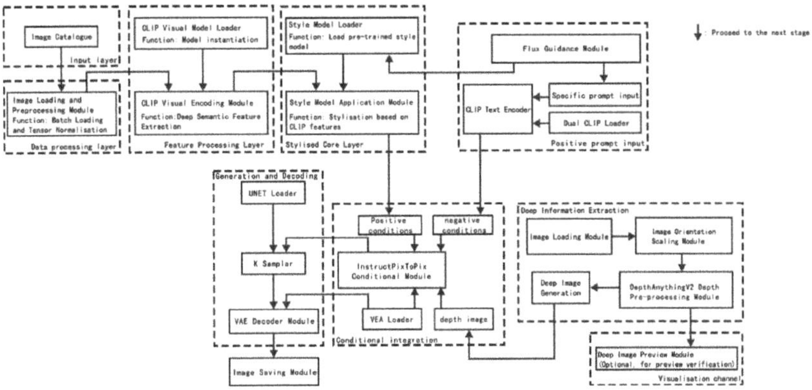

Figure 6. Overall workflow. (Source: made by authors.)

First, regarding AI model selection, this research workflow does not train models from scratch but instead integrates top-tier open-source foundational models, such as Flux Redux and Flux Depth. These models have already undergone pre-training on massive, high-quality, rigorously cleaned, and curated internet image-text pair datasets.

Second, in workflow design, we innovatively implemented multi-module collaboration. The deep information extraction module allows users to directly input architectural model images, reflecting the input image's inherent 3D geometric structure. This significantly reduces the uncertainty inherent in AI-generated images.

4. Experiments and Results

4.1. Experimental Setup

The data used for testing primarily consists of architectural images. All architectural photographs originate from publicly available urban architectural image databases. These databases encompass real-world architectural scenes from diverse cities and styles, effectively reflecting visual characteristics under authentic conditions to ensure the test data's authenticity and representativeness.

For variable configuration, image complexity was hierarchically categorized through comprehensive evaluation. This involved calculating the information entropy (H), edge density (E), and color complexity (C) for each image. A composite metric (S) was then computed using the formula $S = 0.4H + 0.3E + 0.3C$. Based on these calculations, complexity levels were divided into three tiers.

The weight distribution $S = 0.4H + 0.3E + 0.3C$ is based on the following logic: In image processing, information entropy (H) is widely regarded as the core metric for measuring the overall information content and uncertainty of an image. It carries the most significant weight in emphasizing that the total information contained in the original image is the primary factor influencing processing difficulty. Second, edge density (E) directly reflects the structural complexity of an image, while color complexity (C) relates to the richness and variation of hues. In architectural facades, the complexity of structure (e.g., decorative lines, window divisions) and the diversity of materials (color and texture) collectively contribute to the visual perception of "complexity." Assigning equal weight to both dimensions reflects a balanced emphasis on the "form" and "material" aspects of architectural facades.

https://doi.org/10.3390/buildings16030494

It is imperative to note that the "complexity" measured in this study specifically refers to the visual computational complexity of architectural facade images. This metric primarily quantifies the technical challenges faced by style transfer models when processing input images with varying information density and levels of detail.

The complexity definition employed in this experiment does not aim to replace the architectural ontology complexity derived from multidimensional constraints, such as function, structure, environment, and context, within architectural theory. This study focuses on the performance of AI models in visual generation tasks. Therefore, using image complexity as a control variable for grouped experiments is technically sound and practical.

4.2. Experimental Results

To ensure data authenticity, all 24 selected architectural images were uniformly cropped to a consistent size before complexity calculations. The final experimental data ranged from 3.6 to 15.23. Based on these results, eight images with values between 9.96 and 15.23 were classified into the high-complexity group, eight images between 8.12 and 9.91 into the medium-complexity group, and eight images between 3.6 and 8.08 into the low-complexity group (Table 1). Style transfer was performed on each group, and similarity measurements were conducted between the transferred images and their originals.

Table 1. Complexity Metrics for Different Image Datasets.

ID	Information Entropy Complexity (H *)	Edge Density (E *)	Color Complexity (C *)	Comprehensive Score (S *)
1	7.16	0.418	0.245	8.96
2	7.52	0.291	0.261	9.08
3	7.24	0.368	0.207	7.68
4	7.21	0.392	0.279	9.91
5	7.53	0.31	0.261	9.14
6	7.49	0.404	0.279	9.96
7	7.49	0.342	0.287	10.01
8	7.47	0.443	0.214	8.12
...	...	...	...	...
24	7.32	0.392	0.286	10.12

* $H = -\sum(p(i) \times \log_2(p(i)))$. * E = (Number of edge pixels)/(Total number of pixels). * C = (Number of Unique Colors)/(Total Number of Pixels). * S = 40%H + 30%E + 30%C.

Subsequent image similarity calculations reveal that, unlike conventional architectural facade style transfer, this experiment focuses on transferring buildings with altered core structures, forms, and layouts. Consequently, traditional similarity metrics, such as structural similarity, become inadequate. This necessitates a new evaluation framework—one that shifts from assessing "pixel or structural similarity" to evaluating "the inheritance of architectural semantics and stylistic genes." The evaluation framework elevates from the "geometric level" to the "semantic and imagery level." It assesses explicitly four aspects: shape contextual similarity (S1), key-point matching accuracy (K), texture feature similarity (T), and color distribution similarity (C). The overall similarity score is calculated as: Comprehensive Score(S2) = 0.35S1 + 0.25K + 0.25T + 0.15C.

The weighting distribution for the composite score (S2) = 0.35S1 + 0.25K + 0.25T + 0.15C is primarily based on the following logic: Shape Contextual Similarity (S1) is the core of this evaluation system. As demonstrated earlier, generative AI style transfer may alter a building's fundamental form. Yet, a structure's "essence" and "recognizability" are most fundamentally embodied in its basic shape, silhouette, and volumetric relationships. Therefore, Shape Contextual Similarity is assigned the highest weight. Key point matching (K) and texture feature similarity (T) evaluate two important yet slightly secondary

https://doi.org/10.3390/buildings16030494

dimensions—structural stability and surface similarity, respectively. Thus, they are assigned equal weight. Color Distribution Similarity (C) is intentionally assigned a lower weight in the evaluation system. Compared to shape and structure, color alterations have a relatively minor impact on the "identity" of the building itself, and color changes tend to be minimal before and after style migration.

Additionally, due to AI processing uncertainties, transferred architectural images may exhibit excessive background rendering. To ensure consistency in subsequent measurements, transferred images undergo standardized preprocessing to eliminate background interference, providing only the main building structure remains (Table 2). Following the preparatory steps, data were measured, and statistical analysis was conducted (Table 3). Comparing results with the prior complexity grouping, the high-complexity group achieved an average comprehensive semantic inheritance score of 78.7125, while the medium-complexity group scored 78.65. The slight difference (approximately 0.06) indicates relatively similar semantic inheritance after style transfer in medium- and high-complexity buildings. However, the low-complexity group achieved an average comprehensive semantic inheritance score of 82, significantly higher than those of the high- and medium-complexity groups. This suggests that style transfer in low-complexity buildings better preserves semantic and stylistic characteristics, resulting in higher inheritance rates (Table 4).

Table 2. Comparison of Preprocessed and Unprocessed Images.

Original Image	Transferred Image	Preprocessed Image

Table 3. Similarity Metrics for Transferred Image.

ID	Shape Context Similarity (S1 *)	Key-Point Matching (K *)	Texture Feature Similarity (T *)	Color Gradation Similarity (C *)	Comprehensive Semantic Inheritance Score (S2 *)
1	88	50	99	63	77.5
2	89	50	99	53	76.35
3	95	50	99	73	81.45
4	92	50	99	74	80.55
5	88	50	99	66	77.95
6	87	50	99	71	78.35
7	90	50	99	66	78.65
8	91	50	99	81	81.25
...	...	...	...	...	...
24	93	50	99	71	80.45

* $S1 = 1 - \chi^2(P_1, P_2) = 1 - ½\sum(P_1(i,j) - P_2(i,j))^2/(P_1(i,j) + P_2(i,j))$. * $K = \sum\_(i,j)\ wij \cdot \exp(-d\_(ij)^2/2\sigma^2)$. * $T = \cos(\theta) = (V_1 \cdot V_2)/(|V_1||V_2|)$. * $C = 1 - ½\sum|H_1(i) - H_2(i)|$. * S2 = 35%S1 + 25%K + 25%T + 15%C

https://doi.org/10.3390/buildings16030494

Table 4. Image Similarity Metrics for Different Complexity Groups.

Complexity Group	Data ID	Complexity Range	Average Similarity Score
High Complexity Group	18 11 22 10 20 24 7 6	9.96–15.23	78.7125
Medium Complexity Group	4 15 17 5 2 1 14 8	8.12–9.91	78.65
Low Complexity Group	16 23 3 19 12 9 21 13	3.6–8.08	82

Based on the findings of this experiment, low-complexity buildings feature simpler structures, fewer details, and more uniform textures. This enables AI models to more readily capture core semantic features (such as basic shapes and color distributions), thereby reducing information loss during transfer. Conversely, high- and medium-complexity buildings contain more intricate details, complex edges, and color variations, increasing the model's learning difficulty. This may result in incomplete preservation of key semantic information during transfer, particularly when structural alterations are significant. The model may overemphasize local features while neglecting overall style inheritance.

Moreover, the evaluation system in this experiment emphasizes semantic levels. The diversity of high-complexity buildings may make key-point matching and shape context similarity calculations more susceptible to noise interference (e.g., residual background rendering artifacts). In contrast, the uniformity of low-complexity buildings facilitates stable metric computation. Additionally, regarding the two evaluation framework models mentioned above, the weighting data represent reasonable and interpretable initial settings derived from theoretical analysis and preliminary experiments. However, their "optimality" is not absolute. For instance, the keypoint matching score (K) consistently measures 50 across all datasets, indicating that this metric may not fully capture variations in complexity and requires further optimization.

4.3. Ablation Studies

To systematically evaluate the role of each module in the style transfer task, this study conducted targeted controlled experiments, primarily focusing on the deep information extraction module and the positive prompt input module. Specifically, the core function of the deep information extraction module is to parse the geometric layout and spatial structure of the scene from the source image, providing crucial structural constraints for subsequent generation. The core function of the positive prompt input module is to assist in generating the transferred image using textual descriptions.

The experiments will focus on observing and comparing performance differences in multiple aspects of the generated images. Through this controlled experimental design—where all other generation module parameters and conditions remain constant while manipulating only the depth information and forward prompt input modules—we can effectively analyze the relationship between variations in image generation outcomes and the absence of depth information. Alternatively, we can examine the control capabilities of textual guidance over the semantic content and stylistic details of generated images (Table 5), thereby dissecting the specific roles of these two modules in maintaining the structural integrity of generated images.

https://doi.org/10.3390/buildings16030494

Table 5. Ablation Experiment Results.

Experimental Variables	Experimental Flowchart	Experimental Results
No Variables	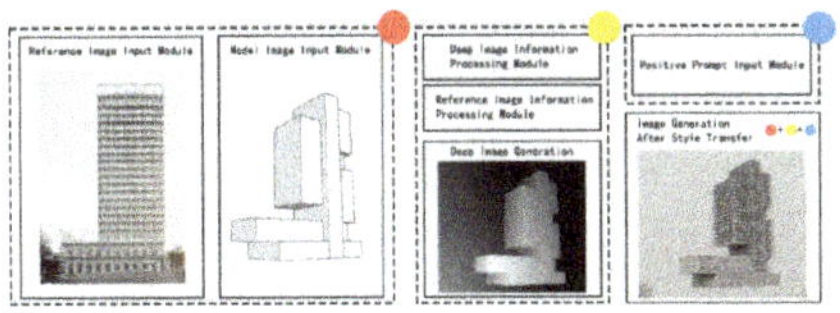	
Only the Deep Information Extraction Module is Disabled	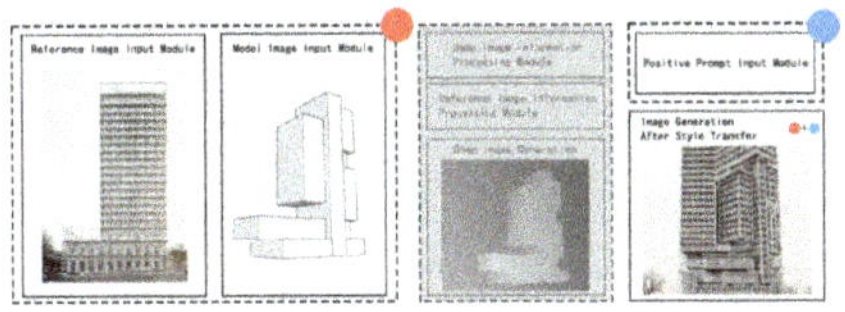	
Disable only the positive enhancement word input module	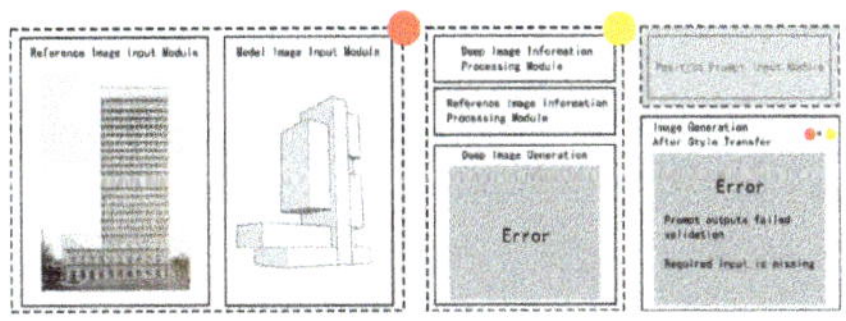	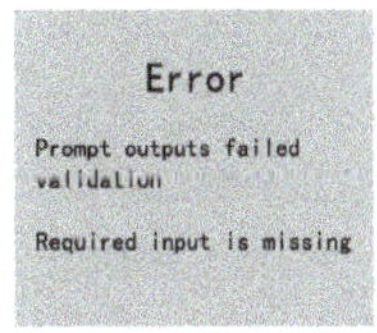

Experimental results demonstrate that the forward prompt input module is an indispensable component of the Flux Redux model's generation workflow. Disabling this module results in the loss of conditional information relied upon by the model, ultimately causing the generation process to fail.

Turning off the depth information extraction module entirely causes the workflow's core functionality to revert to the text-to-image generation mechanism based on the Flux Redux model. Under this configuration, the final generated images become highly dependent on two key input elements. One is the style reference image, whose color patterns, texture features, and compositional elements are analyzed by the algorithm and transferred into the generated image. The other element is the forward prompt. Using semantic encoding techniques, concepts, scenes, attributes, and other textual information are converted into feature vectors processable by the image generation model. This unidirectional relationship implies that when deep information is absent, the spatial structure, perspective relationships, and three-dimensional layering of the generated image are not incorporated into the model's computational process. Consequently, an image generation paradigm dominated by two dimensional feature transfer and semantic mapping emerges.

4.4. Experimental Analysis and Discussion

Experimental results clearly demonstrate a negative correlation between the initial complexity of architectural images and the semantic inheritance rate after style transfer. The low-complexity group achieved the highest average inheritance score (82), while the medium- and high-complexity groups scored similarly and significantly lower (78.65 and 78.7125, respectively). This finding indicates that a structurally simple, less detailed architecture enables AI models to capture better and preserve their core stylistic essence. Furthermore, ablation experiments preliminarily validate the critical role of the deep information extraction module in constraining the three-dimensional spatial structure of generated images, as well as the indispensable conditional guidance provided by the forward prompt input module within the current workflow.

https://doi.org/10.3390/buildings16030494

4.4.1. Limitations of the Study

Although this study has achieved specific results, several limitations remain:

First, the experiment used only 24 architectural images, which may limit the generalizability of the findings. Furthermore, variables such as building type, regional style, and size were not systematically considered. Future experiments should expand the sample database to cover a broader range of architectural types and styles, thereby enhancing the robustness of the conclusions.

Second, the accuracy of the evaluation system requires further refinement. The assessment framework developed in this study remains preliminary, with specific metrics failing to demonstrate sufficient discriminative power in the experimental data. This indicates their limited effectiveness in the evaluation process, necessitating optimization and further research.

Third, while the experiment observed a negative correlation between complexity and preservation, explanations for the underlying mechanisms remain speculative. For instance, the assertion that "high complexity hinders model learning" lacks a direct analysis of the model's internal feature representations. Techniques such as feature visualization could be considered to more deeply reveal differences in attention distribution when AI models process architectural images of varying complexity.

4.4.2. Scope and Boundary Conditions of the Study

Experiments revealed that to obtain optimal experimental data, this study faces certain limitations in its applicability and requires specific conditions to be met.

For instance, in terms of applicability, low-complexity architecture, such as minimalist modern building facades, yields the best style transfer results, as their uniform structure facilitates semantic preservation.

Regarding boundary conditions, input images must possess precise geometric contours; blurred or heavily occluded architectural photos may result in failed style extraction. Furthermore, the workflow involves serialized multi-module operations, and ablation experiments indicate high inter-module dependencies—failures in individual modules can cause the entire workflow to collapse.

4.4.3. Applications of Research in Architecture

The research currently presented primarily focuses on visual generation techniques for architectural images, yet it also demonstrates practical applications in architectural design, historic buildings, and architectural education.

Gao et al. highlighted in their study of Jiangmen's overseas Chinese heritage architecture that digital processing and intelligent analysis of architectural imagery are crucial for cultural heritage preservation and restoration [17]. The AI image style transfer technology examined in this research leverages style learning and transfer to provide robust generative capabilities for virtual restoration of damaged historical structures, thereby expanding the application boundaries of AI technology in architectural heritage.

In prior research, Jin et al. developed an AI-embedded teaching model for a 9-week graduate course [18]. This study clarified AI's core role in architectural education—as a creative generation tool—providing educational value support for our current technical application. Within architectural design education, this research serves as a cognitive tool for generating spatial style. It enables students to create and compare facades across multiple styles, helping them intuitively grasp the relationships among diverse styles and building volumes. This approach both encourages bold formal exploration and deepens understanding of architectural essence.

https://doi.org/10.3390/buildings16030494

Regarding the design process, research by Bagasi et al. explicitly positions AI image generation tools as core technologies for early design stages, directly demonstrating the practicality and efficiency gains of AI image style transfer techniques during architectural creative generation [19]. The workflow generated by this research can serve as a "creative accelerator" for early conceptual development. Designers can input images of volumetric models and specify one or more style reference images to rapidly generate architectural facade possibilities that simultaneously meet defined volumetric and stylistic objectives. Crucially, the workflow's deep information extraction module parses fundamental spatial hierarchies and front-back relationships from simple block model images. This ensures generated outputs transcend flat texture mapping, instead attempting three-dimensional style mapping that provides a more rational visual foundation for subsequent model refinement.

Furthermore, as Cai et al. prospectively noted in their research, integrating artificial intelligence with 3D printing [20] enables the digital models of architecturally designed solutions—generated through style transfer and embodying both aesthetic innovation and performance optimization—to interface directly with 3D printing construction systems. This facilitates the precise and efficient materialization of virtual styles into physical architecture.

5. Conclusions

This study successfully establishes a new paradigm for evaluating semantic inheritance in architectural style transfer and preliminarily validates that image complexity is a key factor influencing style transfer outcomes. Although improvements are possible in sample size, metric sensitivity, and theoretical depth, the research framework, problem awareness, and preliminary conclusions provide valuable insights for interdisciplinary research at the intersection of architectural design and digital technology. It is important to emphasize that the core contribution of this study lies in the innovation of visualization methods. Its structural relevance is applicable only to specific scenarios such as early-stage conceptual design, educational instruction, or visualization analysis, and is not intended to replace actual structural engineering practices.

Future research will advance in the following directions: First, expand the experimental sample size and systematically introduce building types, architectural styles, and other factors as control variables to conduct multi-factor variance analysis, ensuring that the research findings apply to a broader range of architectural contexts. Second, optimize the evaluation metric system, with the primary task being to refine the "keypoint matching accuracy" metric to enhance the scientific rigor of this evaluation framework. Third, conduct a combined qualitative and quantitative analysis of the image results generated by ablation experiments to assess the role of each module within the workflow rigorously. Fourth, based on the finding that "low complexity facilitates inheritance," explore applying semantic abstraction or simplification preprocessing to high-complexity architectural images before style transfer. Observe whether this enhances semantic inheritance after style transfer, potentially providing a practical technical pathway for digital style innovation in complex historical and landmark buildings.

Author Contributions: Conceptualization, C.Q.; Methodology, C.Q.; Software, C.Q.; Validation, C.Q.; Formal analysis, C.Q.; Investigation, C.Q.; Data curation, C.Q.; Writing—original draft, C.Q.; Writing—review & editing, C.X.; Visualization, C.Q.; Supervision, C.X. All authors have read and agreed to the published version of the manuscript.

Funding: This research received no external funding.

Data Availability Statement: The original contributions presented in this study are included in the article. Further inquiries can be directed to the corresponding author.

Conflicts of Interest: The authors declare no conflicts of interest.

https://doi.org/10.3390/buildings16030494

References

1. Gu, S.J.; Wang, R.X.; Wu, Y.F.; Xu, X.Y.; Yan, C.; Gao, T.Y.; Yuan, F. Exploration of AI-Heuristic Architectural Generative Design Process Based on FUGenerator Platform. In *Proceedings of the 2023 National Conference on Digital Technology in Architectural Education and Research*; Tongji University: Shanghai, China, 2023; pp. 441–444.
2. Huang, X.R.; Wang, Y.D.; White, M.; Zhang, B. Logic and Black Box:Prospecting the Integration of Artificial Intelligence and Computer-Aided Technology in Future Architectureand Urban Design. *Urban Archit.* **2022**, *19*, 1–6+18.
3. Komatina, D.; Miletić, M.; Mosurović Ružičić, M. Embracing Artificial Intelligence (AI) in Architectural Education: A Step towards Sustainable Practice? *Buildings* **2024**, *14*, 2578. [CrossRef]
4. Zhang, J. An Example of Creating a Dragon Year Promotion ID by Invoking the SVD Animation Model via ComfyUI. *Video Prod.* **2024**, *30*, 68–70.
5. Singh, A.; Jaiswal, V.; Joshi, G.; Sanjeeve, A.; Gite, S.; Kotecha, K. Neural style transfer: A critical review. *IEEE Access* **2021**, *9*, 131583–131613. [CrossRef]
6. Johnson, J.; Alahi, A.; Fei-Fei, L. Perceptual losses for real-time style transfer and super-resolution. In *European Conference on Computer Vision*; Springer International Publishing: Cham, Switzerland, 2016; pp. 694–711.
7. Cai, G.; Lou, Y.; Lu, F. AI enhancing prefabricated aesthetics and low carbon coupled with 3D printing in chain hotel buildings from multidimensional neural networks. *Sci. Rep.* **2025**, *15*, 13229. [CrossRef] [PubMed]
8. Dumoulin, V.; Shlens, J.; Kudlur, M. A learned representation for artistic style. *arXiv* **2016**, arXiv:1610.07629.
9. Cai, G.; Sun, L.; Liu, D.; Xu, B.; Mo, Z. Potential of indoor room 3D ratio in reducing carbon emissions by prefabricated decoration in chain hotel buildings via multidimensional algorithm models for robot in-situ 3D printing. *J. Build. Eng.* **2025**, *101*, 111757. [CrossRef]
10. Lin, H.; Huang, L.; Chen, Y.; Zheng, L.; Huang, M.; Chen, Y. Research on the Application of CGAN in the Design of Historic Building Facades in Urban Renewal—Taking Fujian Putian Historic Districts as an Example. *Buildings* **2023**, *13*, 1478. [CrossRef]
11. Meng, J.; Fang, X.; Xu, J.; Zhang, Z. Research on the Innovative Application of Song Dynasty Boundary Painting in Interior Soft Decoration Design Based on AIGC. *Buildings* **2025**, *15*, 1067. [CrossRef]
12. Chen, G.; Tong, Y.; Wu, Y.; Wu, Y.; Liu, Z.; Huang, J. Reconstruction of Cultural Heritage in Virtual Space Following Disasters. *Buildings* **2025**, *15*, 2040. [CrossRef]
13. Yan, W.; Wang, T.; Zhang, C. Renewal Design of Architectural Facade Features in the Shantou Xiaogongyuan Historic District Based on Deep Learning. *Buildings* **2025**, *15*, 4404. [CrossRef]
14. Chen, Z.; Zhang, N.; Xu, C.; Xu, Z.; Han, S.; Jiang, L. Typological Transcoding Through LoRA and Diffusion Models: A Methodological Framework for Stylistic Emulation of Eclectic Facades in Krakow. *Buildings* **2025**, *15*, 2292. [CrossRef]
15. Duan, W.; Rao, J.; Zhao, J.; Tao, N.; Chen, J. AI-Based Pre-Renewal Design for Historic Building Facades: An AIGC–LoRA Framework with Collaborative Assessment. *Buildings* **2025**, *15*, 4212. [CrossRef]
16. Wang, J.; Shi, Y.; Chen, X.; Lan, Y.; Liu, S. Teaching with Artificial Intelligence in Architecture: Embedding Technical Skills and Ethical Reflection in a Core Design Studio. *Buildings* **2025**, *15*, 3069. [CrossRef]
17. Gao, L.; Wu, Y.; Yang, T.; Zhang, X.; Zeng, Z.; Chan, C.K.D.; Chen, W. Research on Image Classification and Retrieval Using Deep Learning with Attention Mechanism on Diaspora Chinese Architectural Heritage in Jiangmen, China. *Buildings* **2023**, *13*, 275. [CrossRef]
18. Jin, S.; Tu, H.; Li, J.; Fang, Y.; Qu, Z.; Xu, F.; Liu, K.; Lin, Y. Enhancing Architectural Education through Artificial Intelligence: A Case Study of an AI-Assisted Architectural Programming and Design Course. *Buildings* **2024**, *14*, 1613. [CrossRef]
19. Bagasi, O.; Nawari, N.O.; Alsaffar, A. BIM and AI in Early Design Stage: Advancing Architect–Client Communication. *Buildings* **2025**, *15*, 1977. [CrossRef]
20. Cai, G.; Liu, D.; Wu, Z. Multidimensional algorithms-based carbon efficiency model of building geometric 3D ratios for prefabricated 3D printing design and construction. *npj Clean Energy* **2025**, *1*, 5. [CrossRef]

https://doi.org/10.3390/buildings16030494

buildings

MDPI

Article

Research Progress and Frontier Trends in Generative AI in Architectural Design

Yingli Yang [1], Yanxi Li [2,*], Xuefei Bai [1], Wei Zhang [1] and Siyu Chen [1]

1 School of Art and Design, Guangdong University of Finance and Economics, Guangzhou 510320, China; yangyingli@gdufe.edu.cn (Y.Y.)
2 School of Humanities and Arts, Heyuan Polytechnic, Heyuan 517000, China
* Correspondence: liyanxi19971004@163.com

Abstract

In recent years, with the rapid advancement of science and technology, generative artificial intelligence has increasingly entered the public eye. Primarily through intelligent algorithms that simulate human logic and integrate vast amounts of network data, it provides designers with solutions that transcend traditional thinking, enhancing both design efficiency and quality. Compared to traditional design methods reliant on human experience, generative design possesses robust data processing capabilities and the ability to refine design proposals, significantly reducing preliminary design time. This study employs the CiteSpace visualization tool to systematically organize and conduct knowledge map analysis of research literature related to generative AI in architectural design within the Web of Science database from 2005 to 2025. Findings reveal the following: (1) International research exhibits a trend toward interdisciplinary convergence. In recent years, research in this field has grown rapidly across nations, with continuously increasing academic influence; (2) Research primarily focuses on technological applications within architectural design, aiming to drive innovation and development by providing superior, more efficient technical support; (3) Generative AI in architectural design has emerged as a prominent international research focus, reflecting a shift from isolated design to industry-wide integration; (4) Generative AI has become a core global architectural design topic, with future research advancing toward full-process intelligent collaboration. High-quality knowledge graphs tailored for the architecture industry should be constructed to overcome data silos. Concurrently, a multidimensional evaluation system for generative quality must be established to deepen the symbiotic design paradigm of human–machine collaboration. This significantly enhances efficiency while reducing the iterative nature of traditional methods. This study aims to provide empirical support for theoretical and practical advancements, offering crucial references for practitioners to identify business opportunities and policymakers to optimize relevant strategies.

Keywords: generative AI; architectural design; intelligence; visualization; CiteSpace

1. Introduction

The application of generative AI in architectural design represents a multidisciplinary concept dedicated to integrating generative AI with architectural practice. The application of artificial intelligence in architectural design has brought unprecedented opportunities for creativity and efficiency, allowing architects to explore new forms and functions with the help of generative models [1,2]. Conceptually, it emphasizes the convergence of Building

https://doi.org/10.3390/buildings16020388

Information Modeling (BIM), the Internet of Things (IoT), and generative AI large models. Generative AI is positioned as a creative partner, collaborating to stimulate innovative thinking and fostering disruptive concepts through cross-disciplinary integration [3]. Architects can use these AI applications to explore various design possibilities, enhance creativity, and receive instant feedback for iterative improvement [4,5]. Specifically, generative AI in architectural design operates across five key dimensions: First, the theoretical framework dimension explores how generative AI reshapes fundamental architectural cognition-shifting from the "form-function" dualism to a "multi-subjectivity" paradigm and investigating novel spatial paradigms. Second, the educational innovation dimension focuses on transforming designer competencies and educational reform in the generative AI era, shifting emphasis from skill transmission to cultivating innovation literacy. At the same time, artificial intelligence has traditionally been used as a digital design tool, but its role as a teaching aid to enhance creative thinking has not been fully explored [6]. Third, the technical methodology dimension centers on core technologies, encompassing AI model evolution and integration with design workflows, with the core being the technological shift from machine learning to generative AI. Fourth, the performance optimization dimension leverages generative AI to enhance building performance for sustainability, with its core function being the simulation of green buildings to achieve integrated "generation-simulation-optimization". Fifth, the practice integration dimension centers on reshaping the architecture industry's technical ecosystem through generative AI integration. Its core lies in breaking information silos by combining with digital technologies, while its frontier drives the industry's evolution from "technology empowerment" to "user experience" to ensure efficient implementation of complex designs.

While academia has extensively researched generative AI's applications in architectural design, several shortcomings persist: First, in technical maturity, the reliability and precision of generated outputs remain insufficient, potentially yielding non-compliant, unsafe, or impractical design solutions. Additionally, integration with traditional workflows remains cumbersome, often creating information silos [7]. Second, data governance faces challenges due to inconsistent standards and diverse formats across design, construction, and supplier data, resulting in scarce and uneven-quality industry datasets [8]. Third, there is a shortage of interdisciplinary professionals proficient in both civil engineering/architecture and AI, compounded by lagging interdisciplinary programs and curriculum development in higher education. Fourth, application limitations: Generative AI is predominantly used in early-stage processes like visualization and concept generation within the construction industry, while deeper applications in project management and cost control remain scarce [9]. Despite high industry interest, few construction firms have scaled the integration of generative AI into their operations [10]. Unlike previous studies that focused on theoretical discussions, case summaries, or technical reviews, this research employs systematic bibliometric analysis. Utilizing the CiteSpace visualization tool and drawing data from the Web of Science database, it retrieves and filters relevant academic literature in the field of architectural design from 2005 to 2025, constructing a network diagram through visualization tools. Through node size, color, and connecting lines, the evolution logic of the field is clearly reconstructed. While existing studies often focus on specific technical branches or short-term trends, this research systematically organizes and reveals the overall evolutionary path, knowledge structure clusters, and international collaboration networks of generative AI research in architectural design through long-term, holistic visualization. This addresses the shortcomings of existing reviews in analyzing dynamic evolution and structural relationships. By systematically mapping the research progress and emerging trends in generative AI for architectural design, this study aims to fill potential gaps in integrating technology with architectural design scenarios. It provides new insights with a more structured and evolutionary perspective for both theoretical research and practical exploration in this field.

2. Research Methods and Data Collection

2.1. Research Methods

CiteSpace is a pioneering bibliometric and scientific knowledge mapping software developed by Professor Chen Chaomei, an internationally renowned expert in scientometrics and information visualization. It is not merely a simple "visualization tool," but rather an integrated research system combining complex network analysis, clustering algorithms, and temporal analysis. Its purpose is to assist scholars in extracting clear, profound patterns of knowledge development from vast, chaotic literature data. Its core philosophy treats a knowledge domain as an evolving complex system. Within this system, "nodes" can represent academic papers, authors, institutions, or keywords, while the "connections" between them symbolize relationships such as co-occurrence, co-citation, or collaboration. CiteSpace employs advanced algorithms to compute and render these relationships, ultimately generating dynamic, multi-layered visual maps.

This study employs bibliometric methods and utilizes the knowledge map analysis software CiteSpace (version 6.4.R1) to conduct a visual analysis of literature data. The research focuses on the knowledge structure, evolutionary trajectory, and collaborative networks of generative artificial intelligence within the field of architectural design. The specific analysis spans the time period from 2005 to 2025, sliced by year. Network nodes represent countries, institutions, and keywords to reveal the field's macro-level landscape. Colors in the diagram denote years, with shades transitioning from blue to red as years increase. To ensure clarity and highlight core structures, redundant connections were pruned using the Pathfinder algorithm. For co-citation analysis, the g-index (k = 25) serves as the threshold for screening highly cited literature. This parameter balances representativeness and influence, representing a robust and widely adopted setting in related studies. It ensures the core literature included in the analysis possesses sustained and significant academic impact while reflecting the field's fundamental knowledge base.

It should be noted that while the above parameter selection follows common practices in bibliometric analysis, it may still exert specific influences on results. For instance, annual slicing may fail to reveal micro-fluctuations within shorter cycles; the Top 50 node threshold may inadequately reflect certain low-frequency yet promising emerging concepts. To mitigate such potential biases, this study will complement overall trend analysis with techniques like keyword outbreak detection to keenly capture emerging frontiers. All other analysis options retain the software's default configurations to ensure the standardization of the analytical process and the reproducibility of results. Overall, the parameter settings in this study aim to strike an optimal balance between scientific rigor and operational feasibility, systematically and objectively revealing the panoramic landscape of knowledge evolution in this field.

2.2. Data Collection

This study employs the Web of Science (WOS) Core Collection database as its retrieval source, primarily based on the following considerations: First, the WOS Core Collection has become the gold standard data source for bibliometric research, ensuring the reliability and academic validity of analytical results while effectively guaranteeing the scholarly authority and international representativeness of the analyzed sample [11]. Second, its comprehensive citation indexing facilitates in-depth co-citation and collaboration network analysis, aligning with this study's objective of knowledge map construction. Scopus and Google Scholar were excluded because the former has a relatively diverse coverage

https://doi.org/10.3390/buildings16020388

scope, while the latter contains a high proportion of non-core literature, both of which are detrimental to systematic metric analysis.

Web of Science (WOS) is one of the world's largest citation databases and the primary database supporting current knowledge graph analysis. This study utilized the Web of Science Core Collection as the foreign literature data source, focusing on document types such as "Article", " Early Access", "Proceeding Paper" AND "Review Article" with the themes "Topic = Generative AI" AND "architectural design". A total of 183 documents were collected. Manually excluded documents irrelevant to the theme, such as "Artificial Intelligence, Lymphoid Neoplasms, and Prediction of MYC, BCL2, and BCL6 Gene Expression Using a Pan-Cancer Panel in Diffuse Large B-Cell Lymphoma", yielding 151 valid documents. Subsequently, a search using the keywords "Topic = AI" AND "architectural design" yielded 833 documents. Irrelevant documents, such as "Kamino: A Scalable Architecture to Support Medical AI Research Using Large Real World Data", were manually excluded, resulting in 657 valid documents. After deduplicating the 151 valid documents from the first search with the 657 valid documents from the second search, 598 valid documents were obtained. These 598 valid documents were authored by 261 authors from 217 distinct institutions. The collection process is shown in Figure 1.

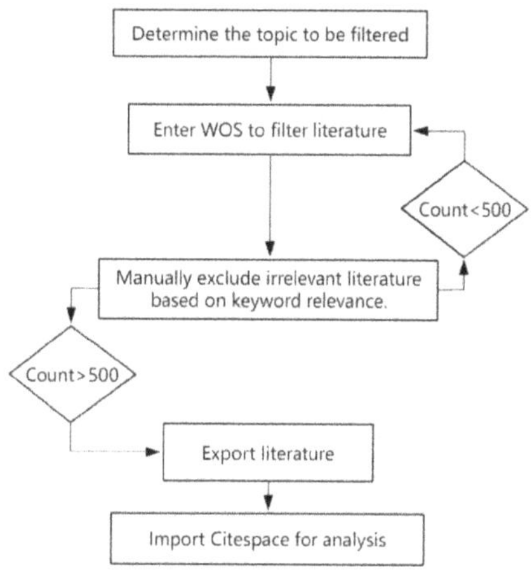

Figure 1. Data Collection Flowchart.

3. Literature Feature Analysis

3.1. Quantitative Characteristics

The temporal variation in publication volume reflects research progress and evolutionary trends [12]. To clearly understand international research outputs on generative AI in architectural design, this study utilized CiteSpace's visualization capabilities to statistically organize the publication dates of retrieved sample literature. Analysis yielded a quantitative trend chart. Figure 2 displays the annual publication volume from the WOS English database between 2005 and 2025. The line graph indicates that publication volumes remained at relatively low levels in the early years, with very gradual growth and minimal fluctuations before 2017, typically maintained between 0 and 5 articles. From 2017 onward, the growth trajectory accelerated sharply, with the curve exhibiting rapid expansion—particularly near 2025—indicating that research in this field has entered a period of intense exploration. This suggests the presence of high-caliber scholars and institutions abroad maintaining sustained focus on this domain [13]. Why did 2017 mark a turning point in this field's research? According to the literature, this shift is closely tied

https://doi.org/10.3390/buildings16020388

to relevant policies issued by China. In 2017, authors led by Ashish Vaswani published the paper "Attention Is All You Need" [14]. This paper introduced a revolutionary neural network architecture that fundamentally transformed sequence data processing. Compared to traditional RNN/LSTM models, the Transformer could handle longer sequences, resolving the "long-text forgetting" issue. Its powerful parallel computing capabilities significantly boosted model training efficiency, becoming the technical cornerstone for all subsequent large generative models (GPT, DALL-E, etc.). Concurrently, Foster + Partners established its Applied Research and Development (ARD) group in 2017, applying AI/machine learning to design assistance, knowledge dissemination, and business insights. They developed models that generate images from natural language. Subsequently, "Deep Learning Architect: Classification for Architectural Design Through the Eye of Artificial Intelligence" proposed a deep learning-based architectural style classification method, achieving a breakthrough in core technology and laying the foundation for intelligent classification and generation in architectural design [15]. Subsequently, in 2018, Achlioptas, P. published "Learning Representations and Generative Models for 3D Point Clouds". This paper explores geometric data represented as point clouds, introducing deep autoencoder (AE) networks with advanced reconstruction quality and generalization capabilities [16]. It conducts an in-depth study of various generative models, including GANs operating on raw point clouds, significantly improved GANs trained in the fixed latent space of AEs, and Gaussian mixture models (GMMs). This enables intelligent generation from point cloud data to architectural forms, offering new technological pathways for architectural design. In 2023, "Integrating BIM and AI for Smart Construction Management: Current Status and Future Directions" proposed that the integration of BIM and AI will demonstrate new added value in handling construction projects characterized by complexity and uncertainty [17]. It provided an in-depth analysis of the current status and future trends of AI utilization throughout the entire BIM project lifecycle. In Web of Science (WOS), research development in this field transitioned from a gradual start to rapid growth. Publication volume (orange line) shows an overall upward trend. Early publications fluctuated at low levels, indicating an initial exploratory phase with minimal academic output. From 2017, publication volume began a modest increase, and surged dramatically to approximately 200 papers by 2025, reflecting an extremely rapid growth phase. The maturation of technologies (such as multimodal models and open-source tools) after 2020 significantly enhanced the practical applicability of architectural design research, ultimately driving the substantial increase in WOS publications by 2025. Overall, while international research started slowly, it accelerated rapidly in later years. In recent times, generative AI in architectural design has rapidly advanced to become a mainstream frontier in academic focus. Simultaneously, it has gradually entered the public eye, attracting widespread attention [2]. As research depth expands and application scenarios continue to materialize, this field has begun to show clear signs of specialization and diversification.

3.2. Distribution of Key Authors and Research Institutions

3.2.1. Author Collaboration Network Analysis

Co-occurrence analysis identifies core authors within a field and measures collaboration intensity among them [18]. Publication volume is represented by node size. The larger the node, the greater the volume of posts. Network density refers to an indicator measuring the closeness of connections between nodes in a network, defined as the ratio of actual connections to the number of possible connections. Collaboration relationships are depicted as connections, where denser lines signify stronger ties [19]. As shown in Figure 3.

https://doi.org/10.3390/buildings16020388

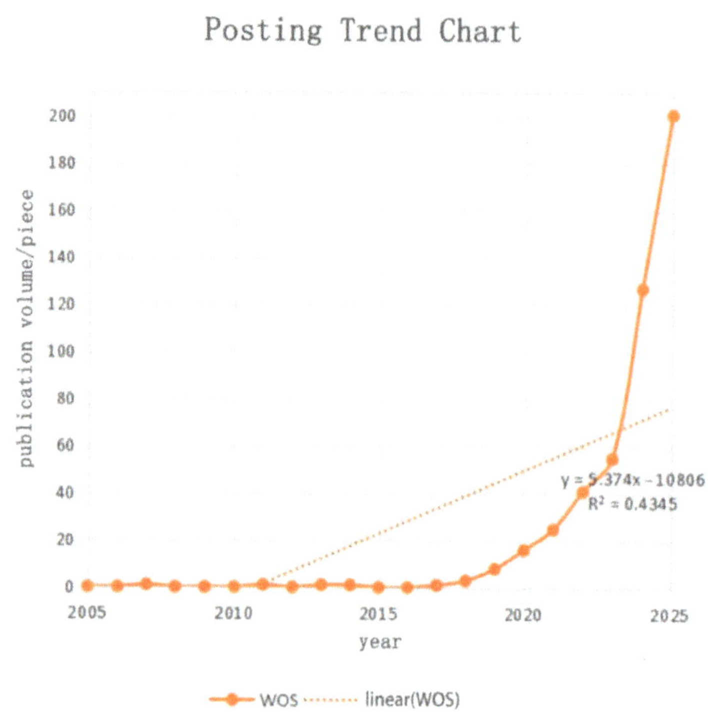

Figure 2. Annual Publication Volume Trend (2005–2025).

Figure 3 shows a visualization of the Web of Science sample literature, revealing 261 network nodes and 290 connections, with a network density of 0.0085. Prominent large nodes indicate key authors in foreign generative AI literature for architecture: Lu Shuai (4 publications), Lu Weisheng (4 papers), Yin Jun (4 papers), Yoon Sungmin (4 papers), Alanne Kari (3 papers), Darko Amos (3 papers), Li Jiteng (3 papers), and Zeng Pengyu (3 papers). In terms of connectivity, collaboration among international authors is relatively close. The dense connections in the network diagram indicate that a relatively mature academic collaboration community has formed in this field abroad, with frequent interactions among different scholars. Particularly prominent is the research team centered around Hong Tianzhen and Liang Jiadong, which represents a cohesive cluster within the current collaboration network. The formation of such clusters also facilitates the integration of different research perspectives, driving cross-disciplinary innovation in technologies and methodologies within the field.

The team centered around Lu Shuai's AI research primarily focuses on automating the generation and modification of residential layouts, aiming to enhance efficiency in architectural design processes [20]. Targeting the complex and highly customized process of residential layout design, they employ generative AI to automate layout generation and dynamic modification. They have also established collaborative interaction channels between users and designers, proposing the Text2 FloorEdit framework that decomposes design tasks into three components: Window, Door, and Wall Generation (WD-Net) for detailed floor plan creation at lower training costs, enabling precise translation of ordinary users' spatial needs into design parameters [21,22]. This enhances collaboration between users and designers, advances architectural automation, and ultimately boosts design efficiency and automation adoption. Concurrently, Lu Shuai's team introduced an AI-based framework integrating generative models, energy prediction, and evolutionary

https://doi.org/10.3390/buildings16020388

optimization. This approach reduced energy consumption by 17.5% compared to the best baseline model, demonstrating its effectiveness in lowering residential energy usage [23].

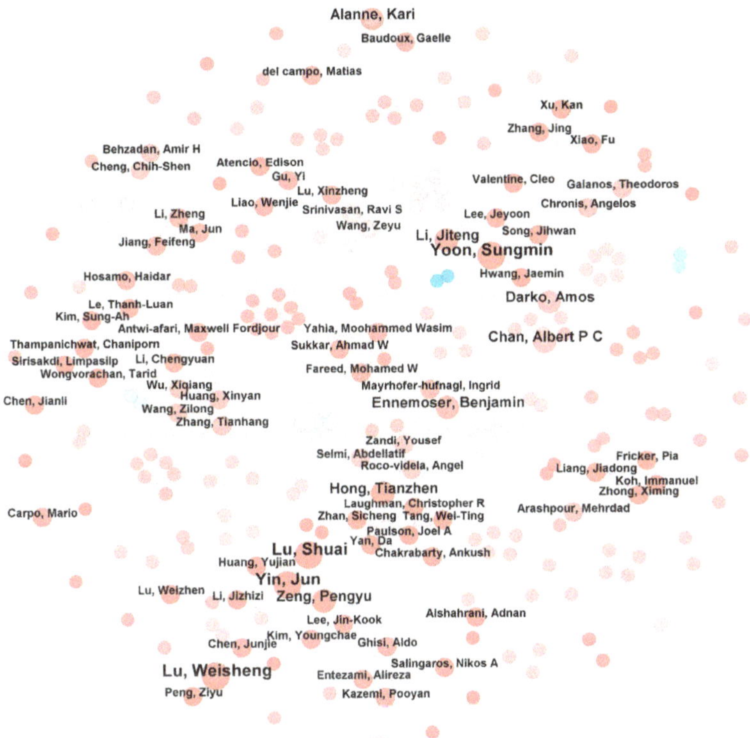

Figure 3. Author Collaboration Network.

The team led by Lu Weisheng focuses on researching AI agent based conceptual models for Indoor Environmental Informatics (IEI), intelligent exploration in building operations and maintenance, and continuous indoor environmental management via AI agents [24]. Their AI agent conceptual model enables real-time collection, analysis, and regulation of indoor environmental parameters, proposing a segmentation-free, Derivative-Free Optimization (DFO) method. This approach transforms finished BIM generated from 2D images into an optimization problem for fitting BIM components under architectural and topological constraints. By connecting fitted components to existing semantic sources, it enriches BIM semantics, partially alleviating the dilemma of information scarcity versus overload in BIM development [25]. Lu Weisheng's team discovered that user attitudes toward intelligent building systems are neither constant nor necessarily tied to perceptions of utility and usability. Managers can tactically apply pressure to enhance organizational technology acceptance. These findings, combined with an expanded TAM theory, lay the groundwork for developing a universal theoretical framework for technology adoption in the construction industry [26]. The team further explored computational BIM algorithms that control information to facilitate CWM decision-making. By linking computational BIM operations to multiple mainstream procurement models, they identified the platforms where computational BIM can advance more efficiently and effectively [27]. Grounded in ubiquitous computing and universal intelligent object theory, they defined the panoramic and interconnected characteristics that distinguish SCOs from traditional building objects, thereby explicitly specifying SCOs' core attributes, computational applications, and representations. The team's research is closely integrated with Building Information Modeling (BIM) and the Internet of Things (IoT) [28].

3.2.2. Institutional Collaboration Network Analysis

The presentation format for institutional collaboration networks mirrors that of author networks. As shown in Figure 4 for the Web of Science sample literature network, foreign literature comprises 217 nodes and 222 links, with a network density of 0.0095. Core institutions in related fields include Hong Kong Polytechnic University, Tsinghua University, Texas A&M University System, Universidad Politécnica de Madrid, University College London, University of London, Egyptian Knowledge Bank (EKB), Islamic Azad University, Southeast University–China, Sungkyunkwan University (SKKU), Tongji University, and University of Hong Kong, occupying pivotal positions within the institutional collaboration network.

Figure 4. Institutional Collaboration Network.

From the overall research institution map, most institutions have established only one-way or limited collaborations with a few core institutions, failing to form a comprehensive, dense collaborative network. However, the connection strength and interaction frequency within the core institution clusters are relatively high. Institutions like Islamic Azad University and Egyptian Knowledge Bank (EKB) have formed a cross-institutional collaborative team with strong cooperative capabilities. Nevertheless, some nodes still exhibit independent and fragmented states, indicating significant room for improvement in future collaborative exchanges among research institutions in this field. Overall, research on generative AI in architectural design follows a pattern dominated by universities, supplemented by research institutes and associations.

https://doi.org/10.3390/buildings16020388

3.3. Analysis of Major Country Distribution and Disciplinary Distribution

As shown in Figure 5, the Web of Science country co-occurrence map reveals 185 connections among 69 countries, with a network density of 0.0789. Most countries maintain unidirectional collaborations with only 1–2 core nations, failing to form a comprehensive and tightly knit cooperative network. Collaboration is predominantly concentrated among a few leading countries, while peripheral nations exhibit weak connectivity.

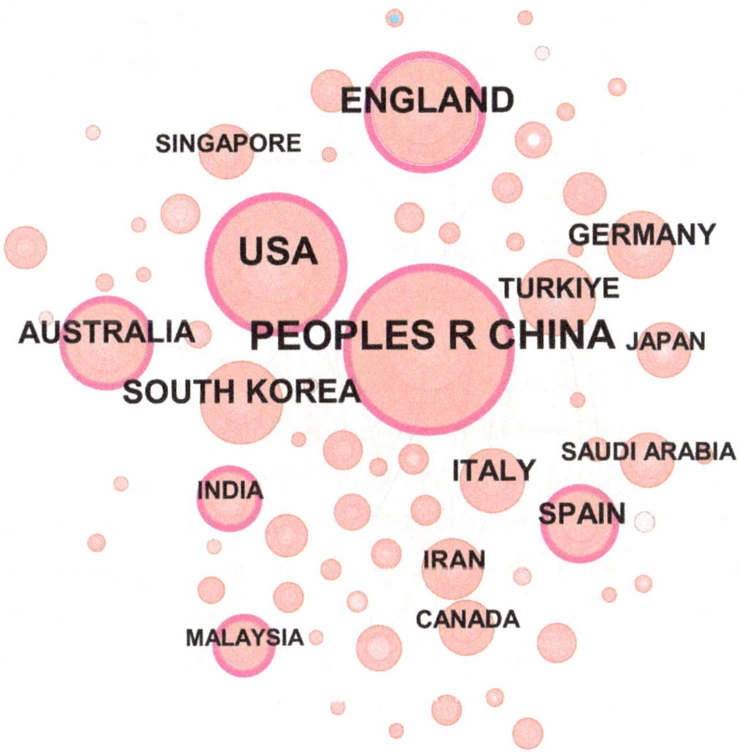

Figure 5. National Institutional Collaboration Map.

According to Table 1's institutional collaboration data, the top ten countries by publication volume are: China, the United States, the United Kingdom, South Korea, Australia, Germany, Italy, Spain, Turkey, and Iran. China leads in total publications, followed closely by the United States and the United Kingdom. The top three countries by centrality are the United Kingdom, China, and the United States. Germany (6th in publication volume) and Italy (7th) both exhibit centrality values of merely 0.01. Here, intermediary centrality is defined for each node in the network. It measures the likelihood that any shortest path in the network traverses the node. The year indicates when a country's counted papers (or contribution volume) first exceeded zero. Nodes with high intermediary centrality may lie between two large clusters or subnetworks, hence the term "intermediary." German research primarily focuses on developing artificial intelligence technologies, with collaborations predominantly involving North American institutions. Italian research emphasizes the preservation and restoration of traditional architecture, though its collaborative scope remains relatively limited [29,30]. Consequently, despite producing some outcomes, these nations have not deeply integrated into the global collaborative network. This results in

https://doi.org/10.3390/buildings16020388

weak international connectivity and influence within the field. Thus, despite high publication volumes, both countries exhibit limited influence in generative AI research for architectural design and urgently require enhancement.

Table 1. National Institution Collaboration Matrix.

Number	Country	Count	Quantity	Year
1	PEOPLES R CHINA	130	0.23	2006
2	USA	91	0.33	2015
3	ENGLAND	65	0.54	2007
4	SOUTH KOREA	43	0.03	2013
5	AUSTRALIA	37	0.17	2019
6	GERMANY	30	0.01	2018
7	ITALY	30	0.01	2012
8	SPAIN	30	0.12	2016
9	TURKIYE	25	0.04	2011
10	IRAN	20	0.06	2020
11	JAPAN	17	0	2011
12	SINGAPORE	17	0.09	2020
13	CANADA	16	0.05	2019
14	INDIA	16	0.2	2015
15	SAUDI ARABIA	16	0.1	2015
16	MALAYSIA	14	0.12	2020
17	FRANCE	12	0.08	2017
18	EGYPT	11	0.03	2020
19	SWEDEN	10	0	2021
20	TAIWAN	10	0	2023
21	BRAZIL	9	0	2015
22	PORTUGAL	9	0	2022
23	AUSTRIA	8	0.05	2022
24	FINLAND	8	0	2021
25	U ARAB EMIRATES	8	0.01	2024
26	BELGIUM	7	0.02	2021
27	THE NETHERLANDS	6	0.04	2013
28	CHILE	5	0.01	2021
29	DENMARK	5	0.01	2022
30	NEW ZEALAND	5	0	2023
31	NORWAY	5	0	2024
32	SOUTH AFRICA	5	0.01	2022
33	VIETNAM	5	0	2021
34	GREECE	4	0.05	2021
35	MEXICO	4	0	2021
36	RUSSIA	4	0	2023
37	THAILAND	4	0	2024
38	ARGENTINA	3	0	2016
39	JORDAN	3	0	2022
40	NIGERIA	3	0	2021
41	PAKISTAN	3	0	2016
42	POLAND	3	0	2023
43	QATAR	3	0.01	2022
44	SWITZERLAND	3	0	2022
45	TUNISIA	3	0	2021
46	BAHRAIN	2	0	2007
47	HUNGARY	2	0	2021
48	IRAQ	2	0.01	2021
49	KAZAKHSTAN	2	0.02	2021
50	KUWAIT	2	0	2025
51	ROMANIA	2	0	2023
52	SERBIA	2	0	2023
53	ALGERIA	1	0	2025
54	COLOMBIA	1	0	2025
55	CZECH REPUBLIC	1	0	2023

https://doi.org/10.3390/buildings16020388

Table 1. *Cont.*

Number	Country	Count	Quantity	Year
56	ETHIOPIA	1	0	2022
57	INDONESIA	1	0	2020
58	IRELAND	1	0	2024
59	LEBANON	1	0	2020
60	LITHUANIA	1	0	2024
61	MONTENEGRO	1	0	2024
62	MOROCCO	1	0	2025
63	PALESTINE	1	0	2024
64	PHILIPPINES	1	0	2025
65	SLOVAKIA	1	0	2022
66	SLOVENIA	1	0	2021
67	SRI LANKA	1	0	2023
68	UKRAINE	1	0	2025
69	YEMEN	1	0	2025

3.4. Co-Citation Analysis of Literature, Authors, and Journals

Co-citation analysis serves as a method to evaluate the influence of academic literature. Quantifying citation frequencies of documents, authors, and journals, it identifies research hotspots. Literature clustering analysis can be employed to discern similarities and differences in research themes. This study utilizes co-citation analysis to identify significant research outcomes in generative AI within architectural design, providing references for academic evaluation and research direction selection.

3.4.1. Document Co-Citation Analysis

Document co-citation analysis reflects a study's academic influence, research quality, and recognition within the scholarly community, serving as a key indicator for evaluating scholarly contributions. This study conducted co-citation analysis on documents published between 2005 and 2025, using one-year intervals as the temporal unit. The co-citation network comprises 598 nodes and 1558 links (see Figure 6a). The purple circle represents centrality; the thicker the circle, the greater the centrality. In the co-citation analysis graph of Web of Science publications, the top publications with 15 or more co-citations are ranked as follows: Baduge SK (2022, AUTOMAT CONSTR, V141, P0, DOI 10.1016/j.autcon.2022.104440) [31]. Abioye SO (2021, J BUILD ENG, V44, P0, DOI 10.1016/j.jobe.2021.103299) [32]. Pan Y (2021, AUTOMAT CONSTR, V122, P0, DOI 10.1016/j.autcon.2020.103517) [33]. Rombach R, (2022, PROC CVPR IEEE, V0, P10674, DOI 10.1109/CVPR52688.2022.01042) [34]. Goodfellow I (2020, COMMUN ACM, V63, P139, DOI 10.1145/3422622) [35]. Nauata N (2020, IMG PROC COMP VIS RE, V12346, P162, DOI 10.1145/3422622) [36]. Darko A (2020, AUTOMATIC CONSTRUCTION, V112, P0, DOI 10.1016/j.autcon.2020.103081) [37]. Chaillou Stanislas (2020, ARCHITECTURAL INTELLIGENCE, V0, P117, DOI 10.1007/978-981-15-6568-7_8) [38]. Hu RZ (2020, ACM Transactions on Graphics, V39, P0, DOI 10.1145/3386569.3392391 [39]. Liao WJ, (2024, AUTOMATIC CONSTRUCTION, V157, P0, DOI 10.1016/j.autcon.2023.105187) [2].

Among these, Goodfellow I (2020) [35] serves as the theoretical cornerstone of the field, representing pioneering research on generative adversarial networks. This work established the core generative paradigm for the entire field, enabling the creation of realistic and diverse architectural images, floor plans, and more from random noise, laying the technical foundation for all subsequent applications. Subsequently, Rombach R (2022) [34] served as the technical engine and application inflection point for the field, representing a breakthrough in latent diffusion models. This technology drastically reduces computational costs for high-quality image generation by performing denoising in a low-dimensional latent space. Combined with natural language understanding capabilities, it enables an "AIGC" workflow that directly generates and manipulates architectural designs via text prompts, thereby igniting large-scale application and innovation of generative AI in architectural design.

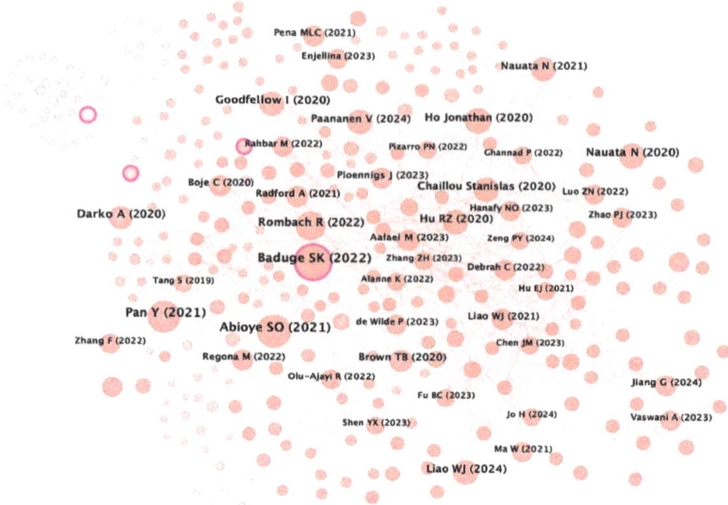

(a)

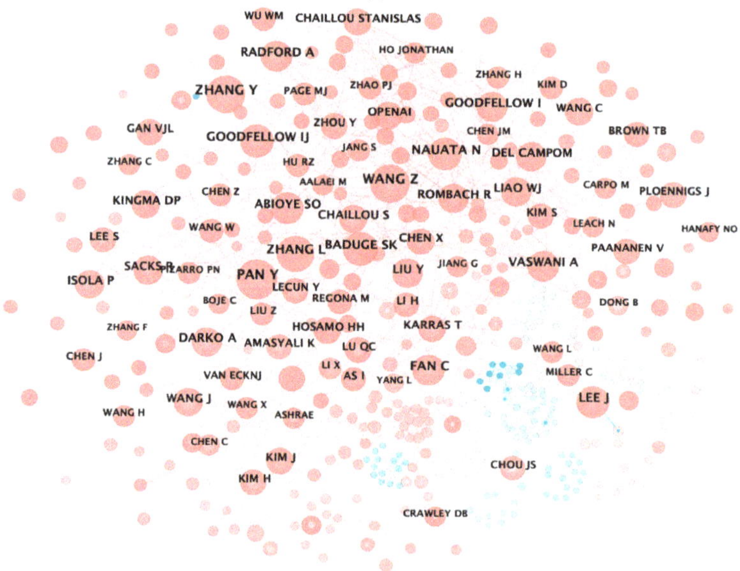

(b)

Figure 6. *Cont.*

https://doi.org/10.3390/buildings16020388

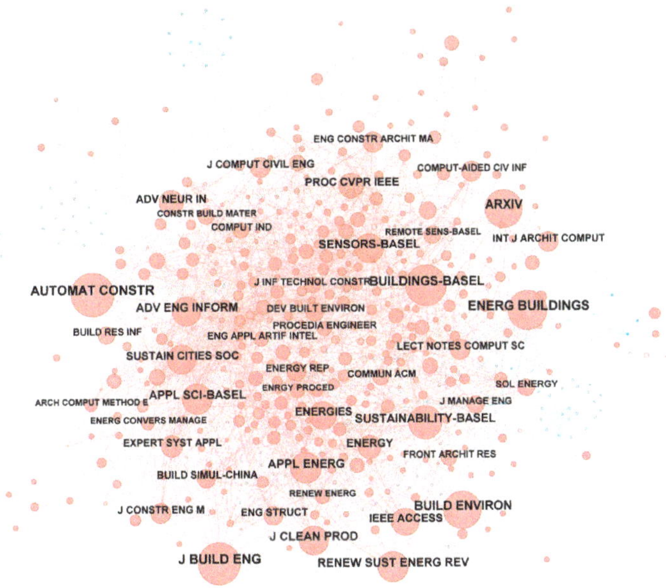

(c)

Figure 6. (**a**) Cited Literature. (**b**) Cited Authors. (**c**) Cited Journals.

3.4.2. Co-Citation Analysis of Authors

Co-citation analysis illuminates relationships and influence among authors in academic research, aiding the identification of core scholars, potential collaboration directions, and interdisciplinary trends within a field. This paper employs the CITESpace tool to construct an author co-citation network. Figure 6b shows authors with co-citation frequencies exceeding 25, listed in descending order: PAN Y, WANG Z, ZHANG Y, ZHANG L, GOODFELLOW IJ, BADUGE SK, FAN C, ABIOYE SO, and LEE J. This indicates that these publications and authors have exerted significant influence on the development of generative AI in architectural design, with their research findings receiving high recognition.

3.4.3. Journal Co-Citation Analysis

Core journals play a vital role in the development of academic fields. The nodes in Figure 6c, representing the number of published papers, display a dual-layer overlay of citing and cited journals for 598 relevant publications. This illustrates citation relationships within the research field, as well as interdisciplinary integration and mutual influence across different disciplines. Journals with co-citation frequencies exceeding 150 are listed in descending order: AUTOMAT CONSTR (284 times), J BUILD ENG (249 times), ENERG BUILDINGS (246), BUILD ENVIRON (223), BUILDINGS-BASEL (222), ARXIV (200), SUSTAINABILITY-BASEL (167), RENEW SUST ENERG REV (164), APPL ENERG (163 times), and APPL SCI-BASEL (150 times). Specifically, as the top journal in this field, the high citation count of AUTOMAT CONSTR signifies that generative AI research is at the forefront of the digital and intelligent transformation in architecture. This journal focuses on BIM, digital construction, robotics, and related directions. As a key technology driving the next wave of automation, generative AI research is deeply embedded within the end-to-end digital workflow spanning design, construction, and operations. It represents the highest level of pursuit for process innovation and technological integration within the field. The high citation count of J BUILD ENG indicates that generative AI research also places significant emphasis on practical engineering applications and performance

https://doi.org/10.3390/buildings16020388

implementation. This journal emphasizes specific engineering challenges in building materials, structures, energy efficiency, and environmental protection. Its extensive citations demonstrate that generative AI has transcended conceptual generation, deeply integrating into performance-driven design domains like structural optimization, energy efficiency calculations, and sustainability analysis. It is dedicated to solving practical engineering challenges in building safety, comfort, and environmental sustainability. In summary, these journals exhibit high attention and in-depth research on generative AI in architectural design, making them essential long-term tracking targets for scholars in this field.

4. Research Hotspot Analysis

Keywords provide a rapid overview of an article's core research content—its essence, condensed substance, and research direction. Analyzing keywords reveals research hotspots and frontiers, with Citespace's keyword co-occurrence and clustering effectively mapping these areas.

Research hotspots represent the focal points of current studies, a group of intrinsically connected, relatively numerous papers addressing popular issues or themes within a specific period. The frequency distribution and centrality of literature keywords can be used to study the development trends and research hotspots in a field. Meanwhile, keyword emergence maps utilize the rate of change in co-occurrence frequency to reflect the fundamental characteristics of evolving research frontiers.

4.1. Keyword Analysis

This study employs the co-occurrence analysis method with keywords as node types using CiteSpace software to identify research hotspots. The time scale was set from 1 January 2005, to 1 November 2025, with a time interval of 1. The data extraction target was set to TOP50, and the clipping method was Pathfinder. The resulting keyword knowledge map is shown in Figure 7. Among the keywords, "performance", "design" and "artificial intelligence (AI)" had the highest frequency and largest nodes, followed by "BIM", "artificial intelligence" and "architectural design", "artificial intelligence (AI)" exhibit the highest frequency and largest node size, followed by "bim", "artificial intelligence" and "architectural design".

Centrality refers to the ability to act as a mediator within a relational network. Generally, nodes with a centrality value greater than 0.1 are considered to occupy relatively important positions within the network structure. According to software statistics (Table 2), eight keywords in WOS literature exhibit centrality exceeding 0.1: "performance" (0.19), "design" (0.18), "artificial intelligence (AI)" (0.15), "bim" (0.14), "artificial intelligence" (0.13), "architectural design" (0.12), "big data" (0.12), "behavior" (0.12), and "neural networks" (0.10). Among these keywords, "artificial intelligence (AI)" represents the methodological cornerstone of the research, specifically referring to generative models exemplified by deep learning, generative adversarial networks (GANs), and diffusion models. These serve as novel "creativity engines," fundamentally transforming the technical paradigm of architectural design. "design" specifically denotes the entire intelligent design process from concept generation to scheme refinement. It is no longer solely the domain of human designers but has evolved into an iterative process of "human–machine dialog and collaborative exploration." Here, artificial intelligence handles massive data processing and generates vast alternative solutions, while designers steer the direction, inject creativity, and make final decisions; "Performance" defines the value orientation and optimization goals of this research domain. It emphasizes that the generative process and final solutions must transcend formal aesthetics, deeply integrating and optimizing the building's comprehensive performance across energy efficiency, structural integrity, lighting, com-

https://doi.org/10.3390/buildings16020388

fort, materials, and other aspects. This indicates that these keywords possess the highest mediating centrality in generative AI research within architectural design, signifying their significant role as connectors and intermediaries within the entire relational network. They represent the focal points of research in this field.

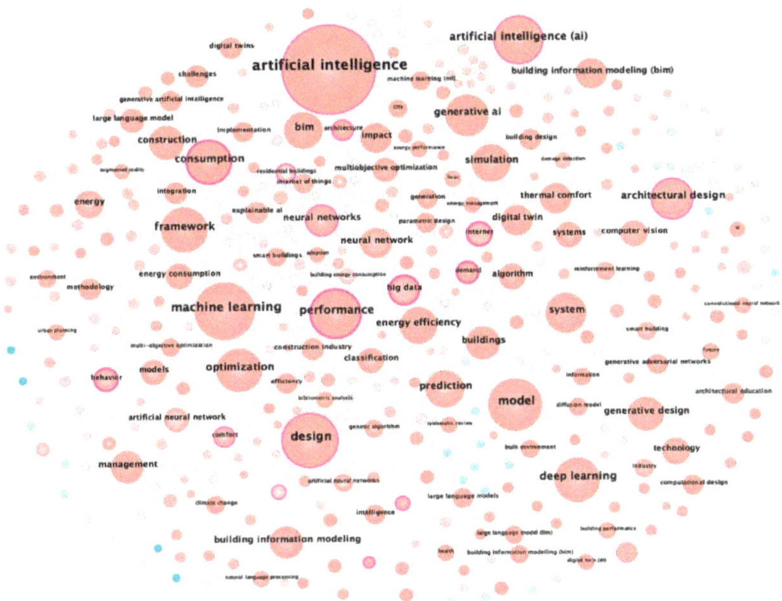

Figure 7. Literature Keyword Map.

Table 2. Keywords with Centrality Greater than 0.1.

Rank	Count	Centrality	Year	Keywords
1	41	0.19	2016	performance
2	52	0.18	2011	design
3	36	0.15	2020	artificial intelligence (AI)
4	24	0.14	2020	bim
5	135	0.13	2015	artificial intelligence
6	14	0.12	2016	big data
7	27	0.12	2013	architectural design
8	8	0.12	2019	behavior
9	17	0.10	2015	neural networks

4.2. Burst Analysis

Keyword burst maps reveal sudden drops or surges in citation frequency, reflecting major shifts in research focus. To track inflection points in generative AI research trends within architectural design, CiteSpace's Burstness feature was employed to detect keyword bursts. By analyzing key terms associated with generative AI's emergence in architectural design, Figure 8a illustrates the citation burst intensity of 24 keywords in academic literature from 2013 to 2025. The red line indicates the burst period of a keyword, during which its citation frequency significantly increases, signifying widespread academic attention to the topic. The blue curve represents the time span from a keyword's emergence to its disappearance, with light blue indicating it has not yet begun and dark blue signifying the keyword is still active. Figure 8b shows that, based on the distribution characteristics of the red and blue curves, keyword outbreaks can be divided into three stages.

https://doi.org/10.3390/buildings16020388

In chronological order: fuzzy logic, knowledge representation, building envelope, case-based reasoning, artificial neural network, residential buildings, comfort, building energy consumption, artificial neural networks (ANNs), support vector machine, classification, behavior, neural networks, artificial neural networks, models, optimization, neural network, damage detection, internet of things, generative adversarial networks, computer vision, explainable AI, architectural design, natural language processing. Among these, the keyword "fuzzy logic" maintained the longest research intensity period, spanning from 2013 to 2021. This was followed by "artificial neural network" and "comfort," both active from 2015 to 2021. Additionally, the keyword "residential buildings" showed sustained research activity from 2015 to 2022. Keywords such as "damage detection", "internet of things", "generative adversarial networks", "computer vision", "explainable AI", "architectural design", and "natural language processing" have maintained research momentum from 2023 to the present, exhibiting strong growth trends.

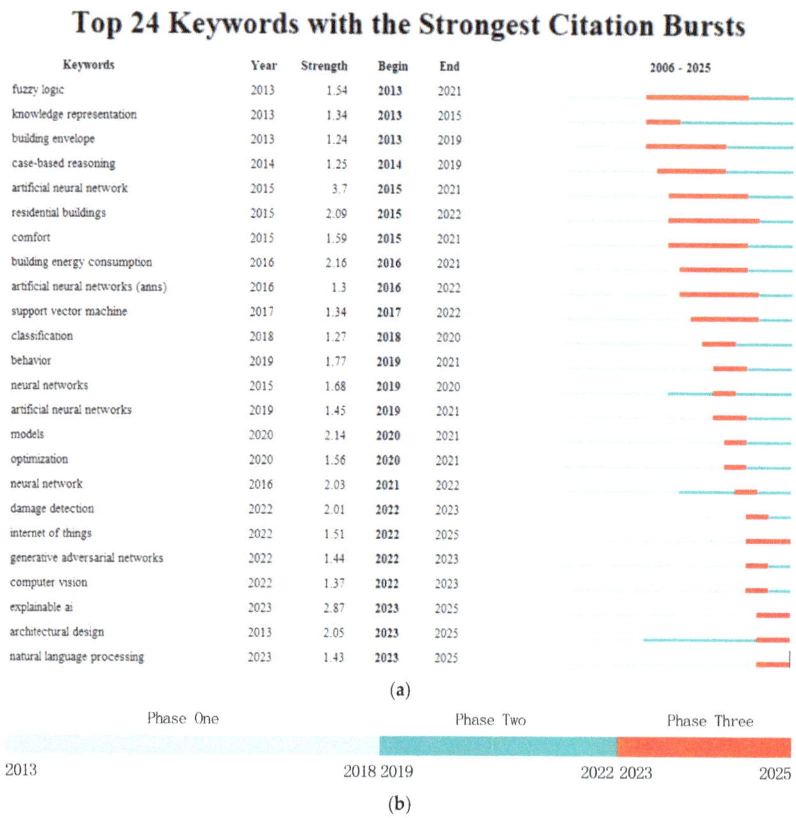

Figure 8. (**a**) Keyword Emergence Map of Literature. (**b**) Literature keywords highlight the timeline.

Phase One (approx. 2013–2018): Research during this period exhibited distinct problem-driven and foundational methodological characteristics. Keywords that emerged early on, such as "fuzzy logic," "knowledge representation," and "case-based reasoning," indicate that researchers were focused on introducing traditional AI methods into the architectural domain to address uncertainties in design and the reuse of experience. Subsequently, the research focus rapidly shifted toward specific building performance and energy consumption issues. Keywords like "building envelope", "comfort", and "building energy consumption" became prominent. Concurrently, the emergence of "artificial neural

https://doi.org/10.3390/buildings16020388

networks" and their variants signaled researchers' adoption of more advanced machine learning models to address these performance optimization challenges, laying the technical foundation for subsequent research.

Phase Two (approx. 2019–2022): This period represented a critical phase of deepening core technologies and broadening application scenarios. On one hand, the persistent prominence of keywords like "optimization" and various "neural networks" indicated that deep learning methods had become mainstream, shifting the research focus from solving single-performance problems to more complex multi-objective optimization and precise prediction. On the other hand, the emergence of "generative adversarial networks" and "computer vision" around 2022 marked a turning point. This signaled a shift in research focus from "analysis and optimization" to "creation and generation," propelling generative AI from laboratories to the forefront of architectural design. The emphasis now centers on automating the generation of design drawings and forms.

Phase Three (approx. 2023–2025): Current research frontiers exhibit distinct trends toward multimodal fusion, explainability, and cross-technology integration. The recent emergence of "explainable AI" directly addresses the "black box" challenges of generative models, emphasizing the reliability and transparency of generated outcomes—an essential requirement for technological maturity and responsible application. The re-emergence of "architectural design" as a core domain term, now appearing alongside "natural language processing", strongly signals a shift toward a new paradigm of human–machine interaction through natural language and cross-modal generation. Architects can now directly guide AI in design through language. Combined with earlier technologies like sketch generation and image control, a highly intelligent, naturally interactive design partner role is taking shape.

5. Analysis of Research Frontiers

Research frontiers play a crucial role in defining research directions, referring to potential research questions and a set of emerging dynamic concepts within a field. In CiteSpace, clusters can be formed by a group of closely related documents based on their interconnections, accurately depicting frontier focal points [40].

Cluster Analysis

This study employs the LLR (Log-Likelihood Rate) algorithm for cluster analysis, categorizing keywords to clearly reveal research trends and hotspots. This approach identifies thematic categories of generative AI research within architectural design. labels automatically assigned based on dominant keywords [41]. Figure 9 presents the clustering results for 598 documents. By consolidating similar clusters across different time periods (one year per segment), this study identified nine primary clusters: #0 Building Information Modeling, #1 Generative AI, #2 Computer Vision, #3 Multi-Objective Optimization, #4 Digital Twins, #5 Machine Learning, #6 Energy Savings, #7 Network Analysis, #8 building envelope, #9 artificial intelligence. The Web of Science literature clustering achieved Modularity = 0.5252 and Mean Silhouette = 0.9201, both exceeding 0.5, indicating good clustering reliability. Among these, Cluster 7 (network analysis) involves evaluation methods for other clusters, while Clusters 6 (energy savings) and 8 (building envelope) address specific architectural design issues. Therefore, these three clusters are excluded from this study. The remaining six clusters are examined in detail. Additionally, Clusters 1 (generative AI) and 9 (artificial intelligence) belong to the same domain and will be discussed together.

The primary keywords extracted using the LLR algorithm are summarized in Table 3. Content analysis of valid literature based on LLR keywords reveals that Building Information Modeling (BIM) has the highest citation count among these clusters, indicating it is a research hotspot in this field. Since its emergence, artificial intelligence has been

https://doi.org/10.3390/buildings16020388

increasingly applied in architectural design. Sustained research in recent years indicates continuous refinement of AI applications within this domain. The following sections provide detailed discussions of each cluster.

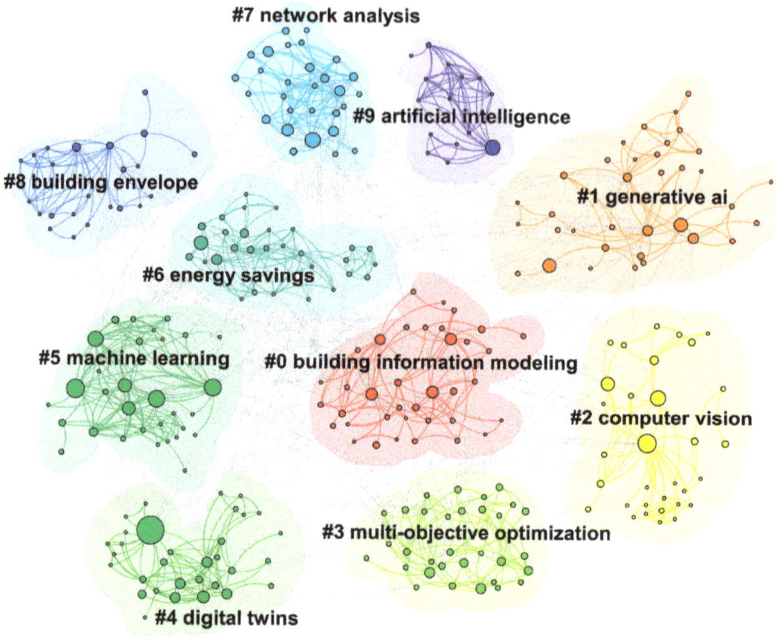

Figure 9. Keyword Cluster Map.

Table 3. Keywords Extracted from Literature.

WOS Clustering Label	The Main Keywords Extracted by Using the LLR (log-Likelihood Rate) Algorithm
#0	building information modeling (10.38, 0.005); modular building (10.18, 0.005); conversational ai (10.18, 0.005); machine learning (7.12, 0.01); 4D bim (6.53, 0.05)
#1	generative ai (11.61, 0.001); large language models (11.13, 0.001); building design (10.3, 0.005); open ai gym (10.3, 0.005); prompt engineering (10.3, 0.005)
#2	computer vision (30, 1.0×10^{-4}); deep learning (18.46, 1.0×10^{-4}); convolutional neural network (8.96, 0.005); residential building (8.96, 0.005); design studio (8.96, 0.005)
#3	multi-objective optimization (10.41, 0.005); life cycle assessment (6.75, 0.01); computational design (6.75, 0.01); machine learning (6.71, 0.01); simulation tools (5.2, 0.05)
#4	digital twins (14.44, 0.001); artificial intelligence (12.02, 0.001); energy efficiency (9.67, 0.005); reinforcement learning (8.73, 0.005); digital twin (7.67, 0.01)
#5	machine learning (20.09, 1.0×10^{-4}); sustainable construction (8.25, 0.005); carbon emissions (8.25, 0.005); explainable artificial intelligence (8.25, 0.005); energy management (8.17, 0.005)
#6	energy savings (13.07, 0.001); aaron hertzmann (8.7, 0.005); generative adversarial network (gan) (8.7, 0.005); robert pepperell (8.7, 0.005); anil seth (8.7, 0.005)
#7	network analysis (9.34, 0.005); digital transformation (9.34, 0.005); large language model (7.7, 0.01); systematic literature review (slr) (7.7, 0.01); smart buildings (7.7, 0.01)
#8	building envelope (15.14, 1.0×10^{-4}); artificial neural network (7.69, 0.01); energy efficiency in buildings (7.55, 0.01); ai algorithms (7.55, 0.01); algorithm (7.55, 0.01)
#9	artificial intelligence (ai) (17.52, 1.0×10^{-4}); artificial intelligence (16.47, 1.0×10^{-4}); m casey rehm (10.16, 0.005); machine learning (7.17, 0.01); wenjia guo (5.07, 0.05)

https://doi.org/10.3390/buildings16020388

(1) Building Information Modeling (BIM)

Building Information Modeling, abbreviated as BIM, integrates information across a building's entire lifecycle through the creation of virtual 3D models. This approach aims to enhance construction efficiency, reduce costs, and conserve energy.

Ploszaj-Mazurek, M and Rynska, E emphasize that current BIM technology is crucial for promoting whole-life-cycle building assessment and enhancing environmental performance. The authors integrated BIM, IFC, AI, and carbon footprint calculation into a single tool. This tool generates material quantities from IFC models, calculates a building's carbon footprint, and solicits improvement suggestions from AI [42]. Park, C and Lee, YG highlight users as a critical factor in architectural design tasks. They developed a real-time BIM-based human behavior simulation system, enabling architects to simulate user behavior at any design stage and provide essential behavioral insights for design proposals [43]. Jungsik Choi and Sejin Lee emphasize the importance of enhancing BIM quality to elevate architectural design standards. Through case studies of BIM data quality control, the research reconfigures derived quality control objectives to establish a quality inspection standard [44]. Mikyoung Kim, Seungyeul Ji, and Hanjong Jun facilitate BIM project management by synchronizing BIM data across different computers and providing BIM model viewers and data via the web, thereby supporting collaboration in BIM projects [45]. Omar Bagasi, Nawari O. Nawari, and Adel Alsaffar pointed out that the integration of Building Information Modeling (BIM) and Artificial Intelligence (AI) is reshaping the Architecture, Engineering, and Construction (AEC) industry, exploring the technical and ethical challenges faced by AI-BIM integration, and further proposing strategies to support its practical application [46]. Bagasi, O.; Dagmar Kutá, and Michal Faltejsek explored the role of Artificial Intelligence (AI) in the transformation of BIM environments, analyzing the integration of AI and BIM from both theoretical and practical perspectives, and evaluating its potential value to the construction industry [47]. Jinyi Li, Zhen Liu, Guizhong Han, Peter Demian, and Mohamed Osmani proposed that BIM and AI-driven deep digitalization of the AEC industry has promoted the rise of "smart cities" [48].

Although generative AI interventions offer numerous conveniences for architectural design, AI remains in its developmental stage. BIM still faces significant issues and challenges during promotion and application. For instance, usage costs are relatively high. Professional BIM software like Autodesk Revit (https://www.autodesk.com.cn/products/revit/overview, accessed on 18 December 2025) or ArchiCAD (https://www.archweb.com/zh-CN/Archicad-23/, accessed on 18 December 2025) requires expensive subscriptions. Additionally, high-performance computers and servers capable of smoothly running these programs necessitate substantial hardware investments. Beyond this, challenges include steep learning curves, talent shortages, and software interoperability issues. To effectively address these, we must recognize that BIM is not merely software but a process. In application, we can adopt subscription models based on demand, enhance pre-design planning, and optimize usage efficiency. In education, we need to intensify talent cultivation, establish "industry-academia collaboration" platforms, and translate knowledge into practice to drive the continuous advancement of Building Information Modeling.

(2) Artificial Intelligence (AI)

Artificial Intelligence (AI) is a technology that simulates human intelligence. It processes problems by mimicking human thinking and decision-making capabilities through computers, emphasizing judgment and prediction. If AI is considered a "whole," generative AI represents a crucial branch within it. Compared to general AI, generative AI places greater emphasis on design and creation, with common technologies including Generative Adversarial Networks (GANs).

https://doi.org/10.3390/buildings16020388

Özorhon, G, Gelirli, DN, Lekesiz, G and Müezzinoglu, C, they investigated and discussed the artificial intelligence involved in architectural design. Their research indicates AI significantly aids students in project development and provides foundational spatial solutions [49]. Yangluxi Li, Huishu Chen, Peijun Yu, and Li Yang propose that AI enhances efficiency, innovation, and sustainability in architectural design. Their analysis of the FU Generator platform and interactive architecture further substantiates this argument, offering direction for future AI development in architectural design [50]. Feiran Chen, Mengran Mai, Xinyi Huang, and Yinghan Li applied AI to traditional Chinese architecture. They identified limitations in current AI applications for traditional structures and enhanced AI matching accuracy and sustainability by extracting samples from traditional architectural styles, thereby strengthening AI's applicability in indigenous architectural contexts [51]. Federica Joe Gardella, Luciana Mastrolia, Francesca Moro, and Marta Rossi examined AI tool implementation in practice using large corporations as case studies [52]. Carlo Deregibus linked emerging trends like AI and social media to future architectural design significance by referencing two conceptual frameworks: systems theory and a priori phenomenology [53]. Nashaat, B, and Elzeni, MM explored the application of AI in computational architectural design processes, proposing AI-enabled computational design workflows [54]. Montenegro, N examined the potential of text-to-image AI systems in architectural design education and further discussed how AI could shape future architectural practices [55]. Adam Fitriawijaya and Taysheng Jeng illustrated the practical integration of generative AI and blockchain technology in architectural design, which can bring more transparent, secure, and efficient outcomes in the early stages of design [3].

Currently, the application prospects of AI in architectural design are vast, yet numerous challenges persist at both technical and cultural levels. For instance, data quality varies significantly. Generative AI requires vast amounts of high-quality, standardized data for training, but architectural data is fragmented within the field. Different stakeholders possess disparate datasets in inconsistent formats, potentially yielding unreliable or erroneous results. Additionally, issues such as poor model interpretability, copyright division, and human–machine collaboration remain.

(3) Computer Vision

Computer vision, a branch of artificial intelligence, employs computers and related devices to simulate biological vision. By processing captured images or videos, it extracts relevant information. Applications such as facial recognition, medical image analysis, and image retrieval all utilize computer vision technology, profoundly transforming how we interact with the world.

Vijayakumar, A and Vairavasundaram, S propose that accuracy in object detection remains the most fundamental and challenging problem in computer vision. Their comprehensive analysis of various YOLO versions demonstrates the model's exceptional precision and speed. In architectural design, it provides data-driven insights by analyzing massive datasets—including lighting, weather, and foot traffic—to inform design solutions. In practical applications, it enables real-time monitoring of foot traffic and biometric identification to enhance building security [56]. Isik, M. similarly notes that feature detection and matching form the foundational components of computer vision. The research improves algorithm performance by comparing descriptions from different images to identify corresponding key points for analysis [57]. Fabian Jarrin, Yasuko Koga, and Diego Thomas pointed out that architectural practice is continuously evolving through digital fabrication technologies, exploring the challenges of quantifying the complexity of architectural facade design, and investigating whether methods combining virtual reality (VR) and computer vision (CV) can effectively measure facade complexity and align with user perception [58]. Nan-Ching Tai focused on visual perception issues under binocular vision,

https://doi.org/10.3390/buildings16020388

using three-dimensional stereoscopic display technology to construct computer-generated environments with perceptual realism, verifying whether reliable computer-generated environments can be used for potential applications of anticipated luminance contrast in architectural design [59]. Rongbing Mu, Yue Cheng, and Haoxuan Feng applied computer vision and machine learning technologies to develop an automatic optimization method for low-carbon architectural landscape spaces, improving the efficiency and accuracy of landscape space optimization [60].

Currently, computer vision still faces significant challenges on its path toward understanding the visual world. For instance, poor generalization ability. Computer vision also relies heavily on massive datasets. Models that perform well on specific training data often degrade in real-world applications due to variations in photo angles, lighting, backgrounds, and other factors. Additionally, high computational costs and privacy concerns pose significant issues.

(4) Multi-Objective Optimization

Multi-objective optimization seeks balanced solutions when multiple conflicting goals exist within an optimization problem. In architectural design, multi-objective optimization seeks optimal solutions across building cost, energy consumption, comfort levels, structural integrity, and other factors.

Ding Yang, Danilo Di Stefano, Michela Turrin, Sevil Sariyildiz, and Yimin Sun note that current multi-objective optimization methods are increasingly applied to conceptual architectural design. To treat poorly defined conceptual designs as an iterative exploration process, restructuring the optimization problem is essential. This research introduces an innovative SBMOO method that enables simultaneous dynamic Re-OPF and interactive Re-OPF, making it suitable for conceptual architectural design exploration [61]. Fei Guo, Shiyu Miao, Sheng Xu, Mingxuan Luo, Jing Dong, and Hongchi Zhang focus on applying multi-objective optimization to building sustainability. The study highlights how rapid urbanization profoundly alters urban climates, intensifies extreme weather events, exacerbates urban heat island effects, and drives global carbon emissions upward. Multi-objective optimization has emerged as an effective tool to balance diverse building performance metrics and adapt flexibly to external meteorological conditions [62]. Ramadan, LA, El Mokadem, A, and Badawy, N proposed a multi-objective optimization framework for parametric architectural form generation, covering energy performance, digital construction, and aesthetic considerations [63]. Ginnia Moroni, Eric Forcael, and Cristian Berrios conducted a comprehensive review of research on multi-objective optimization in architectural design, identifying its key features and assessing the relationships between variables and objectives [64]. Zhao Wang proposed a multi-objective architectural space generation method based on lightweight AIGC, providing an efficient and sustainable intelligent tool for early-stage architectural design, thereby promoting the digital and green transformation of the architecture industry [65].

Its application in architectural design revolutionizes traditional decision making, once reliant on experience and personal preferences, by enabling systematic, quantifiable design choices. Of course, numerous challenges persist, such as the difficulty in quantifying certain objectives. While multi-objective optimization can quantify aspects like building energy consumption, cost, and materials, it struggles to quantify humanistic values like aesthetic appeal and cultural expression. This often leads to prioritizing technical aspects during optimization, resulting in designs that are technically efficient but lack humanistic and artistic value. Additionally, issues like high investment costs and design homogenization remain.

https://doi.org/10.3390/buildings16020388

(5) Digital Twins

The application of digital twins paints a promising future for the construction industry, yet practical implementation still faces numerous challenges. For example: data silos. The architectural design process involves multiple industry sectors, each utilizing distinct software that generates different formats and data types.

Joel Lehmann, Andreas Lober, Tim Häußermann, Alessa Rache, Lisa Ollinger, Hartwig Baumgärtel, and Julian Reichwald note that escalating environmental and geopolitical challenges are complicating supply chains, presenting significant new issues for industry and society. Simultaneously, as society evolves, human-centered and sustainable approaches have become critical focal points that cannot be ignored. This research supports meeting humanity's growing future demands by establishing digital twins [66]. Marcus Wiens, Gernot Steindl, Carlotta Tubeuf, Felix Birkelbach, Juliane Burfeind, and Tobias Meyer introduced DigiWind, a scalable digital twin platform specifically designed for the wind energy sector. This research aims to define the fundamental requirements and architectural design for such platforms, offering a solution for the construction industry to develop, manage, and integrate digital twins in wind energy. This enhances the performance and operational efficiency of wind farms [67]. Ipek Ozkaya proposed that digital twin technology in the construction field provides an effective solution for the entire lifecycle management of intelligent manufacturing and its operation and maintenance [68]. Franca Rocco di Torrepadula, Alessandra Somma, Alessandra De Benedictis, and Nicola Mazzocca focused on the architectural integration of digital twins in smart ecosystems and proposed an ecosystem architecture that supports digital twins [69]. Gernot Steindl, Martin Stagl, Lukas Kasper, Wolfgang Kastner, and Rene Hofmann pointed out that digital twins have become a research hotspot for realizing the Industry 4.0 vision and proposed a technology-independent General Digital Twin Architecture (GDTA) [70].

The application of digital twins paints a promising future for the construction industry, yet numerous challenges persist in practical implementation. For instance: data silos. The construction design process involves multiple industry sectors, each utilizing different software that generates disparate formats and data. This creates data silos, resulting in incomplete and inaccurate twin models. Additionally, issues such as overly stringent technical requirements and inconsistent standards remain.

(6) Machine Learning

Machine learning is a multidisciplinary field encompassing deep learning, algorithms, generative design, and more. It learns patterns and rules from vast datasets to make new predictions and judgments. This represents a crucial and effective approach to achieving artificial intelligence.

Weisheng Zhang, Yue Wang, Zongliang Du, Chang Liu, Sung-Kie Youn, and Xu Guo proposed a machine learning-assisted topology optimization method for artistically styled architectural design. By utilizing different convolutional layers within a CNN (Convolutional Neural Network) based on the VGG-19 (Visual Geometry Group) model, they constructed artistic content and styles spanning low to high levels of abstraction [71]. Ali Shehadeh and Odey Alshboul harness the synergistic potential of virtual reality (VR) and machine learning (ML) to enhance graphical modeling in engineering and architectural design, significantly boosting efficiency while reducing the iterative nature of traditional approaches [72]. Yansong Wang and Xian Hu focus on rural construction, delving into image recognition technologies like machine learning for architecture and proposing application and development strategies for rural architectural design. The research employs the HOG algorithm to extract image contour features, using a trained SVM classifier to categorize these features, ultimately enabling architectural recognition [73]. Tamke, M., Nicholas, P.,

https://doi.org/10.3390/buildings16020388

and Zwierzycki, M. describe how machine learning can be applied throughout the design and manufacturing process, developing relationships between design, performance, and learning, referencing recent architectural research [74]. Fatemeh Mostafavi, Casper van Engelenburg, Seyran Khademi, and Georg Vrachliotis highlighted recent advancements in machine learning (ML) in the field of architectural design, which have driven the development of automatic floor plan generation technology. However, for larger and more complex floor plans, critical evaluation of ML-based floor plan generation has not kept pace [75]. Tiemen Strobbe, Francis Wyffels, Ruben Verstraeten, Ronald De Meyer, and Jan Van Campenhout focused on addressing the issue of automatically identifying architectural schemes in specific architectural styles or design corpora, proposing the use of one-class Support Vector Machines (SVM) combined with graph kernel techniques [76]. Lindenthal, Thies, and Johnson, Erik B. combined traditional feature-based pricing models with the classification results of architectural styles from both human experts and machine learning (ML) classifiers, indicating that properties with unclear style characteristics weaken the impact of architectural style on price and found no evidence of differentiated price effects for revival or contemporary styles in new buildings [77].

Although machine learning-based generative AI has made significant strides in architecture, product design, and computing, it still faces challenges such as "the reliability of generated outputs", "the accuracy of training data," and "the maturity of human–machine collaboration mechanisms". In the architectural domain, machine learning-based approaches remain immature, currently confined to highly regularized structures like apartment buildings. These structures exhibit strong summarizability in construction methods, architectural forms, and structural systems, yielding generative outputs with high reliability and implementability. However, applying such methods to larger-scale and more complex architectural dimensions proves challenging. Therefore, future pathways for machine learning-empowered design innovation should focus on enhancing the reliability of generated results, diversifying database construction, and refining human–machine co-creation interaction mechanisms to broaden its application scope. Concurrently, we must integrate humanistic considerations into machine learning to align with esthetic and functional human needs, thereby driving the data- and algorithm-driven transformation of the design industry.

6. Trend Analysis and Future Outlook

To visually illustrate the evolution and development trends of keywords across different time periods, a Time zone visualization of keywords was performed using CiteSpace. Combined with the clustering timeline diagram in Figure 10, this approach analyzes the research characteristics of generative AI in architectural design from 2005 to 2025 (as of November 2025), identifies dominant research trends across phases, and examines the temporal span of literature related to specific research themes, thereby exploring the dynamic evolution of this field. Figure 11 displays the overall Time-zone map of generative AI research in architectural design from Web of Science. The horizontal axis represents years, illustrating keyword evolution over time. This clearly reveals research trends in generative AI within architectural design. A higher concentration of literature within a specific time period indicates that researchers were focusing on that area at that time. The connecting lines between time periods indicate continuity, with the number of lines representing the strength of the connection between two periods.

https://doi.org/10.3390/buildings16020388

Figure 10. Keyword Cluster Timeline Map.

Figure 11. Literature Time Zone Map.

6.1. Research on Current Development Status

The Web of Science keyword time zone map in Figure 11 reveals that international research on generative artificial intelligence in architectural design began relatively late. The period from 2005 to 2010 represents an initial gap, with no significant relevant literature in this field. From 2011 to 2014, related literature emerged, but with few nodes and connections, also termed the slow development phase. Notably, no significant relevant literature appeared in 2012. From 2015 to 2020, the number of nodes increased, with keywords such as artificial intelligence, neural networks, performance, and BIM gaining prominence. This period thus represents the exploratory phase of international research in this field, indicating growing attention to the integration of generative AI with architectural design. From 2021 to 2022, the number of keyword nodes surged sharply, including terms like machine learning, optimization, models, systems, framework, deep learning, construction, digital twin, thermal comfort, technology, and the construction industry. This phase represents a period of stable growth, with expanding technological integration and application scenarios. Although publication volume slightly declined in 2023 with smaller nodes, the number of nodes did not decrease significantly. Subsequently, 2024 showed a sharp growth trend, with research hotspots concentrated on buildings, impact, methodology, climate change, efficiency, large language models, challenges, building information modeling, and building performance. Overall, the period from 2020 to 2025 represents a rapid development phase.

https://doi.org/10.3390/buildings16020388

characterized by numerous nodes, dense connections, expanded research domains, and enriched content.

6.2. Development Research Trends

By integrating Web of Science keyword clustering results with the phased evolution characteristics of the time zone map, three core research trends can be identified. These trends follow a progressive logic, evolving from deepening technical connections to expanding application scenarios, ultimately culminating in a clear value orientation:

At the technological level, the cross-integration of generative AI with core architectural technologies has become a mainstream research direction, breaking the isolation of traditional technology applications. Early "BIM + AI" represented a tool-based integration where "BIM stored data while AI performed analysis separately"; Today's "Generative AI + BIM + Digital Twin" forms a closed-loop chain of data, generation, validation, and optimization. For instance, generative AI generates solutions based on BIM component libraries, while digital twins simultaneously construct virtual models for real-time daylighting/energy simulations. Computer vision automatically identifies compliance gaps in solutions, and multi-objective optimization algorithms iterate designs—creating a system-level collaborative workflow. At the application level, traditional design relies on designers' subjective interpretation of user needs during the early stages, employing a serial trial-and-error approach of "one version, one review, one revision." Generative AI, however, integrates urban population data, regional functional planning, and operational data from similar projects. It automatically converts ambiguous requirements into quantifiable spatial parameters, significantly shortening the cycle from needs assessment to initial design drafts. Simultaneously, it generates multiple differentiated solutions while concurrently performing "compliance verification + performance simulation + cost estimation" in parallel. It can even dynamically adjust generation logic based on client feedback, reducing the time cost of the proposal phase by over 60%. At the value level, designers' roles shift from "scheme drafters" to "AI logic definers," requiring mastery of "how to input precise design constraints to AI" and "how to evaluate the humanistic value of AI proposals." The industry's demands for designers' "technical literacy + interdisciplinary cognition" have significantly increased.

6.3. Future Directions

Against the backdrop of driving digital transformation and achieving intelligent upgrades in the construction industry, generative AI has emerged as a significant research direction in architectural design. Drawing upon existing domestic technical practices and insights from cutting-edge international applications, China's research in generative AI for architectural design requires continued advancement in the following areas:

(1) Strengthen training of large generative AI models. Currently, large models for architectural design remain confined to mimicking visual forms and surface features, necessitating a deeper understanding of spatial relationships, functional requirements, and regulatory compliance. To overcome this limitation, we can construct specialized databases and establish multidimensional evaluation systems. For instance, systematically integrating relevant architectural literature, design codes, construction drawings, and other materials—while incorporating classic cases spanning diverse regions, schools, and styles—can provide models with extensive foundational expertise. Establish a triple evaluation mechanism of "form-function-performance," incorporating architectural esthetic standards alongside spatial algorithms and performance metrics. This ensures generated models achieve formal beauty and practical functionality while minimizing energy consumption and maximizing performance.

https://doi.org/10.3390/buildings16020388

(2) A robust ethical and accountability framework must be established. Current generative AI faces risks regarding solution safety, copyright ownership, and data privacy, necessitating regulatory guidance through established rules and systems. For instance, a multidimensional copyright attribution system should be developed. When generative AI serves as an auxiliary tool, designers retain primary copyright; when AI independently generates solutions, copyright is distributed among multiple parties. Simultaneously, generated designs must undergo copyright registration, and critical information—such as source data and instructions—should be systematically archived to provide evidence for subsequent liability determinations. On another front, mechanisms for data privacy and algorithmic transparency must be established. Data involving user privacy, corporate core solutions, and sensitive site-specific information should be protected. Algorithmic transparency mechanisms help address the "black box" dilemma in decision-making by requiring developers to periodically disclose portions of foundational model training data. This ensures processes are traceable, auditable, and subject to scrutiny, enabling better human–machine collaboration.

(3) Interdisciplinary exchange should be strengthened to cultivate versatile talents capable of meeting contemporary demands. China currently faces significant shortcomings in its interdisciplinary talent development models. Among domestic architecture schools, less than 10% of faculty possess both architectural expertise and proficiency in generative AI applications. Most institutions' "AI + Architecture" courses exhibit polarized teaching approaches: either delivered solely by computer science instructors focusing exclusively on technical principles, or taught by architecture faculty who learn on the job, focusing solely on tool operation demonstrations. Neither approach facilitates deep integration between technical logic and architectural design requirements. Concurrently, inherent flaws in the traditional education system—such as rigid disciplinary barriers and disconnect between technical instruction and design practice—hamper its ability to effectively respond to the impact and transformation brought by emerging technologies like generative AI to the architecture industry. Therefore, accelerating the transformation and upgrading of architectural education is imperative, requiring coordinated advancement across three dimensions: restructuring knowledge frameworks, innovating teaching models, and deepening practical mechanisms. At the knowledge structure level, an interdisciplinary framework integrating "architecture + computer science + humanities and arts" should be established, enabling students to master design principles while applying AI algorithms and humanistic esthetics. Regarding teaching models, industry-academia-research collaboration should be deepened through joint curriculum development and laboratory construction. Students must hone human–machine collaboration skills and accumulate practical experience within real project contexts. For practical mechanisms, beyond foundational training in campus labs and joint courses, opportunities for corporate internships and participation in actual engineering projects should be expanded. Students should be encouraged to explore cutting-edge AI-driven architectural topics in their graduation projects. This approach will cultivate a new generation of architects who master technology and lead design innovation.

(4) Strengthen international collaboration to facilitate knowledge exchange. In the globalized context, enhancing international cooperation and experience sharing is crucial for advancing generative AI research in architectural design. For instance, establishing international partnerships, actively participating in relevant organizations, launching global research initiatives, hosting high-level academic conferences, and building international databases can foster cross-border dialogue on critical issues like

https://doi.org/10.3390/buildings16020388

data security, copyright ownership, and algorithmic transparency. This collaborative approach will help build consensus and collectively address the challenges facing architectural design in the AI era.

In summary, research on generative AI in architectural design represents a dynamic and promising field. Future studies should continuously advance in theoretical refinement, technological breakthroughs, and practical exploration, providing enhanced theoretical support and practical guidance for the integration of generative AI into the architectural design industry. Simultaneously, researchers must maintain an open and critical stance. We should not only actively validate the benefits of the technology but also objectively evaluate the effectiveness and limitations of generative AI in architectural design. This approach prevents the degradation of design thinking and the decline of architectural culture stemming from overreliance on tools and technology. Only through continuous critical examination and self-reflection can we ensure technological progress remains aligned with the fundamental aspirations of architectural design, thereby driving the ongoing advancement of generative AI.

7. Conclusions

This paper employs CiteSpace to conduct a visual analysis of Web of Science literature on generative AI in architectural design from 2005 to 2025. It first examines publication trends alongside author collaboration networks, institutional collaboration networks, and national distribution. Second, based on keyword network knowledge graphs, it discusses the characteristics, research stages, clusters, emergent themes, and trends in foreign scholars' studies. It analyzes research hotspots and trends in generative AI within architectural design abroad, summarizes core content and future research directions, and draws the following conclusions:

(1) Research on generative AI in architectural design has become a hot topic in international academia. However, scholars from different regions exhibit distinct characteristics in research directions and focal points due to variations in their research backgrounds and societal demands. Overseas scholars predominantly explore how to refine the technical application of generative AI in architectural design from perspectives such as digital modeling, information technology, building systems, design methodologies, and machine learning, thereby driving innovation in architectural design. Research hotspots primarily focus on the technological innovations, economic impacts, and developmental challenges in architectural design brought about by generative AI. Scholars are particularly interested in how generative AI can more effectively and efficiently assist architectural design at the technical level.

(2) From 2005 to 2025, research on generative AI in architectural design progressed through three distinct phases, each exhibiting unique characteristics. The first phase (2013–2018) was the embryonic stage, characterized by limited research scope, scale, and perspectives. The second phase (2019–2022) marked stable growth, featuring abundant empirical findings and deepening research frameworks. The third phase is the rapid growth stage (2023–2025), where research focuses intensify on "safety" and "automation," interdisciplinary collaboration strengthens, research methodologies diversify, and innovative, forward-looking outcomes emerge continuously.

(3) Annual publication statistics from Web of Science (WOS) show a consistent upward trend. The R^2 value indicates that publication growth does not follow a strictly linear pattern; simple annual progression does not fully determine output volume. Actual academic productivity is likely influenced by complex factors, including policy implementation, technological breakthroughs, societal hot topics, and global issues. Regarding publications in international journals, China, the United States, South

https://doi.org/10.3390/buildings16020388

Korea, the United Kingdom, and Australia are the primary research nations in this field. In the collaboration network among authors, interactions between foreign authors are relatively scarce, with most operating in a dispersed manner, demonstrating a preference for independent research. From the institutional collaboration network perspective, foreign institutions predominantly follow a "university-centric" research model, with research institutions like The Hong Kong Polytechnic University providing robust support to the field.

(4) Future research will transcend the current rudimentary stage of efficiency tools, advancing toward a new paradigm of deep symbiosis with human designers. The core focus lies in developing "design thinking partners" capable of comprehending complex contextual environments, ethical constraints, and multidimensional performance metrics. These partners will move beyond mere command execution to proactively propose disruptive solutions, continuously performing multi-objective optimization during design refinement. This achieves a true closed-loop process from form generation to performance integration. Simultaneously, technology will deeply integrate and reshape the entire industry chain. Research will focus on constructing generative models that span the full building lifecycle, enabling them to integrate real-time environmental data, occupant behavior, and material supply chain information. This will achieve adaptability and evolvability throughout a building's generation, construction, operation, and renovation phases.

The aforementioned keywords, research content, and trends not only reflect the diversity and complexity of generative AI research in architectural design but also indicate that various countries and regions are actively addressing challenges arising from its application. Through ongoing research, they are refining corresponding theories and practices, providing crucial references for future development.

This study deepens our understanding of generative AI, advances its application in architectural design, and offers empirical support for related theoretical development. The findings contribute to analyzing key issues and developmental trends in generative AI within architectural design, offering scientific grounds for policy formulation and design development. Simultaneously, they provide practical references for the ongoing refinement of generative AI. However, this study retains several limitations: (1) Database coverage is restricted to the Web of Science Core Collection, excluding non-indexed literature (e.g., regional databases, gray literature); (2) Keyword-based retrieval strategies may omit synonym variants, and the exclusive screening of English-language literature may introduce language bias; (3) Manual screening involves subjectivity, potentially leading to bibliometric bias. Therefore, the current conclusions primarily reflect mainstream trends. To conduct a more comprehensive study in this field, it is necessary to integrate multiple databases, expand the semantic retrieval scope, incorporate broader information sources, and ensure data comprehensiveness and accuracy.

Author Contributions: Y.Y.: conceptualization, methodology, formal analysis, investigation, writing—original draft preparation, visualization; Y.L.: conceptualization, supervision, writing—review and editing; X.B.: investigation, data curation, writing—review and editing; W.Z.: formal analysis, data curation, investigation, writing—review and editing; S.C.: investigation, writing—review and editing. All authors have read and agreed to the published version of the manuscript.

Funding: This research received no external funding.

Data Availability Statement: The original contributions presented in this study are included in the article. Further inquiries can be directed to the corresponding author.

Conflicts of Interest: The authors declare that they have no known competing financial interests or personal relationships that could have appeared to influence the work reported in this paper.

https://doi.org/10.3390/buildings16020388

References

1. Zheng, H.; Yuan, P.F. A generative architectural and urban design method through artificial neural networks. *Build. Environ.* **2021**, *205*, 108178. [CrossRef]
2. Liao, W.; Lu, X.; Fei, Y.; Gu, Y.; Huang, Y. Generative AI design for building structures. *Autom. Constr.* **2024**, *157*, 105187. [CrossRef]
3. Fitriawijaya, A.; Jeng, T. Integrating multimodal generative AI and blockchain for enhancing generative design in the early phase of architectural design process. *Buildings* **2024**, *14*, 2533. [CrossRef]
4. Milošević, J.; Đukanović, L.; Zivkovic, M.; Žujović, M.; Gavrilović, M. Automated Compositions: Artificial Intelligence Aided Conceptual Design Explorations in Architecture. In Proceedings of the 9th International Scientific Conference on Geometry and Graphics MONGEOMETRIJA 2023, Novi Sad, Serbia, 7–10 June 2023; pp. 103–115.
5. Almaz, A.; El-Agouz, E.; Abdelfatah, M.; Raafat, I. The Future Role of Artificial Intelligence (AI) Design's Integration into Architectural and Interior Design Education is to Improve Efficiency, Sustainability, and Creativity. *Civ. Eng. Archit.* **2024**, *12*, 1749–1772. [CrossRef]
6. Lee, S.; Kang, S. The influence of generative AI with prompt engineering on creative design in architectural education. *J. Asian Archit. Build. Eng.* **2025**, 1–16. [CrossRef]
7. Brewer, J.; Patel, D.; Kim, D.; Murray, A. Navigating the challenges of generative technologies: Proposing the integration of artificial intelligence and blockchain. *Bus. Horiz.* **2024**, *67*, 525–535. [CrossRef]
8. Abanda, F.H.; Jian, N.; Adukpo, S.; Tuhaise, V.V.; Manjia, M.B. Digital twin for product versus project lifecycles' development in manufacturing and construction industries. *J. Intell. Manuf.* **2025**, *36*, 801–831. [CrossRef]
9. Davidson, E.; Wessel, L.; Winter, J.S.; Winter, S. Future directions for scholarship on data governance, digital innovation, and grand challenges. *Inf. Organ.* **2023**, *33*, 100443. [CrossRef]
10. Ye, Y.; Hao, J.; Hou, Y.; Wang, Z.; Xiao, S.; Luo, Y.; Zeng, W. Generative AI for visualization: State of the art and future directions. *Vis. Inform.* **2024**, *8*, 43–66. [CrossRef]
11. Huang, Z.; Zhan, X.; Jin, J.; Wang, X.; Jin, Q. Fatty Acids as Nutritional Therapy for NAFLD: A Bibliometric Analysis of Research Trends and Future Directions. *Foods* **2025**, *14*, 4277. [CrossRef]
12. Zhang, T.H. Will the increase in publication volumes "dilute" prestigious journals' impact factors? A trend analysis of the FT50 journals. *Scientometrics* **2021**, *126*, 863–869. [CrossRef] [PubMed]
13. Cocho-Bermejo, A. Artificial intelligence and architectural design before generative AI: Artificial intelligence algorithmics approaches 2000–2022 in review. *Eng. Rep.* **2024**, *7*, e70114. [CrossRef]
14. Vaswani, A.; Shazeer, N.; Parmar, N.; Uszkoreit, J.; Jones, L.; Gomez, A.N.; Kaiser, Ł.; Polosukhin, I. Attention Is All You Need. *Adv. Neural Inf. Process. Syst.* **2017**, *30*.
15. Kang, J.; Lee, H. Deep Learning Architect: Classification for Architectural Design through the Eye of Artificial Intelligence. In *International Conference on Computers in Urban Planning and Urban Management*; Springer International Publishing: Cham, Switzerland, 2017.
16. Achlioptas, P.; Diamanti, O.; Mitliagkas, I.; Guibas, L. Learning Representations and Generative Models for 3D Point Clouds. In Proceedings of the 35th International Conference on Machine Learning, Stockholm, Sweden, 10–15 July 2018.
17. Pan, Y.; Zhang, L.M. Integrating BIM and AI for Smart Construction Management: Current Status and Future Directions. *Arch. Comput. Methods Eng.* **2023**, *30*, 1081–1110. [CrossRef]
18. Muppidi, S.; Reddy, K.T. Co-occurrence analysis of scientific documents in citation networks. *Int. J. Knowl.-Based Intell. Eng. Syst.* **2020**, *24*, 19–25. [CrossRef]
19. Lou, W.; Qiu, J.P. Semantic information retrieval research based on co-occurrence analysis. *Online Inf. Rev.* **2014**, *38*, 4–23. [CrossRef]
20. Li, J.M.; Luo, Y.L.; Lu, S.; Zhang, J.Y.; Wang, J.; Guo, R.Z.; Wang, S.M. ChatDesign: Bootstrapping generative floor plan design with pre-trained large language models. In Proceedings of the 29th International Conference of the Association for Computer-Aided Architectural Design Research in Asia, CAADRIA 2024, Singapore, 20–26 April 2024; Gardner, N., Herr, C.M., Wang, L.T.H., Khan, S.A., Eds.; Association for Computer-Aided Architectural Design Research in Asia (CAADRIA): Hongkong, China, 2024; Volume 1, pp. 99–108.
21. Zeng, P.Y.; Yin, J.; Zhang, M.; Lu, J.; Zhang, Y.C.; Lu, S. Unified residential floor plan generation with multimodal inputs. *Autom. Constr.* **2025**, *178*, 178. [CrossRef]
22. Zeng, P.Y.; Gao, W.; Li, J.Z.; Yin, J.; Chen, J.L.; Lu, S. Automated residential layout generation and editing using natural language and images. *Autom. Constr.* **2025**, *174*, 106133. [CrossRef]
23. Zeng, P.Y.; Yin, J.; Huang, Y.J.; Zhong, J.; Lu, S. AI-based generation and optimization of energy-efficient residential layouts controlled by contour and room number. *Build. Simul.* **2025**, *18*, 2777–2805. [CrossRef]

https://doi.org/10.3390/buildings16020388

24. Wang, J.; Lu, W.S.; Liu, D.D.; Xue, F. The heterogeneity of BIM objects in different construction contexts. In Proceedings of the Construction Research Congress 2018: Construction Information Technology, New Orleans, LA, USA, 2–4 April 2018; pp. 210–220.
25. Xue, F.; Lu, W.S.; Chen, K. Automatic generation of semantically rich as-built building information models using 2D images: A derivative-free optimization approach. *Comput.-Aided Civ. Infrastruct. Eng.* **2018**, *33*, 926–942. [CrossRef]
26. Liu, D.D.; Lu, W.S.; Niu, Y.H. Extended technology-acceptance model to make smart construction systems successful. *J. Constr. Eng. Manag.* **2018**, *144*, 04018037. [CrossRef]
27. Lu, W.; Webster, C.; Chen, K.; Zhang, X.; Chen, X. Computational Building Information Modelling for construction waste management: Moving from rhetoric to reality. *Renew. Sustain. Energy Rev.* **2017**, *68*, 587–595. [CrossRef]
28. LNiu, Y.H.; Lu, W.S.; Chen, K.; Huang, G.G.; Anumba, C. Smart construction objects. *J. Comput. Civ. Eng.* **2016**, *30*, 04016007. [CrossRef]
29. Rhee, J.; Veloso, P. Type investigation in the form of high-rise building using deep neural network. In *Computer-Aided Architectural Design: Design Imperatives: The Future Is Now*; Gerber, D., Pantazis, E., Bogosian, B., Nahmad, A., Miltiadis, C., Eds.; Springer: Singapore, 2021; Volume 1465, pp. 175–189.
30. Memon, S.A.; Shehata, W.; Rowlinson, S.; Sunindijo, R.Y. Generative Artificial Intelligence in Architecture, Engineering, Construction, and Operations: A Systematic Review. *Buildings* **2025**, *15*, 2270. [CrossRef]
31. Baduge, S.K.; Thilakarathna, S.; Perera, J.S.; Arashpour, M.; Sharafi, P.; Teodosio, B.; Shringi, A.; Mendis, P. Artificial intelligence and smart vision for building and construction 4.0: Machine and deep learning methods and applications. *Autom. Constr.* **2022**, *141*, 104440. [CrossRef]
32. Abioye, S.O.; Oyedele, L.O.; Akanbi, L.; Ajayi, A.; Delgado, J.M.D.; Bilal, M.; Akinade, O.O.; Ahmed, A. Artificial intelligence in the construction industry: A review of present status, opportunities and future challenges. *Build. Eng.* **2021**, *44*, 103299. [CrossRef]
33. Pan, Y.; Zhang, L. Roles of artificial intelligence in construction engineering and management: A critical review and future trends. *Autom. Constr.* **2021**, *122*, 103517. [CrossRef]
34. Rombach, R.; Blattmann, A.; Lorenz, D.; Esser, P.; Ommer, B. High-Resolution Image Synthesis with Latent Diffusion Models. In Proceedings of the 2022 IEEE/CVF Conference on Computer Vision and Pattern Recognition (CVPR), New Orleans, LA, USA, 18–24 June 2022; pp. 10674–10685. [CrossRef]
35. Goodfellow, I.; Pouget-Abadie, J.; Mirza, M.; Xu, B. Generative adversarial networks. *Commun. ACM* **2020**, *63*, 139–144. [CrossRef]
36. Nauata, N.; Chang, K.-H.; Cheng, C.-Y.; Mori, G.; Furukawa, Y. House-gan: Relational generative adversarial networks for graph-constrained house layout generation. In *European Conference on Computer Vision*; Springer International Publishing: Cham, Switzerland, 2020.
37. Darko, A.; Chan, A.P.; Adabre, M.A.; Edwards, D.J.; Hosseini, M.R.; Ameyaw, E.E. Artificial intelligence in the AEC industry: Scientometric analysis and visualization of research activities. *Autom. Constr.* **2020**, *112*, 103081. [CrossRef]
38. Chaillou, S. Archigan: Artificial intelligence × architecture. In *Architectural Intelligence: Selected Papers from the 1st International Conference on Computational Design and Robotic Fabrication (CDRF 2019)*; Springer Nature: Singapore, 2020.
39. Hu, R.; Huang, Z.; Tang, Y.; Van Kaick, O.; Zhang, H.; Huang, H. Graph2plan: Learning floorplan generation from layout graphs. *ACM Trans. Graph. (TOG)* **2020**, *39*, 118:1. [CrossRef]
40. Li, H.; Li, B. The state of metaverse research: A bibliometric visual analysis based on CiteSpace. *J. Big Data* **2024**, *11*, 14. [CrossRef]
41. Sanchez-Cortez, L.; Salvador-Gutierrez, B.; Pantoja-Carhuavilca, H.; Tinoco-Gomez, O.; Montaño-Pisfil, J.; Chávez-Sánchez, W.; Gutiérrez-Tirado, R.; Poma-García, J.; Santos-Mejia, C.; Vara-Sanchez, J. Pico-Hydropower and Cross-Flow Technology: Bibliometric Mapping of Scientific Research and Review. *Water* **2025**, *17*, 3524. [CrossRef]
42. Płoszaj-Mazurek, M.; Ryńska, E. Artificial Intelligence and Digital Tools for Assisting Low-Carbon Architectural Design: Merging the Use of Machine Learning, Large Language Models, and Building Information Modeling for Life Cycle Assessment Tool Development. *Energies* **2024**, *17*, 2997. [CrossRef]
43. Lee, Y.G. Developing a Design Supporting System in the Real-Time Manner for Low-Energy Building Design Based on BIM. In *HCI International 2016—Posters' Extended Abstracts. HCI 2016. Communications in Computer and InformationScience*; Stephanidis, C., Ed.; Springer: Cham, Switzerland, 2016; Volume 618.
44. Choi, J.; Lee, S.; Kim, I. Development of Quality Control Requirements for Improving the Quality of Architectural Design Based on BIM. *Appl. Sci.* **2020**, *10*, 7074. [CrossRef]
45. Kim, M.; Ji, S.; Jun, H. BIM-based File Synchronization and Permission Management System for Architectural Design Collaboration. *J. Asian Arch. Build. Eng.* **2017**, *16*, 511–518. [CrossRef]
46. Nawari, N.O.; Alsaffar, A. BIM and AI in Early Design Stage: Advancing Architect–Client Communication. *Buildings* **2025**, *15*, 1977. [CrossRef]
47. Kutá, D.; Faltejsek, M. The Role of Artificial Intelligence in the Transformation of the BIM Environment: Current State and Future Trends. *Appl. Sci.* **2025**, *15*, 9956. [CrossRef]

https://doi.org/10.3390/buildings16020388

48. Li, J.; Liu, Z.; Han, G.; Demian, P.; Osmani, M. The Relationship Between Artificial Intelligence (AI) and Building Information Modeling (BIM) Technologies for Sustainable Building in the Context of Smart Cities. *Sustainability* **2024**, *16*, 10848. [CrossRef]
49. Özorhon, G.; Gelirli, D.N.; Lekesiz, G.; Müezzinoğlu, C. AI-assisted architectural design studio (AI-a-ADS): How artificial intelligence join the architectural design studio? *Int. J. Technol. Des. Educ.* **2025**, *35*, 1999–2023. [CrossRef]
50. Li, Y.; Chen, H.; Yu, P.; Yang, L. A Review of Artificial Intelligence in Enhancing Architectural Design Efficiency. *Appl. Sci.* **2025**, *15*, 1476. [CrossRef]
51. Chen, F.; Mai, M.; Huang, X.; Li, Y. Enhancing the Sustainability of AI Technology in Architectural Design: Improving the Matching Accuracy of Chinese-Style Buildings. *Sustainability* **2024**, *16*, 8414. [CrossRef]
52. Gardella, F.J.; Mastrolia, L.; Moro, F.; Rossi, M. The Impact of AI on Small Architecture Firms. *Technol. | Archit. + Des.* **2025**, *9*, 72–83. [CrossRef]
53. Deregibus, C. The Key Role of Keywords in Architectural Design: A Systemic Framework. *Architecture* **2025**, *5*, 32. [CrossRef]
54. Nashaat, B.; Elzeni, M.M. Reviewing AI in architectural computational design: Applications, opportunities, and the AI-ACD workflow for improved design integration. *Int. J. Archit. Comput.* **2025**, 14780771251405443. [CrossRef]
55. Montenegro, N. Integrative analysis of Text-to-Image AI systems in architectural design education: Pedagogical innovations and creative design implications. *J. Arch. Urban.* **2024**, *48*, 109–124. [CrossRef]
56. Vijayakumar, A.; Vairavasundaram, S. YOLO-based Object Detection Models: A Review and its Applications. *Multimedia Tools Appl.* **2024**, *83*, 83535–83574. [CrossRef]
57. Isik, M. Comprehensive empirical evaluation of feature extractors in computer vision. *PeerJ Comput. Sci.* **2024**, *10*, e2415. [CrossRef]
58. Jarrin, F.; Koga, Y.; Thomas, D. VR and computer vision based facade complexity analysis for building design. *J. Asian Arch. Build. Eng.* **2025**, *25*, 363–387. [CrossRef]
59. Tai, N.C. Computational Investigation of the Illusory Depth Effect of Luminance Contrast on a Three-Dimensional Architectural Scene with Binocular Vision. *J. Asian Arch. Build. Eng.* **2015**, *14*, 331–338. [CrossRef]
60. Mu, R.; Cheng, Y.; Feng, H. Research on automated optimization of low-carbon architectural landscape spaces based on computer vision and machine learning. *Int. J. Low-Carbon Technol.* **2025**, *20*, 146–153. [CrossRef]
61. Yang, D.; Di Stefano, D.; Turrin, M.; Sariyildiz, S.; Sun, Y. Dynamic and interactive re-formulation of multi-objective optimization problems for conceptual architectural design exploration. *Autom. Constr.* **2020**, *118*, 103251. [CrossRef]
62. Guo, F.; Miao, S.; Xu, S.; Luo, M.; Dong, J.; Zhang, H. Multi-Objective Optimization Design for Cold-Region Office Buildings Balancing Outdoor Thermal Comfort and Building Energy Consumption. *Energies* **2025**, *18*, 62. [CrossRef]
63. Ramadan, L.A.; El Mokadem, A.; Badawy, N. Multi-objective optimization framework based on review of form-finding of architectural parametric forms. *Int. J. Archit. Comput.* **2024**, 14780771241299595. [CrossRef]
64. Moroni, G.; Forcael, E.; Berrios, C. Variables and objectives in multi-objective optimization for the integration of architecture, structure, and environmental impact: A literature review. *Arch. Sci. Rev.* **2024**, *68*, 534–550. [CrossRef]
65. Wang, Z. A Lightweight AIGC-Based Multi Objective Architectural Space Generation Method. Collaborative Optimization of Thermal Comfort and Ventilation Performance. *Int. J. Heat Technol.* **2025**, *43*, 1151–1163. [CrossRef]
66. Lehmann, J.; Lober, A.; Häußermann, T.; Rache, A.; Ollinger, L.; Baumgärtel, H.; Reichwald, J. The Anatomy of the Internet of Digital Twins: A Symbiosis of Agent and Digital Twin Paradigms Enhancing Resilience (Not Only) in Manufacturing Environments. *Machines* **2023**, *11*, 504. [CrossRef]
67. Wiens, M.; Steindl, G.; Tubeuf, C.; Birkelbach, F.; Burfeind, J.; Meyer, T. DigiWind-An Open-Source Digital Twin Framework for Wind Energy Systems. *IEEE Access* **2024**, *12*, 84046–84063. [CrossRef]
68. Ozkaya, I. Architectural Concerns of Digital Twins. *IEEE Softw.* **2022**, *39*, 3–6. [CrossRef]
69. di Torrepadula, F.R.; Somma, A.; De Benedictis, A. Mazzocca Smart Ecosystems and Digital Twins: An Architectural Perspective and a FIWARE-Based Solution. *IEEE Softw.* **2025**, *42*, 38–46. [CrossRef]
70. Steindl, G.; Stagl, M.; Kasper, L.; Kastner, W.; Hofmann, R. Generic Digital Twin Architecture for Industrial Energy Systems. *Appl. Sci.* **2020**, *10*, 8903. [CrossRef]
71. Zhang, W.; Wang, Y.; Du, Z.; Liu, C.; Youn, S.-K.; Guo, X. Machine-learning assisted topology optimization for architectural design with artistic flavor. *Comput. Methods Appl. Mech. Eng.* **2023**, *143*, 116041. [CrossRef]
72. Shehadeh, A.; Alshboul, O. Enhancing Engineering and Architectural Design Through Virtual Reality and Machine Learning Integration. *Buildings* **2025**, *15*, 328. [CrossRef]
73. Wang, Y.; Hu, X. Machine learning-based image recognition for rural architectural planning and design. *Neural Comput. Appl.* **2025**, *37*, 8073–8082. [CrossRef]
74. Tamke, M.; Nicholas, P.; Zwierzycki, M. Machine Learning for Architectural Design: Practices and Infrastructure. *International J. Archit. Comput.* **2018**, *16*, 123–143. [CrossRef]
75. Mostafavi, F.; van Engelenburg, C.; Khademi, S.; Vrachliotis, G. Floor plan generation: The interplay among data, machine, and designer. *Int. J. Archit. Comput.* **2025**, *23*, 370–386. [CrossRef]

https://doi.org/10.3390/buildings16020388

76. Strobbe, T.; Wyffels, F.; Verstraeten, R.; De Meyer, R.; Van Campenhout, J. Automatic architectural style detection using one-class support vector machines and graph kernels. *Autom. Constr.* **2016**, *69*, 1–10. [CrossRef]
77. Lindenthal, T.; Johnson, E.B. Machine Learning, Architectural Styles and Property Values. *J. Real Estate Financ. Econ.* **2025**, *71*, 353–384. [CrossRef]

https://doi.org/10.3390/buildings16020388

buildings

MDPI

Review

Artificial Intelligence Empowering the Transformation of Building Maintenance: Current State of Research and Knowledge

Yaqi Zheng [1], Boyuan Sun [2], Yiming Guan [2] and Yufeng Yang [3,*]

1 Department of Program in Urban Regeneration, Korea University, Seoul 02841, Republic of Korea; zheng0624@korea.ac.kr
2 School of Architecture, Central Academy of Fine Arts, Beijing 100102, China; 11250900039@cafa.edu.cn (B.S.); 12230500048@cafa.edu.cn (Y.G.)
3 Intelligent Design Laboratory, School of Fine Arts, Central China Normal University, Wuhan 430079, China
* Correspondence: yufengyang@ccnu.edu.cn

Abstract: With the acceleration of urbanization and the continuous expansion of building stock, building maintenance plays a critical role in ensuring structural safety, extending service life, and promoting sustainable development. In recent years, the application of artificial intelligence (AI) in building maintenance has expanded significantly, markedly improving detection accuracy and decision-making efficiency through predictive maintenance, automated defect recognition, and multi-source data integration. Although existing studies have made progress in predictive maintenance, defect identification, and data fusion, systematic quantitative analyses of the overall knowledge structure, research hotspots, and technological evolution in this field remain limited. To address this gap, this study retrieved 423 relevant publications from the Web of Science Core Collection covering the period 2000–2025 and conducted a systematic bibliometric and scientometric analysis using tools such as bibliometrix and VOSviewer. The results indicate that the field has entered a phase of rapid growth since 2017, forming four major thematic clusters: (1) intelligent construction and digital twin integration; (2) predictive maintenance and health management; (3) algorithmic innovation and performance evaluation; and (4) deep learning-driven structural inspection and automated operation and maintenance. Research hotspots are evolving from passive monitoring to proactive prediction, and further toward system-level intelligent decision-making and multi-technology integration. Emerging directions include digital twins, energy efficiency management, green buildings, cultural heritage preservation, and climate-adaptive architecture. This study constructs, for the first time, a systematic knowledge framework for AI-enabled building maintenance, revealing the research frontiers and future trends, thereby providing both data-driven support and theoretical reference for interdisciplinary collaboration and the practical implementation of intelligent maintenance.

Keywords: artificial intelligence; building maintenance; predictive maintenance; ibliometrics; knowledge mapping

1. Introduction

In the context of accelerating global urbanization, the unprecedented growth in both newly constructed and existing buildings has placed building maintenance and management at the forefront of ensuring structural safety, extending service life, and promoting sustainable development [1]. Effective routine maintenance can reduce equipment downtime, prolong the lifespan of components, and enhance operational safety, thereby playing a critical role in preserving the functionality and value of buildings [2]. However, current maintenance practices still rely predominantly on traditional approaches such as manual inspections and reactive repairs [3,4]. These methods are not only labor-intensive and time-consuming but also heavily dependent on the subjective judgment and experience of personnel, making it difficult to achieve comprehensive and consistent inspection results [5]. As maintenance tasks become increasingly complex, teams are confronted with challenging environments and insufficient equipment, particularly in high-risk areas such as façade inspections and work at height, which further exacerbates safety hazards. Against this backdrop, the limitations of traditional maintenance approaches in terms of efficiency, precision, and safety assurance have become increasingly evident, rendering them insufficient to meet the growing demands of modern cities for building safety and sustainable management [4,6]. In response to these challenges, the application of emerging technologies in the field of building maintenance has become an important focus in both academic research and engineering practice [7]. In recent years, the rapid advancement of digital technologies, particularly artificial intelligence (AI), has opened new possibilities for innovation in maintenance strategies [8]. AI technologies enable automated monitoring of building conditions, intelligent diagnostics, and predictive maintenance, significantly improving the efficiency and accuracy of maintenance operations, reducing operational costs, and effectively preventing major safety incidents [9]. Therefore, against the backdrop of traditional maintenance methods being unable to meet the demands of modern building management, systematically advancing the application of intelligent technologies in building maintenance can not only significantly enhance efficiency and safety but also provide essential support for achieving high-quality and sustainable development in the industry [10,11].

Artificial intelligence (AI) has emerged in recent years as a rapidly advancing frontier technology, encompassing core subfields such as machine learning, deep learning, computer vision, and natural language processing [12]. Machine learning endows computers with the ability to improve performance through data-driven learning, while deep learning, through multi-layer neural networks, has greatly enhanced the accuracy of pattern recognition and prediction [13]. Computer vision enables machines to "interpret" the content of images and videos [14]. Breakthroughs in these core AI technologies have laid the foundation for their application in buildings maintenance. In recent years, scholars both domestically and internationally have conducted extensive research on building structural inspection, equipment operation and maintenance, and energy efficiency management, achieving notable progress [15]. For example, in the field of structural health monitoring, researchers have proposed automated defect recognition methods based on computer vision [16]. By utilizing unmanned aerial vehicles (UAVs) to capture images of building façades and applying deep learning-based image segmentation algorithms, these methods enable rapid and accurate identification and quantitative assessment of surface defects such as concrete cracks [17]. Such AI-powered inspection systems can also be integrated with Building Information Modeling (BIM) platforms or intelligent operation and maintenance systems to automatically generate façade health diagnostic reports and maintenance recommendations, thereby assisting facility managers in implementing timely preventive maintenance [18]. In the area of mechanical and electrical equipment operation and maintenance, AI-driven building management platforms integrate Internet of Things (IoT) sensor data with predictive analytics models to continuously monitor the operational status of key equipment such as heating, ventilation, and air conditioning (HVAC) systems and elevators, allowing for early prediction of potential failure trends [19,20].

Compared with traditional manual inspections and reactive repair methods, artificial intelligence (AI) offers significant advantages for building maintenance [21]. By leveraging sensors and intelligent algorithms, AI enables round-the-clock automated monitoring, greatly improving the timeliness and accuracy of hazard detection [22]. Through the processing of large volumes of historical and real-time data, AI enhances maintenance decision-making while reducing the costs of manual inspections and operations [23]. Applications such as deep learning-based image analysis have substantially improved both the efficiency and accuracy of defect identification [24]. Moreover, the integration of AI with technologies such as the Internet of Things (IoT) and Building Information Modeling (BIM) has accelerated the digital and intelligent transformation of building management systems, enabling performance optimization, energy reduction, and fault prediction [25]. Consequently, AI has become a central driving force for innovation and upgrading in the field of building maintenance [26].

Current research in this field primarily focuses on predictive maintenance of building facilities, intelligent anomaly detection and defect recognition, multi-source data integration and big data management systems, as well as intelligent operation and maintenance (O&M) management supported by data-driven decision-making [27,28]. These efforts have collectively advanced the automation and intelligence of building maintenance technologies [8]. However, the existing literature largely lacks systematic bibliometric and visualization studies that comprehensively examine the overall knowledge structure, thematic hotspots, and technological evolution of the field. As a result, no global and structured cognitive framework has yet been established, and it remains difficult to capture the latest developments in knowledge innovation and technological breakthroughs in a timely manner. To address this gap, this study adopts a bibliometric approach based on the Web of Science (WOS) Core Collection, selecting the relevant literature from the past 25 years concerning the application of artificial intelligence in building maintenance. The research systematically reviews the knowledge landscape, thematic hotspots, and developmental trajectory of the field. Using visualization tools such as ArcGIS 10.8.1 and VOSviewer (v1.6.20), we construct knowledge maps and research trend visualizations to comprehensively reveal the dynamic evolution of research hotspots and emerging frontiers. The specific objectives of this study are as follows:

RQ1: To reveal the evolution of research themes, the distribution of core authors and institutions, and the clustering characteristics of knowledge networks through bibliometric and scientometric analyses, thereby outlining the overall research landscape of the field.

RQ2: To critically examine the technical, managerial, and governance challenges encountered in the application of AI, identify major research topics, and explore their underlying knowledge structures.

RQ3: To analyze research hotspots and frontiers, while also identifying inherent limitations in existing studies and potential directions for future research.

This study systematically reviews the application of artificial intelligence (AI) in building maintenance through bibliometric analysis, encompassing technologies such as machine learning, deep learning, and computer vision. It summarizes the practices and potential of these technologies in predictive maintenance, health monitoring, defect detection, and energy efficiency optimization. From a systemic perspective, the research explores the integration of AI with intelligent operation and maintenance platforms, highlighting its value in proactive maintenance, risk prevention, and collaborative optimization. The ultimate aim is to provide valuable insights for future academic research and industry practice, thereby advancing the safety, cost-effectiveness, and sustainability of building maintenance.

2. Literature Review

Building maintenance refers to a series of planned, executed, supervised, and continuously improved activities aimed at ensuring the safety, reliability, energy efficiency, and regulatory compliance of a building and its mechanical and electrical systems throughout its life cycle [29,30]. During their service life, buildings inevitably undergo varying degrees of aging and performance degradation. Such deterioration may stem from untreated structural defects in the early construction phase, insufficient maintenance during operation, or long-term exposure to natural environmental factors such as wind and rain erosion or seismic activity [7]. Therefore, systematic maintenance activities are both necessary and urgent to slow the aging process and preserve the safety and functional continuity of buildings [4,31].

In general, building maintenance strategies can be categorized into planned maintenance and unplanned maintenance [4,32]. Planned maintenance is carried out under a predefined maintenance scheme, in which maintenance cycles and component replacement schedules are determined based on the operational conditions, performance requirements, and service life of equipment [33,34]. Its core objective is to prevent failures under normal operating conditions, maintain system stability and reliability, and reduce life-cycle costs [35]. This type of maintenance is characterized by its systematic and controllable nature, facilitating rational resource allocation and transparency in operation and maintenance processes [36]. In contrast, unplanned maintenance lacks prior arrangement and is typically triggered by user complaints, unexpected failures, or system malfunctions [37]. It is highly reactive, requires immediate response, and often incurs unpredictable maintenance costs and downtime losses, exerting substantial impacts on the continuity of building operations [7].

Within the framework of planned maintenance, preventive maintenance and corrective maintenance constitute two fundamental strategies [38]. Preventive maintenance centers on "proactive control," aiming to reduce the likelihood of failures and slow performance degradation through early intervention [39]. It can be further divided into three forms. First, scheduled maintenance is performed at fixed time intervals, irrespective of the actual condition of the components [33]. Although simple to implement, it may lead to excessive maintenance [40]. Second, condition-based maintenance relies on periodic or continuous monitoring to collect equipment operation data and assess its health status, enabling more targeted maintenance decisions [41,42]. Third, predictive maintenance utilizes performance degradation models or trend analyses to estimate potential failure times and conduct interventions before failures occur, thereby optimizing maintenance timing while minimizing operational interruptions [43].

In contrast, corrective maintenance is performed after a component or system failure has occurred, with the aim of restoring functionality [44]. This strategy may take place as planned maintenance, in which repairs are deliberately scheduled once performance declines to a certain level, or as unplanned emergency maintenance, which involves addressing unpredictable failures immediately to prevent further damage [45,46]. Depending on risk levels and available resources, unplanned corrective maintenance may be classified into deferred repair—applied to non-critical components—and immediate repair, which is required for critical systems where safety and operational continuity are of paramount importance [47]. The specific technical processes of building maintenance are presented in Figure 1.

A substantial body of research has proposed integrated information technology solutions for the full life cycle of building projects, including Computerized Maintenance Management Systems (CMMS), Building Information Modeling (BIM), and Supervisory Control and Data Acquisition (SCADA) platforms [48–50]. These systems primarily rely on documentation and predefined rules, however, and demonstrate limited adaptability to complex and dynamically changing environments [51]. Maseda et al. (2021) developed a data-driven anomaly detection framework based on SCADA that can identify potential failures in advance without relying on fixed alarm thresholds, and automatically trigger maintenance work orders, thereby reducing downtime and improving equipment availability, demonstrating a feasible pathway toward predictive maintenance [52]. The review by Abideen et al. (2022) indicates that BIM is primarily used for information integration and model visualization during the operation and maintenance (O&M) phase, while its application in decision-making and value assessment remains insufficient, largely due to issues such as inconsistent information standards, difficulties in cross-departmental coordination, and the absence of dynamic update mechanisms [53]. Rui Calejo Rodrigues et al. (2023) proposed a CMMS selection model for Integrated Maintenance Systems (IMS), emphasizing that the system should support equipment asset registers and performance monitoring, and be capable of integration with BIM and IoT to enhance preventive maintenance and resource management efficiency [54].

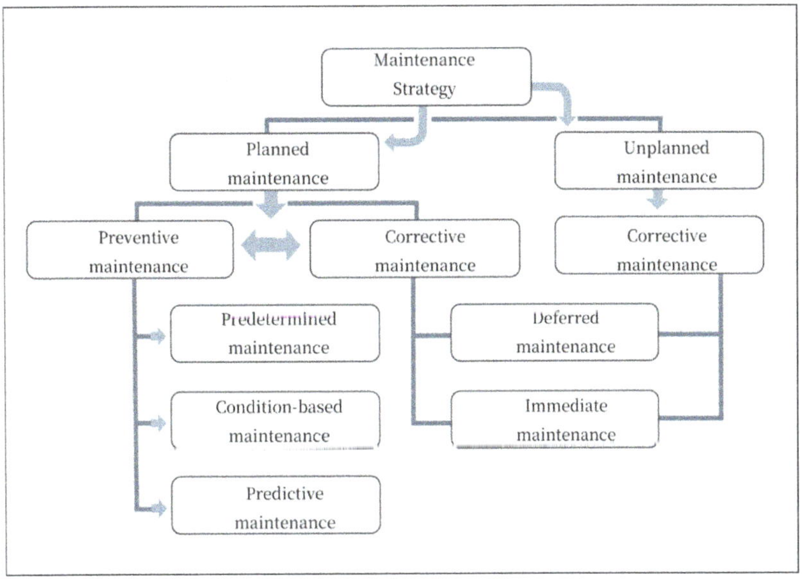

Figure 1. Technical Processes of Building Maintenance.

However, in practical building maintenance, CMMS is often still limited to log recording and work order tracking, with underutilized functional potential [55].With the advancement of IoT and big data technologies, artificial intelligence (AI) methods—including machine learning, computer vision, reinforcement learning, and large-scale language models—have increasingly been incorporated into the field of building maintenance [56]. By automatically learning from massive multi-source datasets, these methods enable higher degrees of automation and predictive capability, shifting maintenance management from reactive mode to proactive prevention [57]. Matteo Interlando et al. (2024) integrated Vision Transformers with ConvNeXt for multi-class façade defect recognition, significantly improving fine-grained detection accuracy and enhancing inspection efficiency and automation [58]. Jihan Zhang et al. (2025) proposed an AI-enabled digital twin framework that integrates UAV data acquisition, deep learning-based defect detection, and GeoBIM semantic registration to achieve automated exterior wall defect recognition and centimeter-level 3D localization; field validation on high-rise buildings in Hong Kong demonstrated scalability and positional accuracy within the 1–5 cm range, offering an extensible digital twin solution for structural health monitoring and intelligent maintenance in complex built environments [59]. Siliang Chen et al. (2025) introduced a human–AI collaborative large-model framework for O&M of building energy systems, achieving complex fault diagnosis through few-shot learning and knowledge enhancement, with an accuracy of 96.3%, demonstrating the scalability of large models in energy operation scenarios [60]. Bo Pang et al. (2025) combined infrared thermography with an improved YOLOv8 to enhance the identification of concealed defects such as cracks and water seepage, improving detection accuracy by 18.7% and effectively overcoming the limitations of visible-light inspection [61]. Zhou Xiaoling et al. (2026) developed an automated multi-type façade defect detection system based on UAV imagery and deep learning. By constructing a multi-scale segmentation model using EfficientUNet++, the system achieved high-precision identification of cracks, spalling, water leakage, and glass breakage across different façade materials (including concrete, ceramics, masonry, and glass), verifying strong generalizability and robustness, and providing a scalable AI framework for intelligent façade inspection and urban infrastructure maintenance [62].

3. Materials and Methods

3.1. Data Sources and Retrieval Strategy

The data used in this study were obtained from the Web of Science (WOS) Core Collection database, provided by Clarivate Analytics. As one of the most widely used multidisciplinary literature retrieval tools worldwide [63]. WOS covers a broad range of disciplines, including natural sciences, social sciences, arts, and humanities [64]. With its comprehensive coverage, powerful search capabilities, and timely data updates, WOS ensures high-quality bibliographic support for this research.

To ensure the systematicity and accuracy of the literature retrieval process, this study followed the principles of "comprehensiveness, representativeness, and consistency" in designing the expanded set of keywords. The artificial intelligence-related terms were selected to encompass major research directions and algorithmic categories within AI, while the building maintenance-related terms corresponded to application domains such as building operation and maintenance, facility inspection, and structural health monitoring. In terms of logical structure, the retrieval strategy was constructed using Boolean operators: the OR operator was applied to expand synonymous and semantically similar terms, and the AND operator was used to define the intersection between AI-related technologies and the building maintenance domain, thereby ensuring thematic focus and relevance in the search results. The literature retrieval was conducted in Web of Science on 15 June 2025, using the Topic Search (TS) method, which matches terms within the title, abstract, author keywords, and KeyWords Plus fields. This approach ensured both the comprehensiveness and representativeness of the retrieved publications. The search query was designed as follows: TS = (("artificial intelligence" OR "machine learning" OR "deep learning" OR "neural network*" OR "computer vision" OR "expert system*" OR "intelligent system*" OR "AI") AND ("building maintenance" OR "facility maintenance" OR "structural maintenance" OR "building inspection" OR "facility inspection" OR "structural inspection" OR "infrastructure maintenance" OR "infrastructure inspection")). The time span for the search was set from 1 January 2000 to 1 June 2025, in order to comprehensively capture the major developmental stages and the latest progress in the field of artificial intelligence applied to building maintenance research.

The initial search yielded a total of 503 relevant publications. To improve the accuracy of the analysis, the results were further restricted to English-language documents, reducing the number to 499. Subsequently, based on document type, only Articles and Proceeding Papers were retained as high-quality sources, leaving 460 papers. Finally, an independent screening and verification were conducted by a researcher: studies highly relevant to the topic were retained, while those that were tangential, incomplete, or duplicate were excluded. This process resulted in a final dataset of 423 highly relevant publications, which served as the basis for the subsequent bibliometric analysis. The detailed retrieval and screening process is illustrated in Figure 2.

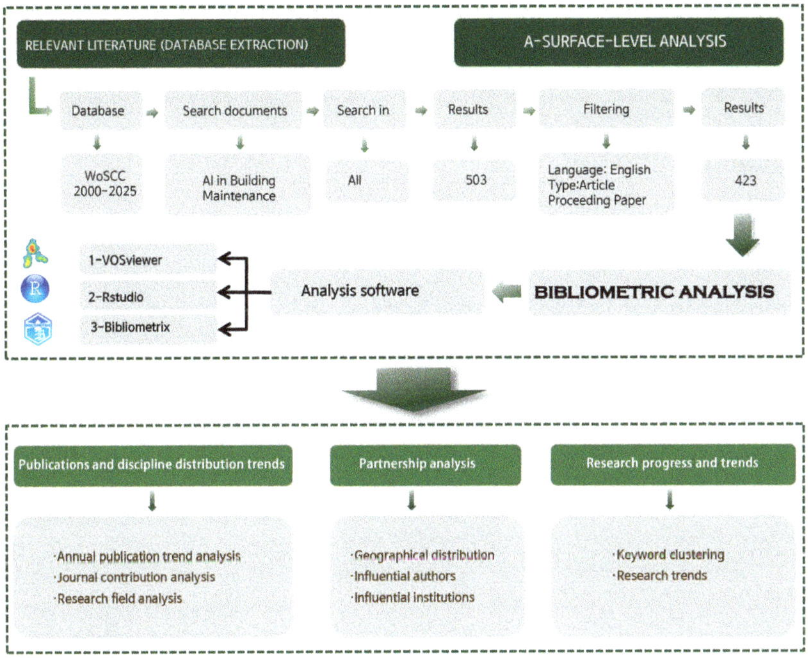

Figure 2. Research Workflow.

3.2. Data Analysis

This study is based on 423 publications related to artificial intelligence and building maintenance, and conducts a systematic bibliometric analysis using Microsoft Office Excel 2021, ArcGIS 10.8.1, VOSviewer (v1.6.20), RStudio, and Bibliometrix v5.0.1 (R package). Specifically:

Microsoft Office Excel 2021 was used for basic statistical analyses, including annual publication trends, core journal distribution, research category statistics, and publication outputs by major authors and institutions. The extracted results provided foundational data support for subsequent visualization.

VOSviewer (v1.6.20) was employed to construct national collaboration networks, institutional collaboration networks, author collaboration networks, and keyword co-occurrence networks.

For the keyword analysis, the minimum occurrence threshold was set to 6, resulting in 91 high-frequency keywords selected from a total of 1795.

For the author collaboration analysis, the minimum publication threshold was set to 2, yielding 75 authors visualized from a pool of 1542.

For the institutional collaboration analysis, the minimum publication threshold was set to 3, resulting in 51 institutions visualized from 663.

In each network, nodes represent countries, institutions, authors, or keywords; node size reflects publication output or occurrence frequency, and connecting lines represent collaborative or co-occurrence relationships.

RStudio was used to generate the thematic map and the keywords distribution over time using the Bibliometrix package. The thematic map was configured with Field = All Keywords, Number of Words = 250, Minimum Cluster Frequency = 5, Number of Labels = 3, and the Walktrap clustering algorithm. The time range for the keyword temporal distribution analysis was set to 2005–2025.

ArcGIS 10.8.1 was primarily used to create the geographic distribution map of publications. A choropleth mapping approach was employed to visualize publication outputs across countries or regions, thereby illustrating the spatial distribution characteristics of research activities.

4. Results

4.1. Publication and Disciplinary Distribution Trends

Analyzing the annual publication trend provides researchers with a systematic understanding of the developmental dynamics of the field, thereby offering data support for formulating informed research strategies [65]. Figure 3 illustrates the annual number of publications in the field of artificial intelligence applications in building maintenance from 2000 to 2025. Over the past 25 years, the annual output has exhibited distinct stages of growth. Between 2000 and 2016, the annual number of publications remained at a consistently low level, typically ranging from 0 to 1 per year, with a cumulative total of only 12 publications by 2016. Research during this period was relatively fragmented, and the overall development of the field progressed at a slow pace.

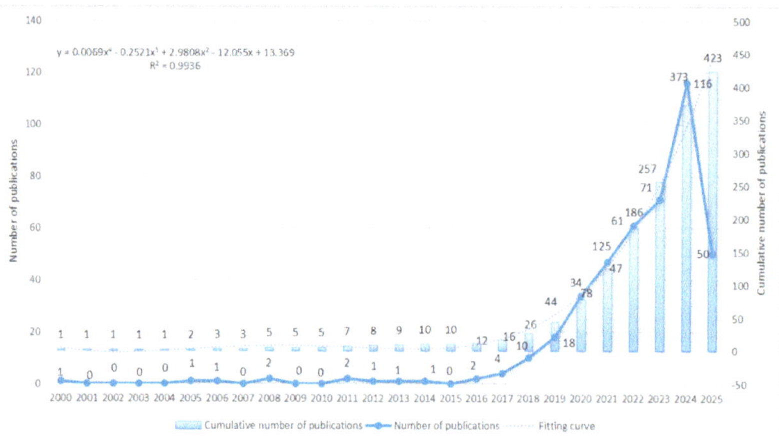

Figure 3. Annual Publication Trend.

Since 2017, the annual number of publications has shown steady growth. In 2017, 2018, and 2019, the numbers reached 4, 10, and 18, respectively, with the growth rate gradually accelerating. After 2020, the annual output increased sharply, rising to 34 in 2020, 47 in 2021, 61 in 2022, and 71 in 2023. By 2024, the annual number of publications peaked at 116. The cumulative total also grew rapidly, reaching 257 by 2023 and 423 by 2025. The relatively smaller number for 2025 can be attributed to the fact that data collection was completed in May. A polynomial regression analysis of the annual publication trend yielded a coefficient of determination (R^2) of 0.9936, indicating an excellent fit and a highly consistent relationship between publication output and time. Overall, 2017 marked a turning point in research activity, with the field entering a phase of accelerated development after 2020 and reaching its peak in 2024. The rapid increase in publication volume reflects the deepening integration of AI technologies into the domain of building maintenance and the continuously rising research interest in this area.

The impact factor (IF) is one of the most important international indicators for evaluating the academic influence of journals. It not only reflects a journal's status and role in scientific communication but also serves as a critical measure of its academic value. Likewise, the number of citations a paper receives is a key metric for assessing its quality and scholarly significance, as highly cited studies often indicate substantial influence and contribution to the field. Our analysis identified a total of 423 relevant publications distributed across 81 journals. Table 1 lists the top ten core journals in terms of publication output within this research domain.

Table 1. Top Ten Core Journals by Publication Volume.

Ran	Institutions	NP	NC	IF (2024)
1	Journal Of Building Engineering	26	420	7.4
2	Buildings	25	165	3.1
3	Automation In Construction	13	1004	11.5
4	Applied Sciences-Basel	13	400	2.5
5	Energy And Buildings	10	388	7.1
6	Energies	10	139	3.2
7	Sustainability	9	54	3.3
8	Building And Environment	7	242	7.6
9	Ieee Access	7	108	3.6
10	Sensors	6	241	3.5

Note: NP = Number of Publications; NC = Number of Total Citations; IF = Impact Factor.

In terms of publication volume, the Journal of Building Engineering ranks first with 26 papers, accounting for 6.1% of the total, far surpassing other journals and highlighting its prominent influence in the field of building maintenance and intelligent applications. Following closely, Buildings and Automation in Construction occupy the second and third positions with 25 and 13 papers, respectively, underscoring their central roles in research on building technologies and automation. Notably, Automation in Construction has an impressive impact factor (IF) of 11.5 and leads by a wide margin with 1004 citations, indicating not only a considerable number of publications but also exceptionally strong academic influence. This positions the journal as an authoritative platform for research on the application of artificial intelligence in building maintenance.

From the perspective of impact factor (IF2024), in addition to Automation in Construction and Building and Environment, both the Journal of Building Engineering and Energy and Buildings also demonstrate strong performance, each with an IF exceeding 7.0. This further confirms the international influence of these journals within the field. Overall, the top ten journals not only include mainstream outlets in architecture and construction but also encompass interdisciplinary fields such as intelligence and sustainable development. This highlights the multidimensional research hotspots of artificial intelligence applications in building maintenance and provides valuable references for future journal selection and research planning.

By conducting a statistical analysis of research categories for publications on artificial intelligence in building maintenance indexed in the Web of Science, we found that studies in this field are primarily concentrated in engineering and computer science. As shown in Table 2, the top five research categories are Construction Building Technology (146 papers), Engineering Civil (137 papers), Engineering Electrical Electronic (47 papers), Engineering

Multidisciplinary (45 papers), and Energy Fuels (44 papers). The high publication volume within these categories indicates that research on AI in building maintenance is developing along a composite trajectory: with civil engineering and building technology as the core, driven by computer science, and extending into multiple domains such as energy, materials, and environment. This trend not only provides a solid disciplinary foundation for the deep integration of AI and building maintenance but also establishes both theoretical and practical support for future cross-disciplinary collaborative innovation.

Table 2. Top 20 Research Categories.

Rank	Web of Science Categories	NP	Rank	Web of Science Categories	NP
1	Construction Building Technology	146	11	Computer Science Theory Methods	23
2	Engineering Civil	137	12	Engineering Mechanical	19
3	Engineering Electrical Electronic	47	13	Telecommunications	19
4	Engineering Multidisciplinary	45	14	Engineering Industrial	17
5	Energy Fuels	44	15	Physics Applied	17
6	Computer Science Artificial Intelligence	40	16	Instruments Instrumentation	16
7	Computer Science Interdisciplinary Applications	39	17	Chemistry Multidisciplinary	15
8	Computer Science Information Systems	31	18	Automation Control Systems	13
9	Materials Science Multidisciplinary	26	19	Environmental Sciences	13
10	Green Sustainable Science Technology	24	20	Engineering Manufacturing	12

Note: NP = Number of Publications.

4.2. Geographic Distribution of Publications

Analyzing the global publication landscape of artificial intelligence in building maintenance from a national perspective helps to reveal the level of research engagement and the leading forces in the international academic community [66]. Figure 4 illustrates the global distribution of research outputs in this field. Specifically, China ranks first with 80 publications, accounting for 18.7% of the total, followed by the United States with 52 publications (12.1%). Italy ranks third with 37 publications, while Canada (26), the United Kingdom (25), and India (24) closely follow. These countries, supported by well-established research systems, international collaboration networks, and policy incentives, have achieved rapid application and transformation of AI technologies in the field of building maintenance.

From the perspective of global regional distribution, Asian countries (such as China, India, Vietnam, Malaysia, South Korea, and Singapore) and European countries (including Italy, the United Kingdom, Germany, France, Poland, Sweden, and Norway) are particularly active, while North America (the United States and Canada) and Oceania (Australia) also demonstrate strong competitiveness. This distribution pattern indicates that research on artificial intelligence in building maintenance has moved beyond the traditional dominance of developed countries, evolving toward a new trend of global multi-centered development and regional collaboration. With the continuous iteration of relevant technologies and the deepening digital transformation of the construction industry, it is anticipated that more emerging economies will actively participate in international academic exchange and technological innovation, thereby driving the geographical distribution of this field toward greater diversification and dynamism.

To further reveal the global academic collaboration landscape of artificial intelligence in building maintenance, this study employed VOSviewer to construct a country-level scientific collaboration network (see Figure 5). In the network, each node represents a country, with node size indicating the number of publications produced by that country in this field. The links between nodes represent collaborative relationships between countries, with thicker lines signifying stronger and more frequent cooperation.

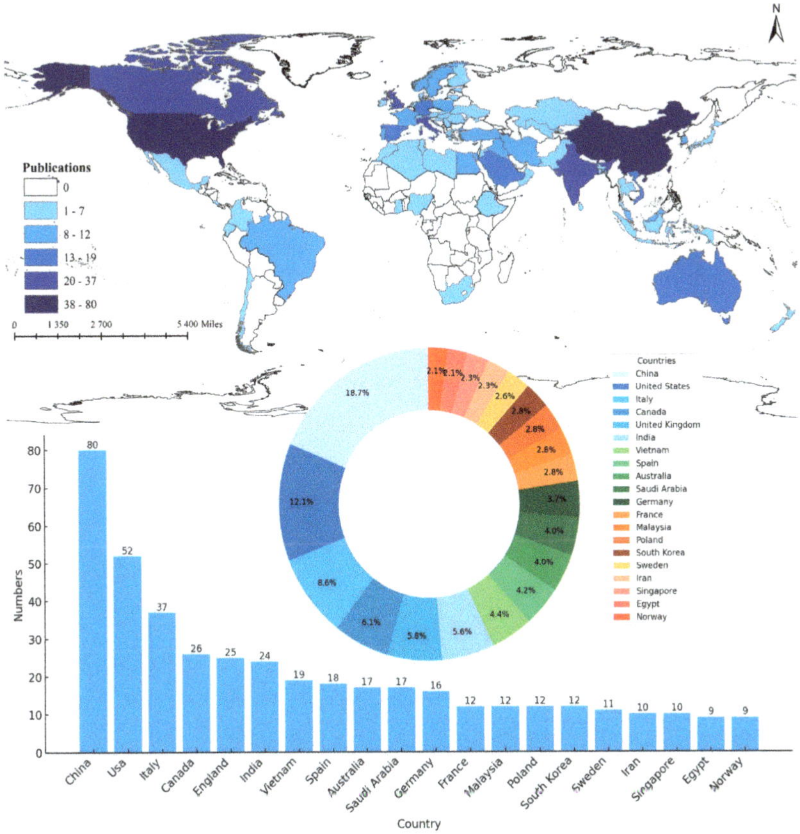

Figure 4. National Publications and Geographic Distribution.

As shown in Figure 5, the collaboration network exhibits characteristics of multi-centered and clustered structures. Core-node countries are closely interconnected, whereas peripheral countries are generally engaged in limited or single-link collaborations. China, the United States, Italy, the United Kingdom, and Canada occupy central positions in the network, with larger node sizes and dense linkages, underscoring their dominant and pivotal roles in international academic cooperation. China not only leads in publication output but also demonstrates remarkable strength and breadth in international collaborations, maintaining partnerships with nearly all major research-active countries. This highlights its high level of openness and international influence. In contrast, countries such as Ukraine, Qatar, Nigeria, Greece and Lebanon display relatively limited collaboration, characterized primarily by unidirectional or small-scale partnerships. This reflects their comparatively lower level of participation in research on AI in building maintenance, which may be constrained by factors such as academic resources, policy environments, or institutional capacity.

Overall, international collaboration in this field demonstrates a globalized pattern centered on major technological powers such as China, the United States, and European countries, characterized by multi-regional cooperation and multi-level interconnections. Looking ahead, as international cooperation mechanisms continue to improve and the research capacity of emerging countries strengthens, the global collaboration network is expected to become more open and balanced, thereby providing a solid foundation of international cooperation for the sustainable development of this field.

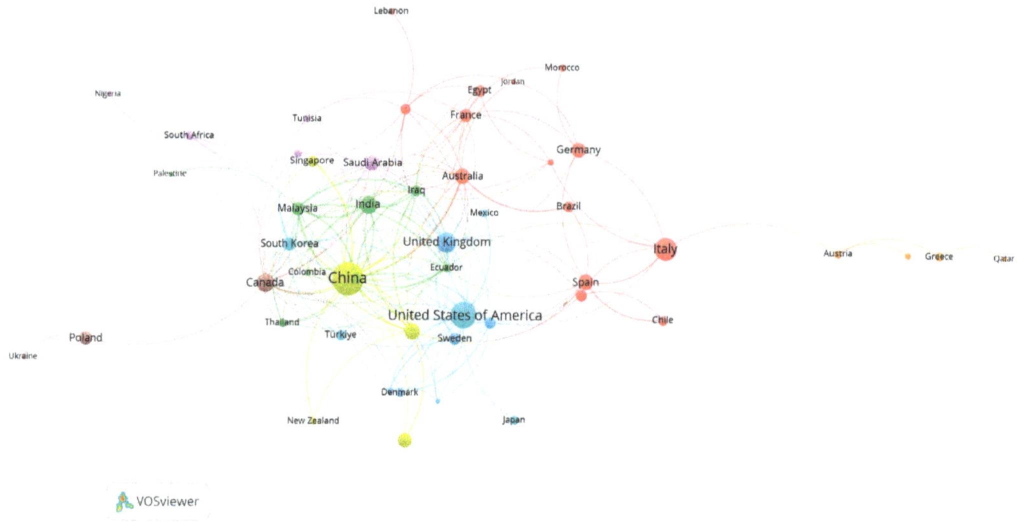

Figure 5. International Collaboration Network of Countries.

4.3. *Influential Authors*

The analysis of prolific authors and core institutions helps to clarify the primary contributors in the field of artificial intelligence applied to building maintenance, the distribution of academic resources, and potential directions for future collaboration. Table 3 presents the top ten authors in terms of publication output. In terms of the number of publications, Hoang Nhat-Duc ranks first with seven papers, demonstrating sustained research commitment and high activity in this domain, followed by Andrés J. Prieto (five papers) and Cao Minh-Tu (four papers). With respect to total citations, Hoang Nhat-Duc likewise leads with 175 citations, indicating that his research work has received widespread academic attention and recognition. Notably, although Shen Weiming has published only four papers, his total citation count reaches 101 and his H-index is 58—the highest among all listed authors—reflecting the exceptional academic value and influential impact of his contributions.

Table 3. Top 10 Authors by Publication Output.

Rank	Author	NP	NC	H-Index
1	Hoang, Nhat-Duc	7	175	45
2	Andrés J Prieto	5	44	15
3	Cao, Minh-Tu	4	73	20
4	D'Orazio, Marco	4	7	8
5	Gunay, Burak	4	101	16
6	Shen, Weiming	4	101	58
7	Yang, Chunsheng	4	101	22
8	Bernardini, Gabriele	3	7	25
9	Di Giuseppe, Elisa	3	7	20
10	El-Rayes, Khaled	3	13	32

Note: NP = Number of Publications; NC = number of total citations; H-Index = Hirsch Index.

In summary, while publication quantity reflects an author's level of research activity to some degree, citation count and the H-index provide a more accurate measurement of academic influence. The findings indicate that Hoang Nhat-Duc performs strongly in both research output and citation impact, whereas Shen Weiming achieves significant influence with comparatively fewer publications. The work of such high-impact scholars plays a crucial guiding role in advancing theoretical development and methodological innovation within this field.

To reveal the research collaboration patterns and academic network structure in the field of artificial intelligence applied to building maintenance, this study employed VOSviewer to visualize author publication data and generate an author collaboration network (Figure 6). In the figure, each node represents an author, while the edges indicate collaborative relationships. The colors differentiate distinct collaboration clusters, and the density and positioning of the edges reflect the closeness of collaboration and the core structure of the academic network.

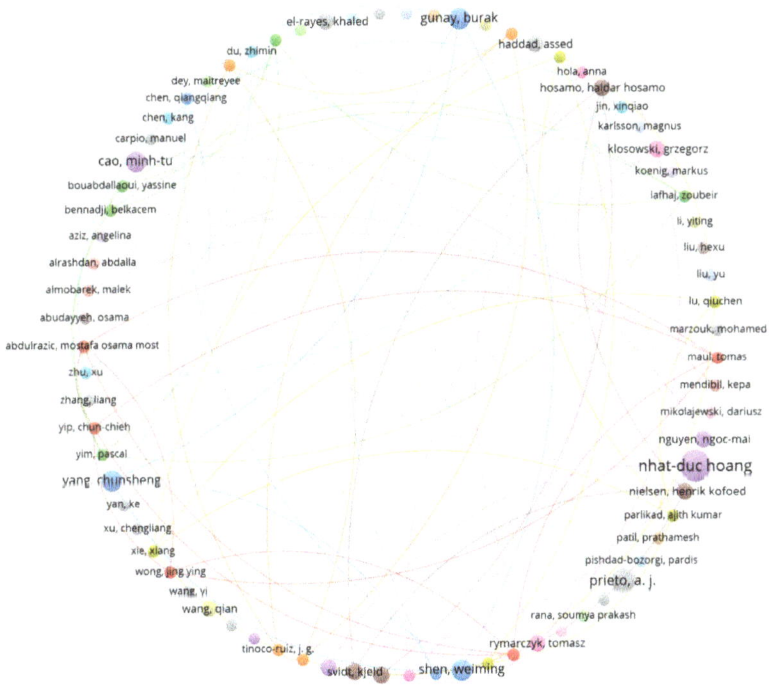

Figure 6. Author Collaboration Network Analysis.

From the overall structure of the collaboration network, the relationships exhibit a "multi-core with weak interconnections" pattern, meaning that several scholars have formed relatively stable collaboration clusters, while connections between these clusters remain comparatively limited. Hoang Nhat-Duc, Shen Weiming, and Gunay Burak occupy central positions in the network, demonstrating broad collaborative reach and strong academic influence, indicating their organizational capacity and cohesive roles in advancing the field. For instance, Hoang Nhat-Duc forms a stable collaborative cluster with Prieto A. J. and Nguyen Ngoc-Mai, showing both high productivity and strong scholarly impact. Shen Weiming's collaboration network is highly concentrated, involving close cooperation with scholars such as Svidt Kjeld and Tinoco-Ruiz J. G., highlighting his significant role in facilitating international academic collaboration. Additionally, Gunay Burak exhibits dense network linkages with collaborators such as El-Rayes Khaled and Haddad Assed, reflecting active efforts to expand collaborative boundaries within the research domain.

In conclusion, core authors form academic communities through stable collaborative relationships, playing a central role in driving the coordinated development of research and enhancing the influence of the field. Future research should further strengthen cross-team and cross-regional collaboration to promote knowledge integration and methodological innovation, thereby advancing the deeper application of artificial intelligence technologies in building maintenance.

4.4. Institutional Contributions

Institutional analysis helps to reveal the research landscape and core strengths within the field, providing insight into the academic influence and collaboration networks of leading institutions [67]. This information serves as a valuable reference for future research planning and opportunities for inter-institutional collaboration. Table 4 lists the top ten institutions ranked by publication volume, providing a clear overview of the research distribution and academic influence of each institution.

Table 4. Top 10 Institutions by Publication Output.

Rank	Institutions	NP	NC	AC
1	Duy Tan University	11	231	21
2	The Hong Kong Polytechnic University	10	127	13
3	Islamic Azad University	7	71	10
4	National University of Singapore	6	372	62
5	Shanghai Jiao Tong University	6	163	27
6	University of Illinois at Urbana-Champaign	6	20	3
7	Università Politecnica delle Marche	6	29	5
8	Chongqing University	5	69	14
9	The Hong Kong University of Science and Technology	5	360	72
10	Pontificia Universidad Católica de Chile	5	20	4

Note: NP = Number of Publications; NC = number of total citations; AC = Average Citations per Publication.

In terms of publication output, Duy Tan University ranks first with 11 articles, indicating sustained research investment and a high level of scholarly activity in this field. It is followed by The Hong Kong Polytechnic University (10 articles) and Islamic Azad University (7 articles), both of which demonstrate strong organizational capacity and research productivity in advancing studies on intelligent building maintenance. However, when considering average citation impact, The Hong Kong University of Science and Technology (AC = 72) and the National University of Singapore (AC = 62) emerge as the most influential institutions. Although these two institutions have published only five and six articles, respectively, their total citation counts reach 360 and 372, far surpassing other institutions. This indicates that their research outputs possess high scholarly value and citation influence, reflecting their critical role in theoretical innovation and methodological advancement within the field.

To further elucidate the international collaboration landscape in the application of artificial intelligence to building maintenance, this study employed VOSviewer to visualize institutional publication data. By setting the minimum publication threshold at three papers per institution, an institutional collaboration network map was generated (Figure 7). In this network, each node represents an institution, with node size corresponding to publication volume, connecting lines indicating collaborative relationships, and colors distinguishing different collaboration clusters.

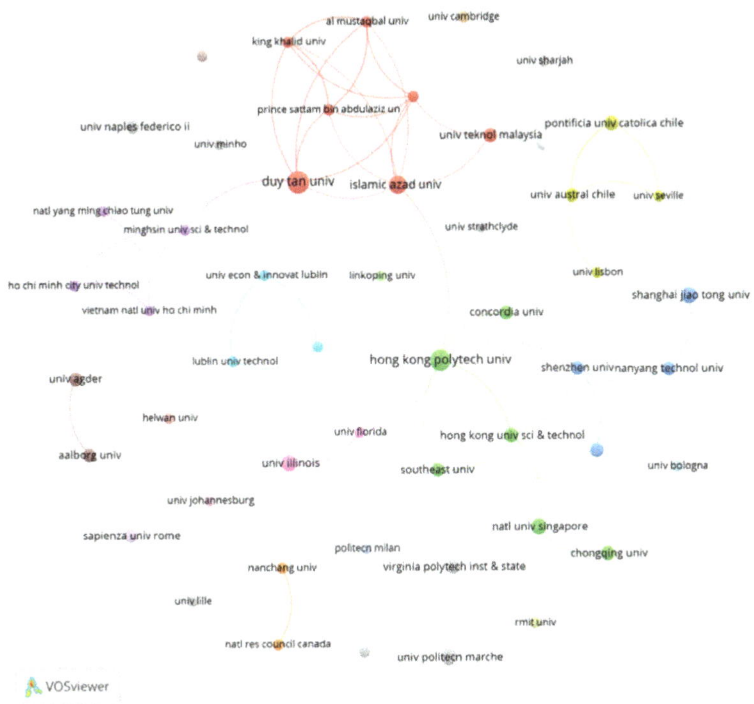

Figure 7. Institutional Collaboration Network Map.

From the overall network structure, institutional collaboration exhibits a distinctly multi-centered cluster pattern. High-output institutions tend to form close cooperative networks around specific regions or research alliances. Duy Tan University and Islamic Azad University form the most prominent cluster (highlighted in red), demonstrating a strong regional synergy between institutions in the Middle East and Southeast Asia. This cluster includes multiple universities from Arab and Asian regions, such as King Khalid University and Prince Sattam Bin Abdulaziz University, characterized by high collaboration strength and dense network connectivity. The Hong Kong Polytechnic University occupies a core position within the green cluster, linking with institutions such as The Hong Kong University of Science and Technology, Shenzhen University, and Southeast University. This indicates that universities in mainland China and Hong Kong have established relatively stable regional cooperation alliances in research on intelligent building maintenance. In contrast, some institutions—such as the University of Illinois and Univ Politecn Marche—are relatively isolated within the network, reflecting limited collaboration ties and suggesting that research resources in this field remain unevenly distributed across regions.

In summary, institutional collaboration in the field of artificial intelligence for building maintenance currently shows a trend centered in Asia, characterized by multi-polar linkages and a gradually increasing degree of internationalization. Strengthening cross-regional research collaboration and knowledge exchange will be essential for enhancing the global level of intelligent building maintenance, accelerating the translation of research outcomes into practice, and promoting broader technological application.

4.5. Knowledge Structure and Clustering Characteristics

By analyzing high-frequency keywords, it is possible to gain deeper insights into the main research directions, technical approaches, and research hotspots in the field [68]. In this study, VOSviewer software was employed to conduct a co-occurrence network analysis of the keywords extracted from the core literature (Figure 8). The co-occurrence relationships among high-frequency keywords effectively map the implicit knowledge structure of the discipline and reveal the evolutionary trajectory of research hotspots [69]. A total of 1795 high-frequency keywords were identified, and four major thematic clusters were generated using the Louvain clustering algorithm. These clusters maintain relative thematic independence while simultaneously achieving close coupling within a complex network structure through cross-domain nodes.

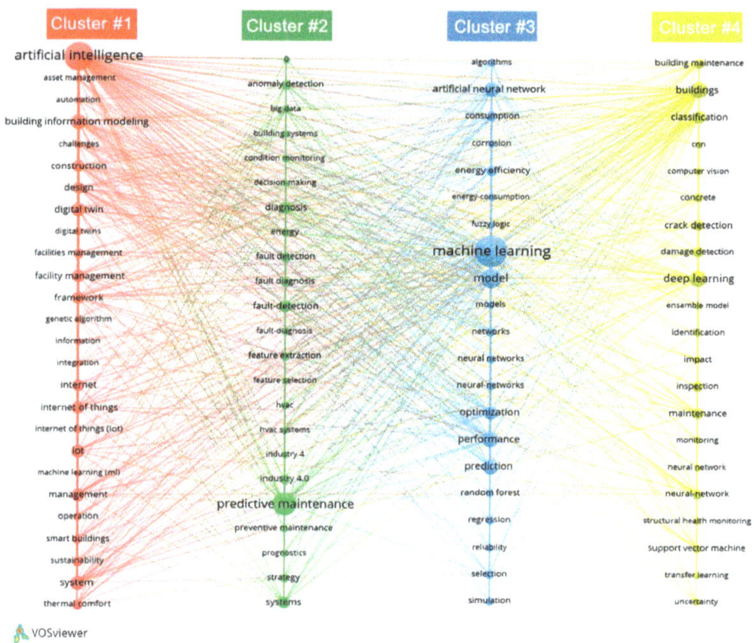

Figure 8. Keyword co-occurrence network diagram (in the figure, each node represents a high-frequency keyword, and the connecting lines indicate the co-occurrence relationship of keywords within the same document. Different colors represent clusters of keywords automatically identified by the system, the size of each node reflects the frequency of keyword occurrence, and the density of the connecting lines indicates the strength of co-occurrence with other keywords).

Cluster 1: Intelligent Construction and Digital Twin Integration

This cluster centers on the keywords "artificial intelligence," "building information modeling (BIM)," "digital twin," and "internet of things (IoT)," revealing a trend toward deep integration of digitalization and intelligent technologies in the building sector. Current research is evolving from the digitalization of isolated subsystems to the real-time interconnection of multiple platforms, with BIM and digital twins serving as central carriers for building life-cycle data, while IoT provides continuous sensing and data input [70,71]. The introduction of artificial intelligence enables a closed-loop workflow of "perception–analysis–decision–execution," shifting maintenance practices from experience-driven to knowledge-driven paradigms. The key drivers behind this trend include increasing industry demands for energy efficiency optimization and operational reliability, as well as policy initiatives supporting smart city development [72]. However, challenges remain regarding data standardization, interface interoperability, and model compatibility, which constrain large-scale implementation. Future research is expected to focus on open platforms, semantic interoperability, and knowledge graph development, thereby establishing a unified data foundation for life-cycle intelligent maintenance management.

Cluster 2: Predictive Maintenance and Health Management

This cluster is characterized by the keywords "predictive maintenance," "fault detection," and "condition monitoring," reflecting the ongoing shift in building maintenance from passive response to proactive prediction. By integrating multi-source sensor data with machine learning algorithms, predictive maintenance can identify potential equipment failures in advance, enabling optimal resource allocation and risk prevention [73]. The emergence of this research direction is driven by the dual pressures of increasing building facility aging and rising operational costs [4]. The research frontier is evolving from single-equipment health assessment toward system-level operational optimization, focusing on the integrated balance among energy consumption, service life, and reliability. Future breakthroughs are expected to lie in the development of generalized and interpretable intelligent models that can be deployed robustly in complex building environments [74]. Moreover, through deep coupling with BIM and IoT, it will be possible to establish a truly "self-sensing, self-diagnosing, and self-decision-making" building maintenance system.

Cluster 3: Algorithmic Innovation and Performance Evaluation

This cluster centers on the keywords "machine learning," "optimization," and "performance," reflecting the ongoing evolution of algorithmic frameworks and the deepening of intelligent maintenance methodologies. The progression from early statistical learning to deep neural networks, reinforcement learning, and transfer learning has continuously improved prediction accuracy and operational efficiency [75]. In recent years, the research focus has shifted from model selection to performance optimization and enhanced interpretability, with emerging directions including uncertainty quantification, multi-task learning, and knowledge fusion [76]. The significance of algorithmic innovation lies in advancing intelligent maintenance from being merely "usable" to becoming truly "trustworthy." Future development should strengthen research on model transparency and computational sustainability, and explore lightweight, deployable AI models that can support real-time decision-making and large-scale application in building operation and maintenance.

Cluster 4: Deep Learning Driven Structural Detection and Automated Operation and Maintenance

This cluster centers on the keywords "deep learning," "CNN," "computer vision," and "structural health monitoring (SHM)," highlighting the extensive application of visual intelligence in structural inspection and condition monitoring. Automated detection methods based on convolutional neural networks and point cloud analysis enable efficient, non-contact crack identification, damage localization, and surface anomaly detection, significantly improving the real-time performance and accuracy of building safety monitoring [77]. The emergence of this research direction has benefited from advancements in high-resolution imaging and multimodal sensing technologies; however, limited availability of annotated data and insufficient model generalization continue to restrict large-scale deployment in complex real-world environments [78]. Future studies are expected to integrate generative AI, active learning, and knowledge distillation techniques to develop sustainable learning frameworks for structural health monitoring, achieving an intelligent transition from localized inspection to holistic operational management.

Together, the four clusters constitute the core knowledge structure of artificial intelligence in the field of building maintenance: Cluster 1 provides the data and platform foundation; Cluster 2 reflects the shift in maintenance logic toward intelligent prediction; Cluster 3 offers algorithmic support; and Cluster 4 drives practical application and deployment. The overall research evolution follows the trajectory of digital integration → intelligent sensing → algorithmic empowerment → automated operation and maintenance. The general trend indicates that the field is transitioning from tool development toward a new stage of systematic, sustainable, and autonomous intelligent maintenance. Future research should further strengthen integration across clusters, build an open and shared data ecosystem, and establish interpretable intelligent decision-making systems to enable adaptive management across the full life cycle of buildings.

4.6. Thematic Evolution and Trend Analysis

To reveal the developmental trajectory of research themes and the migration of hotspots in the field of artificial intelligence for building maintenance, we employed RStudio and the bibliometrix visualization tool to systematically construct the keyword temporal distribution heatmap (Figure 9A) and the thematic structure evolution diagram (Figure 9B). The keyword temporal distribution heatmap (Figure 9A) provides an intuitive representation of the emergence and evolution of various research themes. Meanwhile, the thematic structure evolution diagram (Figure 9B), based on the dual dimensions of centrality and density, delineates the structural characteristics and evolutionary pathways of core, foundational, and emerging themes within the field.

4.6.1. Dynamic Evolution of Keyword Temporal Distribution

As shown in Figure 9A, the temporal visualization of keywords clearly reveals the dynamic evolutionary trajectory of research themes in this field. Between 2005 and 2015, related studies primarily focused on foundational themes such as "Fault Diagnosis" and "Feature Extraction." During this period, keywords appeared infrequently and were dispersed, indicating that the penetration of artificial intelligence technologies into building maintenance was still at an early stage. After 2016, with the introduction of emerging technologies such as machine learning and deep learning, both the variety and density of keywords gradually increased. In particular, keywords such as "Machine Learning," "Artificial Intelligence," and "Deep Learning" began to appear with high frequency after 2018, reflecting the consolidation of the technological foundation of the field and a gradual shift in research attention toward the practical application of intelligent algorithms. Moreover, compound keywords such as "Digital Twin," "IoT," and "Energy Efficiency" have risen significantly since 2020, demonstrating the advancement of building maintenance research from single-technology approaches toward multi-technology integration and system-level intelligence.

From the distribution on the right side of the heatmap, it is evident that during the period from 2021 to 2025, the concentration of keywords further intensified, with high-frequency themes represented by "Digital Twins," "Predictive Artificial Intelligence Maintenance," and "Anomaly Detection." This indicates that the discipline has entered a new stage characterized by the deep integration of technologies and application scenarios. The vertical hierarchical changes in the keywords clearly illustrate the shift in research focus from fundamental data sensing and monitoring toward advanced objectives such as full life-cycle intelligent operation and maintenance, energy efficiency management, and system-level optimization. This trend not only reflects the upgrading of industry demands but also signals that the future of building maintenance will increasingly rely on the diversified empowerment of artificial intelligence.

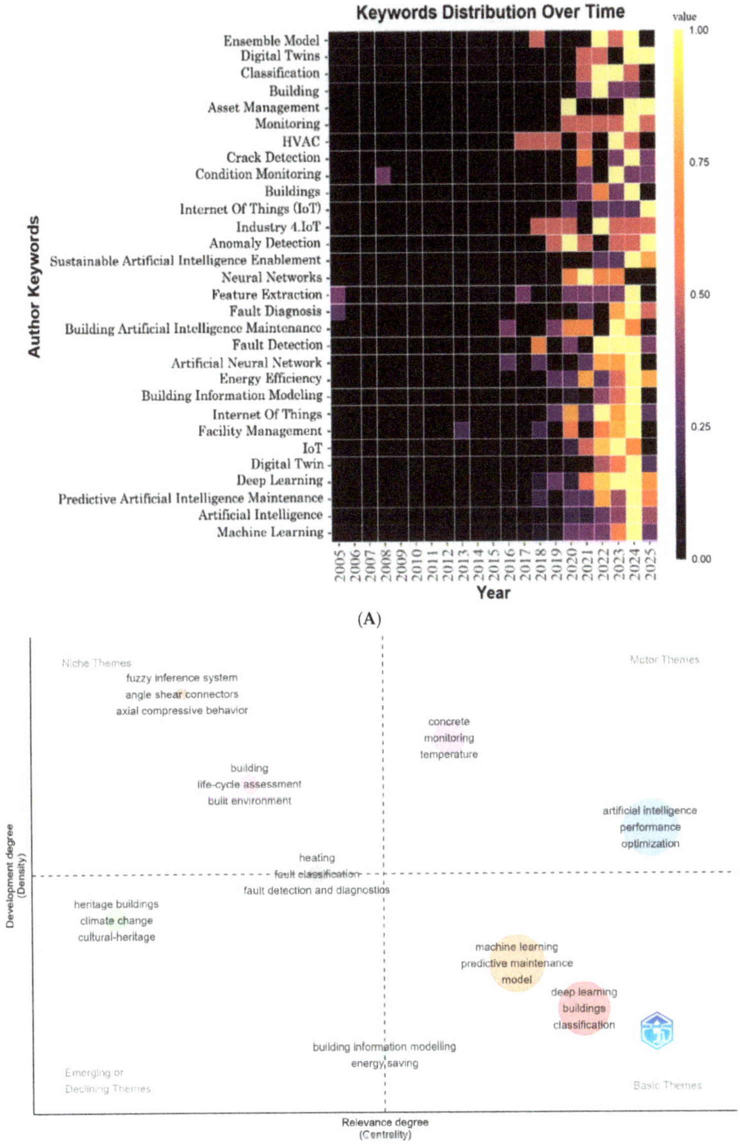

Figure 9. (**A**) Temporal Distribution Heatmap of Keywords (Colors closer to yellow indicate higher keyword occurrence frequency in the corresponding year, while black indicates no co-occurrence record, reflecting the evolution of research focus over time.). (**B**) Thematic Structure Evolution Map (The horizontal axis represents keyword centrality, and the vertical axis represents maturity (density). The upper-right quadrant denotes "motor themes," the upper-left "niche themes," the lower-left "emerging or declining themes," and the lower-right "basic themes." The color and size of the circles indicate the clustering intensity and frequency of each theme).

4.6.2. Thematic Structure and Evolutionary Pathways

Figure 9B presents the thematic evolution structure diagram based on co-occurrence analysis, further revealing the structural relationships and evolutionary directions of different research themes in the field of artificial intelligence for building maintenance. In this two-dimensional coordinate system, the horizontal axis represents thematic relevance (centrality), while the vertical axis represents thematic development (density). The distribution across the four quadrants clearly delineates the hierarchical importance of research themes as well as the emerging frontiers and hotspots.

The upper-right quadrant represents the "Motor Themes," exemplified by keywords such as "artificial intelligence," "performance," and "optimization." These themes exhibit both high centrality and high density, indicating that they are the core driving forces in the field. Themes in this quadrant are not only highly connected and influential but also demonstrate a high level of technological maturity, serving as critical nodes for advancing the intelligence and automation of building maintenance. Related research focuses on performance enhancement, energy optimization, and intelligent decision-making enabled by AI, reflecting the parallel progression of technological innovation and industry application.

The lower-right quadrant represents the "Basic Themes," including keywords such as "machine learning," "predictive maintenance," "deep learning," "buildings," and "classification." These themes exhibit high relevance and moderate development, serving as the methodological and practical foundation for the application of artificial intelligence in building maintenance. Themes in this quadrant often act as bridges for knowledge transfer and innovation expansion, supporting the overall progression of the field from theoretical research to practical implementation.

The upper-left quadrant represents the "Niche Themes," dominated by keywords such as "fuzzy inference system" and "axial compressive behavior," reflecting in-depth exploration within specialized subfields. Although these themes exhibit relatively low centrality, they possess unique advantages in addressing specific technical challenges and application scenarios. As such, they hold non-negligible value in fostering the diversified development of the technological framework within the field.

The lower-left quadrant represents the "Emerging or Declining Themes," including keywords such as "heritage buildings," "climate change," and "cultural heritage," which exhibit relatively low density and centrality. These themes largely correspond to emerging fields or cross-disciplinary explorations, with current research still at an early or transitional stage but demonstrating considerable development potential. With the growing prominence of issues such as green buildings and cultural heritage preservation, these themes are expected to evolve into new hotspots of future growth.

5. Discussion

5.1. Major Research Findings (Knowledge Structure)

By combining the temporal perspective of Figure 9A with the structural perspective of Figure 9B in Section 4.6, the bibliometric analysis of literature on artificial intelligence in building maintenance management reveals that the development of this field is characterized by dynamic complexity, with various research directions and hotspots emerging at different stages. Some new research directions expand upon existing themes, while others deepen the investigation of established issues, indicating that the evolution of research focuses and directions is inherently dynamic. Although several review studies have examined the application of Building Information Modeling (BIM) in operations and maintenance or suggested the use of AI to improve facility management, these works have not succeeded in establishing a coherent knowledge framework, making it difficult for readers to access key information in a concise and intuitive manner. Therefore, it is necessary to construct a new, valuable, and comprehensive theoretical knowledge framework for research on AI in building maintenance, in order to present a holistic view of the field and to provide guidance for future studies.

Figure 10 presents the integrated knowledge framework of AI in building maintenance management research, clearly illustrating the interrelationships among major research themes, the primary stakeholders, and potential future research directions, thereby providing valuable references for subsequent studies.

Publication Analysis: The bibliometric results indicate that the number of publications in this research field has shown a continuous upward trend, with research topics becoming increasingly diverse. This reflects the growing scholarly attention to the field and suggests a promising outlook for future development. Basic publication statistics demonstrate that related studies began to rise significantly in the late 2010s, with a sharp increase in annual publications after 2018. In terms of major academic journals, Journal of Building Engineering, Buildings, and Automation in Construction are the top three in terms of publication volume. The most prominent subject categories include Civil Engineering and Building Technology, Computer Science, and Energy and Environmental Sciences, reflecting the interdisciplinary and cross-domain development trend of the field. multidisciplinary and cross-disciplinary development trajectory.

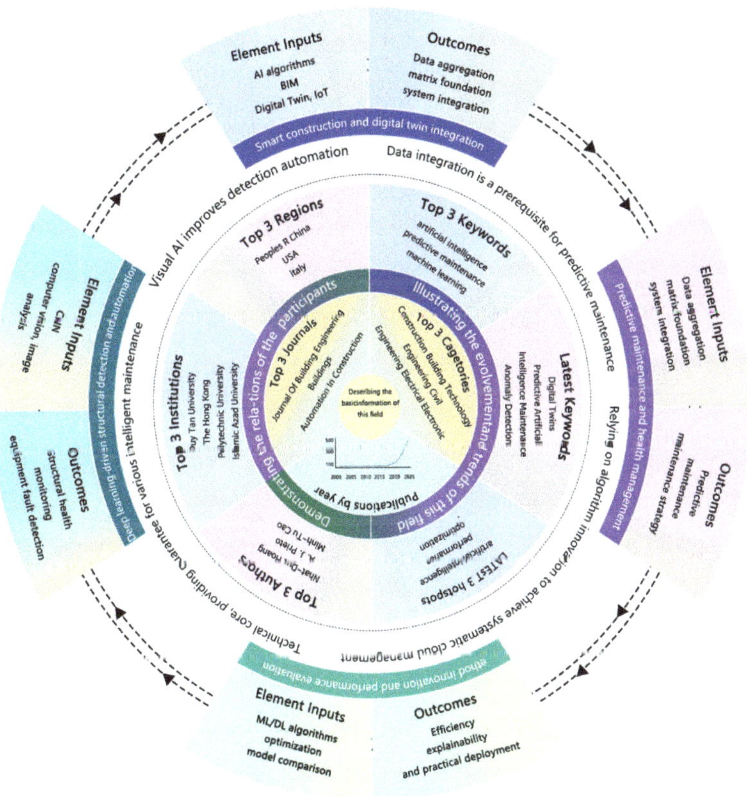

Figure 10. Illustrates the knowledge framework of artificial intelligence research in the field of building maintenance. The framework is presented in a multi-layered circular structure, depicting the primary research elements, key thematic keywords, regional distribution, and interrelationships among research outputs. The outer rings sequentially display information on the evolution of research hotspots.

Collaboration Analysis: The collaboration network reveals the dynamic cooperative relationships among research institutions, countries/regions, and authors. Scholars from all major countries are actively engaged in this field, and the leading contributors have been identified to provide references for researchers seeking potential collaborators and appropriate research domains. China, the United States, Italy, Canada, and the United Kingdom are the countries with the closest collaboration. More specifically, the leading institutional participants include Duy Tan University, The Hong Kong Polytechnic University, and Islamic Azad University. Among individual scholars, Nhat-Duc Hoang, A. J. Prieto, and Minh-Tu Cao are identified as the most active collaborators.

The outer ring displays the set of research keywords: the keyword co-occurrence patterns reveal the major research hotspots and evolutionary trends in the field, outlining the core themes and potential turning points across different periods. The analysis indicates that the thematic structure is primarily organized around four core clusters: "intelligent construction and digital twin integration," "predictive maintenance and health management," "algorithmic innovation and performance evaluation," and "deep learning–driven structural inspection and automated operation and maintenance," together forming the central lexical network of this research domain.

5.2. *Thematic Evolution Trends and Emerging Hotspots*

By integrating the keyword evolution pathways with the thematic strategic diagram, it becomes evident that research themes in the field of artificial intelligence for building maintenance management are undergoing a progressive upgrade—from "passive monitoring" to "proactive prediction," and further toward "intelligent decision-making" and "system integration." In recent years, "digital twins," "energy efficiency management," and "multi-technology integration" have emerged as new research hotspots, while novel application scenarios such as green sustainability, cultural heritage, and climate-adaptive buildings have continuously surfaced. The emergence of these hotspots has not only advanced the levels of intelligence and coordination in building maintenance but has also provided more efficient and scientifically grounded solutions for energy conservation and emission reduction, as well as for the preservation of historic buildings, thereby driving the industry toward greater intelligence and sustainability.

5.2.1. Integration with Digital Twin Technology

With the integration of Artificial Intelligence (AI) and Digital Twin technology, building maintenance is shifting from traditional reactive monitoring toward proactive prediction [19,79]. A systematic review by Siyuan Chen et al. indicates that AI techniques such as supervised learning, deep learning, and reinforcement learning can enhance digital twin models to enable fault prediction and data-driven maintenance strategies [80]. For example, Chen et al. (2023) note that digital twins provide accurate equipment condition recognition and failure forecasting for predictive maintenance (PdM), supporting a transition from "passive response" to "active prevention." This shift allows for optimization of maintenance scheduling, reduction of downtime, and improvement in system reliability and economic efficiency. However, current industrial applications are still constrained by challenges including scalability, heterogeneous data integration, and model deployment complexity [81]. Future research should continue to advance standardized modeling frameworks, mechanisms for multi-source data fusion, model interpretability, and real-time reliability validation, in order to facilitate the practical implementation of digital twin technology within building maintenance [82].

5.2.2. Energy Efficiency Management

AI is playing an increasingly significant role in building energy efficiency management [83]. According to the review by Bajwa et al. (2024), AI technologies have been widely applied in Smart Building Management Systems (SBMS) for HVAC optimization, lighting control, and renewable energy forecasting, achieving substantial energy savings (with consumption reductions of 20–50%) [84]. Algorithms such as reinforcement learning and deep learning have demonstrated greater flexibility than traditional rule-based approaches by adaptively adjusting control strategies in response to real-time sensor data. Future development requires the establishment of unified frameworks and strengthened interdisciplinary collaboration, along with expanded experimental validation and real-world project implementation, to promote the industrial-scale application of AI in building energy management [85]. By integrating dynamic control with predictive maintenance, energy management can evolve from a narrow focus on "energy-saving" to a more comprehensive approach encompassing "energy efficiency, emission reduction, and sustainable operation [86]."

5.2.3. Multi-Technology Integration

The trend of intelligent building maintenance is increasingly characterized by the integrated coordination of multiple technologies. Under the combined influence of BIM, the Internet of Things (IoT), 5G, virtual/augmented reality (VR/AR), and AI-driven robotics, building maintenance capabilities are being significantly expanded to form comprehensive intelligent maintenance solutions [87,88]. Wong et al. (2025) note that IoT enables real-time data acquisition and monitoring in buildings, VR/AR and digital twins support immersive visualization and simulation, and AI algorithms provide predictive analytics for maintenance decision-making through machine learning [89]. Specifically, Building Information Modeling (BIM), as a collaborative digital platform, ensures shared situational awareness of building conditions across departments; meanwhile, AI techniques such as generative design can search for optimal solutions under constraints such as cost and energy efficiency [90–93]. The combination of these tools enables scenario-based simulation and continuous condition sensing, thereby optimizing maintenance pathways and enhancing sustainability.

The integration of robotics and AI provides solutions for building inspection in inaccessible or hazardous environments [94,95]. Maryam Kouzehgar et al. proposed a self-reconfigurable façade-cleaning robot equipped with a deep convolutional neural network (CNN) for crack detection. Using a CNN model implemented in TensorFlow, the system analyzes real-time video captured from the robot's onboard camera to achieve high-precision (approximately 90%) automatic crack identification and obstacle avoidance, significantly improving the safety and automation level of high-rise glass curtain wall cleaning [96]. The convergence of diverse technologies is thus a core pathway for the systematic development of intelligent maintenance. By integrating BIM, IoT, AI, and robotic automation into a unified platform, fragmented maintenance operations can be transformed into a coordinated and intelligent system [97].

5.2.4. Applications of Artificial Intelligence in Green Buildings

Under the dual-carbon policy targets, AI-empowered green and sustainable buildings have become an emerging application scenario in building maintenance [98,99]. Modern green buildings not only aim for energy conservation but also emphasize full life-cycle sustainability and occupant health and comfort [100]. AI technologies are increasingly applied to optimize the operational performance of green buildings [101]. For example, Prabhu Rajaram et al. employed algorithms such as KNN, SVM, Random Forest, XGBoost,

AdaBoost, and Naive Bayes to enhance energy efficiency and reduce carbon footprints. Their approach effectively addressed challenges such as inaccurate energy consumption prediction, limited real-time indoor environmental optimization, and insufficient decision support in sustainable smart buildings, achieving coordinated improvements in both energy performance and occupant comfort [102]. Furthermore, AI has been utilized to support green building design and certification decision-making. Mansouri et al. used machine learning models (including decision trees, support vector regression, and XGBoost) to predict LEED green building certification scores, with the XGBoost model achieving the best performance ($R^2 \approx 0.81$), demonstrating the effectiveness of machine learning for LEED credit score prediction [103]. Future research may focus on developing intelligent design assistants aligned with dual-carbon objectives, embedding carbon simulation into BIM platforms, and leveraging generative and reinforcement learning approaches to promote proactive carbon reduction throughout the building life cycle [104].

5.2.5. Applications of Artificial Intelligence in Cultural Heritage Preservation

For historical buildings and cultural heritage, the introduction of AI technologies provides new approaches to monitoring and risk management in this traditionally conservation-oriented field [105]. As heritage structures are exposed to long-term climate change and environmental pollution, traditional manual inspections often fail to detect slow and subtle deterioration processes in a timely manner [106]. By deploying IoT sensor networks in historic buildings to monitor environmental and structural parameters such as temperature, humidity, stress, strain, and air quality, and applying AI-based data analysis, real-time early warning and predictive assessment can be achieved [107]. Neeraparng Laohaviraphap et al. report that IoT enables continuous acquisition of environmental and structural condition data from heritage sites, while AI performs pattern analysis and risk prediction, shifting heritage conservation from passive restoration to proactive prevention. For example, when abnormal moisture levels are detected in wall materials, AI can predict risks such as mold growth or structural weakening and issue early alerts [105]. At the same time, deep learning-based image recognition technologies enable automated identification of surface cracks and weathering, facilitating fine-grained monitoring. Sida Xu and Haonan Chen et al. integrated digital twin technology with an improved YOLO object detection algorithm by using UAV panoramic scanning to construct a high-precision 3D model of the Nanjing Qixia Temple Pagoda. They then applied YOLO within the digital twin environment to detect crack and corrosion damage, enabling multi-angle and cross-season automated defect identification, achieving significantly higher accuracy and efficiency compared with traditional manual inspection [108]. AI is transforming cultural heritage preservation into a proactive and scientific process: through intelligent monitoring and simulation-based analysis, it enhances the foresight, precision, and coverage of conservation actions, thereby greatly improving the capacity for long-term preservation of valuable cultural assets [109].

5.2.6. Applications of Artificial Intelligence in Climate-Resilient Architecture

With the increasing frequency of extreme weather events and natural disasters, building resilience and climate adaptability have become emerging focal points, in which AI plays a critical role [110]. By applying machine learning to meteorological, atmospheric, and structural response data, AI-driven disaster prediction and response capabilities have rapidly advanced in extreme event scenarios [111]. For example, Alessandro Rocchi et al. analyzed data from 331 municipalities in the Emilia–Romagna region of Italy and, after data standardization and PCA dimensionality reduction, applied K-means clustering to classify the region into four risk categories. They then incorporated variables such as seismic acceleration, building condition, and population exposure to generate comprehensive risk labels, providing a scientific basis for government prioritization of disaster prevention interventions [112]. According to a review by Vagelis Plevris, AI has already been employed in fields such as earthquake early warning, structural health monitoring, damage assessment, and multi-hazard risk mapping, improving early warning accuracy and enhancing structural resilience. However, challenges remain, including uncertainties in climate models, insufficient reliability of intelligent systems under extreme conditions, and difficulties in integrating such systems into regulatory frameworks [113]. Future research directions may explore dynamic building envelopes (e.g., adaptive shading systems, deployable or morphable structural assemblies) and real-time climate control strategies enabled by edge computing, thereby coupling AI-based predictive analytics with automated building control [114,115]. This integration is expected to enhance the capability of buildings to autonomously anticipate, respond to, and recover from extreme climate impacts [116].

5.3. Challenges and Major Issues

Overall, the engineering application of artificial intelligence in building maintenance management still faces numerous challenges. First, data sources related to building maintenance are highly diverse yet lack unified standards [117]. Data quality is uneven, formats are inconsistent, and effective sharing remains difficult. For instance, large volumes of building operation and maintenance data have not been systematically collected or stored, and there are almost no open-access databases for building maintenance (with only a few available for energy consumption) [118]. This severely constrains the efficient integration of heterogeneous multisource data and limits the widespread adoption of AI models. Second, current AI models often struggle to adapt to variations across building types, climatic conditions, and operational scenarios, reflecting insufficient generalization capability [119]. Due to the scarcity and fragmentation of training datasets, the transferability of models across climate zones and different building structures remains limited [120]. On the one hand, differences in operating conditions between regions and buildings lead to poor performance of "well-trained" models when applied in new environments [121]. On the other hand, the "black-box" nature of AI systems renders their decision-making processes opaque, making it difficult for building operators to understand the rationale behind maintenance decisions [122]. This lack of interpretability undermines user trust in AI-generated recommendations and constitutes a significant barrier to real-world adoption in facility maintenance [123]. Third, many existing building maintenance platforms, sensor monitoring networks, and Building Information Modeling (BIM) systems operate in isolation without effective integration, restricting AI from accessing comprehensive and standardized data support. Moreover, different generations of equipment and software often lack standardized interfaces; for example, legacy building systems are not interoperable with modern IoT edge nodes, preventing seamless data exchange [11]. Such information silos increase the complexity of AI deployment, necessitating extensive customized development to bridge disparate systems. Finally, disparities across countries and regions in terms of resource investment, policy support, and technology transfer capacity further exacerbate the uneven global development of intelligent building maintenance.

In the future, efforts should focus on advancing data standardization, enhancing the generalizability and interpretability of models, and optimizing pathways for system integration. At the same time, policies and resource allocation should be reinforced in alignment with regional contexts to ensure the high-quality implementation and sustainable development of artificial intelligence technologies in building maintenance management.

6. Conclusions

Over the past decade, the application of artificial intelligence in building maintenance has exhibited rapid expansion and, driven by the maturation of technologies such as the Internet of Things (IoT), digital twins, 5G, and edge computing, is expected to experience even stronger growth momentum in the coming years. With the implementation of predictive maintenance, unmanned visual inspection, intelligent optimization of Building Management Systems (BMS), and coordinated management of energy and carbon, the industry faces increasingly urgent demands for effective evaluation frameworks encompassing accuracy, reliability, interpretability, interoperability, and cost-effectiveness. This study seeks to systematically reveal the latest developmental trajectories and methodological evolution in the "AI–building maintenance" domain by identifying the most influential research sources, research hotspots, and emerging topics. Using scientific mapping techniques, we constructed knowledge graphs that delineate the theoretical foundations of the field—such as condition monitoring and remaining service life prediction, mechanism–data fusion, and the integration of Facility Management (FM) with BIM/digital twins—as well as recent integrated framework studies. The observed trend highlights the growing prominence of intelligent and sustainable maintenance, with a continuous increase in publications across interdisciplinary areas, including facility management, architecture and environmental engineering, computer science, and energy and environmental sciences, as indicated by the content analysis.

We emphasized the dynamic evolution of the role of artificial intelligence in building maintenance: shifting from passive inspection to proactive prediction, and from isolated optimization to system-level integration (BIM/digital twin/IoT/CMMS collaboration). This progression demonstrates its potential to enhance maintenance decision-making, energy efficiency management, and risk control through predictive capabilities and real-time data analytics. The four thematic clusters are highly interrelated and intricately coupled: Cluster 1 provides the data and semantic foundation; Cluster 2 constructs a closed loop of "monitoring–diagnosis–prediction–intervention"; Cluster 3 functions as the engine of methodological innovation and evaluation standardization; and Cluster 4 connects the perception–execution chain and feeds new evidence back into the digital twin for continuous learning. Based on this thematic structure, this paper proposes a comprehensive evaluation framework that integrates artificial intelligence with traditional facility management and reliability engineering approaches. Anchored in a digital twin platform, the framework spans data governance, model development and interpretability, business process orchestration, value assessment, and risk control, thereby establishing an end-to-end management and decision-making cycle. This ensures transparency, traceability, and fairness in technological implementation, while enhancing the feasibility of building maintenance management and fostering social trust.

Several critical gaps remain in the existing literature: insufficient data standardization and interoperability, particularly manifested in semantic inconsistencies across BIM, digital twins (DT), and IoT systems; limited generalization and interpretability of models, with inadequate transferability across building types, climatic conditions, and operational contexts; challenges in large-scale deployment due to technological readiness level (TRL) bottlenecks, where transitions from laboratory validation to practical implementation remain problematic; the absence of open benchmark datasets and reproducibility mechanisms, restricting horizontal comparison and validation of different algorithms; and systemic deficiencies in ethical, privacy, and compliance evaluations, including shortcomings in data governance, auditability, and accountability.

7. Limitations and Future Perspectives

This study employed bibliometric methods, drawing on the core dataset of the Web of Science database, to systematically review research hotspots and developmental trajectories of artificial intelligence in building maintenance management. However, certain limitations remain. First, the analysis was based on a single mainstream database, without incorporating additional sources such as Google Scholar, ProQuest, SpringerLink, and CNKI. This may have resulted in the omission of important studies, thereby affecting the comprehensiveness and representativeness of the findings. Future research should expand the range of databases utilized to better capture global research trends in the field of intelligent building maintenance and to continuously track the latest advancements.

Author Contributions: Y.Z.: Writing—original draft, Writing—review and editing, Visualization, Investigation, Formal analysis, Conceptualization. B.S.: Writing—review and editing, Visualization, Methodology, Resources, Formal analysis, Data curation. Y.G.: Validation, Resources, Investigation, Software, Formal analysis. Y.Y.: Supervision, Project administration, Conceptualization, Writing—review and editing, Funding acquisition. All authors have read and agreed to the published version of the manuscript.

Funding: The authors gratefully acknowledge the financial support of the Fundamental Research Funds for the Central Universities (CCNU23XJ049).

Data Availability Statement: The raw data supporting the conclusions of this article will be made available by the authors on request.

Conflicts of Interest: The authors declare that they have no known competing financial interests or personal relationships that could have appeared to influence the work reported in this paper.

References

1. Zhang, X.; Bayulken, B.; Skitmore, M.; Lu, W.; Huisingh, D. Sustainable urban transformations towards smarter, healthier cities: Theories, agendas and pathways. *J. Clean. Prod.* **2018**, *173*, 1–10. [CrossRef]
2. West, J.; Siddhpura, M.; Evangelista, A.; Haddad, A. Improving equipment maintenance—Switching from corrective to preventative maintenance strategies. *Buildings* **2024**, *14*, 3581. [CrossRef]
3. Pan, Y.; Zhang, L. Integrating BIM and AI for smart construction management: Current status and future directions. *Arch. Comput. Methods Eng.* **2023**, *30*, 1081–1110. [CrossRef]
4. Hauashdh, A.; Jailani, J.; Rahman, I.A.; Al-Fadhali, N. Strategic approaches towards achieving sustainable and effective building maintenance practices in maintenance-managed buildings: A combination of expert interviews and a literature review. *J. Build. Eng.* **2022**, *45*, 103490. [CrossRef]
5. Valença, J.; Morin, K.; Jouen, N.; Olivo, N.; Torres-Gonzalez, M.; Mendes, M.P.; Silva, A. Feeling-BIM: A digital model to support maintenance decisions, based on automatic inspection and dwellers' feelings. *J. Build. Eng.* **2024**, *87*, 108937. [CrossRef]
6. Wei, Y.; Yuan, H.; Li, H. Exploring the contribution of advanced systems in smart city development for the regeneration of urban industrial heritage. *Buildings* **2024**, *14*, 583. [CrossRef]
7. Dzulkifli, N.a.; Sarbini, N.N.; Ibrahim, I.S.; Abidin, N.I.; Yahaya, F.M.; Azizan, N.Z.N. Review on maintenance issues toward building maintenance management best practices. *J. Build. Eng.* **2021**, *44*, 102985. [CrossRef]
8. Foresti, R.; Rossi, S.; Magnani, M.; Bianco, C.G.L.; Delmonte, N. Smart society and artificial intelligence: Big data scheduling and the global standard method applied to smart maintenance. *Engineering* **2020**, *6*, 835–846. [CrossRef]
9. Himeur, Y.; Elnour, M.; Fadli, F.; Meskin, N.; Petri, I.; Rezgui, Y.; Bensaali, F.; Amira, A. AI-big data analytics for building automation and management systems: A survey, actual challenges and future perspectives. *Artif. Intell. Rev.* **2023**, *56*, 4929–5021. [CrossRef]
10. Chen, Y.; Huang, D.; Liu, Z.; Osmani, M.; Demian, P. Construction 4.0, Industry 4.0, and Building Information Modeling (BIM) for sustainable building development within the smart city. *Sustainability* **2022**, *14*, 10028. [CrossRef]
11. Volk, R.; Stengel, J.; Schultmann, F. Building Information Modeling (BIM) for existing buildings—Literature review and future needs. *Autom. Constr.* **2014**, *38*, 109–127. [CrossRef]
12. Jaboob, A.; Durrah, O.; Chakir, A. Artificial intelligence: An overview. In *Engineering Applications of Artificial Intelligence*; Springer: Berlin/Heidelberg, Germany, 2024; pp. 3–22.

13. Waheed, W.; Xu, Q. Data-driven short term load forecasting with deep neural networks: Unlocking insights for sustainable energy management. *Electr. Power Syst. Res.* **2024**, *232*, 110376. [CrossRef]
14. Gong, J.; Caldas, C.H. Computer vision-based video interpretation model for automated productivity analysis of construction operations. *J. Comput. Civ. Eng.* **2010**, *24*, 252–263. [CrossRef]
15. Eleftheriadis, S.; Mumovic, D.; Greening, P. Life cycle energy efficiency in building structures: A review of current developments and future outlooks based on BIM capabilities. *Renew. Sustain. Energy Rev.* **2017**, *67*, 811–825. [CrossRef]
16. Dong, C.-Z.; Catbas, F.N. A review of computer vision–based structural health monitoring at local and global levels. *Struct. Health Monit.* **2021**, *20*, 692–743. [CrossRef]
17. Wang, C.; Antos, S.E.; Triveno, L.M. Automatic detection of unreinforced masonry buildings from street view images using deep learning-based image segmentation. *Autom. Constr.* **2021**, *132*, 103968. [CrossRef]
18. Bouabdallaoui, Y.; Lafhaj, Z.; Yim, P.; Ducoulombier, L.; Bennadji, B. Predictive maintenance in building facilities: A machine learning-based approach. *Sensors* **2021**, *21*, 1044. [CrossRef] [PubMed]
19. Sadri, H. AI-driven integration of digital twins and blockchain for smart building management systems: A multi-stage empirical study. *J. Build. Eng* **2025**, *105*, 112439. [CrossRef]
20. Alijoyo, F.A. AI-powered deep learning for sustainable industry 4.0 and internet of things: Enhancing energy management in smart buildings. *Alex. Eng. J.* **2024**, *104*, 409–422. [CrossRef]
21. Rojek, I.; Jasiulewicz-Kaczmarek, M.; Piechowski, M.; Mikołajewski, D. An artificial intelligence approach for improving maintenance to supervise machine failures and support their repair. *Appl. Sci.* **2023**, *13*, 4971. [CrossRef]
22. Khan, A.; Mishra, S.; Pandey, S.; Sardar, T.; Mishra, A.; Mudgal, M.; Singhai, S. A Critical Review of IoT-Based Structural Health Monitoring for Dams. *IEEE Internet Things J.* **2024**, *12*, 1368–1379. [CrossRef]
23. Turner, C.J.; Emmanouilidis, C.; Tomiyama, T.; Tiwari, A.; Roy, R. Intelligent decision support for maintenance: An overview and future trends. *Int. J. Comput. Integr. Manuf.* **2019**, *32*, 936–959. [CrossRef]
24. Bhatt, P.M.; Malhan, R.K.; Rajendran, P.; Shah, B.C.; Thakar, S.; Yoon, Y.J.; Gupta, S.K. Image-based surface defect detection using deep learning: A review. *J. Comput. Inf. Sci. Eng.* **2021**, *21*, 040801. [CrossRef]
25. Baghalzadeh Shishehgarkhaneh, M.; Keivani, A.; Moehler, R.C.; Jelodari, N.; Roshdi Laleh, S. Internet of Things (IoT), Building Information Modeling (BIM), and Digital Twin (DT) in construction industry: A review, bibliometric, and network analysis. *Buildings* **2022**, *12*, 1503. [CrossRef]
26. Wu, J.; Wang, Q.; Guo, Z.; Peng, C. AI-Driven Green Building Technology Innovation: Knowledge Structure, Evolution Trends, Research Paradigms and Future Prospects. *Buildings* **2025**, *15*, 1754. [CrossRef]
27. Lu, Q.; Xie, X.; Parlikad, A.K.; Schooling, J.M. Digital twin-enabled anomaly detection for built asset monitoring in operation and maintenance. *Autom. Constr.* **2020**, *118*, 103277. [CrossRef]
28. Liu, Z.; Shi, G.; Meng, X.; Sun, Z. Intelligent control of building operation and maintenance processes based on global navigation satellite system and digital twins. *Remote Sens.* **2022**, *14*, 1387. [CrossRef]
29. Motawa, I.; Almarshad, A. A knowledge-based BIM system for building maintenance. *Autom. Constr.* **2013**, *29*, 173–182. [CrossRef]
30. Zhang, H.C.; Kuo, T.C.; Lu, H.; Huang, S.H. Environmentally conscious design and manufacturing: A state-of-the-art survey. *J. Manuf. Syst.* **1997**, *16*, 352–371. [CrossRef]
31. Madureira, S.; Flores-Colen, I.; de Brito, J.; Pereira, C. Maintenance planning of facades in current buildings. *Constr. Build. Mater.* **2017**, *147*, 790–802. [CrossRef]
32. Pampana, A.K.; Jeon, J.; Yoon, S.; Weidner, T.J. Comprehensive maintenance dataset of building facilities for planned preventive and unplanned maintenance in North American universities. *Data Brief* **2024**, *57*, 110845. [CrossRef]
33. Gustavsson, E.; Patriksson, M.; Strömberg, A.-B.; Wojciechowski, A.; Önnheim, M. Preventive maintenance scheduling of multi-component systems with interval costs. *Comput. Ind. Eng.* **2014**, *76*, 390–400. [CrossRef]
34. Raghavan, S.; Chowdhury, B. State diagram-based life cycle management plans for power plant components. *IEEE Trans. Smart Grid* **2015**, *6*, 965–972. [CrossRef]
35. Zio, E. Some challenges and opportunities in reliability engineering. *IEEE Trans. Reliab.* **2016**, *65*, 1769–1782. [CrossRef]
36. Khan, S.; Phillips, P.; Hockley, C.; Jennions, I. No Fault Found events in maintenance engineering Part 2: Root causes, technical developments and future research. *Reliab. Eng. Syst. Saf.* **2014**, *123*, 196–208. [CrossRef]
37. Okoh, P.; Haugen, S. Maintenance-related major accidents: Classification of causes and case study. *J. Loss Prev. Process Ind.* **2013**, *26*, 1060–1070. [CrossRef]
38. Doyen, L.; Gaudoin, O. Modeling and assessment of aging and efficiency of corrective and planned preventive maintenance. *IEEE Trans. Reliab.* **2011**, *60*, 759–769. [CrossRef]
39. Jamarani, A.; Haddadi, S.; Sarvizadeh, R.; Haghi Kashani, M.; Akbari, M.; Moradi, S. Big data and predictive analytics: A systematic review of applications. *Artif. Intell. Rev.* **2024**, *57*, 176. [CrossRef]

40. Moradi, E.; Ghomi, S.F.; Zandieh, M. Bi-objective optimization research on integrated fixed time interval preventive maintenance and production for scheduling flexible job-shop problem. *Expert Syst. Appl.* **2011**, *38*, 7169–7178. [CrossRef]
41. Ali, A.; Abdelhadi, A. Condition-based monitoring and maintenance: State of the art review. *Appl. Sci.* **2022**, *12*, 688. [CrossRef]
42. Bousdekis, A.; Magoutas, B.; Apostolou, D.; Mentzas, G. Review, analysis and synthesis of prognostic-based decision support methods for condition based maintenance. *J. Intell. Manuf.* **2018**, *29*, 1303–1316. [CrossRef]
43. Chen, Q.; Cao, J.; Zhu, S. Data-driven monitoring and predictive maintenance for engineering structures: Technologies, implementation challenges, and future directions. *IEEE Internet Things J.* **2023**, *10*, 14527–14551. [CrossRef]
44. Wang, Y.; Deng, C.; Wu, J.; Wang, Y.; Xiong, Y. A corrective maintenance scheme for engineering equipment. *Eng. Fail. Anal.* **2014**, *36*, 269–283. [CrossRef]
45. Eti, M.C.; Ogaji, S.; Probert, S. Reducing the cost of preventive maintenance (PM) through adopting a proactive reliability-focused culture. *Appl. Energy* **2006**, *83*, 1235–1248. [CrossRef]
46. Endrenyi, J.; Aboresheid, S.; Allan, R.; Anders, G.; Asgarpoor, S.; Billinton, R.; Chowdhury, N.; Dialynas, E.; Fipper, M.; Fletcher, R. The present status of maintenance strategies and the impact of maintenance on reliability. *IEEE Trans. Power Syst.* **2001**, *16*, 638–646. [CrossRef]
47. Khazraei, K.; Deuse, J. A strategic standpoint on maintenance taxonomy. *J. Facil. Manag.* **2011**, *9*, 96–113. [CrossRef]
48. Wang, T.; Chen, H.-M. Integration of building information modeling and project management in construction project life cycle. *Autom. Constr.* **2023**, *150*, 104832. [CrossRef]
49. Meng, Q.; Zhang, Y.; Li, Z.; Shi, W.; Wang, J.; Sun, Y.; Xu, L.; Wang, X. A review of integrated applications of BIM and related technologies in whole building life cycle. *Eng. Constr. Archit. Manag.* **2020**, *27*, 1647–1677. [CrossRef]
50. Hassan, M.M.; Gumaei, A.; Huda, S.; Almogren, A. Increasing the trustworthiness in the industrial IoT networks through a reliable cyberattack detection model. *IEEE Trans. Ind. Inf.* **2020**, *16*, 6154–6162. [CrossRef]
51. Nan, N. Capturing bottom-up information technology use processes: A complex adaptive systems model. *MIS Q.* **2011**, *35*, 505–532. [CrossRef]
52. Maseda, F.J.; López, I.; Martija, I.; Alkorta, P.; Garrido, A.J.; Garrido, I. Sensors data analysis in supervisory control and data acquisition (SCADA) systems to foresee failures with an undetermined origin. *Sensors* **2021**, *21*, 2762. [CrossRef]
53. Abideen, D.K.; Yunusa-Kaltungo, A.; Manu, P.; Cheung, C. A systematic review of the extent to which BIM is integrated into operation and maintenance. *Sustainability* **2022**, *14*, 8692. [CrossRef]
54. Rodrigues, R.C.; Sousa, H.; Gondim, I.A. SMARTS-based decision support model for CMMS selection in integrated building maintenance management. *Buildings* **2023**, *13*, 2521. [CrossRef]
55. Yao, Y.; Patterson, E.A.; Taylor, R.J. The influence of digital technologies on knowledge management in engineering: A systematic literature review. *IEEE Trans. Knowl. Data Eng.* **2023**, *36*, 854–867. [CrossRef]
56. Senevirathna, T.; La, V.H.; Marcha, S.; Siniarski, B.; Liyanage, M.; Wang, S. A survey on XAI for 5G and beyond security: Technical aspects, challenges and research directions. *IEEE Commun. Surv. Tutor.* **2024**, *27*, 941–973. [CrossRef]
57. Qin, J.; Liu, Y.; Grosvenor, R. Multi-source data analytics for AM energy consumption prediction. *Adv. Eng. Inf.* **2018**, *38*, 840–850. [CrossRef]
58. Interlando, M.; Pacifico, M.G.; Novellino, A.; Pastore, V.P. Ensembles of deep neural networks for the automatic detection of building facade defects from images. *IEEE Access* **2024**, *12*, 164953–164965. [CrossRef]
59. Zhang, J.; Zhao, B.; Yang, G.; Zhou, X.; Huang, Y.; Gao, C.; Chen, X.; Chen, B.M. AI-Empowered Digital Twin Modeling for High-Precision Building Defect Management Integrating UAV and GeoBIM. In *Building Simulation*; Tsinghua University Press: Beijing, China, 2025; pp. 1–28.
60. Chen, S.; Liang, X.; Liu, Y.; Li, X.; Jin, X.; Du, Z. Customized large-scale model for human-AI collaborative operation and maintenance management of building energy systems. *Appl. Energy* **2025**, *393*, 126169. [CrossRef]
61. Pang, B.; Wang, F.; Zhang, A.; Zhang, K.; Yang, J. Stochastic analysis and machine learning techniques for maintenance of heritage buildings with multi-source data. *J. Build. Eng.* **2025**, *103*, 111998. [CrossRef]
62. Zhou, X.; Tiong, R.L.K. Defects inspection system for building facades using drones and deep learning method. *Expert Syst. Appl.* **2025**, *298*, 129715. [CrossRef]
63. Zhu, J.; Liu, W. A tale of two databases: The use of Web of Science and Scopus in academic papers. *Scientometrics* **2020**, *123*, 321–335. [CrossRef]
64. Li, K.; Rollins, J.; Yan, E. Web of Science use in published research and review papers 1997–2017: A selective, dynamic, cross-domain, content-based analysis. *Scientometrics* **2018**, *115*, 1–20. [CrossRef] [PubMed]
65. Hassan, W.; Duarte, A.E. Bibliometric analysis: A few suggestions. *Curr. Probl. Cardiol.* **2024**, *49*, 102640. [CrossRef] [PubMed]
66. Wang, K. Mapping the global knowledge landscape of digital transformation in the AEC industry: A scientometric analysis. *Eng. Constr. Archit. Manag.* **2024**. [CrossRef]
67. Hage, J.; Hollingsworth, J.R. A strategy for the analysis of idea innovation networks and institutions. *Organ. Stud.* **2000**, *21*, 971–1004. [CrossRef]

68. Shao, B.; Li, X.; Bian, G. A survey of research hotspots and frontier trends of recommendation systems from the perspective of knowledge graph. *Expert Syst. Appl.* **2021**, *165*, 113764. [CrossRef]
69. Zhang, X.; Xie, Q.; Song, C.; Song, M. Mining the evolutionary process of knowledge through multiple relationships between keywords. *Scientometrics* **2022**, *127*, 2023–2053. [CrossRef]
70. Petri, I.; Amin, A.; Ghoroghi, A.; Hodorog, A.; Rezgui, Y. Digital twins for dynamic life cycle assessment in the built environment. *Sci. Total Environ.* **2025**, *993*, 179930. [CrossRef]
71. Wang, W.; Guo, H.; Li, X.; Tang, S.; Xia, J.; Lv, Z. Deep learning for assessment of environmental satisfaction using BIM big data in energy efficient building digital twins. *Sustain. Energy Technol. Assess.* **2022**, *50*, 101897. [CrossRef]
72. Liu, F.; Li, X. Integrating deep reinforcement learning with evolutionary algorithms for advanced optimization in smart city energy management. *IEEE Access* **2024**, *12*, 177103–177118. [CrossRef]
73. Gawde, S.; Patil, S.; Kumar, S.; Kotecha, K. A scoping review on multi-fault diagnosis of industrial rotating machines using multi-sensor data fusion. *Artif. Intell. Rev.* **2023**, *56*, 4711–4764. [CrossRef]
74. Wang, D.; Liu, L.; Jia, H.; Wang, W.; Zhi, Y.; Meng, Z.; Zhou, B. Review of key problems related to integrated energy distribution systems. *CSEE J. Power Energy Syst.* **2018**, *4*, 130–145. [CrossRef]
75. Zhu, Z.; Lin, K.; Jain, A.K.; Zhou, J. Transfer learning in deep reinforcement learning: A survey. *IEEE Trans. Pattern Anal. Mach. Intell.* **2023**, *45*, 13344–13362. [CrossRef]
76. Lu, J.; Ma, G.; Zhang, G. Fuzzy machine learning: A comprehensive framework and systematic review. *IEEE Trans. Fuzzy Syst.* **2024**, *32*, 3861–3878. [CrossRef]
77. Azimi, M.; Eslamlou, A.D.; Pekcan, G. Data-driven structural health monitoring and damage detection through deep learning: State-of-the-art review. *Sensors* **2020**, *20*, 2778. [CrossRef]
78. Liu, X.; Zhang, J.; Zhou, S.; van der Plas, T.L.; Vijayaraghavan, A.; Grishina, A.; Zhuang, M.; Schofield, D.; Tomlinson, C.; Wang, Y. Towards deployment-centric multimodal AI beyond vision and language. *Nat. Mach. Intell.* **2025**, *7*, 1612–1624. [CrossRef]
79. Yoon, S.; Song, J.; Li, J. AI agent-based intelligent digital twins for building operations and maintenance. *J. Build. Eng.* **2025**, *108*, 112802. [CrossRef]
80. Chen, S.; Bekar, E.T.; Bokrantz, J.; Skoogh, A. AI-enhanced digital twins in maintenance: Systematic review, industrial challenges, and bridging research–practice gaps. *J. Manuf. Syst.* **2025**, *82*, 678–699. [CrossRef]
81. Chen, C.; Fu, H.; Zheng, Y.; Tao, F.; Liu, Y. The advance of digital twin for predictive maintenance: The role and function of machine learning. *J. Manuf. Syst.* **2023**, *71*, 581–594. [CrossRef]
82. Ouedraogo, E.B.; Hawbani, A.; Wang, X.; Liu, Z.; Zhao, L.; Al-qaness, M.A.; Alsamhi, S.H. Digital Twin Data Management: A Comprehensive Review. *IEEE Trans. Big Data* **2025**, *11*, 2224–2243. [CrossRef]
83. Elmousalami, H.; Maxy, M.; Hui, F.K.P.; Aye, L. AI in automated sustainable construction engineering management. *Autom. Constr.* **2025**, *175*, 106202. [CrossRef]
84. Bajwa, A.; Jahan, F.; Ahmed, I. A systematic literature review on AI-enabled smart building management systems for energy efficiency and sustainability. *Am. J. Sch. Res. Innov.* **2024**, *3*, 1–7. [CrossRef]
85. Lee, J.; Ni, J.; Singh, J.; Jiang, B.; Azamfar, M.; Feng, J. Intelligent maintenance systems and predictive manufacturing. *J. Manuf. Sci. Eng.* **2020**, *142*, 110805. [CrossRef]
86. Cai, W.; Lai, K.-h.; Liu, C.; Wei, F.; Ma, M.; Jia, S.; Jiang, Z.; Lv, L. Promoting sustainability of manufacturing industry through the lean energy-saving and emission-reduction strategy. *Sci. Total Environ.* **2019**, *665*, 23–32. [CrossRef]
87. Bhatarai, R.; Banihashemi, S.; Shakouri, M.; Antwi-Afari, M. Integration of Augmented Reality with Building Information Modeling: Design Optimization and Construction Rework Reduction Perspective. *Arch. Comput. Methods Eng.* **2025**, *32*, 2233–2254. [CrossRef]
88. Li, J.; Liu, Z.; Han, G.; Demian, P.; Osmani, M. The relationship between Artificial Intelligence (AI) and Building Information Modeling (BIM) technologies for sustainable building in the context of smart cities. *Sustainability* **2024**, *16*, 10848. [CrossRef]
89. Wong, P.Y.; Lo, K.C.; Long, H.; Lai, J.H. Towards Digital Transformation in Building Maintenance and Renovation: Integrating BIM and AI in Practice. *Appl. Sci.* **2025**, *15*, 11389. [CrossRef]
90. Ali, K.N.; Alhajlah, H.H.; Kassem, M.A. Collaboration and risk in building information modelling (BIM): A systematic literature review. *Buildings* **2022**, *12*, 571. [CrossRef]
91. Papadonikolaki, E.; van Oel, C.; Kagioglou, M. Organising and Managing boundaries: A structurational view of collaboration with Building Information Modelling (BIM). *Int. J. Proj. Manag.* **2019**, *37*, 378–394. [CrossRef]
92. Song, J. Research on energy optimization space design of intelligent buildings based on generative AI technology and 3D modeling. *Microchem. J.* **2025**, *218*, 115500. [CrossRef]
93. Li, Y.; Chen, H.; Yu, P.; Yang, L. A review of artificial intelligence in enhancing architectural design efficiency. *Appl. Sci.* **2025**, *15*, 1476. [CrossRef]
94. Halder, S.; Afsari, K. Robots in inspection and monitoring of buildings and infrastructure: A systematic review. *Appl. Sci.* **2023**, *13*, 2304. [CrossRef]

95. Tian, Y.; Chen, C.; Sagoe-Crentsil, K.; Zhang, J.; Duan, W. Intelligent robotic systems for structural health monitoring: Applications and future trends. *Autom. Constr.* **2022**, *139*, 104273. [CrossRef]
96. Kouzehgar, M.; Tamilselvam, Y.K.; Heredia, M.V.; Elara, M.R. Self-reconfigurable façade-cleaning robot equipped with deep-learning-based crack detection based on convolutional neural networks. *Autom. Constr.* **2019**, *108*, 102959. [CrossRef]
97. Gyrard, A.; de Freitas, E.P.; Serrano, M.; Li, H.; Gonçalves, P.; Quintas, J.; Vermesan, O.; Olivares-Alarcos, A.; Kung, A.; Cavallo, F. Internet of Robotic Things Evolution, Standards and Data Interoperability Best Practices for the Next Generation of Artificial Intelligence-Powered Systems. *J. Field Rob.* **2025**, 1556–4959. [CrossRef]
98. Yan, K.; Zhou, X.; Yang, B. AI and IoT applications of smart buildings and smart environment design, construction and maintenance. *Build. Environ.* **2022**, *229*, 109968. [CrossRef]
99. Huo, D.; Sun, T.; Gu, W.; Qiao, L. Smart Cities with Green Resilience: A Quasi-Natural Experiment Based on Artificial Intelligence. *Smart Cities* **2025**, *8*, 67. [CrossRef]
100. Liang, R.; Ma, H.; Wang, P.; Zhao, L. The applications of building information modeling in the life-cycle of green buildings: A comprehensive review. *Sci. Technol. Built Environ.* **2024**, *30*, 932–958. [CrossRef]
101. Huang, Y. The technological assessment of green buildings using artificial neural networks. *Heliyon* **2024**, *10*, 2405–8440. [CrossRef] [PubMed]
102. Rajaram, P. Data-driven predictive models for sustainable smart buildings. *Results Eng.* **2025**, *27*, 105916. [CrossRef]
103. Mansouri, A.; Naghdi, M.; Erfani, A. Machine Learning for Leadership in Energy and Environmental Design Credit Targeting: Project Attributes and Climate Analysis Toward Sustainability. *Sustainability* **2025**, *17*, 2521. [CrossRef]
104. Humayun, M.; Tariq, N.; Alfayad, M.; Zakwan, M.; Alwakid, G.; Assiri, M. Securing the Internet of Things in artificial intelligence era: A comprehensive survey. *IEEE Access* **2024**, *12*, 25469–25490. [CrossRef]
105. Laohaviraphap, N.; Waroonkun, T. Integrating Artificial Intelligence and the Internet of Things in Cultural Heritage Preservation: A Systematic Review of Risk Management and Environmental Monitoring Strategies. *Buildings* **2024**, *14*, 3979. [CrossRef]
106. Moropoulou, A.; Labropoulos, K.C.; Delegou, E.T.; Karoglou, M.; Bakolas, A. Non-destructive techniques as a tool for the protection of built cultural heritage. *Constr. Build. Mater.* **2013**, *48*, 1222–1239. [CrossRef]
107. Dolińska, N.; Wojciechowska, G.; Bednarz, Ł. Monitoring Environmental and Structural Parameters in Historical Masonry Buildings Using IoT LoRaWAN-Based Wireless Sensors. *Buildings* **2025**, *15*, 282. [CrossRef]
108. Xu, S.; Chen, H. Deep learning and digital twin integration for structural damage detection in ancient pagodas. *Sci. Rep.* **2025**, *15*, 28408. [CrossRef]
109. Mazzetto, S. Integrating emerging technologies with digital twins for heritage building conservation: An interdisciplinary approach with expert insights and bibliometric analysis. *Heritage* **2024**, *7*, 6432–6479. [CrossRef]
110. Singh, S.; Goyal, M.K. Enhancing climate resilience in businesses: The role of artificial intelligence. *J. Clean. Prod.* **2023**, *418*, 138228. [CrossRef]
111. Rehan, H. Enhancing Disaster Response Systems: Predicting and Mitigating the Impact of Natural Disasters Using AI. *J. Artif. Intell. Res.* **2022**, *2*, 501.
112. Rocchi, A.; Chiozzi, A.; Nale, M.; Nikolic, Z.; Riguzzi, F.; Mantovan, L.; Gilli, A.; Benvenuti, E. A machine learning framework for multi-hazard risk assessment at the regional scale in earthquake and flood-prone areas. *Appl. Sci.* **2022**, *12*, 583. [CrossRef]
113. Plevris, V. AI-driven innovations in earthquake risk mitigation: A future-focused perspective. *Geosciences* **2024**, *14*, 244. [CrossRef]
114. Li, Y.; Chen, H.; Yu, P.; Yang, L. A review of artificial intelligence applications in architectural design: Energy-saving renovations and adaptive building envelopes. *Energies* **2025**, *18*, 918. [CrossRef]
115. Marquez-Sanchez, S.; Calvo-Gallego, J.; Erbad, A.; Ibrar, M.; Fernandez, J.H.; Houchati, M.; Corchado, J.M. Enhancing building energy management: Adaptive edge computing for optimized efficiency and inhabitant comfort. *Electronics* **2023**, *12*, 4179. [CrossRef]
116. Montillet, J.-P.; Kermarrec, G.; Forootan, E.; Haberreiter, M.; He, X.; Finsterle, W.; Fernandes, R.; Shum, C. How big data can help to monitor the environment and to mitigate risks due to climate change: A review. *IEEE Geosci. Remote Sens. Mag.* **2024**, *12*, 67–89. [CrossRef]
117. Luo, N.; Pritoni, M.; Hong, T. An overview of data tools for representing and managing building information and performance data. *Renew. Sustain. Energy Rev.* **2021**, *147*, 111224. [CrossRef]
118. Casini, M. Extended reality for smart building operation and maintenance: A review. *Energies* **2022**, *15*, 3785. [CrossRef]
119. Camps-Valls, G.; Fernández-Torres, M.-Á.; Cohrs, K.-H.; Höhl, A.; Castelletti, A.; Pacal, A.; Robin, C.; Martinuzzi, F.; Papoutsis, I.; Prapas, I. Artificial intelligence for modeling and understanding extreme weather and climate events. *Nat. Commun.* **2025**, *16*, 1919. [CrossRef]
120. Fan, C.; Lei, Y.; Sun, Y.; Piscitelli, M.S.; Chiosa, R.; Capozzoli, A. Data-centric or algorithm-centric: Exploiting the performance of transfer learning for improving building energy predictions in data-scarce context. *Energy* **2022**, *240*, 122775. [CrossRef]
121. Du, Z.; Fan, B.; Jin, X.; Chi, J. Fault detection and diagnosis for buildings and HVAC systems using combined neural networks and subtractive clustering analysis. *Build. Environ.* **2014**, *73*, 1–11. [CrossRef]

122. Carabantes, M. Black-box artificial intelligence: An epistemological and critical analysis. *AI Soc.* **2020**, *35*, 309–317. [CrossRef]
123. Mathew, D.E.; Ebem, D.U.; Ikegwu, A.C.; Ukeoma, P.E.; Dibiaezue, N.F. Recent emerging techniques in explainable artificial intelligence to enhance the interpretable and understanding of AI models for human. *Neural Process. Lett.* **2025**, *57*, 16. [CrossRef]

MDPI AG
Grosspeteranlage 5
4052 Basel
Switzerland
Tel.: +41 61 683 77 34

Buildings Editorial Office
E-mail: buildings@mdpi.com
www.mdpi.com/journal/buildings

www.ingramcontent.com/pod-product-compliance
Lightning Source LLC
Chambersburg PA
CBHW061543220926
51850CB00005B/235
* 9 7 8 3 7 2 5 8 7 1 2 4 7 *